Endocytosis

From Cell Biology to Health, Disease and Therapy

NATO ASI Series

Advanced Science Institutes Series

A series presenting the results of activities sponsored by the NATO Science Committee, which aims at the dissemination of advanced scientific and technological knowledge, with a view to strengthening links between scientific communities.

The Series is published by an international board of publishers in conjunction with the NATO Scientific Affairs Division

A Life Sciences B Physics	Plenum Publishing Corporation London and New York
C Mathematical and Physical Sciences D Behavioural and Social Sciences E Applied Sciences	Kluwer Academic Publishers Dordrecht, Boston and London
F Computer and Systems Sciences G Ecological Sciences H Cell Biology I Global Environmental Change	Springer-Verlag Berlin Heidelberg New York London Paris Tokyo Hong Kong Barcelona Budapest

NATO-PCO DATABASE

The electronic index to the NATO ASI Series provides full bibliographical references (with keywords and/or abstracts) to more than 30000 contributions from international scientists published in all sections of the NATO ASI Series. Access to the NATO-PCO DATABASE compiled by the NATO Publication Coordination Office is possible in two ways:

- via online FILE 128 (NATO-PCO DATABASE) hosted by ESRIN, Via Galileo Galilei, I-00044 Frascati, Italy.
- via CD-ROM "NATO-PCO DATABASE" with user-friendly retrieval software in English, French and German (© WTV GmbH and DATAWARE Technologies Inc. 1989).

The CD-ROM can be ordered through any member of the Board of Publishers or through NATO-PCO, Overijse, Belgium.

Series H: Cell Biology Vol. 62

Endocytosis

From Cell Biology to Health, Disease and Therapy

Edited by

Pierre J. Courtoy

Cell Biology Unit
Louvain University Medical School
and International Institute
of Cellular and Molecular Pathology
UCL 75.41
75, av. Hippocrate
1200 Brussels
Belgium

With the assistance of

Alice Dautry-Varsat
Jean Gruenberg
Lucinda Mata
Mark Marsh
Antoinette Ryter
Bo van Deurs

Springer-Verlag
Berlin Heidelberg New York London Paris Tokyo
Hong Kong Barcelona Budapest
Published in cooperation with NATO Scientific Affairs Division

Proceedings of the NATO Advanced Research Workshop on Endocytosis, held at Paris, France, from October 1–5, 1990

Additional material to this book can be downloaded from http://extra.springer.com.

ISBN-13: 978-3-642-84297-9 e-ISBN-13: 978-3-642-84295-5
DOI: 10.1007/978-3-642-84295-5

Softcover reprint of the hardcover 1st edition 1992

Typesetting: Camera-ready by authors
31/3145-5 4 3 2 1 0 – Printed on acid-free paper

CONTENTS

FOREWORD

(Director : Pierre J. COURTOY)

Two years after its first gathering in Oeiras, Portugal, the European Endocytosis Group convened for a second workshop at the Pasteur Institute, Paris, on October 1-5, 1990. The meeting is reported in detail in this volume; a preliminary coverage, based on the overviews of each session, has appeared in the New Biologist (1991, 3:243-252). The three main objectives, to broaden the audience, to present a more comprehensive view of the multiple aspects of endocytosis, from basic biology to health, disease and therapy, as well as to clarify controversial issues, have been largely fulfilled. The Second European Workshop on Endocytosis was attended by more than 100 participants, originating from 18 countries. 59 lectures and 35 posters were presented. In addition, vivid roundtables allowed to thoroughly discuss the dynamics and the regulation of the endocytic apparatus, as well as the role of endocytosis in antigen presentation.

Endocytosis is a general and distinctive property of all eukaryotic cells, including protists, plants and fungi. This process consists of : (1) the invagination of pericellular membrane into a specific pocket; (2) the generation of a vesicle by "pinching off"; (3) its fusion with intracellular intermediates (endosomes); and (4) final delivery after a series of fission-fusion events to various targets such as lysosomes (degradation pathway, exemplified by asialoglycoproteins in hepatocytes), the original pericellular membrane domain (recycling/regurgitation, illustrated by transferrin and most membrane receptors), or a distinct pericellular domain (transcytosis, which can be traced by polymeric IgA in rat hepatocytes). Two additional pathways can be defined, namely sequestration in an endosomal reservoir (exemplified by vitellogenin in xenopus oocytes and glucose transporters in resting adipocytes), and membrane permeation by internalized proteins (e. g. diphteria toxin) or particles (pH-dependent viruses). These pathways are outlined in Fig. 1.

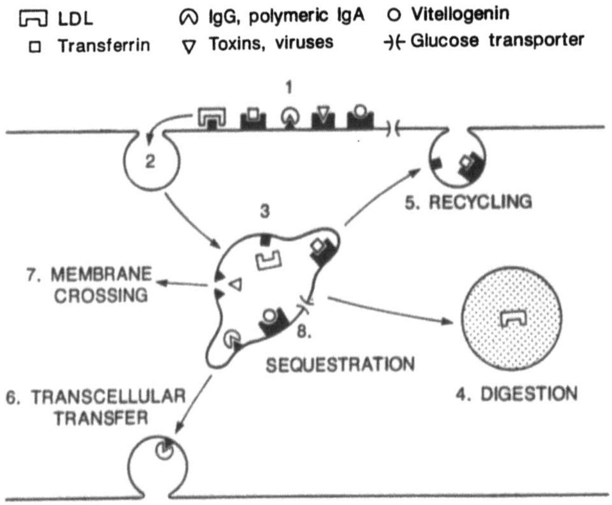

Fig 1 : Pathways of endocytosis from Courtoy (1991) *in* Intracellular Trafficking of Proteins (C. Steer and J. Hanover, eds), Cambridge Univers. Press, pp 107-156.

All components of this network represent the endocytic apparatus. Its proper functioning requires three essential features : (1) mechanisms for selective inclusion into, or exclusion from membrane microdomains in the order of $10^{-2}\ \mu m^2$; (2) an efficient machinery for fission and fusion, the latter having so far received most of the attention; and (3) recognition signals on the intermediate vesicles allowing for their correct addressing.

At the cell surface, molecules which participate in endocytosis are discriminated by lateral mobility, focal clustering and selective incorporation into clathrin-coated or association with the ill-defined clathrin-independent pits. The turn-over of clathrin-coated pits is very fast. Perturbants can block clathrin polymerization on living cells and cell-free assays are now used to unravel the molecular machinery underlining the formation of coated vesicles..

Dynamics and complexity of the endosomal system are evident from video recording and analytical cell fractionation studies, but it is still debated whether observations are best explained with a "maturation" model, for which there is evidence in rapidly growing cells, or with a vesicle-shuttle model, probably prevailing in growth-arrested tissues. Established functions of endosomes include (1) acidification, (2) dissociation of adsorbed constituents from the limiting membrane, (3) selective proteolysis, (4) interchange of disulfide bridges, (5) selective clustering of membrane constituents into outward (tubulo-spherical endosomes) or inward budding membrane expansions (multivesicular bodies), and (6) activation of tyrosine kinase sites associated with some internalized receptors. Cell-free assays are providing new insights on the specificity and the regulation of defined endocytic fusion events.

Lysosomes as an end compartment of three pathways, biosynthetic, endocytic and autophagic, still pose challenging problems of organelle assembly and maintenance. Targeting of soluble lysosomal enzymes involves either a mannose 6-receptor mediated delivery, a sugar-independent adsorptive transport, or the transfer of integral membrane proteins that can be released by a proteolytic cleavage. As more lysosomal membrane proteins are cloned, DNA recombinant technology should rapidly lead to the elucidation of the primary signal(s) of lysosomal addressing. Examples of lysosomal dysfunction due to inborn errors of metabolism, accumulation of xenobiotics such as antibiotics, and possibly labilization of the lysosomal membrane have been presented.

Transcytosis is receiving more attention, for both its contribution to the biogenesis and maintenance of cell polarity, and its involvement in the immune system. Receptor-mediated transepithelial transfer of immunoglobulin A and immunoglobulin G occurs in opposite directions, and contributes to neonatal and mucosal immunity, respectively. These two pathways have been artificially combined in doubly-transfected epithelial cells.

Endocytosis and recycling play a central role in antigen processing by B-lymphocytes and presentation to T-lymphocytes. Indeed, this process involves adsorptive endocytosis of the macromolecule, partial intracellular hydrolysis into fragments, their combination with major histocompatibility complex (MHC) II molecules and final reexposure and presentation. The respective role of endosomes and lysosomes in this process is still debated, but it has become clear that the bulk of MHC II is found in lysosomes.

Although phagocytosis normally terminates most infectious and parasitic diseases, this symposium illustrated various stratagems by which intracellular pathogens escape destruction. It also showed how toxins and viruses may use the endocytic apparatus as a trojan horse to cross the cell membrane and reach the cytosol or the nucleus. Conversely, it emphasized the current attempts to exploit the basic knowledge on endocytosis for drug targeting to activate macrophages or to kill parasites, and for immunotoxins to destroy cancer cells or control auto-immune diseases.

When organizing the meeting, listening to the various presentations, and during the assembly of the proceedings, several reflexions came to my mind. Endocytosis is a respectable field of research. The location of the Second European Workshop at the Pasteur Institute appears quite symbolic, for several topics discussed during the symposium have been pioneered in this very place a century ago or so, by the french Louis Pasteur (bacterial and viral diseases), the russian Ilya Metchnikoff (phagocytosis) and the belgian Jules Bordet (immunology).

It is gratifying to note that endocytosis is studied by a growing number of investigators and is receiving increasing attention by the scientific community, as reflected in the number of Medline entries, which roughly parallels that of oncogenes (Fig 2). This comes to no surprise, since endocytosis is involved in so many aspects of the normal function of our cells and stands at the forefront in our innumerous fights against disease, as exemplified throughout this volume. Indeed, since its acquisition, presumably at the emergence of eukaryotic life, endocytosis has become an essential feature in the general economy of mammalian cells, not only for nutrition, but also to regulate the composition -hence the function- of the cell surface (cell-cell and cell-matrix interactions, transport, signalling to or from the cell...) and possibly to repair altered surface constituents. Endocytosis also compensates for the insertion of membrane from secretory vesicles and could even be required for some transmembrane signals.

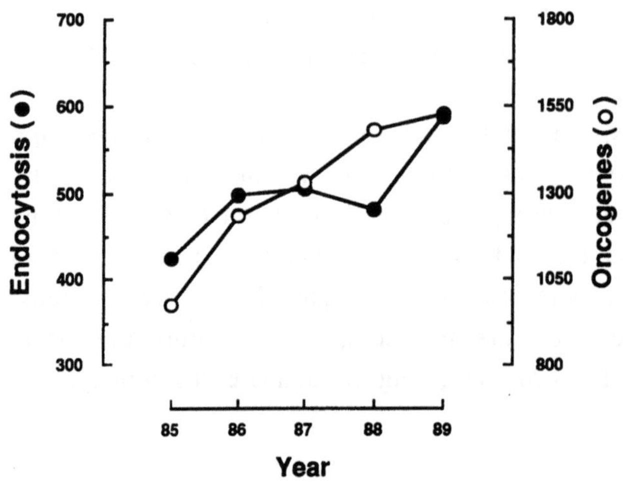

Fig 2 : Number of Medline entries.

The extent of the endocytic traffic is also considerable : the equivalent of the entire cell surface is internalized and recycled within about one hour and the average cell "drinks" the equivalent of its volume every day. Endocytosis could even deal with our origin, if we accept the proposal of Nobel laureate de Duve's developed during his plenary lecture, that the acquisition of endocytosis represented a crucial step for the emergence of eukaryotes from prokaryotes, both to provide independence from the nutritive niche and to accomodate endosymbionts.

However, the price of the success endocytosis enjoys nowadays is exposure to a triple danger : confusion in the terminology, insufficient characterization of complex experimental systems and unwarranted extrapolation from transformed cultured cells to normal tissues. First, due to the multiplicity of approaches, experimental systems and terminology in the various laboratories, we are at risk of building a new Babel tower. The section of keywords lists close to 60 different cell types, mostly transformed, and probably as many probes.

Second, the study of endocytosis is difficult, because the structure of its intricated components is less characteristic than along the secretory pathway, and because specific universal endogenous markers are not yet identified. In addition, intracellular processing of internalized tracers is frequently asynchronous and blocks at intermediate stages are usually not complete. As a result, the extent of overlapping between endosomal compartments in time-defined experiments is not always clear.

Nevertheless, possibly due to an acceleration of the pace of research, systems and fractions are not always carefully defined, and conclusions sometimes ventured without confirmation by available but more time-consuming approaches. Non-linear gradients which enjoy great popularity, expose to the danger of artificial separation of homogenous population of particles, when their median density falls on the flat region of the gradient. The actual composition of subcellular fractions is no longer routinely examined by the analysis of a sufficient number of appropriate markers. Proper quantitation, including indication of the yields in the postnuclear supernatants and recoveries in the fractionation systems, is increasingly missing. Insufficient characterization may lead to confusion, especially at the boundary between late endosomes and lysosomes, referred to as late endosomes, endolysosomes, prelysosomes etc... Shall we end up with a similar complexity along the transcytotic route ? In fact, how many structurally distinct compartments are to be defined in each pathway?

As a third point of caution, transformed cells are convenient to study, but could lead to biased views if conclusions are not validated with non-transformed cells, if not with intact tissues and organs. Vigilance is mandatory but safeguards exist, including regular meetings such as the one reported here, critical evaluation by referees and their insistence on strict controls and quantification, and appreciation of the importance of parallel studies on transformed and non-transformed cells.

In perspective, three avenues are open : (1) to unravel the molecular basis underlining the specificity and the regulation of the complex endocytic traffic and to validate conclusions in non-transformed cells *in situ*; (2) to apply general concepts on endocytosis for a better understanding of its contribution to the physiology of our various tissues; and (3) to exploit emerging tools to fight diseases which subvert the endocytic apparatus or can be defeated by endocytic drug targeting. No doubt endocytosis is a fertile field for future reserarch.

A few words of acknowledgement to conclude. Organizing this meeting would not have been possible without the contribution of a friendly organizing committee, composed of Alice Dautry-Varsat, Jean Gruenberg, Lucinda Mata, Antoinette Ryter and Bo Van Deurs : may they find here the renewed expression of my gratitude. The holding of the meeting heavily depended on a joined funding from EMBO and NATO, and contributions from Novo, Ciba-Geigy and Rhone-Poulenc greatly helped. Michèle Geist kindly assisted assembling this volume. Coworkers showed comprehension for the extra burden. My children Marjorie and Guillaume were particularly supportive and this book is dedicated to them. The demonstration of the dynamism of european investigators in endocytosis and the general appreciation of the usefulness of the meeting are our reward. A third European Workshop on Endocytosis is already scheduled for May 9-13, 1993 in the Netherlands, under the direction of Hans Geuze. The timing is set to alternate with Gordon conferences on lysosomes, held on the other side of the Atlantic Ocean. Significant progress can be expected in between.

I. STRUCTURE, DYNAMICS AND FUNCTIONS OF THE ENDOCYTIC APPARATUS

1.1. SURFACE EVENTS : MOBILITY AND CLUSTERING

(chairperson : Colin HOPKINS)

The first organelle of the endocytic apparatus is the plasma membrane itself. Indeed, lateral mobility of plasma membrane constituents conditions their internalization. **Marc De Brabander** studied the dynamics of membrane proteins on living cells in real time, using the so-called Nanovid microscopy. The phosphatidylinositol-linked Thy-1 antigen shows a fast two-dimensional Brownian motion, with no evidence for backflow related to cell motion. In contast, the transmembrane glycoprotein GP80 assumes a fixed position except at the peripheral lamellae. At the lamellipodia, a steady centripetal migration is observed as waves, if energy is provided and the integrity of the submembranous microfilaments is preserved. These observations underline the control of the plasma membrane by the cytoskeleton and offer another view on membrane dynamics at the leading edge.

Receptor-mediated endocytosis involves clustering and entry in clathrin-coated pits but why some receptors are essentially constitutively internalized, such as LDL receptors in fibroblasts, while the endocytosis of others is mostly triggered by their ligand is still unclear. **Horst Robenek**, used surface replicas to study "en face" the distribution of gold-labeled receptors. His data show that recycled LDL receptors remain as loose agregates or fibroblasts while lipoprotein scavenger receptors are diffusely distributed on mouse peritoneal macrophages, unless they are exposed to their ligands. It seems therefore likely that constitutively internalized receptors spontaneously associate with some components within or below the membrane before they become associated with coated pits. In his contribution to this volume, **Jean-Louis Carpentier**, emphasizes how the clustering of receptors for the C3b intermediate of the complement activation cascade enhances ligand uptake.

One possible signal for clustering in coated pits is the phosphorylation of cytoplasmic domain of receptors. Using cells transfected with various constructs of the insulin receptor prepared by William Rutter, Jean-Louis Carpentier concluded that insulin/insulin receptor complexes are

NATO ASI Series, Vol. H 62
Endocytosis
Edited by P. J. Courtoy

internalized via coated pits only if the endogenous tyrosine kinase can be activated and the receptor be auto-phosphorylated. **Colin Hopkins** showed that the tail-minus transferrin receptors prepared by Ian Trowbridge fail to aggregate in the plane of the membrane and to internalize. There are dramatic differences in the behaviour of low and high expressors. Those two presentations will not be reported further in this volume. In his contribution to the transcytosis session, Ira Mellman describes intriguing differences in two related Fc receptors : FcRII-B1 is poorly internalized, due to a cytoplasmic signal preventing its association wwith coated pits. To the contrary, FcRII-B2 is efficiently internalized in coated pits, thanks to a cytoplasmic domain in which the single tyrosine plays no crucial role. Another example of retention signal, as reported in the session on immunity by Annegret Pelchen-Matthews, is the phosphorylation of CD4 in T cells by the lymphoid-specific tyrosine kinase $P56^{lck}$.

Taken together, these various data begin to clarify the constrains which determine the behavior of migrating surface proteins : free mobility in the plane of the membrane and protein-protein interaction. Phosphorylation in cytoplasmic domains may be involved in either migration or association with the cortical cytoskeleton.

DYNAMIC CYTOMATRIX-MEMBRANE INTERACTIONS INVESTIGATED WITH NANOVID MICROSCOPY

M. De Brabander, R. Nuydens*, H. Geerts* Clinical Research Unit St Bartolomeus, AZ Jan Palfyn, B-2030 Merksem, Belgium.

INTRODUCTION

The cell membrane consisting of the phospholipid bylayer and many different glycoproteins embedded in it is suspected to show a highly dynamic yet regulated behaviour. Impressive reorganizations are known to occur in particular in relation to cell movement and differentiation. While some components may move about relatively freely, local constraints and interactions with underlying cytomatrix elements are suspected to exist while others such as adhesion molecules may become immobilized at specific sites.

Until recently most of our knowledge of the dynamic behavior of cell membrane components was derived by indirect means. Constituents labelled with suitable markers were localized after various time intervals using time lapse fixation. Recently, fluorescent analogs have been introduced and used in combination with video intensification microscopy to localize in living cells relatively large accumulations of molecules (Taylor and Wang, 1980). However, the sensitivity is not sufficient to detect single markers, and bleaching obviates prolonged uninterrupted observation.

In order to be able to follow individual discrete markers for an indefinite time we have developed the nanovid (nanoparticle video) approach (De Brabander et al. 1986). Individual gold particles, are visualized by video enhanced contrast microscopy and electronic image improvement.

In this chapter we describe the use of nanovid microscopy to follow the movement of cell membrane components in real time. Negatively charged cell surface residues were detected by gold particles bearing a positive surface charge.These were prepared by incubating gold sols (De Mey 1984), stabilized by polyethylene glycol (PEG) and bovine serum albumin (BSA), with poly-l-lysine hydrobromide of 4.000 or 240.000 D. Specific cell membrane constituents were followed by gold particles coupled to monoclonal antibodies to GP 80 (Jacobson et al. 1984) and to the Thy-1 antigen (Ishihara et al. 1987) as described (De Mey 1984). Procedures were developed to quantify various parameters by an automatic computer driven system. The approach allowed us to characterize the behavior of different membrane components in relation to the underlying cytocortex. For illustrations we refer to De Brabander et al. 1990.

*Janssen Research Fdn, B-2340, Beerse, Belgium.

NATO ASI Series, Vol. H 62
Endocytosis
Edited by P. J. Courtoy

A. BINDING AND DYNAMIC BEHAVIOR OF POLYLYSINE COATED GOLD PARTICLES.

Gold particles stabilized with BSA and PEG are negatively charged and are repelled from the cell surface. When 4K p-L-l Au40 was added to PtK2 cells an evenly distributed label on the cell surface was obtained Most of the membrane bound particles moved freely on the cell surface in a two dimensional Brownian way. With Nanovid correlation spectroscopy we measured a diffusion coefficient (D) of 0.31±0.11 μ^2/sec. This movement persisted for over an hour. After 30 minutes intercellular contact areas became accentuated by particles assuming a fixed position.

In order to see whether the Brownian motion of the of the gold particles was entirely free or whether local constraints existed we analyzed the mean displacement of the gold particles in time, which showed that the displacement is apparently restrained to microdomains of about 1 um2.

The 240K p-L-l Au40 behaved in an entirely different way. The particles assumed fixed positions immediately after touching the cell surface. However, in time lapse recordings, a constant centripetal flow of particles was seen on free peripheral lamellae. This flow stopped at the transition between the peripheral lamelloplasm and the organelle rich central cytoplasm (endoplasm), and the gold particles accumulated in this region . No flow or aggregation of particles was seen on the remainder of the dorsal cell surface or on cells without free peripheral lamellae. With time gold particles accumulated on intercellular contact areas.

Extraction of labelled cells with Triton (1%) removed most (60%) of the 4K p-L-l Au40 suggesting that this was largely bound to membrane embedded components. Only 18% of the 240K p-L-l Au40 was removed suggesting that this probe was bound to cytocortex attached elements.

B. INFLUENCE OF ATP DEPLETION AND CYTOSKELETAL ALTERATIONS.

240K p-L-l Au40 added to cells pretreated with 10 mM sodium azide did not migrate towards the transition zone and no accumulations appeared on intercellular contacts. The majority of the particles moved in a Brownian way (D=0.126 ± 0.04 μ^2/sec). After perfusion with glucose containing medium, which restores ATP levels, the Brownian motion stopped immediately and the centripetal flow resumed. Addition of sodium azide to cells that already had accumulated gold particles induced Brownian motion and gradual dispersal of the probes over the cell surface. The behavior of the 4K p-L-l Au40 was not altered drastically by sodium azide. Particles continued to move in a Brownian way (D=0.36±0.09 μ^2/sec).No accumulations appeared however on cell junctions.

The tumor promoter TPA (10 ng/ml, 2hr pretreatment) which induces motile behavior, disruption of cell junctions and dissolution of stress fibers in PtK2 cells also abolished the centripetal flow of the 240K p-L-l Au40. Loose accumulations of individually motile particles were formed. The 4K p-L-l Au40 showed Brownian motion as in control cells. Neither of the probes accumulated on intercellular contact areas.Pretreatment of the cells with cytochalasin B (1 or 10µg/ml for 24 hr) also prevented the centripetal flow of the 240K p-L-l Au40 and the accumulation on cell junctions. Instead large aggregates where formed above the cytocentre of the cells. The Brownian motion of the 4K p-L-l Au40 was not altered but particles did not accumulate on cell junctions.

C. RELATION BETWEEN THE CENTRIPETAL FLOW AND MICROFILAMENT ORGANIZATION.

The behavior of the 240K p-L-l Au40 was reminiscent of the rearward movement of large particles on fibroblasts (Abercrombie et al. 1970, Dembo and Harris, 1981) and has been related by some authors to the movement of submembraneous microfilament arcs (Heath 1983). Therefore we investigated whether such an interaction could be relevant to our observations. In cells fixed and immunostained for actin different times after addition of the gold probe a series of microfilament arcs was seen in the lamelloplasm. The accumulation of gold particles in the transition zone was in general closely related to the arc that was closest to the nucleus. In whole mount ultrastructural preparations rows of gold particles were clearly associated with successive arcs. Gold particles had also accumulated on the circumferential bundles of microfilaments that are associated with intercellular contact regions. No relationship existed with microtubules or with stress fibers which are arranged more or less orthogonally to the leading edge. In some cells observed with video enhanced differential interference contrast microscopy the movement of the microfilament arcs could be followed.These structures appeared close to the leading edge and moved centripetally until they reached the transition zone. When 240k p-L-l Au40 was added to these cells the migration of the surface bound particles was coincident with the movement of the arcs. Many particles remained apparently attached to an underlying arc throughout their centripetal movement.

The three compounds that arrested the centripetal migration of the gold particles also affected the actin microfilament integrity as visualized with immunofluorescence. Most notably microfilament arcs could no longer be detected. In cytochalasin B treated cells dense actin aggregates were often seen above the nucleus. These corresponded frequently to accumulations of gold particles.

D. BEHAVIOR OF GP 80 AND THY-1.

Monoclonal antibodies to GP 80 (Jacobson et al. 1984), a major transmembrane glycoprotein of mouse fibroblasts, coupled to 40 nm gold bound to 3T3 cells. Their behavior strongly resembled that of the 240K p-L-l Au40. Most of the particles assumed apparently fixed positions on the membrane. In time lapse recordings, a steady centripetal migration was seen on the dorsal surface of the leading front of the cells. They accumulated in aggregates close to the nucleus. In these cells no clear distinction can be made between the peripheral lamelloplasm and the central endoplasm.

Monoclonal antibodies to Thy-1 coupled to 40 nm gold also bound to 3T3 cells. Their behavior was similar to the 4K p-L-l Au40. The majority of the particles were continuously engaged in Brownian motion on the cell surface.

E. DISCUSSION: DYNAMICS OF CELL MEMBRANE COMPONENTS, CELL MOVEMENT AND MORPHOGENESIS.

The observations show that nanovid microscopy provides a powerfull tool to follow the dynamic behavior of individual membrane components. The gold particles can be followed clearly at high resolution.Quantitative information can be obtained by computer driven analysis. With the labels we have used essentially two types of behavior were detected. The anti Thy1 Au40 and the 4K p-L-l Au40 showed predominently twodimensional Brownian motion. For the Thy-1 antigen, which is a glycoprotein linked to the membrane by a phosphoinositol lipid moiety the observations confirm previous data obtained with Frap analysis (Ishihara et al. 1987). The 4K p-L-l Au40 binds probably to negatively charged cell surface residues. Most of these can be extracted with Triton suggesting that they are not linked to submembraneous cytocortical elements. The lack of effect of sodium azide and of compounds affecting the microfilament system suggest that their mobility is due to thermally driven lipid mediated diffusion. Restraints to this motion appear at the sites of intercellular contact. The diffusion coefficient is within the range of that measured for Thy-1 and lipid analogs (Ishihara and Jacobson, 1987). Apparently the size of the gold marker or the expected polyvalency do not introduce major artefacts. However it has become clear that the diffusion is restricted to local domains in the plasma membrane a long debated phenomenon. Recently Sheetz et al.[1989] came to the same conclusions on the basis of observations using colloidal gold coupled to Con-A.

The behavior of the GP 80 Au40 and 240K p-L-I Au40 is quite different. Both probes apparently bind to membrane components that are not freely diffusible. Mere crosslinking can not be responsible for the slow diffusion rate because the Thy-1 and GP 80 probes should in principle be identical. Rather the observations suggest that the GP 80 and the 240K p-L-I binding sites are linked to submembraneous cortical elements. We do not know whether the transmembrane linkage is constitutive or induced by crosslinking . In particular for the 240K Au40 the long extending polycationic molecules may be suspected to induce extensive crosslinking. An alternative possibility however is that by its multiple binding sites the 240K p-L-I Au40 has a greater probability of binding to at least one transmembraneously linked site in addition to other mobile elements.

The rearward migration of particles on the dorsal surface of cultured cells (Abercrombie et al. 1970) has been taken as evidence for the existence of a rearward flow of membrane lipids inserted at the leading edge of motile cells (Bretscher, 1984). Our observations rather support the concept that submembraneous microfilament arcs (Heath, 1983) carry along certain transmembrane components. This is shown by the coincident migration of the gold particles and the arcs. Moreover the movement of both stops at exactly the same site. Further evidence is provided by the disruptive effects of drugs that disintegrate the arcs. Moreover, detailed analysis of the behavior of the 4-K p-L-I Au 40 and the Thy -1 probe did not detect any superimposed drift on the Brownian motion which should be obvious when a constant lipid flow were occuring.

It is questionable whether the constant rearward migration of membrane elements is connected with cell migration in the epithelial PTK2 cells we have used. Indeed this activity has been observed for several hours without any progression of the leading edge. Moreover treatment with TPA, which induces cell motility, actually disrupts the rearward flow and the microfilament arcs. We propose that, at least in epithelial cells, these activities may rather be involved in morphogenetic events ensuing in close apposition of neighboring cells that become linked by intercellular adhesion molecules.

REFERENCES

Abercrombie, M., Haeysman, J.E.M., Pegrum, S.M. (1970) : the locomotion of fibroblasts in culture. III. Movements of particles on the dorsal surface of the leading lamella. Exp. Cell. Res. 62 : 389-398.

Bretscher,M.(1984) : Endocytosis: relation to capping and cell locomotion. Science 224 : 681-686

De Brabander, M., Nuydens, R., Geuens, G., Moeremans, M., De Mey, J. (1986): The use of submicroscopic gold particles combined with video contrast enhancement as a simple molecular probe for the living cell. Cell Motil. Cytoskel. 6: 105-113

De Brabander, M., Nuydens, R., Geerts, H. (1990) : Dynamic cytomatrix-membrane interactions in cell shape and organization. In: Akkas, N. (ed) : Biomechanics of active movement and deformation of cells.Berlin, Heidelberg. Springer Verlag. pp 215-234.

Dembo, M., Harris, A.K. (1981): Motion of particles adhering to the leading lamella of crawling cells. J. Cell. Biol. 91:528-536
De Mey, J. (1984) : Colloidal gold probes in immunocytochemistry. In: Polak, J.M., Van Noorden, S. (eds) : Immunocytochemistry. Bristol, London, Boston. Wright PSG. pp. 82-112
Heath, J. (1983) : Behavior and structure of the leading lamella in moving fibroblasts. Occurence and centripetal movement of arc shaped microfilament bundles beneath the dorsal cell membrane. J. Cell. Sci. 60 : 331-354
Ishihara, A., Hou, Y., Jacobson, K. (1987) : The Thy-l antigen exibits rapid lateral diffusion in the plasma membrane of rodent lymphoid cells and fibroblasts. Proc. Natl. Acad. Sci. USA 84 : 1290-1293.
Jacobson, K., O'Dell, D., Holifield, B., Murphy, T., August, T. (1984) : Redistribution of a major cell surface glycoprotein during cell movement. J. Cell Biol. 99 : 1613-1623
Sheetz, M.P., Turney, S., Qian, H., Elson, E.L. (1989) : Nanometre-level analysis demonstrates that lipid flow does not drive membrane glycoprotein movements. Nature. 340: 284-288
Taylor, D.L., Wang, Y.L. (1980): Fluorescently labelled molecules as probes of the structure and function of living cells. Nature 284 : 405-410

ACKNOWLEDGEMENTS

We would like to express our gratitude to: G. Geuens, M. Moeremans, J. Leunissen, R. Nuyens, J. De Mey, and K. Jacobson, for their contribution to various parts of the work summarized here. We are indebted to L. Leyssen and G. Jacobs for the photographic work and to K. Donné for typing the manuscript.

ANALYSIS OF RECEPTOR DYNAMICS, AS STUDIED WITH SURFACE REPLICAS

Horst Robenek
Institute for Arteriosclerosis Research,
Department of Cell Biology and Ultrastructure Research,
University of Münster,
Domagkstraße 3, 4400 Münster
Federal Republic of Germany

1. INTRODUCTION

Cells interact in a variety of ways with their environment. One essential type of interaction involves the recognition by cells of a wide variety of polypeptides in the surrounding milieu. At the surface of a cell, these molecules are recognized and bound. This process is receptor-mediated and is the first step in the physiologic signaling and/or internalization of the ligand. Investigations on the complement and distribution of receptors at the cell surface are of great significance for determining how receptors modulate cellular activities. The increased interest in studying cell surface receptors in recent years has sparked the improvement of morphological methods and the development of visual probes for localizing receptor sites in the electron microscope.

2. MATERIALS AND METHODS

2.1 Probes for Visualizing Receptors

Since means for the direct localization of receptors are limited, we still depend on more or less indirect procedures for studying receptor distribution and dynamics. We have employed three principally different approaches for visualizing receptors: The first method was based on the selective binding of gold-labeled ligands to localize directly the corresponding receptors at the plasma membrane surface (Robenek et al., 1982; Robenek and Schmitz, 1985). The second method involved indirect immunocytochemical labeling of receptor-bound ligands and the third method indirect immunocytochemical labeling of receptors (Robenek et al., 1990; Gebhardt and Robenek, 1987). Cells were exposed first to specific antibodies directed against either

NATO ASI Series, Vol. H 62
Endocytosis
Edited by P. J. Courtoy

the ligands or their receptors followed by gold-labeled protein A or secondary antibodies.

2.2 Surface Replication Techniques

Thin sectioning, though indispensable for analyzing the intracellular pathway of ligands after endocytosis, does not provide adequate information on the surface distribution of receptors. Therefore we have used three different platinum-carbon (Pt/C) surface replication techniques which are most effective for analyzing the distribution of receptors at the cell surface (Severs and Robenek, 1983). The advantages of the surface replication techniques are that extensive areas of the cell surface can be surveyed. Combined with colloidal gold markers these techniques permit high resolution visualization and reliable identification of cell surface receptors in cultured cells. They also yield more detailed information about the distribution and dynamics of receptors than any other technique (Robenek, 1989 a,b). The surface replication techniques used are the

a) surface replication technique,
b) lysis-squirting technique, and
c) label-fracture technique (Figure 1).

a) Surface replicas visualize the external plasma membrane surface (Robenek et al., 1982). Surface replicas of gold-labeled fixed cells in monolayers are prepared directly on 1 x 1 cm pieces of plastic cut from culture vessels using a soldering iron. The cells on the plastic pieces are dehydrated and dried in a critical-point drying device and shadowed with platinum at an angle of 38°, followed by carbon at 90°, in a vacuum evaporating unit. The preparations are left overnight in household bleach, which separates the metal replicas from the substrate and removes cellular material. The replicas are washed in distilled water and picked up on copper grids prior to examination in the electron microscope. (Surface replication: Label ➤ Fix ➤ Dry ➤ Pt/C replication ➤ Digestion ➤ Replica cleaning).

b) With the aid of the lysis-squirting technique the protoplasmic aspects of both apical and ventral plasma membranes of cultured cells can be exposed (Nicol et al., 1987). The apical protoplasmic surface is exhibited by tearing away the apical portions of the cell using poly-lysine coated coverlips. The lysis-squirting technique has proven most successful for exposing the protoplasmic surface of the "ventral" plasma membrane. In this case, the cells are lysed in hypotonic buffers and the cytoplasmic components are removed with a stream of buffer from a syringe or Pasteur pipette. The combination of lysis-squirting, critical-point drying, and subsequent replication yields good structural preservation of the plasma membrane. In

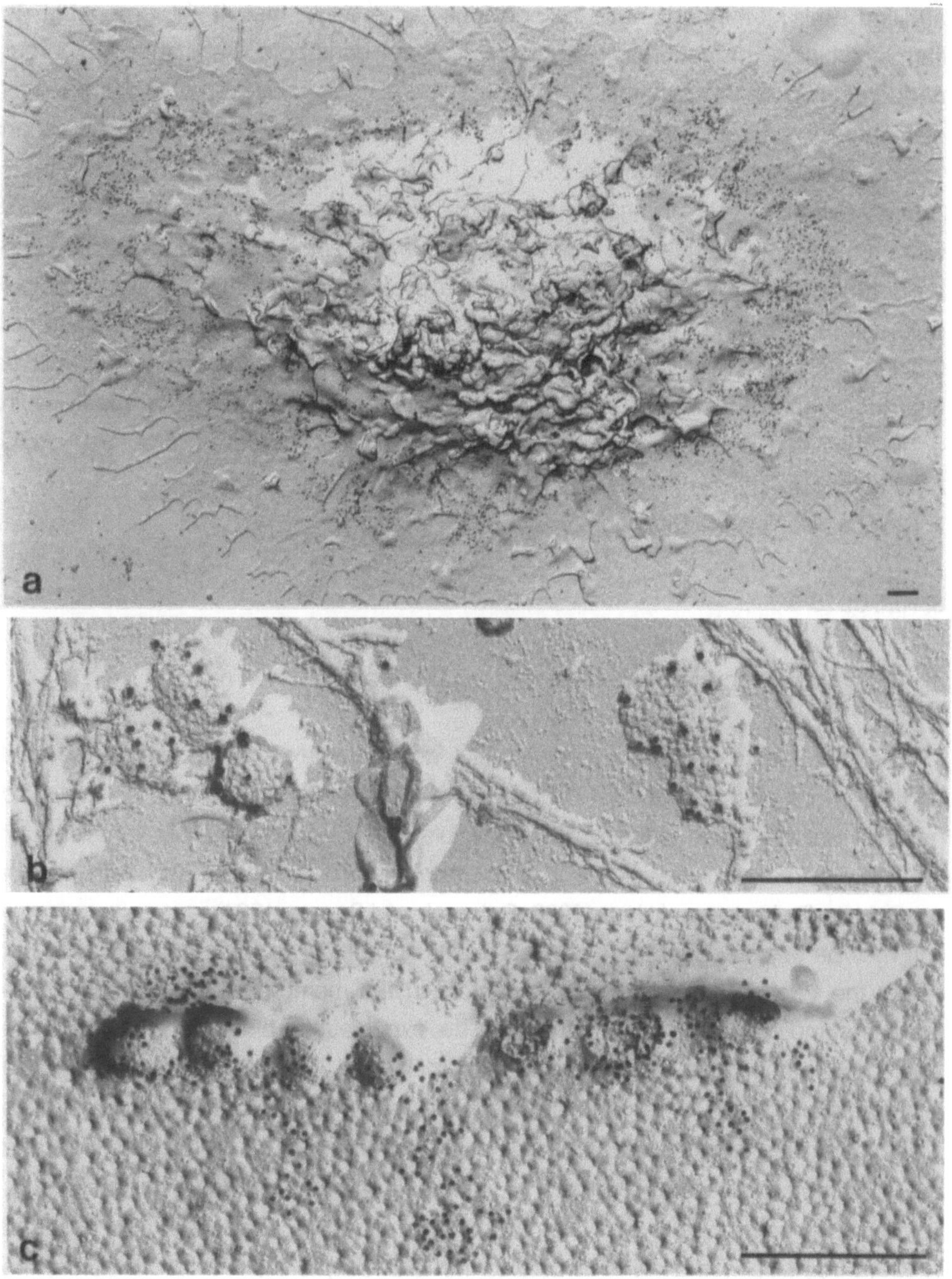

Figure 1: a) Surface replica of a cultured mouse peritoneal macrophage labeled with acLDL-gold complexes.
b) Protoplasmic surface of the plasma membrane of a fibroblast after lysis-squirting labeled with anti-clathrin antibodies followed by protein A-gold.
c) Label-fracture of a fibroblast labeled with LDL-gold complexes. Bar = 0,5 μm.

addition, this method has the obvious advantage that it is amenable to immunolabeling procedures. As such, it may contribute substantially to our understanding of the dynamics of receptors and their relationship to cytoskeletal elements. (Lysis-Squirting: Lysis ➤ Squirting ➤ Label ➤ Fix ➤ Dry ➤ Pt/C replication ➤ Digestion ➤ Replica cleaning).

c) By "label-fracture" plasma membrane E-face structure can be compared directly with corresponding label at the external plasma membrane surface (Pinto da Silva and Kan, 1984). For "label-fracture" the cells are cytochemically labeled using colloidal gold as a marker and then the routine freeze-fracture sequence followed. Where the technique differs crucially from standard freeze-fracture replication is that the replicas are not cleaned in sodium hypochlorite; instead, they are washed extensively in distilled water. (Label-fracture: Label ➤ Fix ➤ Cryoprotect ➤ Freeze ➤ Fracture ➤ Pt/C replication
➤ Replica cleaning (no Digestion).

3. RESULTS

3.1 Distribution, Recycling and Clustering of LDL Receptors

When fibroblasts were incubated with LDL-gold complexes at 4°C for 1h, the gold particles were observed almost exclusively in clusters. The number of solitary gold particles outside these clusters was negligible. The clusters generally consisted of densely packed gold particles and their distribution coincided largely with that of coated pits (Robenek and Hesz, 1983; Robenek and Severs, 1984).

It may be objected that the observed distribution of LDL-gold complexes does not reflect the true distribution of LDL receptors in the membrane, for the complexes represent multivalent ligands that might be expected to bind more efficiently to aggregated rather than to solitary receptor molecules. This was disproved by the following experiments. When we incubated fibroblasts with LDL-gold conjugates in the presence of an excess of unlabeled LDL at 4°C for 1h, both LDL-gold complexes and free LDL molecules were visualized in favourable surface replicas in clusters only. Furthermore, gold clusters equivalent to those already described were also obtained 1) when free LDL was allowed to bind to the LDL receptors a 4°C for 1h before the LDL-receptor complexes were visualized by protein A-gold labeling of antibodies directed against the apolipoprotein B-100 moiety of the LDL molecule and 2) when immunolabeling of LDL receptors themselves was performed using monoclonal antibodies directed against the receptors followed by protein A-gold labeling. From these findings it can be concluded that unoccupied LDL receptors are arranged in the form of clusters in the native cell membrane.

Conflicting views have been proffered concerning how LDL receptors are inserted into the plasma membrane following recycling and how they move to coated pits. Receptors are believed to be inserted into the cell membrane at random locations and to migrate to coated pits either by diffusion or by active flow.

We proposed a new working model for the recycling of LDL receptors to the plasma membrane of cultured fibroblasts and for their clustering in coated pits. We based this concept mainly on our observations from experiments designed to detect only newly inserted receptors. From these observations it can be concluded that after their return to the surface, LDL receptors remain clustered so that they can be incorporated rapidly into newly formed coated pits (Robenek and Hesz, 1983; Robenek et al., 1990).

3.2 Distribution and dynamics of scavenger receptors

Exposure of mouse peritoneal macrophages to acetylated (ac) LDL-gold conjugates for 1h at 4°C yielded preferential binding of the gold label in the intermediate zone of the plasma membrane surface. The gold particles were more or less randomly distributed. When these macrophages were warmed to 37°C, a time-dependent redistribution of gold particles took place. After 4 min. the proportion of gold complexes present in larger aggregates increased considerably and after 8 min. virtually all the gold particles were clustered in coated pits. After 12 - 15 min. the AcLDL-gold complexes had almost completely disappeared from the cell surface.

The following conclusions can be drawn from these results:

1. AcLDL receptors are distributed diffusely on a particular region of the cell surface of cultured mouse peritoneal macrophages.
2. The binding of AcLDL molecules to the receptors induces long-range clustering of the occupied receptors into coated pits. These two results show that both the distribution and dynamics of AcLDL and LDL receptors differ markedly, because LDL receptors in contrast to receptors for AcLDL, are always confined to presumptive coated pit regions, even prior to ligand binding, and do not migrate appreciably in the plane of the membrane.
3. AcLDL is removed rapidly from the plasma membrane and is delivered via coated vesicles to lysosomes for degradation (Robenek et al., 1984).

3.3 Distribution of IgA receptors (Secretory Component, SC)

The distribution of IgA-gold conjugates at 4°C in surface replicas of cultured rat hepatocytes is of a mixed random type. This distribution is intermediate between the exclusively clustered type characteristic of LDL receptors and random scattering. Since direct decoration of SC by anti-SC antibodies or by gold-labeled IgA after fixation yielded virtually the same results, this labeling pattern evidently reflects the distribution of unoccupied receptors. Incubation of hepatocytes at 37°C subsequent to binding of IgA-gold conjugates at 4°C initiated a time-dependent redistribution characterized by progressive coalescence of gold particles into larger clusters.

In cultures exposed to cytochalasin B, the distribution of IgA-gold conjugates in surface replicas after incubation at 4°C was almost the same as that of controls. In contrast, after incubation of cytochalasin B treated hepatocytes at 37°C for 20 to 40 min., huge patches comprising several hundred gold particles were formed preferentially at the periphery of most cells. Cells treated with cytochalasin B and incubated in the absence of IgA for 20 min. at 4°C and then for 40 min. at 37°C before exposure to gold-labeled IgA showed no patch formation. These results indicate that both redistribution as well as patch formation do not occur spontaneously, but are triggered by binding of the ligand to SC (Gebhardt and Robenek, 1987).

Our results on redistribution and patch formation of the IgA receptor can be interpreted by assuming the following series of events:

1. In contrast to the LDL receptor, the IgA receptor is randomly distributed or present in the form of clusters in the hepatocyte plasma membrane. It is tempting to speculate that the dispersed population represents either newly-inserted, recycling receptors, or a dispersion stage following the insertion of preclustered receptors.

2. As a result of binding, the natural tendency of IgA receptor complexes to aggregate is strikingly reinforced due to a conformational change which increases the mutual affinity of these receptor molecules.

3. At this stage, microfilaments may be involved in the control of the redistribution process; in particular they may limit the extent of clustering.

4. The accelerated ligand induced aggregation in turn may enhance the recruitment of clathrin to the clustered ligand receptor complexes to form coated pits as postulated for LDL receptors.

5. Enhanced endocytosis would then be a consequence of the increased number of coated structures.

6. When the control of aggregation by the filamentous system is perturbed experimentally by cytochalasin B, the redistribution process is no longer limited leading to patch formation by unconstrained diffusion within the plasma membrane.

4. CONCLUSIONS

The surface replication techniques are extremely potent and can be employed to study many diverse features of diverse receptor systems.

Depending upon the cell type and the specificity of the receptor, receptors appear either randomly dispersed along the membrane or they appear in clusters. Another distribution pattern is of a mixed random type, intermediary to the extremes of exclusive clustering and to the extremes of random scattering. Examples of random distribution include receptors for EGF, α2-macroglobulin and acetylated low density lipoprotein. In contrast, preclustered receptors include those for maternal IgG and LDL. The IgA-receptor is distributed in the form of clusters and individual, widely-dispersed receptors.

In addition to these characteristic distribution patterns the receptors occupy specific domains at the surface of the plasma membrane. What is responsible for the different initial distribution of these receptors is not known. However, if initially diffuse, it is clear that once the receptors bind ligand most become clustered in coated pits on the membrane. Those that are initially clustered, such as LDL, are predominantly localized over coated regions along the cell membrane.

Supported by Deutsche Forschungsgemeinschaft.

Gebhardt R, Robenek H (1987) Ligand-dependent redistribution of the IgA receptor on cultured rat hepatocytes and its disturbance by cytochalasin B. J Histochem Cytochem 35:301-309

Nicol A, Nermut MV, Doeinck A, Robenek H, Wiegand Ch, Jockusch BM (1987) Labeling of structural elements at the ventral plasma membrane of fibroblasts with the immunogold technique. J Histochem Cytochem 35:499-506

Pinto da Silva P, Kan FWK (1984) Label-fracture: A method for high resolution labeling of cell surfaces. J Cell Biol 99:1156-1161

Robenek H, Rassat J, Hesz A, Grünwald J (1982) A correlative study on the topographical distribution of the receptors for low density lipoprotein (LDL) conjugated to colloidal gold in cultured human skin fibroblasts employing thin section, freeze-fracture, deep-etching, and surface replication techniques. Eur J Cell Biol 27:242-250

Robenek H, Hesz A (1983) Dynamics of low density lipoprotein receptors in the plasma membrane of cultured human skin fibroblasts visualized by colloidal gold in conjunction with surface replicas. Eur J Cell Biol 31:275-282

Robenek H, Severs NJ (1984) Double labeling of lipoprotein receptors in fibroblast cell surface replicas. J Ultrastruct Res 87:149-158

Robenek H, Schmitz G, Assmann G (1984) Topography and dynamics of receptors for acetylated and malondialdehyde-modified low-density lipoproteins in the plasma membrane of mouse peritoneal macrophages as visualized by colloidal gold in conjunction with surface replicas. J Histochem Cytochem 32: 1017-1027

Robenek H, Schmitz G (1985) Receptor domains in the plasma membrane of cultured mouse peritoneal macrophages. Eur J Cell Biol 39:77-85

Robenek H (1989a) Distribution and mobility of receptors in the plasma membrane. In: Freeze-Fracture studies of Membranes, Hui SW (ed) CRC Press Inc, New York 61-86

Robenek H (1989b) Topography and internalization of cell surface receptors as analysed by surface replica and ultrathin sectioning techniques. In: Electron Microscopic Analysis of Cell Dynamics, Plattner H (ed) CRC Press Inc, New York 141-163

Robenek H, Harrach B, Severs NJ (1990) Display of low density lipoprotein receptors is clustered, not dispersed, in fibroblast and hepatocyte plasma membranes. Arteriosclerosis, in press

Severs NJ, Robenek H (1983) Detection of microdomains in biomembranes. An appraisal of recent developments in freeze-fracture cytochemistry. Biochem Biophys Acta 737:373-408

SURFACE DISTRIBUTION AND PATHWAY OF INTERNALIZATION OF C3b RECEPTORS (CR1) IN HUMAN NEUTROPHILS[1]

J.-P. Paccaud and J.-L. Carpentier
Institute of Histology and Embryology
University of Geneva Medical Center
CH-1211 Geneva 4
Switzerland

Neutrophils are implicated in the prevention of bacterial infection. To that end, phagocytes must first sense the presence of bacteria, reach the sites of infection and eventually get in close contact with the microorganisms to ingest them (Elsbach P and Weiss J, 1988). This complex process requires the cooperation of several classes of receptors (adherence, chemoattractants, and phagocytic receptors) present on their surface. Chemoattractant receptors, which bind molecules such as C5a or fMLP, are coupled to phosphoinositide hydrolysis which leads to an increase of cytosolic free calcium $[Ca^{2+}]_i$ and activation of protein kinase C (Snyderman R et al., 1986). In turn, this signalling cascade is thought to be essential for triggering oriented locomotion and secretion of neutrophils (Stossel TP, 1988). Among phagocytic receptors, those for Fc portion of immunoglobulins are endowed with the intrinsic ability to trigger ingestion and metabolic responses towards antibody coated particles. On the other hand, complement receptors such as CR1 (receptors for C3b), promote the binding of complement activating particles (such as bacteria or immune complexes), and their subsequent phagocytosis, synergestically with Fc receptors (Unkeless JC and Wright SD, 1988). In the case of CR1, the aquisition of the phagocytic capability requires its activation by a process thought to involve phosphorylation of the receptor (Changelian PS and Fearon DT, 1986). Full phagocytic capacity of CR1 is obtained when neutrophils are costimulated with fibronectin or laminin (Pommier et al.,1983; Bohnsack et al. 1985). Pharmacological agents such as PMA also mimic this activation process (Kazatchkine MD and Fearon DT, 1990).

The aim of the present study was to get further insight on the cell biology of CR1, first by correlating its distribution at the surface of erythrocytes and human neutrophils with its function, and second by following its internalization pathway and intracellular fate inside human neutrophils.

Because of the low affinity of CR1 for monomeric C3b, the efficient binding of C3b-coated particles to CR1-bearing cells has been suggested to require multivalent interactions between C3b and CR1 (Edberg JC et al., 1987). Hence, the surface

---

[1] This work was supported by the Swiss National Science Foundation, grants 32-25606 and 31-36625.

NATO ASI Series, Vol. H 62
Endocytosis
Edited by P. J. Courtoy

distribution of CR1 might be crucial in cells suceptible to interact with circulating complement-reacted particles, i.e erythrocytes and neutrophils.

Thus, we investigated morphologically and functionnally the state of aggregation of CR1 on the surface of erythrocytes and neutrophils under various conditions of stimulation.

Using monoclonal antibodies against CR1 (E11 and J3D3) tagged with gold-labeled anti-mouse IgG, we localized, by label fracture, CR1 at the surface of human erythrocytes and neutrophils. On erythrocytes, up to 50% of the receptors were organized into clusters containing up to 15 gold particles (Paccaud J-P et al., 1988). The expression of CR1 at the erythrocyte surface is genetically regulated, and varies from 100 to more than 1200 sites per cell (Wong WW et al, 1989). Quantitative analysis performed on different donors revealed a positive correlation between the mean number of CR1 expressed at the surface of erythrocytes and the degree of clustering (Table 1).

Table 1

Donor	CR1/cell	GP [a] per cell	% total GP in clusters ≥3	Nb of clusters≥3 [a] per cell
1	920±23	666±63	48.8	81.8
2	700±18	523±53	46.7	62.4
3	350±19	282±31	36.9	22.8
4	175±20	80±14	28.7	6.7

a GP= gold particles. Mean ± SEM (n=3).
Values reported to the mean erythrocyte surface of 145μm2

(From Paccaud et al., 1988, with permission)

This observation was confirmed functionnaly: erythrocytes bearing a higher number of CR1 per cell bound more, and with greater efficiency opsonized immune complexes than erythrocytes with lower CR1 number (Madi et al., in press). Thus, the surface distribution of CR1 on erythrocytes is such that multivalent binding is favored.

Next we analyzed CR1 distribution on circulating neutrophils stimulated or not with fMLP. By contrast to what is observed on erythrocytes, at the surface of neutrophils only 15 to 20% of gold particles were found in clusters of three or more (Paccaud J-P et al, 1990). In fMLP-stimulated neutrophils (10^{-6} M), the density of gold particles increased proportionnaly to the number of expressed CR1, however the proportion of clustered CR1 did not differ significantly from resting cells (Fig 1).

Figure 1

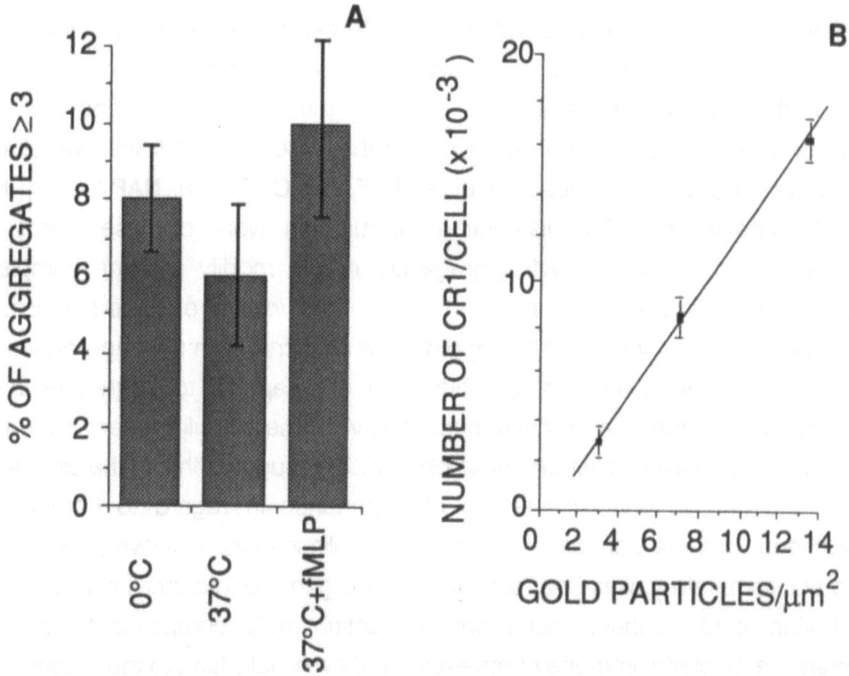

A. Percentage of gold particles in clusters containing 3 or more particles. Mean of 3 different experiments ± SEM.
B. Correlation between the number of CR1 per cell and the number of gold particles per 10µm2.

(Adapted from Paccaud et al., 1990)

These morphological observations were correlated with measurements of C3b-dependent immune complexes binding. To that end, fMLP-stimulated PMN were challenged at 4°C with purified opsonized immune complexes which bound essentially via CR1 (Paccaud J-P et al, 1990). At equal number of receptors, immune complexes adhered less well to neutrophils than to erythrocytes. This functional difference could be attributed to the low level of CR1 clustering in neutrophils as compared to the one observed in erythrocytes. These experiments suggest that, in the circulation, the distribution of CR1 rather than its density on the cell surface is determinant in its ability to efficiently mediate immune adherence reactions. Thus, in the circulation, neutrophils unlikely interact efficiently with C3b-coated particles.

The full phagocytic capacity of neutrophils is expressed in tissues rather than in the blood stream. Several known agonists have been described to upregulate CR1 on human neutrophils, i.e C5a (Fearon DT and Collins LA, 1983), LTB4 (Berger M et al., 1985), TNF (Berger et al, 1988), or NAP-1/IL 8 (Paccaud et al., 1990). However, their potential effect on CR1 distribution remains unknown. To determine whether CR1 present on the neutrophil surface could cluster in the absence of ligand (so as to enhance its adhesion activity prior to encounter the opsonized particles), we studied *in vitro* the effect of some of these agonists (TNF, GM-CSF, C5a, NAP-1/IL-8, fMLP and PMA) on neutrophil's CR1 distribution and function. None of these agents was able to significantly increase CR1 aggregation, nor to modify immune complexes binding efficiency (Paccaud et al, in press). By contrast, *in vivo*, exudated neutrophils (obtained by the skin window technique) did show an increase in CR1 aggregation as compared to circulating neutrophils. Thus CR1 are capable to aggregate in the absence of ligand under the complex combination of the stimuli present during the exudation process. Taken together, these observations suggest that in the circulation, only CR1 present on erythrocytes are of importance in regard to complement regulation and immune complexe transport and elimination. However, neutrophils recruted at inflammatory sites bear activated CR1 organized into small clusters: such a distribution could enhace neutrophil interaction with complement activating substances (i.e. bacteria, immune complexes), and eventually lead to their ingestion.

In the second part of this work, we have examined the mechanisms by which CR1 is internalized and the effect of chemoattractant receptor activation in this process. To that end, CR1 were tagged with ^{125}I-anti-CR1 antibodies and localized by quantitative E.M. autoradiography. These morphological studies allowed us to reach the following conclusions. First CR1 does not concentrate in coated-pits, and is internalized by a macropinocytotic process resembling phagocytosis (Carpentier et al, in press). Thus, internalization of CR1 differ from the uptake of most polypeptide hormones, growth factor and transport protein receptors, which occur through the coated pit route. Second, internalization requires receptor cross-linking, since Fab fragments of anti-CR1 antibody are poorly internalized. That Fc portion of anti-CR1 antibody participate in CR1 internalization can be ruled out, since f(ab)'$_2$ fragments behave similarly to whole antibody. Third, intact microfilaments are necessary to the internalization of CR1. Indeed, treatment with cytochalasin B inhibited CR1 internalization. Fourth, internalized material is found in association with large flattened vacuoles, organized in stacks and probably representing a particular form of endosome before being targetted to lysosomes. Fifth, fMLP increases CR1 internalization without affecting their clustering or their lack of association with coated-pits. We also showed that this fMLP-mediated increase in CR1 internalization was independent of cytosolic free calcium elevation, as blockade of intracellular calcium elevation was without effect on this process. On the other hand, pertussis toxin inhibited the fMLP-induced increase in CR1 internalization, demonstrating that this cross-talking is probably due to second messenger generation, and possibly protein kinase C activation (Fig. 2)

Figure 2:

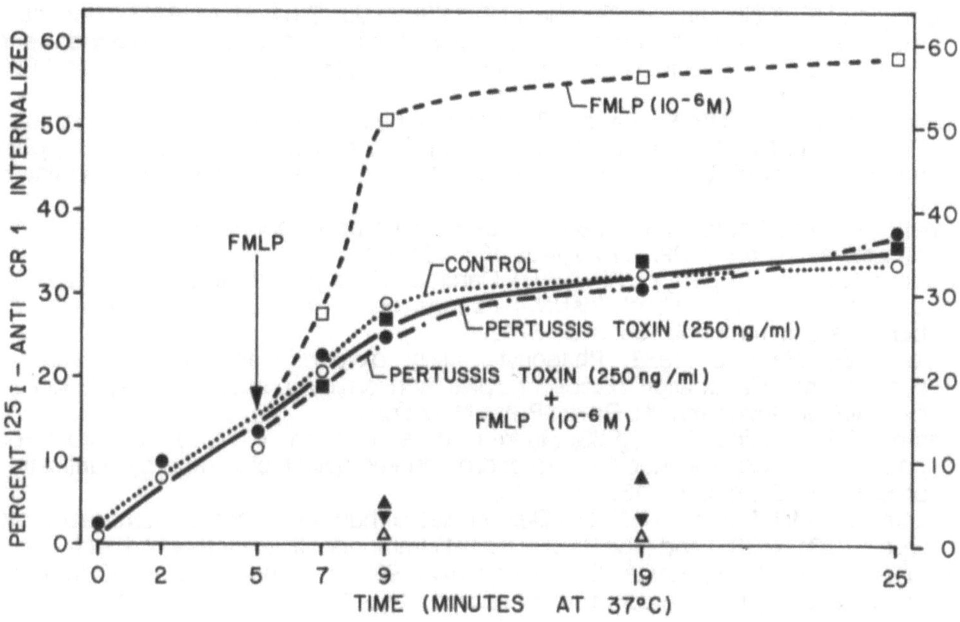

Effect of fMLP in the presence or absence of pertussis toxin (PT) on the internalization of ^{125}I-anti CR1 antibody inhuman neutrophils. Cells were preincubated with 250 ng/ml PT (●,■) or with medium alone (o,□) for 2 hrs at 37°C, washed and further incubated with ^{125}I-anti-CR1 for 1 hrs at 4°C. After 3 washes, cells were incubated at 37°C for various periods of time. When indicated, fMLP (1 µM) was added (□ ,■). The internalization of the radioactive ligand was determined by quantitative EM autoradiography. (From Carpentier et al, in press, with permission).

In conclusion, we have demonstrated first, that the surface distribution of CR1 is the main factor in regulating interactions of erythrocytes, or neutrophils, with complement activating particles: adherence is increased by the clustering of CR1. Thus, in the circulation, only erythrocytes are capable to efficiently interact with such particles, neutrophils being mainly transported by the blood stream until stimulated to leave the intravascular space. Second, when neutrophils have reached inflammatory sites and become activated, crosstalking between chemoattractant receptors and CR1 may further modulate the phagocytotic process, thus couls be of utmost physiological importance in vivo.

REFERENCES:

Berger M, Birx DL, Wetzler EM (1988) Tumor necrosis factor is the major monocyte product that increases complement receptor expression on mature human neutrophils. Blood 71: 151-158.

Berger M, Birx DL, Wetzler EM, O'Shea JJ, Brown EJ, Cross AS (1985) Calcium requirements for increased complement receptor expression during neutrophil activation. J. Immunol. 135: 1342-1348.

Bohnsack HF, Kleinman HK, Takahashi T, O'Shea JJ, Brown EJ (1985) Connective tissue proteins and phagocytic cell function. Laminin enhances complement and Fc-mediated phagocytosis by cultured human macrophages. J. Exp. Med. 161: 912-923.

Carpentier J-L, Lew DP, Paccaud J-P, Gil R, Iacopetta B, Kazatchkine M, Stendhal O, Pozzan T (1990) Internalization pathway of C3b receptors in human neutrophils and its transmodulation by chemoattractant receptors stimulation. Cell Regulation (in press).

Changelian PS, Fearon DT (1986) Tissue-specific phosphorylation of complement receptors CR1 and CR2. J. Exp. Med 163: 101-115.

Edberg JC, Wright E, Taylor RP (1987) Qantitative analysis of the binding of soluble complement-fixing antibody/dsDNA immune complexes to CR1 on human red blood cells J. Immunol. 139: 3739-3747.

Elsbach P, Weiss J (1988) Phagocytic cells: oxygen-independent antimicrobial systems in: Inflammation basic principles and clinical correlates (eds) Gallin JI, Golstein IM, Snyderman R. Raven Press New York.

Fearon DT, Collins LA (1983) Increased expression of C3b receptors on polymorphonuclear leukocytes induced by chemotactic factors and by purification procedures. J. Immunol. 130: 370-375.

Kazatchkine MD, Fearon DT (1990) Deficiencies of human C3 complement receptors type 1 (CR1, CD35) and type 2 (CR2, CD21) Immunodeficiency Rev. 2:17-41.

Madi N, Paccaud J-P, Steiger G, Schifferli JA (1990) Immune complex (IC) binding efficiency of erythrocyte complement receptor type 1 (CR1). Clin. Exp. Immunol. (in press).

Paccaud J-P, Carpentier J-L, Schifferli JA (1988) Direct evidence for the clustered nature of complement receptor type 1 on the erythrocyte membrane. J. Immunol. 141: 3889-3894.

Paccaud J-P, Carpentier J-L, Schifferli JA (1990) Difference in the clustering of complement receptor type 1 (CR1) on polymorphonuclear leukocytes and erythrocytes: effect on immune adherence Eur. J. Immunol. 20: 283-289.

Paccaud J-P, Schifferli JA, Baggiolini M (1990) NAP-1/IL-8 induces up-regulation of CR1 receptors in human neutrophil leukocytes. Biochem. Biophys. Res. Comm. 166: 187-192.

Paccaud J-P, Carpentier J-L, Schifferli JA (1990) Exudation induces clustering of CR1 receptors at the surface of human polymorphonuclear leukocytes. Biochem. Biophys. Res. Comm. (in press).
Pommier CG, O'Shea J, Chused T, Yancey K, Frank MM, Takahashi T, Brown EJ (1984) Studies of the fibronectin receptors of human peripheral blood leucocytes. J. Exp. Med. 159: 137-145.
Snyderman R, Smith CD, Verghese MW (1986) Model for leukocyte regulation by chemoattractant receptors: roles of a guanine nucleotide regulatory protein and polyphosphoinositide metabolism. J. Leuko. Biol. 40: 785-800.
Stossel TP (1988) The mechanical responses of white blood cells in: Inflammation basic principles and clinical correlates (eds) Gallin JI, Golstein IM, Snyderman R. Raven Press New York.
Unkeless JC, Wright SD (1988) Phagocytic cells and complement receptors in: Inflammation basic principles and clinical correlates (eds) Gallin JI, Golstein IM, Snyderman R. Raven Press New York.
Wong WW, Cahill JM, Rosen MD, Kennedy CA, Bonaccio ET, Morris MJ, Wilson JG, Klickstein LB, Fearon DT (1989) Structure of the human CR1 gene. Molecular basis of the structural and quantitative polymorphisms and identificaion of a new CR1-like allele. J. Exp. Med. 169: 847-863.

1.2. CLATHRIN-COATED PITS AND CLATHRIN-INDEPENDENT ENDOCYTOSIS

(Chairperson : Bo VAN DEURS)

As a major door of entry, clathrin-coated pits have received considerable attention, for physiological and practical reasons. Coated pits can be regarded as a rate-limiting stage in receptor-mediated endocytosis. In addition, they are generally easy to identify by conventional electron microscopy, and can be decorated by anti-clathrin antibodies. It is thus no surprise, but still gratifying to note how our understanding of the dynamics, the biochemistry and the molecular biology of clathrin-coated pits has recently progressed.

Jean Davoust exploited confocal laser microscopy to discriminate coated pits at the cell surface from those in the perinuclear region and to provide three-dimensional reconstruction. With the combination of microinjection of fluorescent clathrin into cultured cells and analysis of recovery after fringe pattern fluorescence photobleaching, the dynamics of clathrin triskelion assembly and depolymerization in surface coated pits of control Vero cells was found to be extremely fast, with about 5 cycles per minute. This value has to be compared with current estimations of a one-minute life-time for a coated pit. ATP and pH homeostasis are both required to support the rapid turn-over of clathrin between the plasma membrane and the soluble pool.

Ernst Ungewickell used a clathrin assembly assay to identify clathrin-associated proteins in bovine brain. The adaptin complexes HA_2 (known as plasma membrane adaptor, containing four proteins named α, β, 50 kDa and 16 kDa) and HA_1 (Golgi-specific, composed of β', γ, 47 kDa and 20 kDa) are ubiquitous. Proteins AP180 and auxilin are neurone-specific. In close similarity to the known structure of HA_2, the β' and γ subunits of HA_1 are structured into cores (associated with the small subunits) and appendages, which can be readily separated by mild proteolysis. Appendages are the only phosphorylation sites. In bovine brain coated vesicles, AP180 is the most efficient assembly protein for clathrin heavy chains, followed by HA_2. In contrast, clathrin light chains are inhibitory. According to the current model, adaptors attach to clathrin by their β-type subunit and both cores and appendages are required for a high-affinity interaction. The receptor binding domain(s) has not been identified so far.

NATO ASI Series, Vol. H 62
Endocytosis
Edited by P. J. Courtoy

Margaret Robinson complemented this structural information by molecular biology analysis. She discovered two separate genes for the α subunit. Their N-terminal domains are incorporated in the cores (also called "heads"), a hinge region is rich in proline and glycine and the C-terminal domains correspond to the appendages (also called "ears"). Although one product is neurone-specific, both genes can be expressed in the same cell. The β and β' subunits are highly homologous, but the N-termini of the γ (Golgi) and α (plasmalemmal) adaptins are quite distinct. What determines organelle targeting remains to be elucidated.

Finally, **Bo Van Deurs** called the attention to a distinct, clathrin-independant endocytic pathway. Indeed, when cells are K^+-depleted or acidified, receptor-mediated endocytosis of transferrin is almost arrested whereas the adsorptive tracer ricin and the fluid-phase tracer lucifer yellow are still internalized to a large extent. An ultrastructural assay, designed to detect newly formed endocytic vesicles, revealed two distinct populations of pre-endosomal vesicles, present at approximately the same frequency, one been coated and the other been non coated and somewhat smaller. It is likely that the two pathways are however not of equal importance in every cell type or at every moment. This may represent a new mechanism for the regulation of endocytosis.

THE DYNAMICS OF CLATHRIN COATS IN LIVING CELLS MEASURED BY ADVANCED FLUORESCENCE MICROSCOPY.

JEAN DAVOUST AND PIERRE COSSON (*).

European Molecular Biology Laboratory, Postfach 102209,

D-6900 Heidelberg, Germany

and Centre d'Immunologie INSERM-CNRS de Marseille-Luminy

Case 906, 13288 Marseille, Cedex 9, France. (For correspondence).

(*) Present address is Cell Biology and Metabolism Branch NICHD, Build. 18T NIH, 9000 Rockville Pike, Bethesda MD 20892, USA.

INTRODUCTION

Clathrin coated pits and coated vesicles are responsible for receptor-mediated endocytosis of a wide variety of ligands that bind at the cell surface (Anderson et al., 1977 ; Goldstein et al., 1985). Clathrin coated pits are also involved in the sorting of receptors to be transported to intracellular acidic compartments such as the Mannose-6-phosphate receptor rich pre-lysosomal compartment (Lemansky et al., 1987), the secretory granules in pancreatic endocrine insulin-secreting cells (Orci et al., 1987), or the ACTH-containing vesicles in anterior pituitary cells (Tooze and Tooze, 1986). Clathrin triskelions are composed of three heavy (180 kD) and three light (30 kD) chains. They can assemble in vitro at low pH to form clathrin baskets, the geometry of which has been extensively studied (Crowther and Pearse, 1981 ; Harrison and Kirchhausen, 1983 ; Blank and Brodsky, 1986). Clathrin triskelions are associated with adaptins (Pearse, 1988; Ahle et al., 1988; Ahle and Ungewickell, 1989), which are forming multi-units complexes specific for either plasma membrane or Golgi-derived vesicles (Robinson and Pearse, 1986 ; Robinson, 1987, 1989). In coated vesicles, the adaptins are presumably located between

NATO ASI Series, Vol. H 62
Endocytosis
Edited by P. J. Courtoy

the external shell formed by clathrin triskelions and the cytoplasmic domains of transmembrane proteins with which they might interact (Vigers et al., 1986). It is generally accepted that in addition to clathrin, adaptins are required at the level of the plasma membrane or of Golgi derived membranes to form coated pits that bud into coated vesicles. The coat is then removed presumably by an uncoating ATPase characterized in vitro (Schlossman et al., 1984 ; Brael et al., 1984) which is identical to the hsc 70 heat shock protein (Rothman and Schmid, 1986).

The kinetics of clathrin depolymerization visualized by electron microscopy provided an estimate of less than 2 min for the life time of the clathrin coated vesicles containing internalized LDL molecules (Anderson et al., 1977). Other studies based on a rapid cell lysis followed by quantitation of clathrin in vesicular versus cytoplasmic fractions demonstrated the existence of a pool of free clathrin (Goud et al., 1985 ; Bruder and Wiedeman, 1986). The effect of cytoplasmic acidification and hypertonic media on clathrin lattice morphology has been recently described at the ultrastructural level (Heuser, 1989 ; Heuser and Anderson, 1989). However, these experiments do not tell us what is the half-life time of clathrin coats and what are the amounts of free versus polymerized clathrin under the different experimental conditions. In order to investigate these questions, we used fringe pattern fluorescence photobleaching (Davoust et al., 1982) to measure directly the mobility of fluorescent triskelions microinjected to living cells. Our results indicate that 45 % of the total cellular pool of clathrin can diffuse freely in the cytoplasm and that clathrin coats have a half-life time of 11 seconds. Low cytoplasmic pH, ATP depletion, low temperature or hypertonic media which are known to inhibit membrane transport were found to affect drastically the dynamics of clathrin polymerization in the cytoplasm.

MATERIALS AND METHODS

Isolation and labeling of clathrin triskelions

Isolation of pure rat liver coated vesicles was performed at 4°C essentially as already described (Pearse 1983 ; Daiss and Roth, 1983 ; Simion et al., 1983). For labeling, we used tetra-methyl rhodamine 5-6 iodoacetamide dissolved in DMSO added to the polymerized clathrin coated vesicles as described (Cosson et al., 1990). The labeled vesicles were separated from the free rhodamine and were dialysed against 0.5 M Tris 1 mM EDTA pH=7.7 at 4°C to induce clathrin depolymerization. After pelleting, the supernatant containing depolymerized coat proteins was applied on a Pharmacia CL4B gel filtration column eluted with 20 mM Tris pH=8.0, 20 mM KCl. The rhodamine-labelled clathrin triskelions migrated as a distinct peak shortly after the void volume. To assay the polymerization of clathrin, baskets were obtained by dialysing a sample containing 1 mg/ml of pure rhodamine labeled triskelions and 0.2 mg/ml of pure non labeled assembly proteins (also isolated on a CL4B column) for 24 hours at 4°C against 20 mM MES pH=6.25, 2 mM CaCl2. After pelleting, the coats were observed in electron microscopy using negative staining with 1 % uranyl acetate.

Cell culture, microinjection and Immunofluorescence

Vero cells (African green monkey kidney cells), HepG2 (human hepatoma cells) and AtT20 cells (mouse anterior pituitary cells), Hep G2 cells (human hepatoma cells) were cultured in DME supplemented with 10 % FCS. Microinjection was performed on a Zeiss work station equiped with an IM35 microscope (C. Zeiss Oberkochen FRG) and installed by W. Ansorge, (EMBL Heidelberg). For immunofluorescence staining, we used the monoclonal mouse anti-clathrin heavy chain antibody X22 (kindly provided by F. Brodsky, University of California, San Francisco), followed by an anti-mouse IgG fluorescein-labelled antibody (kindly provided by T. Kreis, EMBL).

The inhibition of cellular metabolism under the different conditions was performed for 15 min. at 37°C prior to the photobleaching experiments as follows : ATP depleted cells were incubated in PBS supplemented with 1 mM CaCl, 0.5 mM MgCl2, 2 mg/ml BSA (Sigma), 10 mM NaN3 and 50 mM deoxyglucose. Less than 1 % of total ATP was present in the ATP depleted cells and normal ATP levels were restored after 30 min of reversion with cell culture medium at 37°C. Low pH treated cells were exposed to DME without bicarbonate supplemented with 20 mM MES and buffered at pH 5.7 (Davoust et al., 1987). High ionic strength was obtained in Hanks'Saline buffer supplemented with 2 mg/ml of BSA and 300 mM of NaCl. Low temperature treated cells were pre-equilibrated for 15 min on ice with Hanks' Saline buffer containing 2 mg/ml BSA. For reversion experiments, the cells were incubated for 30 min in their culture medium in the presence of 5 % CO2 atmosphere.

Fringe pattern fluorescence photobleaching

In order to measure the lateral motion of fluorescent molecules in living cells, a fringe pattern illumination was added to an inverted fluorescence microscope (IM 35, Zeiss, Oberkochen, FRG). The microscope was adapted with a home made temperature controled stage using a vertical stream of regulated flow nitrogen gas. Final temperature of the specimen could be equilibrated from 6 to 40°C with 0.1°C precision. The fluorescent cells are illuminated by vertical planar fringes produced in the cross volume of two interfering beams as described (Davoust et al., 1982). For that reason, the instrument is sensitive to the motion of individual molecules in the focal plane and also above and below this plane ($\pm$ 20 μm). After fringe pattern photobleaching of a single fluorescently labeled cell, the contrast of the striped object is directly measured as a function of time by rapidly scanning the fringe pattern back and forth (Davoust et al. 1982, Lanni and Ware, 1982 ; Koppel and Sheetz, 1983). Other related pattern photobleaching methods using a square grid imaged on fluorescent cellular specimens have been used to estimate the lateral diffusion coefficient of membrane receptors (Smith and McConnell 1978, Koppel and Sheetz 1983, Myles et al. 1984). The fringe pattern photobleaching method cumulates several advantages over spot photobleaching : it averages the lateral mobility signal over a defined field that encompass a whole cell ; low bleaching doses are delivered, here 10 to 20 % of the total fluorescence ; the repetitive distance of the pattern and the exponential decay time constants related to several diffusion components are measured without image processing.

RESULTS and DISCUSSION

Isolation and characterization of fluorescent clathrin

Rat liver coated vesicles were isolated and directly labeled in a polymerized form with rhodamine iodo-acetamide in order to preserve the sites involved in clathrin-clathrin interactions. Labeled coat proteins were then depolymerized and separated by a gel filtration column (Fig 1). The fluorescent clathrin peak (dashed) was concentrated on a DE52 anion exchange column and analyzed by SDS-gel electrophoresis to check its purity. It was composed of pure clathrin labeled on both heavy and light chains. The amount of labeling, deduced from the optical density at 565 nm and a protein assay, was of three rhodamine groups per clathrin molecule, that is, nine rhodamine groups per clathrin triskelion. To check that the polymerization properties of clathrin were unaffected by the labeling and the

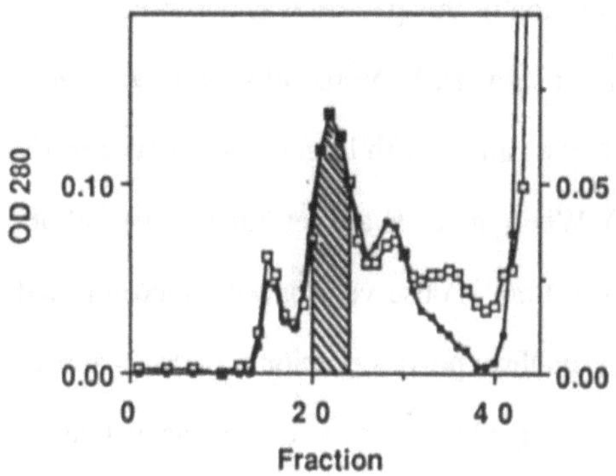

Fig. 1. Isolation of labeled clathrin on a CL4B column. Rat liver coated vesicles were depolymerized, centrifuged at high speed and the supernatant applied on a CL4B gel filtration column. OD_{280} and OD_{565} were measured on each sample. The shaded area points to the fractions that were collected and that contain pure rhodamine labeled triskelions.

isolation protocol, we tested the ability of the final product to polymerize in vitro in the presence of non-labeled assembly proteins (corresponding to peak 3, Figure 1). This fraction contains 100 kD, 50 kD and 15 kD proteins as checked by gel electrophoresis (data not shown). As shown by others, the presence of assembly proteins is necessary to obtain regularly sized clathrin baskets (Vigers et al., 1986). After dialysis for 24 hours against polymerization buffer, 94% of the rhodamine labeled clathrin can be pelleted by airfuge centrifugation at high speed. When visualized in electron microscopy after negative staining the pellet appears composed of typical clathrin baskets (data not shown).

Next, we observed the distribution of fluorescent clathrin microinjected into Vero cells. We used a 0.5 mg/ml of fluorescent clathrin solution and assuming a cytoplasmic protein concentration of 200 mg/ml, there should be approximately 0.5 mg/ml of endogenous clathrin in the cytoplasm (Goud et al., 1985). Since we microinjected less than 10 % of the cytoplasmic volume, the labeled clathrin represents less than 10 % of the endogenous clathrin. Microinjected clathrin gives rise to a typical punctuate pattern indistinguishable from the pattern of total clathrin (microinjected+endogenous) visualized by immunofluorescence using a mouse monoclonal anti-heavy chain X22 kindly provided by F. Brodsky. Such a cell can be visualized by confocal microscopy allowing to discriminate several planes of section where both the plasma membrane in contact with the cell support (fig. 2A) and the perinuclear region of the cell (fig. 2B) can be recorded. This colocalisation indicates that microinjected clathrin incorporates to the endogenous clathrin pool of the cell (color prints not shown). No alteration of this pattern was observed from 2 minutes to 24 hours after microinjection (data not shown). In Vero cells exposed to acid medium the initial fine dotty pattern is replaced by a pattern with larger patches, as already described for BHK cells (Davoust et al., 1987). When the cells are returned to neutral pH using bicarbonate containing medium, the initial pattern is observed for both microinjected and endogenous clathrin (data not shown). Thus, clathrin polymerization properties appear unaffected by labeling and isolation procedure, as judged by an in vitro assay as well as by its behavior when microinjected into living cells.

Photobleaching of microinjected fluorescent clathrin

The principle of the photobleaching technique is to bleach the fluorescence with a laser light in a defined area of a cell and to monitor the recovery of fluorescence from the unbleached to the bleached areas. With the fringe pattern photobleaching device we were using (Davoust et al., 1982), the whole cell is bleached along stripes and we measure the disappearance with time of the contrast between bleached and unbleached stripes. In such experiments a free cytoplasmic protein that diffuses rapidly from bleached to unbleached areas causes a rapid decrease of the initial contrast. On the opposite, a protein engaged in a stable structure cannot diffuse from one area to another, and the contrast created by photobleaching is stable. If the protein is able to depolymerize from a structure with a half-life time larger than the time scale of diffusion, then the contrast disappears with this longer half-life time, independent of the width of the stripes. Thus, fringe pattern photobleaching experiments allow to distinguish a free protein from a protein engaged in a structure and to measure the half-life time of long-lived polymerized structures.

A typical trace from a fringe pattern photobleaching experiment performed on a single cell microinjected with labelled clathrin and maintained at 37°C, is shown on fig. 3. The contrast of the bleached cell is represented as a function of time. Before the bleaching pulse the cell do not contain a periodic fluorescence profile. After bleaching of 10 to 20 % of the

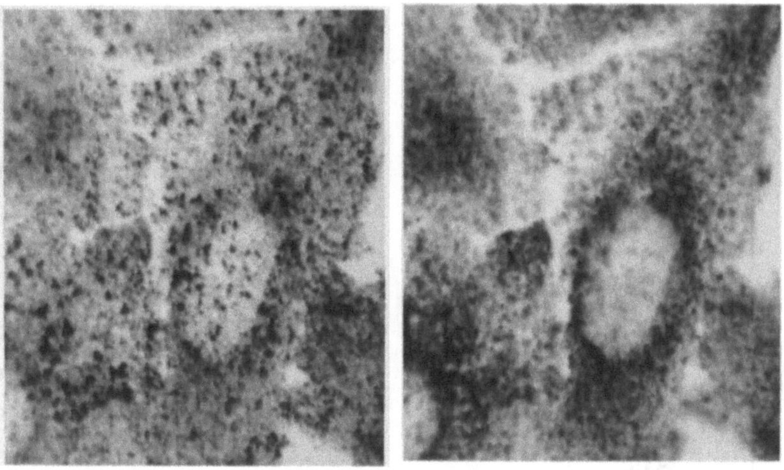

Fig. 2. Observation of cells microinjected with rhodamine-labeled clathrin. Cells microinjected with rhodamine-labeled clathrin were incubated two hours in MEMc in a 5 % CO2 incubator. The dishes were then fixed in 3 % paraformaldehyde and processed for immunofluorescence using monoclonal anti-clathrin X22 as first antibody and a rabbit fluorescein-labeled anti-mouse IgG fraction as second antibody. They were then observed with a LEICA confocal microscope tuned on the fluorescein channel. (a) plane of focus at the cell-substratum surface, (b) plane of focus across the cell nucleus. Total field equals 40x40 μm, the thickness of the optical section is 0.5 μm.

---

fluorophore, fluorescent stripes are generated and give rise to a positive signal which decays with time. Computer analysis revealed that the decaying curve is the sum of two components : a rapidly equilibrating component (t1/2 =0.69 sec ± 0.02) represents 39 ±2 % of the initial contrast. A slower component (t1/2= 11.1 ±0.3 sec) represents 49 ± 1 % of the initial contrast. As observed in all our photobleaching experiments from 10 % to 14 % of the initial contrast does not disappear even after prolonged observation. 12 to 20 cells were analyzed per specimen and the variation in total fluorescence intensity from cell to cell did not lead to significant differences in the decaying curves. We interpret the two main components as follows :

The rapid component corresponds to the diffusion of free clathrin in the cytoplasm. The diffusion coefficient calculated from the 0.7 second half-life time and the interfringe (2 μm) equals 10^{-9} cm^2/sec, about ten times slower than that of free tubulin or of free bovine

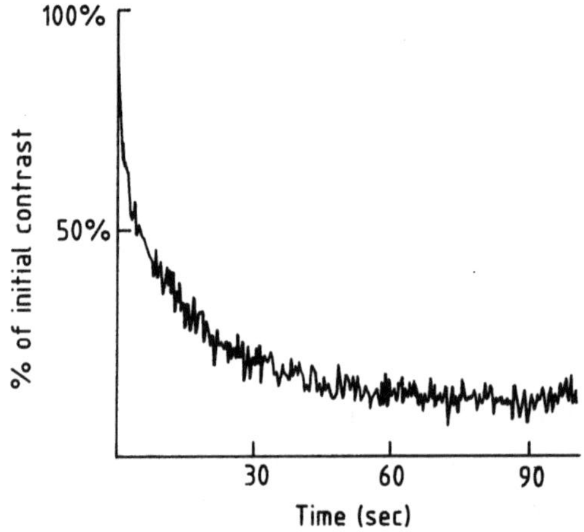

Fig. 3. Mobility of fluorescent clathrin microinjected into living cells.
Photobleaching experiments were performed on living cells 3 to 4 hours after microinjection with rhodamine-labeled clathrin. Photobleaching is performed at time 0 and the contrast is recorded and plotted as a function of time. The photobleaching was performed on cells microinjected with rhodamine-labeled clathrin kept on the stage at 37°C in microchambers.
The contrast of the stripes vanishes as free clathrin diffuses with a half time of less than a sec (needle sharp component) and as clathrin coats depolymerize (slow component with a half time of 11 sec).

---

serum albumin microinjected into Vero cells (Wojcieszyn et al., 1981 ; Jacobson and Wojcieszyn, 1984) and checked also with our instrument. This is not surprising, as clathrin is present in the cell as a triskelion having a molar weight around 600.000 and an extended shape (Ungewickell and Branton, 1981) that probably slowers diffusion. As expected the half life time is dependent on the width of the stripes : it increases to 3 sec when the interfringe is set to 10 μm.

The slower component represents 55% of cellular clathrin. As indicated by its slower relaxation this second component corresponds to clathrin engaged in structures. We interpret the relaxation of the contrast as depolymeryzation of clathrin structures with a half-life time of 11 seconds. However we had to consider the possibility that this relaxation would be caused by movement of whole coated vesicles instead of depolymerization of the coat. This is excluded by the three following experiments. First as compared with the observation of saltatory transport (Gilbert et al., 1985), one would expect vesicles to be transported from time to time with a mean velocity in the range of 0.5-2 μ/min whereas our results would correspond to a complete movement of all the clathrin coats within 11 sec on a distance scale of about 1 μm. Second, the decay curve is insensitive to a 10 μM nocodazole concentration that depolymerize all the microtubules of Vero cells. If we were to measure a vesicular movement, it does not occur along microtubules. Third, when varying the interfringe from 2 to 10 μm we observe identical curves with the same half-life time of 11 seconds for the slow component. If the slow decay of contrast was caused by transport of coated vesicles from unbleached to bleached areas, the time required should increase with the interfringe, at least by 4 or 5 folds as for the free clathrin (see above). On the contrary if the rate-limiting step is depolymerization of clathrin structures, the kinetics is independent upon the interfringe.

Effect on clathrin dynamics of ATP depletion, low pH, low temperature, and high osmotic strength

Using this assay of clathrin dynamics, we tested the effect of conditions known to affect vesicles budding and transport : exposure to low temperature ,low pH, or high osmotic strength, and ATP depletion. We found that these four conditions have a common effect on the stability of the clathrin coats : they strongly stabilize all the clathrin structures contained in the cell (fig. 4) : the slower component of the photobleaching traces does not relax rapidly and it persists even after 3 minutes, indicating that the clathrin-containing structures are now stable. The inhibition of membrane traffic observed in these conditions is probably related to the stabilization of clathrin-containing structures in the cell.

However the various conditions affect differentially the fraction of free clathrin. ATP depletion (to less than 1 % of control values) as well as cytosolic acidification do not affect the pool of free clathrin in living cells (fig. 4 , a and b). On the opposite in cells exposed to low temperature (fig. 4, c), the pool of free clathrin is reduced to 25% of total. Upon exposure to high osmotic strength (fig. 4, d), the effect is even more drastic : less than 10% of the clathrin is freely diffusing in the cytoplasm. These changes in the fraction of free

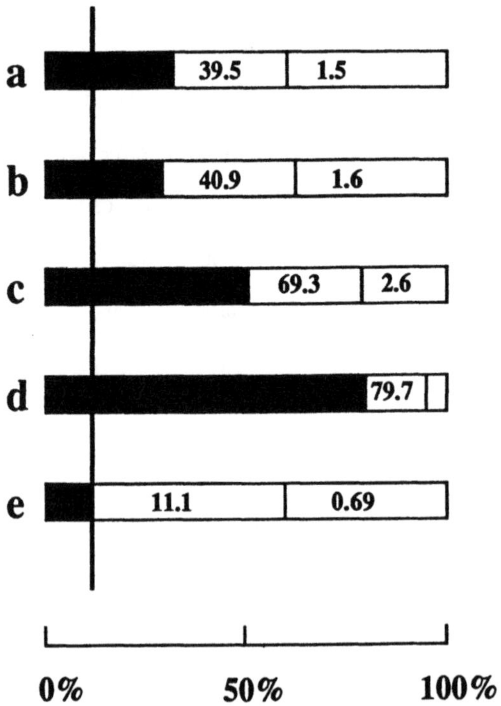

Fig. 4. Effect of temperature, low pH and low ATP and high ionic strength on intracellular clathrin.
Experiments were performed as indicated in figure 3 except that the cell culture conditions or the temperature were varied. The fluorescence recovery curves were fitted by two exponential components plus a constant base line using a spline mode algorithm (Pollerberg et al., 1986). For each experimental condition, the proportion of the separate components is shown as an horizontal bar up to 100% and the number inside the bars represents the half time constant of each component expressed in sec.
The fraction of immobilized rhodamine labeled clathrin is represented in black, the decaying component corresponding to clathrin depolymerization and the rapidly decaying component corresponding to free clathrin are in white. The results are represented as the mean of 12 to 20 single cell measurements per condition. The standard deviations not represented here do not exceed 3 % for the proportion of the separate components and 12 % for half-decay time constants.
Experimental conditions used : (a) Low pH treated cells, DME pH=5.7 at 37°C; (b) ATP depleted cells, (c) cell incubated in Hank's BSA at 6°C, (d) cells exposed to hypertonic media, Hank's BSA + 300 mM NaCl at 37°C and (e) to the control conditions in Hank's BSA at 37°C.

---

versus polymerized clathrin coats are reversible when the cells are returned to the physiological conditions.

We correlated this reduction of free clathrin pool with an increase of clathrin-containing structures visualized in electron microscopy, in conditions where coated pits are still clearly visible on the cell surface. For that we quantitated the surface area and the average size of coated pits at the cell surface of Vero cells incubated in the same conditions as previously described (Cosson et al. 1989). We found that in cells exposed to low temperature the clathrin coats present at the plasma membrane were two times more abundant than in control cells (table I). No effect was observed in cells depleted of ATP or exposed to low pH, when compared to control. In the cells exposed to hypertonic media, the clathrin is polymerized as microcages in the cytoplasm (Heuser and Anderson, 1989) and coated pits and coated vesicles cannot be clearly identified and counted on sections. Altogether, the quantitation by electron microscopy confirms the results obtained by our fluorescence photobleaching measurements : the pool of polymerized clathrin is unaffected at low pH or in ATP-depleting conditions, but is increased at 6°C, and represents more than 90 % of total clathrin in cells exposed to high osmotic strength.

CONCLUSION

Half life time of clathrin coats is 11 sec

We have investigated the kinetics of clathrin depolymerization in living cells microinjected with rhodamine-labeled clathrin and analyzed by pattern Fluorescence Photobleaching. The fluorescent clathrin was fully functional as judged by its ability to reform clathrin baskets in vitro, and it incorporates to the endogenous pool of clathrin when microinjected. We monitored the dynamics of clathrin triskelions in living cells using a fringe pattern photobleaching device built in our group which allows to analyze a whole cell at once. Two components could be clearly distinguished in Vero cells : a rapid decay (45% of initial contrast, t 1/2 equals 0.7 sec) corresponds to rapid diffusion of free clathrin. A slower decay corresponds to progressive release of clathrin molecules from polymerized structures (55% of contrast, t 1/2 = 11 sec). Although we cannot formally exclude that we are actually measuring the exchange of triskelions between stable coats and the pool of free clathrin, it seems to us unlikely that a triskelion could be extracted from the intricated structure of a clathrin coat without destruction of the whole coat. Since we did not find a slower

TABLE I. Size of cell surface coated pits in Vero cells+.

	% of the cell surface *	Diameter in µm *
Control conditions	1.3 ± 0.4	0.40 ± 0.10
ATP depleted cells	0.8 ± 0.3	0.31 ± 0.10
Low pH treated cells	1.6 ± 0.4	0.37 ± 0.07
Cells at 6°C	3.1 ± 0.7	0.36 ± 0.09

* The results are expressed as the mean ± standard deviation. The number of coated pits counted varied from 60 to 100 per condition.
+ Cell culture conditions were identical to those described in figure 4. The cells were then prepared for epon embedding as previously described (Griffiths et al., 1985 ; Cosson et al., 1989). For estimation of cell surface coated pits, 40 to 50 random micrographs were taken of epon section in each conditions. The amount of membrane in coated pits was related to the total amount of plasma membrane using a double lattice grid (D164 ; Weibel, 1979) which allows also the estimation of the mean length of the coated pits.

component, clathrin coats derived either from the Golgi complex or from the plasma membrane have apparently the same dynamics. Given the fact that about half of the clathrin is free within the cytoplasm, its polymerization time must be about 11 seconds as well. Therefore, the whole polymerization-depolymerization cycle takes about 20 seconds. These results are in agreement with the initial results of Anderson et al. (1977) : they followed by electron microscopy a wave of internalization at 37°C of LDL-ferritin prebound at the cell surface at 4°C and set up an upper limit of 2 minutes for the half-time of plasma membrane clathrin coats. The finding that 45 % of total clathrin is free in the cytoplasm is also coherent with a previous study (Goud et al., 1985) where Vero cells were quickly lysed at 20°C and about 50 % of the clathrin was found in a high-speed supernatant, as detected by an enzyme-linked immunoassay. The authors also found that the amount of free clathrin was reduced in highly secreting cells. It is interesting to note that we found the same amounts of free clathrin in cultured Hep 2 cells, AtT20 cells or Vero cells, whereas in HepG2 cells (highly secreting hepatoma cells) the amount of free clathrin was reduced to 25 % of total (data not shown). Using an approach identical to that of Goud et al., Bruder and Wiedeman (1985) found smaller amounts of free clathrin. This is probably due to the fact that their cells were exposed to 4°C prior to lysis. As we show below, the amount of free clathrin is reduced at 4°C.

Effect of low pH, ATP depletion, low temperature, high osmotic strength on clathrin cycle

Next, we have investigated the stability of clathrin coats under various conditions known to inhibit membrane traffic such as cytoplasmic acidification (Davoust et al., 1987 ; Cosson et al., 1989), ATP depletion (Hertel et al., 1986 ; Sullivan et al., 1987), low temperature (6°C) and hypertonic media (Daukas and Zigmond, 1985 ; Heuser and Anderson, 1989). We found that in these various conditions the inhibition of transport was accompanied, and presumably caused, by a stabilization of polymerized coat structures. It must be noted that this applies to all the clathrin coats located at the level of the plasma membrane or not. The formation of plasma membrane or Golgi-derived clathrin-coated vesicles obey to similar mechanisms and it seems logical that low pH, low temperature, ATP depletion or high ionic strength have pleiotropic effects on different pathways of transport. Interestingly the amount of free clathrin was differentially affected. Whereas at low pH or in ATP depleted cells the quantity of free clathrin was unchanged, it was reduced at 4°C concomitantly to an increase in the number of coated pits at the level of the plasma membrane. Free clathrin was also absent when cells were incubated in hypertonic medium. These results are in agreement with the morphological observations reported by the group of J. Heuser (Heuser and Kirchausen, 1985 ; Heuser and Keen, 1988 ; Heuser, 1989 ; Heuser and Anderson, 1989). At low pH the clathrin coats appear normal although more curved and with some microcages polymerizing at the edges. On the contrary when the cells are exposed to hypertonic media, clathrin is seen as microcages scattered on the inner side of the plasma membrane. Our conclusions meet the proposal by the authors that at high osmotic strength most of the clathrin of the cell is polymerized and thus not available for assembly into new coated pits.

As a possible interpretation of the contradicting effects of ATP depletion and exposure to an hypertonic medium, we propose that clathrin nucleation factors are present in limiting amounts and that nucleation itself is an active process requiring ATP. Previous studies showed that binding of coat proteins to membranes is a saturable process in vitro (Unanue et al., 1981; Moore et al., 1987; Mahaffey et al., 1989). A "pre-coated pit" is apparent when clathrin is gently stripped from membranes and supports repolymerization of clathrin on the membranes (Mahaffey et al., 1989). Adaptins and cytoplasmic domains of transmembrane receptors could represent the limiting factors needed for ATP dependant clathrin polymerization. Indeed recent data suggest that cytoplasmic domains of transmembrane receptors can control the formation of coated pits (Iacopetta et al., 1988). When transfecting a cell with increasing amounts of transferrin receptor, the number of coated pits increases at the cell surface. In that respect it is striking that in cells exposed to hypertonic media, the only condition where clathrin is completely polymerized, clathrin forms microcages apparently independent from the cell membrane (Heuser and Anderson 1989). When the need for membrane is bypassed all the clathrin of the cell can polymerize. Adaptins might also be a limiting factor for coated pit formation in vivo. Photobleaching experiments similar to those described here, but using fluorescent adaptins might provide further insights on that subject.

REFERENCES

Anderson, R.G.W., M.S. Brown, and J.L. Goldstein. 1977. Cell. 10:351-364.

Ahle, S., and E. Ungewickell, E. 1988. J. Biol. Chem. 264: 20089-20093.

Ahle, S., A. Mann, U. Eichelsbacher, and E. Ungewickell, E. 1988. EMBO J. 7: 919-929.

Braell, W.A., D.M. Schlossman, S.L. Schmid, and J.E. Rothman. 1984. J. Cell Biol. 99:734-741.

Blank, G.S., F.M. Brodsky. 1986. EMBO J. 5:2087-2095.

Bruder, G., B. Wiedenmann. 1986. Exp. Cell Res. 164:449-462.

Cosson, P., I. de Curtis, J. Pouyssegur, G. Griffiths, and J. Davoust. 1989. J. Cell Biol. 108:377-387.

Cosson, P., R.Peperkok, R.Back, and J. Davoust. 1990. (submitted).

Crowther, R.A., and B.M.F. Pearse. 1981. J. Cell Biol. 91:790-797.

Daiss, J.L., and T.F. Roth. 1983. Methods in Enzymology. 98:337-349.

Daukas, G. and S.H. Zigmond. 1985. J. Cell Biol. 101:1637-1679.

Davoust, J., P.F. Devaux, and L. Leger. 1982. EMBO J. 1:1233-1238.

Davoust, J., J. Gruenberg, and K.E. Howell. 1987. EMBO J. 6:3601-3609.

Gilbert, S.P., R.D. Allen, and R.D. Sloboda. 1985. Nature. 315:245-248.

Goldstein, J.L., M.S. Brown, R.G.W. Anderson, D.W. Russell, and W.J. Schneider. 1985. Ann. Rev. Cell Biol. 1:1-39.

Goud, B., C. Huet, and D. Louvard. 1985. J. Cell Biol. 100:521-527.

Harrison, S.C., and T. Kirchhausen. 1983. Cell. 33:650-652.

Hertel, C., S.J. Coulter, J.P. Perkins. 1986. J. Biol. Chem. 261:5974-5980.

Heuser, J., and T. Kirchhausen. 1985. J. Ultrastructure Res. 92:1-27.

Heuser, J.E., and J. Keen. 1988. J. Cell Biol. 107:877-886.

Heuser, J. 1989. J. Cell Biol. 108:401-411.

Heuser, J.E., and R.G.W. Anderson. 1989. J. Cell Biol. 108:389-400.

Iacopetta, B.J., S. Rothenberger, L.C. Kühn. 1988. Cell. 54:485-489.

Jacobson, K., and J. Wojcieszin. 1984. Proc. Natl. Acad. Sci. USA. 81:6747-6751.

Koppel, D.E. and M.P. Sheetz. 1983. Biophys. J. 43:175-181.

Lanni, F., and B.R. Ware. 1982. Rev. Sci. Instrum. 53:905-908.

Mahaffey, D.T., M.S. Moore, F.M. Brodsky, and R.G.W. Anderson. 1989. J. Cell Biol. 108:1615-1624.

Moore, M.S., D.T. Mahaffey, F.M. Brodsky, and R.G.W. Anderson. 1987. Science. 236:558-563.

Myles, D.G., Primakoff, P. and D.E. Koppel. 1984. J. Cell. Biol. 98:1905-1909.

Orci, L., M. Ravazzola, M.-J. Storch, R.G.W. Anderson, J.-D. Vassalli, and A. Perrelet. 1987. Cell. 49:865-868.

Pearse, B.M.F. 1983. 98:320-326.

Pearse, B. 1988. EMBO J. 7:3331-3336.

Pollerberg, G.E., M. Schachner and J. Davoust. 1986. Nature. 324:462-465.

Robinson, M.S., and B.M.F. Pearse. 1986. J. Cell Biol. 102:48-54.

Robinson, M.S. 1987. J. Cell Biol. 104:887-895.

Robinson, M.S. 1989. J. Cell Biol. 108:833-842.

Rothman, J.E. and S.L. Schmid. 1986. Cell. 46:5-9.

Schlossman, D.M., Schmid, S.L., Braell, W.A., and J.E. Rothman. 1984. J. Cell Biol. 99:723-733.

Simion, F.A., D. Winek, F. Brandan, B. Fleischer, S. Fleischer. 1983. Methods in Enzymology. 98:326-337.

Smith, B.A. and H.M. McConnell. 1978. Proc. Natl. Acad. Sci. USA 75:2759-2763.

Smith, B.A., W.R. Clark and H.M. McConnell. 1979. Proc. Natl. Acad. Sci. USA 76:5641-5644.

Sullivan, P.C., A.L. Ferris, and B. Storrie. 1987. J. Cell. Physiol. 131:58-63.

Tooze, J., and S.A. Tooze. 1986. J. Cell Biol. 103:839-850.

Unanue, E.R., Ungewickell, E., Branton, D. 1981. Cell. 26:439-446.

Ungewickell, E., and D. Branton. 1981. Nature. 289:420-422.

Vigers, G.P.A., R.A. Crowther, and B.M.F. Pearse. 1986. EMBO J. 5:2079-2085.

Weibel, E.R. 1979. Stereological methods.I. Practical methods for biological morphometry. Acad. Press Inc., New York. 318 pp.

Wojcieszyn, J.W., R.A. Schlegel, E.S. Wu, and K.A. Jacobson. 1981. Proc. Natl. Acad. Sci. USA. 78:4407-4410.

ASSEMBLY PROTEINS AND ADAPTORS IN CLATHRIN COATED VESICLES

Ernst Ungewickell, Robert Lindner and Stephan Schröder
Max-Planck-Institute for Biochemistry
D-8033 Martinsried
Germany

INTRODUCTION

Clathrin coated vesicles, which are present in probably all eukaryotes, are the agents for receptor mediated endocytosis and for the transport of lysosomal enzyme receptors from the *trans*-Golgi to the endosomal system (Goldstein et al 1985; Pearse and Bretscher, 1981). In neurons they are conjectured to facilitate also the recycling of synaptic vesicle membrane after neurotransmitter release (Heuser 1989). The functions of the coat components include the binding of specific cargo molecules (receptors), the coordinated assembly of the coat, its reorganization and eventually its dissolution. The emphasis in this report lies on structural and functional aspects of clathrin associated proteins present in coated vesicle populations from bovine brain.

RESULTS

Clathrin associated proteins of bovine brain coated vesicles: Coated vesicle populations from bovine brain are known to contain in their coat the proteins clathrin, the adaptor complexes HA1 and HA2 and the proteins AP180 and auxilin (Pearse & Robinson 1984; Ahle & Ungewickell 1986; Ahle & Ungewickell 1990). The coat proteins AP180 and auxilin were discovered in our laboratory not as the result of an intensive search effort, but in part by chance when they betrayed their existence by their reactivity with monoclonal antibodies raised against crude coat protein extracts. We have now initiated a systematic search for further hitherto unrecognized coat proteins, which

NATO ASI Series, Vol. H 62
Endocytosis
Edited by P. J. Courtoy

might directly or indirectly interact with clathrin. The experimental design of our approach was very simple. Total coat proteins were extracted with 0.5 M Tris from the coated vesicle membrane and the clathrin therein was allowed to assemble by dialysis against 0.1M MES, 0.5 mM $MgCl_2$, 1 mM EGTA, pH 6.5. Under these conditions reassembly of clathrin is strictly dependend on other coat proteins (Keen et al 1979). The clathrin cages were then separated from unassembled clathrin and non-incorporated protein by sucrose gradient centrifugation, harvested by ultracentrifugation and disassembled in 0.5 M Tris, pH 7.0. The solubilized protein was subsequently fractionated by gel-filtration on Superose 6 and the fractions were analysed by SDS-PAGE using different gel methods and immunoblotting procedures for the identification of known clathrin associated proteins and those which might have escaped previous detection. The analysis of the data indicated that with the exception of a very minor 140 kDa polypeptide (~6% of total incorporated protein) all polypeptides could be related to the already known coat proteins (Fig. 1). These include the adaptor complexes HA1 and HA2, AP180 and auxilin. Since clathrin is not capable of binding to 0.5 M Tris-extracted membranes (Unanue

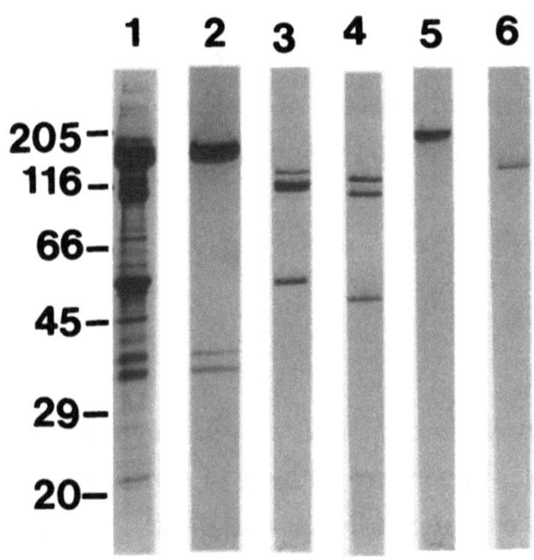

Figure 1: Proteins co-assembling with clathrin into coat structures.
A: Coomassie stained gradient gel (7-15%) according to Laemmli (1970). Note the hitherto undescribed 140 kDa polypeptide.
B: Immunoblot analysis of bound proteins using monoclonal antibodies against AP180 (1); auxilin (2); HA2-α (3); HA1-γ (4)
C: Coomassie stained Laemmli gel, containing 6 M urea. Note the lower electrophoretic mobility of auxilin and HA2-β in this gel system.

et al 1981) it seems unlikely that the membranes contain clathrin binding proteins which might have resisted extraction with Tris and therefore might have escaped detection. Apart from the 140 kDa polypeptide (p140) all the major clathrin associated proteins present in the coat of coated vesicles have now been purified (Fig. 2) and partially characterized in structural and functional terms (see Table I and next section).

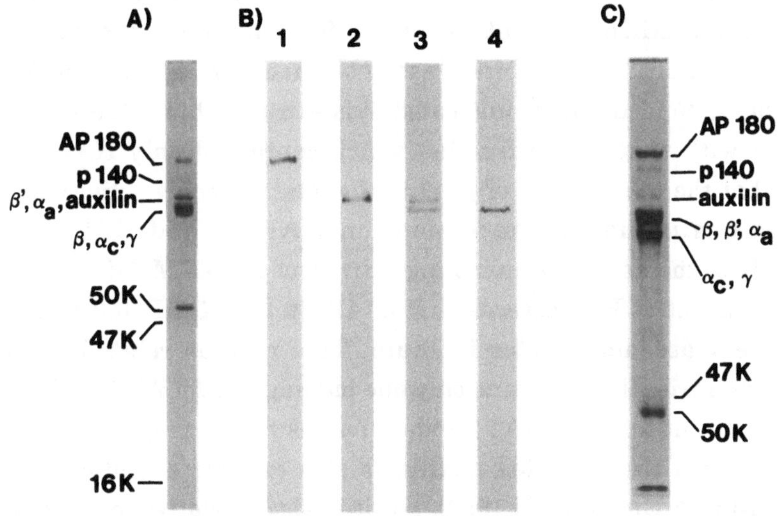

Figure 2: SDS-PAGE of purified coat proteins from bovine brain coated vesicles. Track 1: Protein composition of clathrin coated vesicles; track 2: Clathrin (heavy & light chains); track 3: HA2 adaptor; track 4: HA1 adaptor; track 5: AP180; track 5: Auxilin.

Assignment of assembly activities: To determine which of the proteins that were found to co-assemble with clathrin function as assembly proteins, we analysed clathrin assembly in quantitative terms upon removal of individual coat proteins. This was done by immunoadsorption or fractionation of coat proteins on sucrose gradients or both. This strategy was preferred over the determination of the assembly activity of purified proteins, because the purification regiment may alter this activity. The assembly promoting activity of the unfractionated Tris extract was taken to be 100%. At a clathrin concentration of ~1mg/ml this corresponded usually to 70-80% of assembled clathrin. The removal of the coat protein AP180 by immunoadsorption to immobilized mAb_{AP180} reduced the extent of clathrin assembly relative to an untreated control extract at the same concentration by 55%. The removal of either HA1 or auxilin by adsorption to mAb 100/3 (Ahle et al 1988) or mAb 100/4

(Ahle and Ungewickell 1990), respectively, reduced the assembly of clathrin in either case by at most 5%. This result implies that at least 35% of the assembly activity present in the total Tris-extract is associated with either HA2 and/or the minor p140 component. HA2, HA1 and clathrin have sedimentation coefficients ranging from 8-9S. Thus they are readily separated by sucrose gradient centrifugation from the majority of AP180, auxilin and also from p140 which have sedimentation coefficients that are all well below 4S. Remaining AP 180 and auxilin present in the 8-9S fraction were removed by immunoadsorption. The combined assembly activity of HA1 and HA2 was found to be 40 % of the total untreated Tris-extract. This value is in exellent agreement with that of the fraction which contained only HA2 and p140, implying that the contribution of p140 to the total assembly activity of the Tris extract is negligible. In conclusion, only AP180 and HA2 contribute significantly to the assembly promoting activity of the 0.5 M Tris extract. Since the molar amount of HA2 exceeds that of AP180 by 2.4, AP 180 is clearly the most efficient assembly factor in brain. This may be related to previous findings that a triskelion contains only one binding site for AP 180 (Prasad and Lippoldt 1988) but three for HA2 (Lindner manuscript in preparation). Previous work had shown that purified auxilin is also capable of inducing clathrin assembly (Ahle & Ungewickell 1990), but it is also a relative minor component in coated vesicles and therfore its contribution to the total assembly activity can only be a minor one.

Structure and function of adaptors: In contrast to AP 180 and auxilin which appear to be restricted to cells of neuronal origin, the coat proteins HA1 and HA2 are ubiquitous (Ahle et al 1988). Both proteins are heterotetramers that interact not only with clathrin but also with membranes and certain purified receptor proteins. For this reason they were termed adaptors. The HA1 adaptor is composed of two 105-115 kDa polypeptides, denoted as β′ and γ subunits, along with a 47 kDa and a 20 kDa subunit. Similarly, the HA2 adaptor consists of β (~104 kDa), α (105-112 kDa), 50 kDa and 16 kDa subunits (Ahle et al 1988). The β and β′ polypeptides appear to be very closely related and it has been proposed that the adaptors attach via these subunits to clathrin (Ahle et al 1988). More recently it was shown that the purified β subunit of the HA2 adaptor interacts specificly with clathrin (Ahle and Ungewickell 1989). Immunocytochemical studies showed that membrane-associated HA1 adaptor complex is located only at clathrin coated membranes in the cell's Golgi region, whereas membrane-associated HA2 adaptor is confined to coated pits on the plasma membrane (Robinson 1987; Ahle and Ungewickell 1988). Electron microscopy of the HA2 adaptor complex revealed a brick-shaped molecule with two appendages (Heuser and Keen 1988). The appendages are readily separated from the main body by mild proteolysis and their loss is paralleled by the cleavage of the α and β subunits into 60-70 kDa and 30-41 kDa fragments, respectively (Keen and Beck 1989). We undertook a detailed analysis of the domain structure of the HA1 adaptor using proteolytic dissection and immunological techniques. Like the α and β subunits of HA2, the β′ and γ subunits of the HA1 are structured into 32-44 kDa head (appendages) and 60-70 kDa trunk domains (Fig. 3). The truncated β′ and γ subunits in association with the 47 kDa and 20 kDa subunits form a HA1 adaptor core (Fig. 3). α, β, β′, γ and the 50 kDa subunit are known to be substrates for coated vesicle associated kinases (Morris et al 1990).

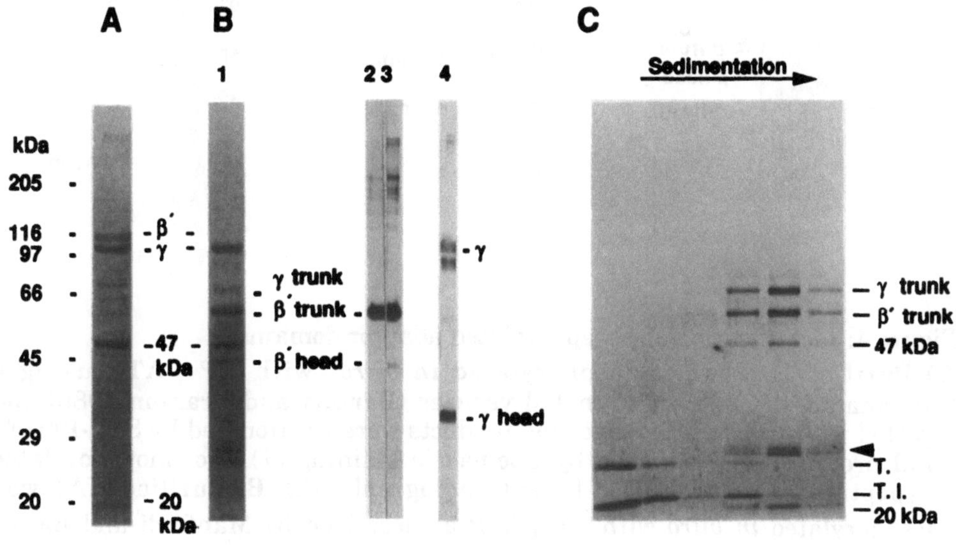

Figure 3: Proteolytic dissection of the HA1 adaptor.
A) Purified HA1, Ponceau stained blot. B) Trypsin digested HA1, Ponceau stained blot (1); immunostaining of β-type subunit with mAb 100/1 (Ahle et al 1988) (2) and monospecific antibodies (Ahle et al 1988) (3); immunostaining with mAb 100/3 against the γ subunit (4). C) Sedimentation analysis of trypsin-digested HA1 complex. Fractions were analyzed by SDS PAGE. T. and T.I. indicate trypsin and soybean trypsin inhibitor, respectively, arrow indicates direction of sedimentation in a 5-30% sucrose gradient, centrifuged for 18 hr at 54.000 rpm in a Beckman SW60 rotor.

Here it is shown that phosphorylation of the large adaptor subunits (α, β, β′and γ) by endogenous casein kinase II is confined to the head region of α, β, β′and γ subunits only (Fig. 4). While our data on the structural organization of the HA2 adaptor confirm and extend those of others (Keen and Beck 1989; Kirchhausen et al 1989), we came to a different conclusion concerning the adaptor domains which are involved in clathrin binding. Under conditions that inhibit aggregation of intact adaptors (10 mM HEPES, 100 mM K_2-tartrate, pH 7.0), but not their interaction with clathrin, we observed release of adaptors from mildly proteolysed coated vesicles as soon as their β or β′subunits were

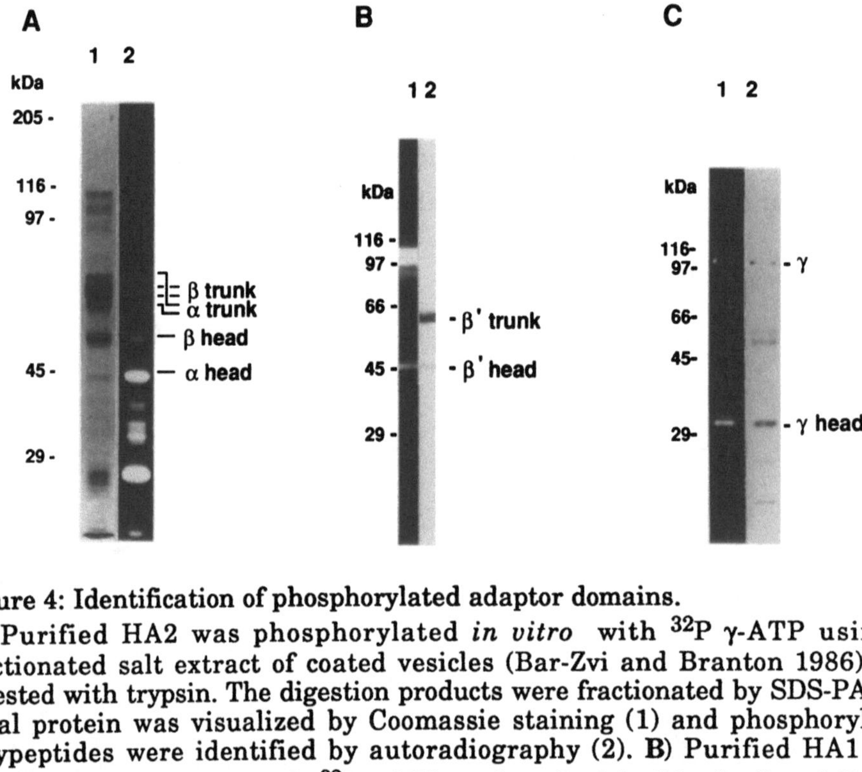

Figure 4: Identification of phosphorylated adaptor domains.
A) Purified HA2 was phosphorylated *in vitro* with ^{32}P γ-ATP using a fractionated salt extract of coated vesicles (Bar-Zvi and Branton 1986) and digested with trypsin. The digestion products were fractionated by SDS-PAGE . Total protein was visualized by Coomassie staining (1) and phosphorylated polypeptides were identified by autoradiography (2). **B)** Purified HA1 was phosphorylated *in vitro* with ^{32}P γ-ATP as described by Manfredi and Bazari (1987). Tryptic digestion products were analyzed by autoradiography (1). The β-type subunit derivatives were identified by immunoblotting with monospecific antibodies (2). **C)** Immunoprecipitation of γ and γ head fragments. Crude adaptor protein fraction was phosphorylated *in vitro* with ^{32}P γ-ATP by endogenous casein kinase II. Trypsin digestion products were immunoprecipitated with immobilized mAb 100/3, directed against the γ subunit of HA1 and subjected to SDS-PAGE and immunoblotting. Autoradiograph of the immunoprecipitate (1) immuno staining of the precipitate with mAb 100/3 (2)

Table I

Properties of Clathrin-associated Coat Proteins from Bovine Brain

Protein	HA2 adaptor	HA1 Adaptor	AP180	Auxilin	p140
Molecular wt.	~280.000[a]	~280.000[a]	120.000[b]	85.000[c]	~140.000
Composition	α; β; 50K;16K[a]	β´; 47K; 20K[a]	monomeric[b]	monomeric[c]	n.d.
Copies per clathrin heavy chain in the coat	0.24	0.03	0.1	0.03	~0.02
rel. assembly activity	35%	≤5%	55%	≤5%	n.d.
Intracellular membrane location	Plasma - membrane[a/d]	Golgi[a]	n.d.	n.d.	n.d.
Tissue	ubiquitous[a]	ubiquitous[a]	neuronal[b]	neuronal[c]	n.d.

a) Ahle et al 1988; b) Ahle and Ungewickell 1986; c) Ahle and Ungewickell 1990
d) Robinson 1987

cleaved into head and trunks. From these observations and the results of reconstitution experiments with truncated adaptors (not shown) we conclude that (i) adaptors attach to clathrin through their β-type (β and β´) subunits and (ii) high affinity interaction between adaptors and clathrin requires the participation of regions from both the head and trunk domains of the β-type subunits.

REFERENCES

Ahle, S. and Ungewickell, E. (1986) Purification and properties of a new clathrin assembly protein. EMBO. J. 5: 3143-3149
Ahle, S., Mann, A., Eichelsbacher, U. and Ungewickell, E. (1988) Structural relationships between clathrin assembly proteins from the Golgi and the plasma membrane. EMBO. J. 7: 919-929.

Ahle, S. and Ungewickell, E. (1989) Identification of a clathrin binding subunit in the HA2 adaptor protein complex. J. Biol. Chem. 264: 20089-20093.

Ahle, S. and Ungewickell, E. (1990) Auxilin, a newly identified clathrin-associated protein in coated vesicles from bovine brain. J. Cell Biol. 111: 19-29.

Bar-Zvi, D. and Branton, D. (1986) Clathrin coated vesicles contain two protein kinase activities. *J. Biol. Chem.* 261, 9614- 9621.

Glickman, J. N., E. Conibear, and B.M.F. Pearse (1989) Specificity of binding of clathrin adaptors to signals on the mannose 6-phosphate/insuline-like growth factor II receptor EMBO J. 8: 1041-1047.

Goldstein, J.L., Brown, M.S., Anderson, R.G.W., Russell, D.W. andSchneider, W.J. (1985) Concepts emerging from the LDL-receptor system. Annu. Rev. Cell Biol. 1: 1-19.

Heuser, J.E., and Keen, J. (1988) Deep-etch visualization of proteins involved in clathrin assembly. J. Cell Biol. 107: 877-886.

Keen, J.H., Willingham, M.C. and Pastan, I.H. (1979)Clathrin coated vesicles: Isolation, dissociation and factor-dependent reassociation of clathrin baskets. 877-886. Cell 16: 303-312.

Keen, J.H. and Beck, K.A. (1989) Identification of the clathrin-binding domain of AP-2. Biochem. Biophys. Res. Commun. 258: 17-23.

Kirchhausen, T., Nathanson, K.L., Matsui, W., Vaisberg, A., Chow, E.P., Burne, C., Keen, J.H. and Davis, A.E. (1989) Structural and functional division into two domains of the large (100- 115 kDa) chains of the clathrin-associated protein complex AP-2.Proc. Natl. Acad. Sci. USA 86: 2612-2616.

Laemmli, U. K. 1970. Cleavage of structural proteins during the assembly of the head of bacteriophage T4. Nature 227: 680- 685.

Manfredi, J.J. and Bazari, W. L. (1987) Purification and characterization of two distinct complexes of assembly polypeptides from calf brain coated vesicles that differ in their polypeptide composition and kinase activities. *J. Biol. Chem.* 262, 12182-12188.

Morris, S.A., Mann, A. and Ungewickell, E. (1990) Analysis of 100-180 kDa phosphoproteins in clathrin-coated vesicles from bovine brain. J. Biol. Chem. 265: 3354-3357.

Pearse, B.M.F. and Robinson, M.S. (1984) Purification and properties of 100kD proteins from coated vesicles and their reconstitution with clathrin. EMBO. J. 3: 1951-1957

Pearse, B.M.F. (1988) Receptors compete for adaptors found in plasma membrane coated pits. EMBO J. 7: 3331-3336.

Prasad, K., and R.E. Lippoldt. (1988) Molecular characterization of the AP180 coated vesicle assembly protein. Biochemistry. 27: 6098-6104.

Robinson, M.S. (1987) 100-kD coated vesicle proteins: Molecular heterogeneity and intracellular distribution studied with monoclonal antibodies. J. Cell Biol., 104: 887-895.

Unanue, E. R., Ungewickell, E., and Branton, D. (1981) The binding of clathrin triskelions to membranes from coated vesicles. Cell 26: 439-446.

Molecular Biology of the Adaptins

Margaret S. Robinson
University of Cambridge, Department of Clinical Biochemistry
Addenbrooke's Hospital, Hills Road, CB2 2QR
England

The adaptins are a family of proteins with molecular weights of ~100,000. They are the major components of adaptors, the protein complexes that are thought to bind to the cytoplasmic domains of selected transmembrane proteins, such as the LDL receptor and the mannose-6-phosphate receptor, and attach them to clathrin, as illustrated in Figure 1.

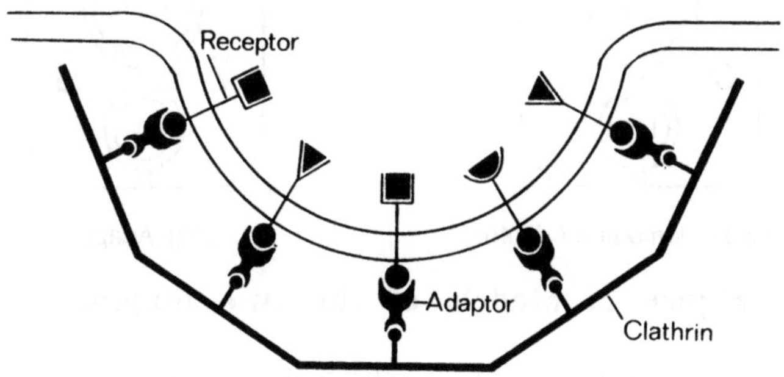

Figure 1. Model of a coated pit.

At least two types of clathrin-coated pits and vesicles exist in most cells. One type, associated with the plasma membrane, is responsible for receptor-mediated endocytosis. A second type, associated with the trans Golgi network, is responsible for the sequestering of newly synthesised lysosomal enzymes bound to the mannose-6-phosphate receptor, diverting such enzymes away from the default pathway and targeting them to prelysosomes. Different adaptors are associated with these two types of coated vesicles (Robinson, 1987; Ahle et al., 1988), and are thought to account for their different specificities (Glickman et al., 1989).

The adaptins have been divided into three classes: α, β, and γ. Each adaptor is a heterotetramer, consisting of one copy of a β-type adaptin, one copy of an α or γ-adaptin, and one copy each of two smaller proteins referred to by their molecular weights. Thus, the plasma membrane adaptor is composed of α-adaptin + β-adaptin + 50K + 17K, while the Golgi adaptor is composed of γ-adaptin + β'-adaptin + 47K + 20K (Ahle et al., 1988). The structure of the plasma membrane adaptor has been examined by electron microscopy of rotary shadowed specimens (Heuser and Keen, 1988). These studies have revealed that the adaptor consists of a central brick-like "head" flanked by two smaller "ears", connected

NATO ASI Series, Vol. H 62
Endocytosis
Edited by P. J. Courtoy

by protease-sensitive hinges. Biochemical characterisation of the proteolysed adaptor (Kirchhausen et al., 1989) has led the model shown in Figure 2. Although the structure of the Golgi adaptor has not yet been studied by electron microscopy, proteolysis studies carried out by Schroeder and Ungewickell (personal communication) indicate that the two adaptors have the same general structure (see Figure 2).

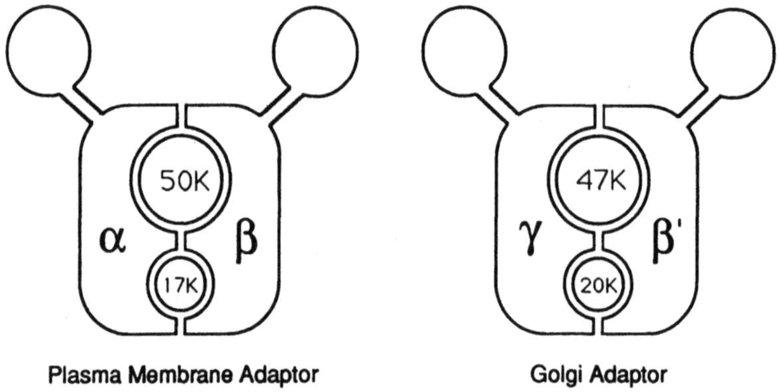

Figure 2. Models of the two adaptors.

The similar susceptibility to proteolysis of the various adaptins, as well as their similar size, suggests that all three classes might be evolutionarily related. With the cloning of α, β, and now γ-adaptins, this hypothesis can be tested by analysing and comparing their sequences.

α-Adaptins

The α-adaptins were the first to be cloned and characterised (Robinson, 1989). There are in fact two distinct α-adaptins with slightly different molecular weights, first observed by SDS gel electrophoresis and Western blotting of brain coated vesicles (Robinson, 1987). The larger of the two appears to be expressed primarily in neuronal tissues, while the smaller is found in virtually all types of cells. Two separate genes encode the two proteins, which show 84% identity at the amino acid level. In situ hybridisation on brain sections indicates that the same cell can co-express both α-adaptins (Robinson, 1989). The relative expression of the two α-adaptins varies from one cell to another, however, raising the possiblility that the two proteins might have different functions: for instance, the neuronal-specific one might be required for synaptic vesicle recycling. It should now be possible to resolve this question by using the sequence information to raise monospecific antibodies against regions of the proteins that differ from each other.

The sequences of the two α-adaptins are consistent with their two-domain structure. In both proteins, there is a region about two-thirds of the way along the molecule which is strikingly rich in proline and glycine residues, typical of hinge regions in other proteins. Evidence from Western blotting using antibodies against the C termini, as well as from sequencing of proteolytic fragments, confirms that this region is the α-adaptin hinge, with the ear fragment corresponding to the C-terminal third of the protein (Camidge and Robinson, unpublished observations).

β-Adaptins

Two distinct cDNAs have also now been cloned for β-adaptin (Kirchhausen et al., 1989; Ponnambalam et al., 1990). One of these appears to encode the β-adaptin subunit of the plasma membrane adaptor, since sequences obtained from the purified protein exactly correspond to sequences deduced from the cDNA. The other β-adaptin shows 83% amino acid identity with the plasma membrane β-adaptin and may be the Golgi adaptor subunit β'. cDNAs encoding the plasma membrane β-adaptin have been sequenced from human fibroblasts as well as from rat lymphocytes, and the two deduced protein sequences have been found to be 100% identical (Ponnambalam et al., 1990). Such a high degree of conservation between species is extremely unusual for a protein of this size: for comparison, the α and γ-adaptins of mouse and cow each contain 10 to 20 amino acid substitutions (Robinson, 1989 and 1990), and suggests that there are extensive interactions between β-adaptin and other proteins. Some of these proteins are presumably the other subunits of the adaptor complex. In addition, purified β-adaptin has been shown to bind to clathrin and is thought to constitute the clathrin binding site of the adaptor (Ahle and Ungewickell, 1989). So far, there is no evidence as to which of the adaptor subunits interacts with the transmembrane receptors found in coated pits.

Like the α-adaptins, the two β-adaptins have a proline and glycine-rich stretch between amino acids 600 and 750, which has been shown to constitute the hinge (Kirchhausen et al., 1989). Interestingly, there appears to be tissue-specific splicing of the plasma membrane β-adaptin so that the hinge region is slightly longer in brain than in lymphocytes or fibroblasts. Thus, at least four coat proteins (α-adaptin, β-adaptin, and the two clathrin light chains) are larger in brain than in other tissues, which may reflect specialisation for a neuronal-specific function.

γ-Adaptin

So far, only one form of γ-adaptin has been identified. The cDNA for this protein has recently been cloned and sequenced, using monoclonal antibodies to screen an expression library (Robinson, 1990). Like the other adaptins, γ-adaptin has a proline and glycine-rich stretch in the expected position, although the C-terminal ear domain is somewhat smaller than those of α or β-adaptins.

How similar are the three classes of adaptins? The Diagon plot shown in Figure 3 reveals that the α and γ-adaptins show substantial sequence homology in their N-terminal and hinge domains, while their ear domains do not appear to be related. An optimal alignment of the two sequences shows that they have an overall amino acid identity of 25.0%. In contrast, Diagon plots of α and β, or of β and γ, do not look impressive. Nevertheless, they share an overall amino acid identity of 17.8% and 15.7% respectively. Thus, it seems likely that all three types of adaptins arose from a single progenitor adaptin gene, although the divergence of α and γ presumably occurred more recently than the divergence of β from the other two. Although virtually nothing is known about non-mammalian adaptins, an antiserum raised against total adaptins from bovine brain coated vesicles was found to cross-react with proteins of ~100kd on Western blots of coated vesicles purified from soybean protoplasts (Mosley and Branton, 1988). It is interesting to note in this regard that plant cells as well as animal cells contain two populations of clathrin-coated vesicles: one associated with the plasma membrane and one associated with the trans Golgi network (Hillmer et al., 1988). The cloning and characterisation of adaptins from other eukaryotes, including plants and fungi, should lead to a better understanding of the evolutionary significance of adaptin divergence, as well as allowing adaptin function to be investigated in genetically tractible organisms.

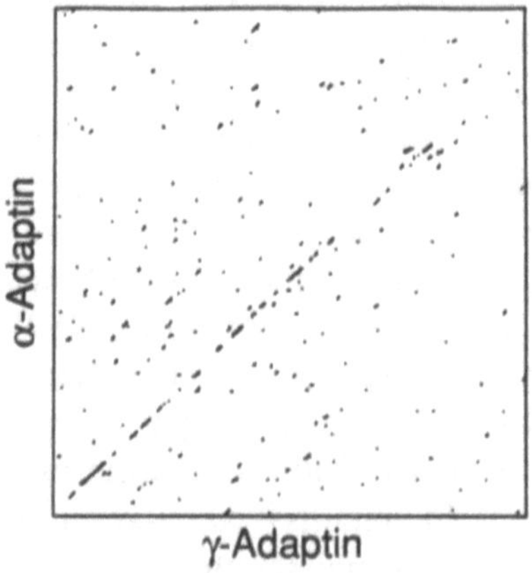

Figure 3. Diagon plot comparing α-adaptin and γ-adaptin.

Adaptins in vivo

Although a straightforward genetic approach is not yet feasible for studying mammalian adaptins, questions about protein function can still be addressed by using in vitro mutagenesis followed by expression in tissue culture cells to investigate the role of adaptins in vivo. A transient expression system has been developed which allows engineered adaptins to be visualised without any background from the cell's endogenous proteins (Robinson, 1990). This system makes use of the monoclonal antibody mAB 100/3, raised by Ahle et al. (1988), which reacts with γ-adaptin from most mammals but not from rodents. Although the complete γ-adaptin cDNA has so far only been cloned from mouse, the epitope for mAb 100/3 has been localised to the hinge region of the protein, allowing the construction of an antigenic mouse-cow chimera. Figure 4a shows Rat 1 fibroblasts transfected with this construct, and demonstrates that one of the cells in the field is expressing detectable antigen. Double labelling the cells with an antiserum against TGN 38, an integral membrane protein of the trans Golgi network (Luzio et al., 1990) (Figure 4b), confirms that the construct has a distribution typical of normal γ-adaptin.

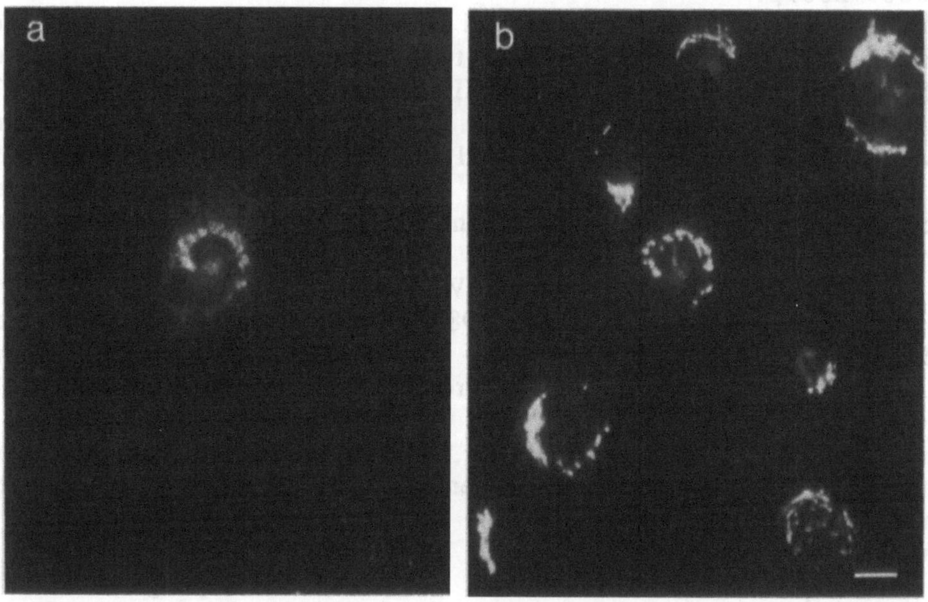

Figure 4. Immunofluorescence localisation of the γ-adaptin chimera (a) and TGN 38 (b) in transfected Rat 1 cells. Bar: 10μm.

Using this system, it will now be possible to address questions about the functions of the various adaptins and their different domains. The C-terminal ear domains of α and γ-adaptins are especially intriguing because of their lack of homology, suggesting that they may be required for an adaptor-specific function. Experiments are in progress to exchange the N-terminal and ear domains between α and γ-adaptins, to try to discover the molecular basis for the differences between the two adaptors, particularly their distinct subcellular distributions and receptor binding properties. Immunofluorescence localisation of such constructs should help to elucidate how the two adaptors are targeted to the correct membrane compartment. In addition, overproducing such constructs to compete out the endogenous pool of adaptins may have profound consequences for the cell, leading to missorting both on the endocytic pathway and on the prelysosomal pathway.

References

Ahle, S., A. Mann, U. Eichelsbacher, and E. Ungewickell. 1988. Structural relationships between clathrin assembly proteins from the Golgi and the plasma membrane. EMBO J. 4: 919-929.

Ahle, S., and E. Ungewickell. 1989. Identification of a clathrin binding subunit in the HA2 adaptor protein complex. J. Biol. Chem. 264: 20089-20093.

Glickman, J.N., E. Conibear, and B.M.F. Pearse. 1989. Specificity of binding of clathrin adaptors to signals on the mannose-6-phosphate/insulin-like growth factor II receptor. EMBO J. 8: 1041-1047.

Heuser, J.E., and J. Keen. 1988. Deep-etch visualization of proteins involved in clathrin assembly. J. Cell Biol. 107: 877-886.

Hillmer, S., H. Freundt, and D.G. Robinson. 1988. The partially coated reticulum and its relationship to the Golgi apparatus in higher plant cells. Eur. J. Cell Biol. 47: 206-212.

Kirchhausen, T., K.L. Nathanson, W. Matsui, A. Vaisberg, E.P. Chow, C. Burne, J.H. Keen, and A.E. Davis. 1989. Structural and functional division into two domains of the large (100- to 115-kDa) chains of the clathrin-associated protein complex AP-2. Proc. Natl. Acad. Sci. USA 86: 2612-2616.

Luzio, J.P., B. Brake, G. Banting, K.E. Howell, P. Braghetta, and K.K. Stanley. 1990. Identification, sequencing and expression of an integral membrane protein of the trans-Golgi network, TGN38. Biochem. J. In press.

Mosley, S.T., and D. Branton. 1988. An antibody against 100- to 116-kDa polypeptides in coated vesicles inhibits triskelion binding. Exp. Cell Res. 174: 511-520.

Ponnambalam, S., M.S. Robinson, A.P. Jackson, L. Peiperl, and P. Parham. 1990. Conservation and diversity in families of coated vesicle adaptins. J. Biol. Chem. 265: 4814-4820.

Robinson, M.S. 1987. 100-kD coated vesicle proteins: molecular heterogeneity and intracellular distribution studied with monoclonal antibodies. J. Cell Biol. 104: 887-895.

Robinson, M.S. 1989. Cloning of cDNAs encoding two related 100-kD coated vesicle proteins (α-adaptins). J. Cell Biol. 108: 833-842.

Robinson, M.S. 1990. Cloning and expression of γ-adaptin, a component of clathrin-coated vesicles associated with the Golgi apparatus. J. Cell Biol. In press.

Clathrin-independent endocytosis

Bo van Deurs[1], Steen H.Hansen[1] and Kirsten Sandvig[2]

[1]Structural Cell Biology Unit, Department of Anatomy, The Panum Institute, University of Copenhagen, DK-2200 Copenhagen N, Denmark.

Cells endocytose plasma membrane components and molecules from their surroundings in at least two different ways: by clathrin coated pits and vesicles and by a mechanism which does not involve clathrin (van Deurs et al., 1989). Whereas our knowledge of clathrin-dependent endocytosis is considerable, studies of the clathrin-independent mechanism(s) have been hampered by lack of specific markers associated with the internalized membrane microdomain, such as transferrin receptors, which are selectively taken up from clathrin coated pits, or the clathrin molecule itself. An example of apparently clathrin-independent endocytosis of a specific membrane domain is the internalization of desmosomes seen, for example, in EGTA-treated epithelial cells (Kartenbeck et al., 1982).

Two rather unusual examples of internalization processes circumventing coated pits and vesicles have recently been reported. Thus, Kartenbeck et al. (Kartenbeck et al., 1989) reported that Simian Virus 40 was endocytosed by small noncoated vesicles ("monopinocytic vesicles") and subsequently delivered to the endoplasmic reticulum without reaching endosomes-lysosomes. Rothberg et al. (1990) found that the glycophospholipid-linked folate receptor internalizes in small, noncoated pits which, however, never pinched off completely from the plasma membrane.

Our present knowledge of clathrin independent endocytosis is to a large extent based on experimental approaches where the internalization of various molecules can be selectively inhibited. Such an inhibition can be carried out in several ways (for a detailed discussion, see van Deurs et al., (1989). For instance, hypertonic media inhibit receptor-mediated endocytosis via coated pits whereas fluid phase endocytosis is not affected (Daukas and

[2] Institute for Cancer Research at the Norwegian Radium Hospital, Montebello, 0310 Oslo 3, Norway

NATO ASI Series, Vol. H 62
Endocytosis
Edited by P. J. Courtoy

Zigmond, 1985; Oka et al., 1989). Heuser and Anderson (1989) found that hypertonic media inhibit formation of clathrin-coated pits. A much used method to inhibit endocytosis from coated pits is potassium depletion in combination with hypotonic shock (Larkin et al., 1983). In this way clathrin is somehow prevented from forming coated pits, and endocytosis of molecules such as LDL and transferrin is strongly inhibited (Larkin et al., 1983; Madshus et al., 1987; Moya et al., 1985). Potassium depletion also inhibits endocytosis of polio virus (Madshus et al., 1987). However, human rhinovirus and the toxic protein ricin are still internalized (Madshus et al., 1987; Moya et al., 1985).

Acidification of the cytosol can be achieved in several ways (for technical details see Sandvig et al., (1989) and inhibits receptor-mediated endocytosis from coated pits by immobilizing or paralyzing the coated pits at the plasma membrane (Heuser, 1989; Sandvig et al., 1987). Nonetheless endocytosis of ricin is only moderately reduced (Sandvig et al., 1987).

Whereas all the above-mentioned methods somehow interfere with the clathrin-dependent pathway without significantly influencing the clathrin-independent pathway, recent experiments showed that it is also possible to selectively modulate clathrin-independent endocytosis (Sandvig and van Deurs, 1990). Thus, it was found that cytochalasin D reduced endocytosis of ricin in Vero cells, and the tumor promotor TPA increased endocytosis of ricin in A431 cells, without changing the uptake of transferrin.

Since specific markers of clathrin independent endocytosis are still lacking, it has been difficult to define and analyze the structural equivivalent of the clathrin-independent endocytic vesicle. Huet et al. (Huet et al., 1980) found that HLA antigen (MHC class I) cross-linked by ferritin-conjugated anti-beta 2 microglobulin was endocytosed via noncoated pits and vesicles in human fibroblasts. Labeling of coated pits was not seen. Using 3T3 fibroblasts, Tran et al. (1987) found that whereas gold-conjugated alpha-2-macroglobulin labeled coated pits and was subsequently internalized via this pathway, cholera toxin-gold did not label coated pits but was endocytosed via small, noncoated pits and vesicles.

We have analyzed ultrastructurally the earliest endocytic - or preendosomal - vesicle populations in HEp-2, A431 and T47D cells (S.H.Hansen, K.Sandvig and B.van Deurs, manuscript submitted). Since such vesicles per definition have detached completely from the plasma membrane but not yet fused to form endosomes, a number of requirements for the basic experimental setup are obvious. First, brief incubations are necessary to exclude any significant "background noise" from tubular endosomal processes and recycling vesicles. After a mere few minutes at 37°C the situation is too complicated to allow detailed analysis of vesicle populations. Secondly, a marker system is needed which makes it possible to readily identify endocytic vesicles and to distinguish truly free (internalized) vesicles from those still connected to the cell surface (in another plane of sectioning). We incubated cells at 4°C with ConA-gold to obtain a marked and uniform labeling of the entire cell surface. The cells were warmed to 37°C to allow endocytosis, followed by replacement of warm medium with fixative within either 30 or 60 s. Then they were incubated with anti-ConA-HRP and further processed for electron microscopy. In this way truly free preendosomal vesicles were easily identified as those containing gold only. In all structures communicating with the cell surface - even though they appeared free in the particular section - a distinct cloud of HRP reaction product was present around each gold particle. This could be confirmed by analysis of thin consecutive sections.

After 30 s we found two distinct preendosomal vesicle populations in the peripheral cytoplasm, one coated and the other noncoated and with a smaller mean diameter than the coated one. The two types of vesicles occurred with approximately the same frequency. After 60 s, the diameter distributions for the two populations were the same. The only difference from 30 to 60 s was the occurrence of a few larger, gold-labeled vesicles resembling endosomes. We also found that a number of larger vacuolar profiles that we routinely would take for endosomes were actually surface-connected.

A Kolmogorov-Smirnov test of the diameter distributions revealed that the two populations were significantly different. Moreover, parallel experiments with an anti-transferrin receptor-gold conjugate showed that only a small fraction of the noncoated vesicles could be due to initially coated vesicles which had shed the coat. Finally, a double-labelling assay excluded the possibility that the noncoated vesicles could represent recycling vesicles rather than preendosomal vesicles, that is, vesicles coming from endosomes rather than from the plasma membrane.

An open question is whether molecules internalized by clathrin-independent endocytosis are delivered to the same endosomes as molecules taken up by coated pits, and whether they are here exposed to the same sorting signals and mechanisms and subsequently follow the same pathways to, for example, lysosomes and the Golgi complex (van Deurs et al., 1989). However, studies by Tran et al. (1987) indicate that gold-conjugated ligands taken up by either mechanism meet in the same endosomal compartment.

Whereas the main function of clathrin-dependent endocytosis is a concentrative internalization of various ligands (transport molecules, hormones, growth factors etc.) binding to transmembrane glycoprotein receptors (e.g., receptors with tyrosine kinase activity; see Ullrich and Schlessinger, 1990), the significance of clathrin-independent endocytosis remains speculative. This endocytic mechanism could be involved in slow, non-concentrative removal of plasma membrane constituents in relation to basic turnover and renewal processes. An uptake rate of just a few per cent per hour could account for a slow turn-over of resident membrane proteins (i.e., proteins not designated for rapid internalization and recycling). Removal of unoccupied receptors or of thyrosine kinase receptors with defective cytoplasmic tails that prevent them from entering into coated pits should be considered. Moreover, a group of non-transmembrane proteins which are anchored to membrane glycolipids (phosphotidylinositol-anchored proteins) seem to be excluded from coated pits. Thus lymphocyte Thy-1 is slowly internalized (Lemansky et al., 1990) although it is excluded from coated pits (Bretscher et al., 1980). Another example of a phosphotidylinositol-anchored protein is the Decay Accelerating Factor of leucocytes which also seems to be internalized in a clathrin-independent way (Tausk et al., 1989). In addition to proteins, membrane glycolipids may also need a slow, continuous internalization, although some are excluded from coated pits. For instance gangliosides are likely to be taken up in this way (Tran et al., 1987). Also, glycolipids with the gal α1-4 gal structure binding Shiga toxin are excluded from coated pits (unless ligand is added) (Sandvig et al., 1989) but could be slowly internalized by a clathrin-independent mechanism in the absence of ligand.

Acknowledgements
Work from our laboratories referred to in this paper has been supported by grants from The Danish and Norwegian Cancer Societies, The Danish

Medical Research Council, The NOVO Foundation, and by a NATO Collaborative Research Grant (CRG 900517).

References

Bretscher, M.S., Thomson, J.N., and Pearse, B.M.F. (1980) Coated pits act as molecular filters. Proc.Natl.Acad.Sci. USA,*77*:4156-4159.

Daukas, G., and Zigmond, S.H. (1985) Inhibition of receptor-mediated but not fluid-phase endocytosis in polymorphonuclear leukocytes. J.Cell Biol.,*101*:1673-1679.

Heuser, J. (1989) Effects of cytoplasmic acidification on clathrin lattice morphology. J.Cell Biol.,*108*:401-411.

Heuser, J.E., and Anderson, R.G.W. (1989) Hypertonic media inhibit receptor-mediated endocytosis by blocking clathrin-coated pit formation. J.Cell Biol.,*108*:389-400.

Huet, C., Ash, J.F., and Singer, S.J. (1980) The antibody-induced clustering and endocytosis of HLA antigens on cultured human fibroblasts. Cell,*21*:429-438.

Kartenbeck, J., Schmid, E., Franke, W.W., and Geiger, B. (1982) Different modes of internalization of proteins associated with adherens junctions and desmosomes: experimental separation of lateral contacts induces endocytosis of desmosomal plaque material. EMBO J.,*1*:725-732.

Kartenbeck, J., Stukenbrok, H., and Helenius, A. (1989) Endocytosis of simian virus 40 into the endoplasmic reticulum. J.Cell Biol.,*109*:2721-2729.

Larkin, J.M., Brown, M.S., Goldstein, J.L., and Anderson, R.G.W. (1983) Depletion of intracellular potassium arrests coated pit formation and receptor-mediated endocytosis in fibroblasts. Cell,*33*:273-285.

Lemansky, P., Fatemi, S.H., Gorican, B., Meyale, S., Rossero, R., and Tartakoff, A.M. (1990) Dynamics and longevity of the glycolipid-anchored membrane protein, Thy-1. J.Cell Biol.,*110*:1525-1531.

Madshus, I.H., Sandvig, K., Olsnes, S., and van Deurs, B. (1987) Effect of reduced endocytosis induced by hypotonic shock and potassium depletion on the infection of Hep 2 cells by picornaviruses. J.Cell.Physiol.,*131*:14-22.

Moya, M., Dautry-Varsat, A., Goud, B., Louvard, D., and Boquet, P. (1985) Inhibition of coated pit formation in Hep-2 cells blocks the cytotoxicity of diphtheria toxin but not that of ricin toxin. J.Cell Biol.,*101*:548-559.

Oka, J.A., Christensen, M.D., and Weigel, P.H. (1989) Hyperosmolarity inhibits galactosyl receptor-mediated but not fluid phase endocytosis in isolated rat hepatocytes. J.Biol.Chem.,*264*:12016-12024.

Rothberg, K.G., Ying, Y., Kolhouse, J.F., Kamen, B.A., and Anderson, R.G.W. (1990) The glycophospholipid-linked folate receptor internalizes folate without endtering the clathrin-coated pit endocytic pathway. J.Cell Biol.,*110*:637-649.

Sandvig, K., Olsnes, S., Brown, J.E., Petersen.O.W., and van Deurs, B. (1989) Endocytosis from coated pits of Shiga toxin: A glycolipid-binding protein from Shigella dysenteriae. J.Cell Biol.,*108*:1331-1343.

Sandvig, K., Olsnes, S., Petersen, O.W., and van Deurs, B. (1987) Acidification of the cytosol inhibits endocytosis from coated pits. J.Cell Biol.,*105*:679-689.

Sandvig, K., Olsnes, S., Petersen, O.W., and van Deurs, B. (1989) Control of coated-pit function by cytoplasmic pH. Meth.Cell Biol.,*32*:365-382.

Sandvig, K., and van Deurs, B. (1990) Selective modulation of the endocytic uptake of ricin and fluid phase markers without alteration in transferrin endocytosis. J.Biol.Chem.,*265*:6382-6388.

Tausk, F., Fey, M., and Gigli, I. (1989) Endocytosis and shedding of the decay accelerating factor on human polymorphonuclear cells. J.Immunol.,*143*:3295-3302.

Tran, D., Carpentier, J.-L., Sawano, F., Gordon, P., and Orci, L. (1987) Ligands internalized through coated or noncoated invaginations follow a common intracellular pathway. Proc.Natl.Acad.Sci. USA,*84*:7957-7961.

Ullrich, A., and Schlessinger, J. (1990) Signal transduction by receptors with tyrosine kinase activity. Cell,*61*:203-212.

van Deurs, B., Petersen, O.W., Olsnes, S., and Sandvig, K. (1989) The ways of endocytosis. Int.Rev.Cytol.,*117*:131-177.

1.3. DYNAMICS OF ENDOSOMES

(Chairperson : Hans GEUZE)

Endosomes are intermediate organelles between primary endocytic vesicles and various subcellular compartments to which internalized constituents are destined The major pathway in most cells is membrane recycling and fluid regurgitation. The second pathway leads to degradation in lysosomes. In polarized cells, a third pathway, usually referred to as transcytosis, results in the transcellular transfer. Thus, endocytic traffic is complex and endosomes are crucial in sorting, segregation and addressing. The traffic is also quite fast : destinations are generally reached in a time scale of a few minutes to less than an hour. The traffic is finally intense : the equivalent of the entire cell surface is internalized and recycled within half an hour to a few hours; the average cell also "drinks" the equivalent of its volume every day, most of which is regurgitated. This high dynamics combined with the plasticity of endosomes has so far prevented us to reach a consensus on their nature. Three approaches contribute to our understanding of the dynamics of endosomes. The visualization of labeled endosomes in living cells provides a dynamic framework. Subcellular fractionation yields quantitative measurements of the transfer between endocytic compartments within the cell. It also provides purified fractions for analysis of protein composition and for identification of regulatory molecules, based on cell-free assays.

Colin Hopkins presented a low-light video recording fluorescence microscopy of living cells having internalized labeled transferrin or anti-EGF receptor antibodies, and discussed his recently published concept of a continuous endosomal reticulum. The concept is based on the observation that the endosome appears to be made of a continuous tubular system, along which varicosities (identified by electron microscopy as multivesicular bodies) move towards the cell center. Discontinuities and refusion are however frequently observed. Electron microscopy also shows sorting by partition between the limiting membrane (transferrin) ant the content (internal vesicles bearing EGF receptors). The major conclusion is that the endosome is probably much more continuous than generally appreciated.

Gareth Griffith presented his personal views of the ultrastructural complexity of the degradative pathway. He proposed that the endocytic pathway consists of two pre-existing, three-dimensionally complex structures, the early endosome and the late, pre-lysosomal compartment.

NATO ASI Series, Vol. H 62
Endocytosis
Edited by P. J. Courtoy

These two stations are highly dynamic structures, their components showing frequent homologous fusion but not with other organelles. Endocytic carrier vesicles and classical lysosomes, respectively, bud therefrom. Both pre-lysosomes and lysosomes contain the bulk of lysosomal membrane glycoproteins and acid hydrolases, but only the pre-lysosomal compartment contains mannose 6-phosphate/IGFII receptors, phosphorylated hydrolases, rab7 and the regulatory subunit of the cAMP-dependent protein kinase (type II). These views consider that the cell is equipped with two major internal sorting stations, the biosynthetic transGolgi network/reticulum and the degradative pre-lysosomal compartment.

This hypothesis is supported by the observations of **Brian Storrie** on fused mouse and rat cells. He followed the fate after fusion of endogenous membrane markers and pre-endocytozed tracers and showed the rapid and extensive lateral intermixing of pre-lysosomes and of endosomes originating from the two parental cells. The exchange of membrane and content can be viewed as a means to maintain organelle homogeneity. Along the degradative pathway, this would promote mixing of enzymes and substrate and optimize intracellular degradation.

A different view on the biogenesis of lysosomes was advocated by **Robert Murphy**. He proposed that endosomes spontaneously mature into dense lysosomes (possibly upon removal of Na^{+}/K^{+}-ATPase whose action antagonizes electrogenic proton pumps), and get continuously equipped with lysosomal enzymes during this maturation. In cultured cells, internalized fluorogenic substrates become rapidly accessible to acid hydrolases and the pH at which tracers are exposed gradually decreases. Further, when endosomes isolated from CHO cells are incubated with ATP in a cell-free system, the density of endocytic tracers apparently increases, up to that of hexosaminidase in total postnuclear preparations. If confirmed by appropriate controls, this mechanism could account for the formation of dense lysosomes (a dominant aspect in rapidly growing cells) with which late endosomes could fuse (a prevailing behaviour in growth-arrested cells).

NEW INSIGHTS INTO THE ENDOCYTIC PATHWAY

Colin R. Hopkins and Karen M. Miller
Department of Biochemistry,
Imperial College,
LONDON SW7 2AZ.

Since 1983 our laboratory has carried out a series of studies on the endocytic pathway in epithelioid cells. In the initial work which described the uptake and recycling of transferrin receptors (TfnR) we identified the endosome as containing networks of branching tubular cisternae and vacuoles (Hopkins and Trowbridge 1983). A most important insight into function was provided by the finding that while a pulse of transferrin is internalised and recycled (t1/2 = 7 mins) continuous incubation brings about linear uptake for up to 40 mins. Later studies showed that the system to which transferrin has access is indeed large and widely distributed, extending from just below the plasma membrane into the pericentriolar area . Within the pericentriolar area the most prominent TfnR-containing elements are multivesicular bodies with long, often branching, tubular arms (Hopkins 1983). Subsequent studies showed that epidermal growth factor (EGF) induced EGFR to move through the full extent of this TfnR-containing system and to be delivered to the TfnR-negative lysosomes in the immediate vicinity of pericentriolar endosomal elements (Miller et al 1986). In collaboration with de Brabander we then used video microscopy to show that TfnR containing endosomes required an intact microtubular cytoskeleton to move from the peripheral cytoplasm to the pericentriolar area (de Brabander et al 1985). Finally, in a series of cell fractionation studies, we were able to purify TfnR-containing endosomes and show that EGF remains undegraded while being processed through them (Beardmore et al 1987, Futter and Hopkins 1989). Most recently, in a related study on lymphocytes we showed that the major change in protein composition along the endocytic pathway occurs in the transition from the TfnR-containing endosome to the lysosome (Beaumelle and Hopkins 1990). This study found significant amounts of mannose-6-phosphate receptor (M6PR) in the TfnR containing endosome.

NATO ASI Series, Vol. H 62
Endocytosis
Edited by P. J. Courtoy

Together these studies suggest that TfnR-containing elements comprise a functionally distinct population within the endocytic pathway. However because they are extremely pleiomorphic and often appear to be free vesicles it has become important to establish the extent to which these variously shaped structures are interrelated. This information should allow us to decide (1) if TfnR-containing endosomes comprise a single, discrete organelle and (2) if the sorting mechanisms for ligand/receptor complexes in this compartment depend upon the formation/selective fusion of free cytoplasmic vesicles.

Further progress using conventional ultrathin sections was hindered because TfnR-containing elements have a complex 3-dimensional form and frequently contain elements within the 30 to 40 nm range. 3D reconstructions from sections require every tubule (even those in grazing, tangential section) to be identified if the extent of interconnection is to be realistically estimated. Morphometry of elements with a diameter at or below the thickness of the section are of very limited value (Weibel 1979). Finally since the the lysosomal part of the pathway has been shown to be plastic in form and difficult to preserve with conventional fixation methods (Heuser 1989) analyses relying exclusively upon embedded sections must be treated with circumspection.

We have therefore developed a method for looking at living cells in which the endocytic pathway is labelled to steady state with a continuous flux of tracer. Loading with a pulse of tracer or using temperature changes to manipulate trafficking profoundly effects the distribution of tracer and is likely to preferentially load compartments with low through-put. The results and experimental procedures of this study have been recently published in detail (Hopkins et al 1990) and for the purpose of this discussion we shall, therefore, only outline our main findings; these are :

1. When H.Ep-2 cells are incubated at 37ºC for 60 mins or longer in subsaturating levels (10^{-9}M) of Texas Red-Tfn confocal and video microscopy indentifies areas of branching reticulum within the cytoplasm.
2. Rapid flows of tracer (≤1µm/sec) are observed within the reticulum.

3. At intervals the tubules have varicosity-like expansions which are an integral part of the reticulum and can move along the tubule; (much like a swelling is seen to move down the length of an anaconda after it has swallowed a pig).

4. When labelled EGFR are induced to internalise by EGF they proceed as a pulse through the TfnR-containing network. They concentrate within the moving varicosities and within these structures they are carried to the pericentriolar area.

5. Electron microscopy shows that the varicosities are multivesicular bodies (MVB) and that while TfnR are localised on their limiting membranes the EGFR become concentrated on the bolus of internal vesicles which eventually fills the MVB lumen.

We propose that the continuity of the system we have observed allows the TfnR-containing elements to be identified as a discrete organelle. Our observations also indentify MVB as an integral part of this organelle operating as a mobile sorting station which removes selected subsets of ligand/receptor complexes into internal vesicles. We think this organelle should be referred to as the "endosome".

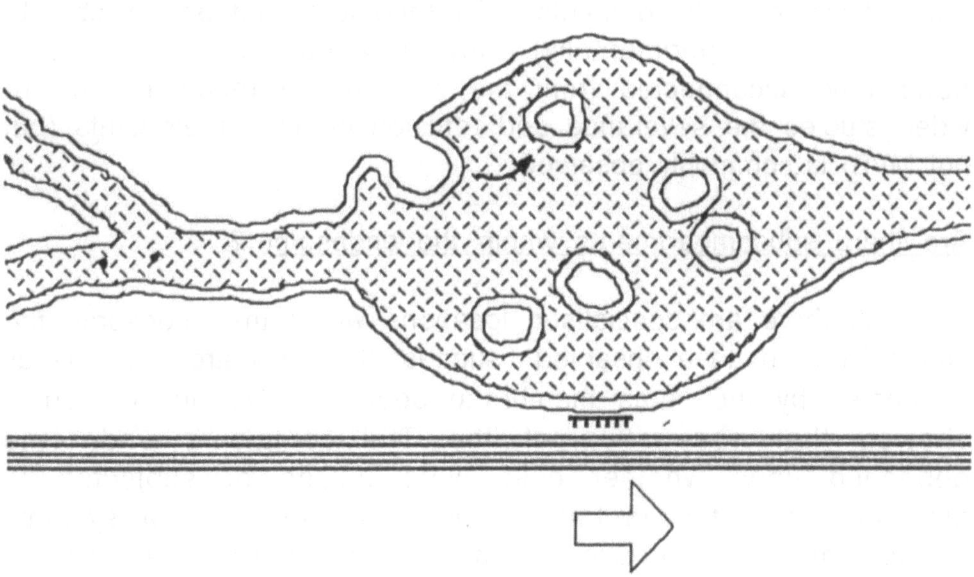

Suggested mechanism for movement of MVB through reticulum

In the context of previous published studies our recent observations leave a number of questions to be resolved. They include the following :

<u>How can the well documented low pH of the endosome be maintained in such an open system</u> ?

Perhaps the first observation to make in this regard is that most of the published evidence for low pH in the TfnR-containing endosome has been obtained either from fixed or fractionated cells. Alternatively manipulations (eg. low temperature) have been used which arrest the flux of traffic through the system. In these circumstances continuity will be severed and pH values may be artificially reduced.

Nevertheless there is good evidence (Fe++ dissociation etc.) that the environment within the endosome is reduced by as much as one pH unit. It is possible that this is achieved by proton pumps being concentrated within localised segments (such as the MVB) in which microenvironments of low pH can be created. To serve most of the known functions of low pH these would have to arise only in the immediate vicinity of the limiting membrane. Alternatively if such segments exist (and it should be possible to examine this experimentally) they may be capable of being transiently disconnected from the rest of the reticulum. In our video studies the severance and refusion of tubular elements (but not MVB) is commonly observed.

<u>Are there subcompartments within the endosome</u> ?

Although there are specialised locations within the endosome the continuities we have observed suggest that they are probably all connected by the same membrane boundary. Our initial studies showed that penetration of the TfnR-containing endosome depended upon whether cells were pulsed or subjected to prolonged incubation in tracer. Recent studies in other systems shows that newly introduced transferrin can quickly gain access to compartments loaded up to 15 minutes previously (Salzman et al 1988, Dunn et al 1989, Stoorvogel et al. this volume). Several reports describe the transferrin tracer recycling from the pericentriolar elements (Dunn et al 1989, Stoorvogel et al 1989).

While it is difficult to reconcile these observations with a sequentially traversed, linear series of compartments ('early' to 'late') they can be readily accommodated by a network with extensive continuity and through which TfnR move at a rapid rate.

In previous studies (including ours) operational definitions for "early" and "late" endosome were made but they did not always refer to the same compartments. For example, in a recent cell fractionation study in which "early" and "late" endosomes were shown to have distinctive protein compositions (Schmid et al. 1988) "early endosomes" were identified by their content of transferrin after an incubation of 30 to 45 min at 37ºC. This treatment would be expected to load the entire TfnR-containing endosome to steady state. The "early endosome" of this study should thus include all of the labelled structures we identified by video microscopy as the "endosome".

The Endosome-Lysosome transition

EGF/EGFR complexes leave the TfnR-containing endosome in the pericentriolar area where an extensive reticulum of lysosomal elements exists (Miller et al 1986). The available cell fractionation studies indicate that major changes in protein composition occur in the endocytic pathway soon after the TfnR-containing endosome is traversed (Schmid et al 1988,Beaumelle and Hopkins 1990) and it is somewhere within this vicinity that the widely used 20ºC block (Dunn et al 1980) which prevents transfer to lysosomes operates. Incubations at 20ºC are not without their side effects and this treatment probably arrests transport at more than one site in the pathway. Nevertheless, several studies demonstrate that significant concentrations of M6PR, lysosomal enzymes and lysosomal membrane proteins (lgp 120, endolyn-78) occur on the TfnR-containing side of the 20ºC block (Griffiths et al 1988, Stoorvogel et al 1989,Croze et al 1989, Beaumelle and Hopkins 1990). Confusingly some reports call these elements "late endosomes", others "prelysosomes".

Beyond the 20ºC block M6PR-negative elements containing endocytosed tracer begin to emerge. It should be noted,however that this can take very much longer (2 to 24 hrs) (Griffiths et al 1988) than it takes for physiological ligands like EGF and ASGP to be degraded (15 to 60 mins) (Dunn et al 1980,Felder et al 1990).

Unfortunately it is not possible to identify lysosomes simply on the basis that they contain acid hydrolase activity because several studies have detected active enzymes (cathepsins) in elements formed within minutes of ligand uptake. i.e. probably within the endosome (Diment and Stahl 1985 Roederer et al 1987). However it should be noted that we and others (Dunn et al 1980, Futter and Hopkins 1989), when following degradation of trafficking ligands like EGF and ASGP have detected degradation only beyond the 20ºC block.

In view of the above complexities and until more information is available the simplest solution would seem to be to reserve the term "lysosome" for all elements beyond the TfnR-containing endosome.

It remains to be seen if the lysosome, like the endosome, is a single organelle in which structural and functional remodelling is a continuum or if there is sequential processing in separate vesicular compartments. Until this issue is resolved there appears to be no reason why a simple division of the endocytic pathway into endosomes and lysosomes as originally described (Helenius et al 1983,Hopkins 1983) should not be adhered to.

BEARDMORE, J., Miller, K., Howell, K. and Hopkins, C.R. Isolation of an endocytic compartment from A431 cells using a density modification procedure employing a receptor- specific monoclonal antibody complexed with colloidal gold. J.Cell Sci. 87. 495 - 506. 1987.

BEAUMELLE, B.D., Gibson, A. and Hopkins, C.R. Isolation and preliminary characterization of the major membrane boundaries of the endocytic pathway. J. Cell Biol. 111. 1811-1823. 1990

CROZE, E., Ivanov, I.E., Kreibich, G., Adesnik, M., Sabatini, D.D. and Rosenfeld, M.G. Endolyn-78, a membrane glycoprotein present in morphologically diverse components of the endosomal and lysosomal compartments: implications for lysosome biogenesis. J. Cell Biol.108. 1597-1613. 1989.

DE BRABANDER, M., Nuydens, R., Geerts, H. and Hopkins, C.R. Dynamic behaviour of the transferrin receptor followed in living epidermoid carcinoma (A431) cells with nanovid microscopy. Cell Motil Cytoskeleton 9. 30-47. 1988

DIMENT, S. and Stahl, P. Macrophage endosomes contain proteases which degrade endocytosed protein ligands J. Biol. Chem. 260. 15311-15317. 1985

DUNN, W.A. Hubbard, A.L. and Aronson, N.N. Low temperature selectively inhibits fusion between pinocytic vesicles and lysosomes during heterophagy of asialofetuin by the perfused rat liver. J.Biol. Chem.255.5971-5978.1980

DUNN, K.W., McGraw, T.E. and Maxfield, F.R. Iterative fractionation of recycling receptors from lysosomally destined ligands in an early sorting endosome. J. Cell Biol. 109. 3303-3314. 1989

FELDER, S., Miller, K., Moehren, G., Ullrich, A., Schlessinger, J. and Hopkins, C.R. Kinase activity controls the sorting of the epidermal growth factor receptor within the multivesicular body. Cell 61. 623-634. 1990

FUTTER, C. and Hopkins, C.R. Subfractionation of the endocytic pathway; isolation of compartments involved in the processing of internalised epidermal growth factor-receptor complexes. J. Cell Sci. 94 685-694. 1989.

GRIFFITHS, G., Hoflack, B.,Simons, K.,Mellman, I. and Konrfeld, S. The mannose 6-phosphate receptor and the biogenesis of lysosomes. Cell 52. 329-341.1988

HELENIUS, A., Mellman, I., Wall, D. and Hubbard A. Endosomes. TIBS 8. 245-250.1983

HEUSER, J. Changes in lysosomes shape and distribution correlated with changes in cytoplasmic pH. J. Cell Biol. 108. 855-864. 1989.

HOPKINS, C., Gibson,A., Shipman, M., and Miller, K. Movement of internalised ligand-receptor complexes along a continuous endosomal reticulum. Nature 346 335-339 1990

HOPKINS, C.R. and Trowbridge, I.S. Internalisation and processing of transferrin and the transferrin receptor in human epidermoid carcinoma A431 cells. J. Cell Biol. 96. 508-521. 1983.

HOPKINS, C.R. Intracellular routing of transferrin and transferrin receptors in epidermoid carcinoma cells. Cell 34, 696-705. 1983 (a).

HOPKINS, C.R. Importance of the endosome in intracellular traffic. Nature 304. 684-685. 1983 (b).

MILLER, K., Beardmore, J., Kanety, H., Schlessinger, J. and Hopkins, C.R. Localisation of the epidermal growth factor (EGF) receptor in EGF stimulated epidermoid carcinoma (A431) cells. J. Cell Biol. 102. 500 - 509. 1986

ROEDERER, M., Bowser, R. and Murphy, R.F. Kinetics and temperature dependence of exposure of endocytosed material to proteolytic enzymes and low pH: evidence for a maturation model for the formation of lysosomes. J. Cell. Physiol. 131. 200-209. 1987

SALZMAN, N.H. and Maxfield, F.R. Intracellular fusion of sequentially formed endocytic compartments. J. Cell Biol. 106. 1083-1091. 1988

SCHMID, S. L., Fuchs, R., Male Ph and Mellman, I. Two distinct subpopulations of endosomes involved in membrane recycling and transport to lysosomes. Cell 52. 73-83. 1988

STOORVOGEL, W., Geuze, H.J., Griffith, J.M., Schwartz, A.L. and Strous, G.J. Relations between the intracellular pathways of the receptors for transferrin, asialoglycoprotein, and mannose 6-phosphate in human hepatoma cells. J. Cell Biol. 108. 2137-2148. 1989

Weibel,E.R. Stereological Methods, Practical methods of biological Morphometry, Academic Press Inc New York.1979

THE COMPARTMENTS OF THE ENDOCYTIC PATHWAY

Gareth Griffiths, European Molecular Biology Laboratory,
Postfach 10.2209, D-6900 Heidelberg, F.R.G.

A few years ago we proposed a four compartment model of the endocytic pathway (Griffiths et al., 1988). According to this scheme, following the initial internalization step by clathrin coated vesicles, material destined for the lysosomes would sequentially pass through the early endosomes, the endosome carrier vesicle (ECV), the prelysosomal compartment (PLC or late endosome) and finally the lysosomes themselves. In this review I shall summarize the recent work of our group and of our collaborators that supports this model. I emphasize that the arguments presented here should be taken more as our personal view of this pathway rather than an attempt to make a comprehensive summary of the whole field (see Kornfeld and Mellman, 1989; Gruenberg and Howell, 1989 for reviews).

The early endosome compartment

A few years ago the early endosome was mostly considered as a simple, tadpole-like structure consisting of a vesicle and attached tubules (see, for example, Rome, 1985). In all our studies of early endosomes in many different kinds of cultured cell, such as BHK, NRK, peritoneal macrophages and MDCK cells, a striking feature has been the three-dimensional complexity of its organization (Griffiths et al., 1989; 1990; McDowall et al., 1989; Parton et al., 1989). As summarized in Fig. 1 this organelle has a number of characteristic features. In light of the earlier model a surprising finding has been the extensive cisternal regions, which, with few exceptions (such as Hopkins, 1983) were not described in earlier studies that emphasized more the tubular projections (e.g. Geuze et al., 1983). In cells such as chicken erythroblasts, which express high concentrations of transferrin (Tf) and transferrin receptors (TfR), the Tf/TfR rich early endosome has very prominent cisternal regions which often resemble the Golgi complex (Killisch, Steinlein, Römisch, Beug and Griffiths, submitted for publication). Overall, this structural organization is much more consistent with the idea of a preexisting early endosome, rather than of a transient structure formed by the fusion of uncoated clathrin-coated vesicles (Helenius et al., 1983). Our combined stereological and biochemical data on endocytosis in BHK cells also argued strongly for a preexisting organelle (see Griffiths et al., 1989 for details).

NATO ASI Series, Vol. H 62
Endocytosis
Edited by P. J. Courtoy

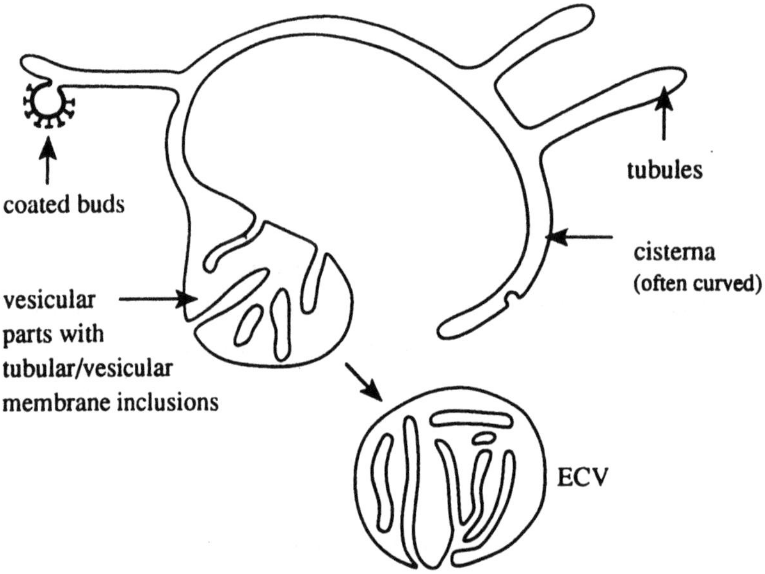

Figure 1. Schematic model showing the main structural features of the early endosome compartment and the ECV's.

The endosome carrier vesicle (ECV)

As summarized in Fig. 1 the early endosome has a characteristic vesicular structure which is continuous with the tubulo-cisternal domains. We have proposed that these large vesicles (ECV) bud off the "fixed" early endosome compartment and move along microtubules to the next station of the pathway, the PLC (Gruenberg et al., 1989). The data in that study argued that some of these vesicles were both structurally and functionally distinct from early endosomes in BHK cells. The key morphological observation was that a plasma membrane marker (the G protein of vesicular stomatitis virus implanted by low pH fusion and marked with antibodies and gold) was internalized sequentially into the early endosomes and the ECV. If the cells were pre-treated with nocodazole to disrupt the microtubule network the gold marker accumulated predominantly in ECV's. When these cells were subsequently allowed to internalize horseradish peroxidase (HRP) for 5 min (a time long enough for the reaction product to fill the early endosome compartment in these cells - Griffiths et al., 1989) the HRP reaction product labelled early endosomes but did not enter most of the ECV's. In parallel, biochemical studies showed that whereas the early endosomes fuse freely with themselves, the ECV's fused neither with themselves nor with early endosomes (Gruenberg et al., 1989). Further, the ECV can fuse with a more distal late endosome, in contrast to early endosomes (Bomsel et al., 1990).

A striking feature of the ECV is the presence of densely-packed membrane structures within their lumen. The consensus in the field has been that these internal membranes represent vesicles that have budded into the lumen; further, that these vesicles and their contents are targeted for degradation (e.g. Hopkins et al., 1990; Warren, 1990). Based on these ideas, as well as their appearance in epoxy sections these structures are widely referred to as multivesicular bodies (MVB's). We do not concur with this model for a number of reasons. First, in cryo-sections, in which preservation can be expected to be better than after embedding in epoxide resins, the bulk of the internal membranes are clearly tubular rather than vesicular. This is also the impression obtained from extensive freeze fracture studies (Dr. Dan Friend, University of California, San Francisco, personal communication). Second, we can consistently detect the presence of two receptors that are clearly recycling and not being destined for degradation on the internal membranes of the ECV. These are the cation-independent mannose 6-phosphate receptor (CI-MPR) in many cell types and Tf/TfR in erythroblasts. Both these receptors have long half lives and, at steady state, only a minute fraction will be degraded. In contrast, recent studies on the distribution of the invariant chain of the major histocompatibility complex in a human melanoma cell line, in which the ECV's are particularly numerous, show that the bulk of these molecules are associated with the outer peripheral membrane of the ECV (Pieters, Lipp, Bakke, Horstmann, Griffiths and Dobberstein - submitted). This result was obtained with antibodies against both the NH_2 (cytoplasmic) and COOH (lumenal) domains of this membrane protein. Paradoxically, the invariant chain *is* destined for degradation and has a half life of less than 4 hrs. While we still do not understand the significance of these localizations they suggest that the current ideas on the MVB's are grossly oversimplified. Another objection one can raise against the term MVB is that profiles of vacuoles containing internal vesicles or vesicle-like structures can often be seen in early endosomes (see Fig. 1) as well as in the MPR-enriched PLC. Thus, it seems likely that the term MVB has been used to describe different functional compartments in different studies.

Our current model of the ECV is that the tubules represent continuities with the outer membrane and that the apparent budding events seen in conventional plastic sections reflect, in part, some of these invaginations. We believe, however, that the organization of the delicate tubules is often destroyed after routine plastic embedding. The structure of isolated ECV's is currently being investigated (without fixation) by cryo EM in vitreous ice by the groups of Gruenberg and Fuller (EMBL) in an attempt to resolve this issue.

The ECV is kinetically an intermediate between the early endosome and the PLC. This idea has been given strong support from *in vitro* studies using a polarized MDCK cell system in which two distinct ECV's, derived from early apical and early basolateral endosomes respectively, appear to fuse with the PLC (Bomsel et al., 1990; see also Parton et al., 1989; Bomsel et al., 1989).

Two, as yet unpublished, observations point to another interesting aspect of the ability of the ECV to fuse with the PLC. The first comes for the mouse peritoneal macrophages treated with wheat germ agglutinin (WGA) for many hours. Under this condition the lectin binds to a major cell surface protein (Rabinowitz and Gordon, 1989) and is internalized into the endocytic pathway. When these cells are subsequently fed BSA-gold particles from the medium the amount of gold which is delivered into the PLC of these cells (see below) is greatly reduced after 30-60 mins when compared to cells not treated with WGA. Instead the gold accumulates predominantly in typical ECV structures (Rabinowitz et al., submitted). The second set of experiments comes from the human melanoma cell line referred to above (Zachgo, Dobberstein and Griffiths, unpublished data). If these cells are treated overnight with the protease inhibitor leupeptin (150 μg/ml), added to the medium, the number of ECV's increases significantly. This increase was apparent by fluorescence microscopy after immunolabelling with antibodies against the invariant chain and could be quantified at the EM level using a stereological approach. Accordingly, the number of ECV structures increased from an average of ≈ 280 per cell (without leupeptin) to about 580 per cell (with leupeptin). We then investigated whether the delivery of internalized HRP to the PLC was altered after leupeptin treatment. The cells were allowed to internalize 5 mg/ml HRP for 10 hrs with or without leupeptin in the medium. While the concentration of HRP in the ECV's, determined by quantitating gold labelling of anti-HRP, was not significantly different the concentration of the marker in the PLC was reduced by half in the presence of leupeptin. These data agree with earlier biochemical studies of rat hepatocytes by Tolleshaug and Berg (1981) who found that in the presence of this inhibitor internalized ^{125}I-asialofetuin was not degraded and accumulated in low density vesicles rather than being transported to dense lysosomes.

The conclusion from both the above experiments is that a treatment which is restricted to the *lumen* of the endocytic pathway (WGA or leupeptin) somehow affects the ability of the ECV to be transported to, or more likely, to "dock" with the *cytoplasmic* side of the membrane of the PLC. An intriguing possibility is that these treatments somehow modulate the PLC in some way so that it then regulates (e.g. by a feedback mechanism) the ability of the ECV to dock/fuse with the PLC. By regulating the ability of ECV to fuse with the PLC the cell could control the delivery of material into the later, mostly degradative, stages of the pathway and the ECV would have the potential to serve as storage sites. That some form of regulation might exist would also be consistent with the observation that, whereas some cells such as the human melanoma cell line have relatively large numbers of ECV, in other cells such as MDBK (Griffiths et al., 1990b) or MDCK cells (Parton et al., 1989) these structures are difficult to find. In the MDCK cells, in fact, convincing evidence for their presence came only after treatment with nocodazole *in vivo* or from *in vitro* preparations (Bomsel et al., 1990).

The prelysosomal compartment (PLC) and the lysosomes

Our interest in the endocytic pathway first focused on this compartment with the finding that the bulk of the CI MPR was found there (Griffiths et al., 1988); similar findings were made independently by Geuze et al. (1988). A subsequent quantitative study showed that about 90% of the gold labelling this receptor co-localizes with the PLC (the rest being distributed over the plasma membrane, early endosomes, ECV and the TGN; Griffiths et al., 1990a). Fig. 2 summarizes our current view of this compartment and its relationship to the terminal compartment, the MPR-negative (dense) lysosomes. In many, but not all, cell types we have examined the lumen of the PLC contains extensive tubulo-cisternal membranes (reminiscent of those in the ECV, above) which are heavily enriched in the CI MPR. Although images of continuities between the outer and inner membranes are mostly equivocal the fact that the whole intracellular pool of the CI MPR is cycling between different intracellular compartments and the plasma membrane (Kornfeld and Mellman, 1989) argues strongly that functional continuities must exist.

The most striking images of the PLC we have seen come from the mouse peritoneal macrophages (Rabinowitz, Gordon and Griffiths, submitted). In these cells the PLC forms an extensive tubulo-cisternal network which fills uniformly with gold markers between 30 and 60 mins after internalization from the medium. In contrast to other cells we have examined, the CI MPR is restricted in its localization to some domains of this structure that are positioned in close proximity to the Golgi complex. This agrees with a recent immunofluorescent study by Tassin et al. (1990) who found a restricted perinuclear localization of this receptor as compared to the more extensive structures that label with endocytic markers. It is probable that the structures we refer to as PLC in our study represent the "tubular lysosome" structures that have been extensively studied by Swanson et al. (1987). The use of the term "lysosome" for these structures opens up a fascinating aspect of the relationship between the PLC and lysosomes, which is more than simple semantics or a question of definition. In the two recent reviews by Gruenberg and Howell (1989) and Kornfeld and Mellman (1989) the lysosome was defined as the terminal compartment of the pathway that is free of MPR's, has the highest density on gradients such as percoll and, in most cases, the bulk of acid hydrolases. While the precise functional differences and the nature of the interrelationships between the PLC and the lysosomes is far from clear, there appears no question that the two compartments are different. The similarities and differences are summarized in Table 1.

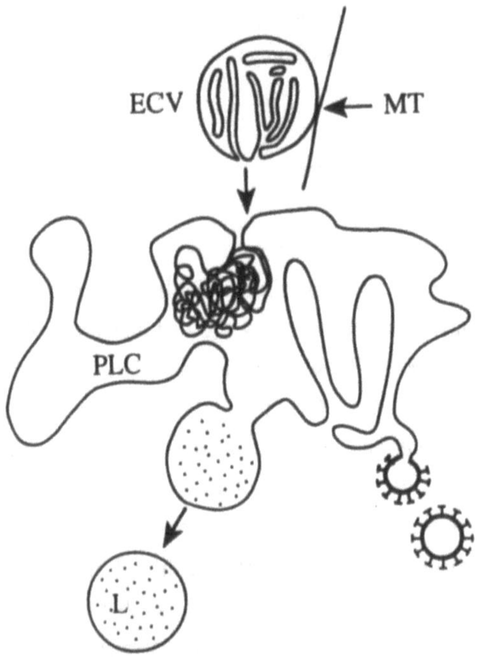

Figure 2. Schematic model showing the main features of the PLC and its relationship to ECV and lysosomes.

As we have recently argued (Griffiths et al., 1990a), it seems likely that the PLC is equivalent to the "light" lysosomes which have been separated from "dense" lysosomes in a multitude of studies using gradient systems such as percoll (see Storrie, 1988 for review). Of particular interest were the studies of Rome et al. (1979) and Rome and Cairn (1981) who showed that the "light" and "dense" lysosomal fractions of human fibroblasts contained similar concentrations of acid hydrolases and, significantly, *both* fractions were capable of fully digesting *in vitro* ^{35}S labelled mucopolysaccharide that had been taken up from the culture medium (a process requiring the action of nine different hydrolases). Further, the *rate* of degradation *in vitro* was the same for both fractions. If these *in vitro* data are representative of the *in vivo* situation and if the PLC and "light" lysosomes are indeed the same compartment, it presents us with an intriguing dilemma: why do some substrates get efficiently degraded while the bulk of the CI MPR (as well as other receptors) avoid degradation? The simplest explanation would be that these receptors molecules are fairly resistant to degradation; since the bulk of hydrolases are in the dense lysosomes (in most cell types) only the relatively small fraction of these recycling molecules that reach this compartment become degraded whereas more sensitive molecules would become degraded in the PLC. While this argument may partially explain the differences between the two sets of structures it is hard to reconcile with the data of Rome et al. (1989) showing that the "light" lysosomes (which we argue represent the PLC) and the "heavy" lysosomes contain similar *amounts* of hydrolase and show a similar *rate* of degradation of mucopolysaccharide. An alternative possibility to explain why the MPR is not significantly degraded in the PLC might be that the receptor molecules may be sequestered in some domains of the compartment that are not freely accessible to the hydrolases.

How do lysosomes form from the PLC? Despite the intense interest in endocytosis over the past decade this crucial question in lysosome biogenesis has received little attention. As for the early endosome to ECV step two extreme models could be envisaged to describe the formation of (dense) lysosomes. First, the PLC somehow "matures" into lysosomes; second, the lysosomes bud off the PLC in analogous fashion to the budding of secretion granules from the trans Golgi network (TGN) and to the proposed budding of ECV from early endosomes. We favour the latter hypothesis for many reasons. First, as for early endosomes, the PLC has a complex three-dimensional organization. Second, when cells are allowed to internalize long-lived endocytic markers such as gold for many hours followed by a "chase" in medium without gold for periods up to many days these markers are invariably distributed in *both* PLC and lysosomes (Griffiths et al., 1988; 1990). This is very hard to reconcile with a simple "maturation" model. These data, plus the fact that the density of the membrane protein lgp120 in the membranes of the PLC and lysosomes are quantitatively similar (Griffiths et al., 1990), rather argues for some kind of dynamic equilibrium between the two compartments. Our working model is that the formation of a lysosome vesicle may be mechanistically equivalent to the formation of a "regulated" secretory granule in secretory cells (Burgess and Kelly, 1987). In the latter it has been proposed that two phenomena are required: first, self-aggregation of secretory proteins to form a dense core and second, binding of the dense aggregates to putative membrane receptor(s) that somehow facilitate the formation of a granule. If similar (putative) receptors were required for lysosome budding they would clearly be different from the MPR's since, first, they would have to bind to dephosphorylated hydrolases and second, they would need to bind their ligands at a relatively low pH. While this idea is admittedly speculative there is some evidence in the literature both for low pH-induced aggregation of lysosomal enzymes (Buckmaster et al., 1988) as well as for acid hydrolase binding to isolated "lysosomal" membranes that is enhanced at low pH (Hohman and Bowers, 1984).

In conclusion we view the endocytic pathway as consisting of two preexisting, three-dimensionally-complex compartments, the early endosome and the PLC (late endosome). These compartments are also functionally complex because they would, like the TGN, need to form vesicles that would be directed to more than one pathway. They are highly dynamic and appear to be able to fuse with themselves but not with other compartments. In contrast, the ECV and the lysosomes are regarded as relatively simple vesicles (notwithstanding the lumenal membranes in the ECV) that bud off the early endosomes and PLC, respectively. Both move in unidirectional fashion along microtubules and, in the simplest model, both would be able to fuse only with the PLC. For the ECV the evidence for their fusion with the PLC is accumulating; for the lysosomes, the postulated retrograde fusion step is highly speculative and is based on the data mentioned above suggesting an equilibrium between the two compartments. We believe that *in vitro* studies using both biochemical and morphological techniques offer the best approach to address this problem in future studies.

List of References (Table 1)

1. Storrie, 1988.
2. Kornfeld and Mellman, 1989.
3. Griffiths et al., 1988
4. Geuze et al., 1988
5. Griffiths et al., 1990a.
6. Parton et al. 1989.
7. Sahagian and Neufeld, 1983.
8. Bleekemolen et al., 1988.
9. Johnson, Griffiths, Hoflack and Kornfeld, submitted.
10. Ludwig, Griffiths and Hoflack, submitted.
11. Killisch, Steinlein, Römisch, Beug and Griffiths, submitted.
12. Deng, Griffiths and Storrie, submitted.
13. Chavrier et al., 1990.
14. Parton et al., in preparation.
15. Griffiths et al., 1990b.

Table 1. Summary of main similarities and differences between PLC and lysosomes

	PLC	Lysosome	References
Similarities:			
Acid hydrolases	Up to 50% of total hydrolases. In most cultured cells ≈ 10-20%	In most cells contain the bulk of acid hydrolases (> 50%)	1,2
pH	below 5.5	below 5.5 (generally considered to be lower than 5)	1,2
lgp's	+++	+++	3,4
Differences:			
Structure	Complex-tubulo-cisternal-vesicular. Probably single copy compartment. In MDCK cells perinuclear, extending into the basal region.	Simple spherical vesicles. In MDCK cells exclusively in the apical region.	4,5,6
Density in gradients (e.g. percoll)	low density	high density	5,7
Cation-independent MPR	+++	0	3,4
Cation-dependent MPR	++	0	8,9
Phosphorylated hydrolases	+++ to 0, depending on cell type	0	10
Other receptors (Tf/TfR)	+ (the bulk are in early endosomes)	0	11
Kinetics of internalization	Fills between 20-60 mins at 37°	Fill after 60 mins at 37°	5,10
Fusion	Fuses together in rat-mouse cell hybrids to form an extended structure.	Contents intermix in rat-mouse cell hybrids without changing the structure of the lysosomes themselves.	12
Regulatory subunit of the cAMP dependent protein kinase (Type II)	+	0	15
Rab7	+	0	13
Relationship to microtubules (MT)	Tethered on MT; structure disrupts with nocodazole. In MDCK cells treated with low media (pH 6.65) fragments of the PLC move to the base of the cell. Upon adding pH 7.4 medium these fragments move rapidly to a position above the nucleus.	Move in directional fashion along MT. In MDCK cells less affected by low pH treatment. Do not fragment.	12,14

References

Bleekemolen, J.E., Stein, M., von Figura, K., Slot, J.W. and Geuze, H.J. 1988. The two mannose 6-phosphate receptors have almost identical subcellular distributions in U937 monocytes. Eur. J. Cell Biol. 47: 366-372.

Bomsel, M., Prydz, K., Parton, R.G., Gruenberg, J. and Simons, K. 1989. Endocytosis in filter-grown Madin-Darby canine kidney cells. J. Cell Biol. 109: 3243-3258.

Bomsel, M., Parton, R., Kuznetsov, S.A., Schroer, T.A. and Gruenberg, J. 1990. Microtubule- and motor-dependent fusion *in vitro* between apical and basolateral endocytic vesicles from MDCK cells. Cell 62: 719-731.

Buckmaster, M.J., Ferris, A.L. and Storrie, B. 1988. Effects of pH, detergent and salt on aggregation of Chinese-hamster-ovary-cell lysosomal enzymes. Biochem. J. 249: 921-923.

Burgess, T.L. and Kelly, R.B. 1987. Constitutive and regulated secretion of proteins. Ann. Rev. Cell Biol. 3: 243-293.

Chavrier, P., Parton, R.G., Hauri, H.P., Simons, K. and Zerial, M. 1990. Localization of low molecular weight GTP binding proteins to exocytic and endocytic compartments. Cell 62: 317-329.

Geuze, H.J., Slot, J.W., Strous, J.A.M., Lodish, H.F. and Schwartz, A.L. 1983. Intracellular site of asialoglycoprotein receptor-ligand uncoupling: double-label immunoelectron microscopy during receptor-mediated endocytosis. Cell 32: 277-87.

Geuze, H.J., Stoorvogel, W., Strous, G.J., Slot, J.W., Bleekemolen, J.E. and Mellman, I. 1988. Sorting of mannose-6-phosphate receptors and lysosomal membrane proteins in endocytic vesicles. J. Cell Biol. 107: 2491-501.

Griffiths, G., Hoflack, B., Simons, K., Mellman, I. and Kornfeld, S. 1988. The mannose 6-phosphate receptor and the biogenesis of lysosomes. Cell 52: 329-41.

Griffiths, G., Back, R. and Marsh, M. 1989. A quantitative analysis of the endocytic pathway in BHK cells. J. Cell Biol. 109: 2703-2720.

Griffiths, G., Matteoni, R., Back, R. and Hoflack, B. 1990a. Characterization of the cation-independent mannose 6-phosphate receptor-enriched prelysosomal compartment in NRK cells. J. Cell Sci. 93: 441-463.

Griffiths, G., Hollinshead, R., Hemmings, B.A. and Nigg, E.A. 1990b. Ultrastructural localization of the regulatory (RII) subunit of cyclic AMP-dependent protein kinase to subcellular compartments active in endocytosis and recycling of membrane receptors. J. Cell Sci. 96: 691-703.

Gruenberg, J., and Howell, K.E. 1989. Membrane traffic in endocytosis: insights from cell-free assays. Ann. Rev. Cell Biol. 5: 453-481.

Gruenberg, J., Griffiths, G. and Howell, K.E. 1989. Characterization of the early endosome and putative endocytic carrier vesicles *in vivo* and with an assay of vesicle fusion *in vitro*. J. Cell Biol. 108: 1301-1316.

Hohman, T.C. and Bowers, B. 1984. Hydrolase secretion is a consequence of membrane recycling. J. Cell Biol. 98: 246-252.

Hopkins, C.R. 1983. Intracellular routing of transferrin and transferrin receptors in epidermoid carcinoma A431 cells. Cell. 35: 321-330.

Hopkins, C.R., Gibson, A., Shipman, M. and Miller, K. 1990. Movement of internalized ligand-receptor complexes along a continuous endosomal reticulum. Nature 346:335-339.

Kornfeld, S. and Mellman, I. 1989. The biogenesis of lysosomes. Ann. Rev. Cell Biol. 5: 482-525.

Parton, R.G., Prydz, K., Bomsel, M., Simons, K. and Griffiths, G. 1989. Meeting of the apical and basolateral endocytic pathways of the MDCK cell in late endosomes. J. Cell Biol. 109: 3259-3272.

Rabinowitz, S. and Gordon, S. 1989. Differential expression of membrane sialoglycoproteins in exudate and resident mouse peritoneal macrophages. J. Cell Sci. 93: 623-630.

Rome, L.H. 1985. Curling receptors. Trends in Biochem. Sci. 10: 151.

Rome, L.H. and Cairn, L.R. 1981. Degradation of mucopolysaccharide in intact isolated lysosomes. J. Biol. Chem. 256: 10763-10768.

Rome, L.H., Garvin, A.J., Ailieta, M.M. and Neufeld, E.F. 1979. Two species of lysosomal organelles in cultured fibroblasts. Cell 17: 143-153.

Sahagian, G.G. and Neufeld, E. 1983. Biosynthesis and turnover of the mannose 6-phosphate receptor in cultured chinese hamster ovary cells. J. Biol. Chem. 258: 7121-7128.

Storrie, B. 1988. Assembly of lysosomes: perspectives from comparative molecular cell biology. Int. Rev. Cytol. 111: 53-105.

Swanson, J., Bushnell, A. and Silverstein, S.C. 1987. Tubular lysosome morphology and distribution within macrophages depend on the integrity of cytoplasmic microtubules. Proc. Natl. Acad. Sci. USA 84: 1921-1925.

Tassin, M.-T., Lang, T., Antoine, J.-C., Hellio, R. and Ryter, A. 1990. Modified lysosomal compartment as carrier of slowly and non-degradable tracers in macrophages. Eur. J. Cell Biol. 52: 219-228.

Tolleshaug, H. and Berg, T. 1981. The effect of leupeptin on intracellular digestion of asialofetuin in rat hepatocytes. Exp. Cell Res. 134: 207-217.

Wang, R.-H., Colbaugh, P.A., Kao, C.-Y., Rutledge, E.A. and Draper, R.K. Impaired secretion and fluid-phase endocytosis in the END4 mutant of chinese hamster ovary cells. J. Biol. Chem., in press.

Warren, G. 1990. Trawling for receptors. Nature 346: 318-319.

Protein Exchange Within the Lysosome and Pre-Lysosome Compartment: A Mechanism for Maintaining Organelle Functionality?

B. Storrie and Y. Deng
Department of Biochemistry and Nutrition
Virginia Polytechnic Institute and State University
Blacksburg, VA 24061
USA

INTRODUCTION

Numerous reports have appeared in the literature regarding heterogeneity within the vacuolar apparatus of mammalian cells and, in particular, the lysosomal compartment. The preceding article in this volume by T. Berg is one example. We have approached the question of lysosomal heterogeneity versus homogeneity from a different standpoint. We have reasoned that, although there may well be transient heterogeneity within the lysosomal population, the maintenance of organelle functionality must require a balanced repertoire of enzymes within each lysosomal unit. Hence we have reasoned that there must be mechanisms for maintaining lysosomal homogeneity and, as a consequence, functionality. Experimentally, we have approached this question by asking if *in vivo* there is a rapid exchange of molecules within the mixed lysosomal population of fused mammalian cells (see, Fig. 1). In our experiments, donor and recipient cell lysosomes have been either content labeled by the accumulation of endocytized solute or gold particles or distinguished by virtue of antigenically distinct lysosomal membrane proteins (lmps). The distribution of markers or biochemical products

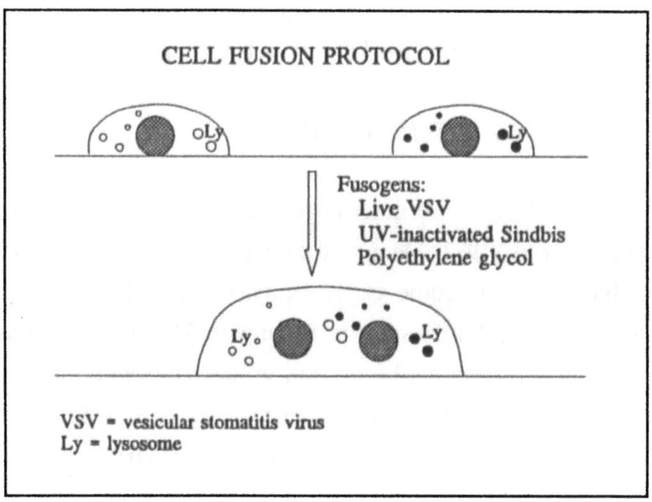

Figure 1: Creation of a Mixed Syncytoplasm

NATO ASI Series, Vol. H 62
Endocytosis
Edited by P. J. Courtoy

resulting from organelle intermixing was scored at various times after cell fusion. In some experiments, the distribution of endocytized markers or lmps have also been scored relative to the 300 Kda, cation-independent, mannose 6-phosphate receptor (MPR), a marker for the pre-lysosome compartment (PLC, Geuze et al., 1988; Griffiths et al, 1988; Griffiths et al., 1990). Lysosomal glycoproteins are present in similar concentrations, albeit perhaps not in equal amounts, in the MPR-enriched PLC and in MPR-negative lysosomes (Griffiths et al., 1990).

The major outcome of this work has been to show that both the PLC and the lysosomes from the parent cells intermix extensively after cell fusion (Ferris et al., 1987; Deng and Storrie, 1988; Deng, Griffiths and Storrie, submitted for publication). Intermixing of lysosomal membrane proteins is a rapid process with a $t_{1/2}$ of 30 min; the structures remain as small, punctate bodies distributed throughout the syncytoplasm (Deng, Griffiths and Storrie, submitted for publication). In contrast, the initially separate units of the PLC coalesce in the fused cells with a $t_{1/2}$ of 1 h to form large, pre-lysosomal complexes associated with individual nuclear clusters. Based on these differences we have concluded that the mechanism(s) of intermixing with respect to the two organelles is different. For the PLC, intermixing is likely to be the direct outcome of organelle coalescence. These results suggest that organelles are capable of self-recognition.

In the present report, we summarize studies on several aspects of lysosomal content exchange and discuss working models through which such exchange may occur.

MATERIALS AND METHODS

Homologous and heterologous cell fusions were performed as previously described (Ferris et al., 1987; Deng and Storrie, 1988; Deng, Griffiths and Storrie, submitted for publication). In some experiments, CHO cells fed [^{14}C]sucrose (15 μC/ml) for 18 h followed by a 2 h chase were fused with 3T3 cells which had been incubated in parallel with yeast invertase (0.5 mg/ml, grade III, Sigma Chemical Co., St. Louis). In other experiments, CHO cells were fed sucrose for 18 h followed by a 1-2 h chase to produce sucrosomes, sucrose swollen lysosomes (DeCourcy and Storrie, 1990), and fused with 3T3 cells containing organelles labeled with Texas red dextran in a parallel uptake protocol. Texas red dextran localization relative to sucrosomes in cell syncytia was determined by fluorescence microscopy. For ATP depletion, cell syncytia were incubated with 50 Mm 2-deoxyglucose and 1 Mm sodium cyanide. This produced >80% depletion of cellular ATP levels using the luciferin-luciferase assay. The frequency of tubular or reticular lysosomes in CHO, HeLa, NRK, and 3T3 cells was assessed in cell cultures which had been pre-fed fluorescent dextrans for 18 h.

RESULTS

Content Exchange Between Lysosomes is a Rapid Process with Minimal Time Lag

Previous results had suggested that, upon cell fusion, there might be a significant 30-45 min, temperature dependent lag in lysosomal content interchange (Ferris et al., 1987; Deng and Storrie, 1988). Based on the temperature sensitivity of the lag period, we had suggested that a "priming" process might be occurring during this period (Ferris et al., 1987). The previous experiments were done chiefly by scoring the transfer of endocytized contents from donor lysosomes into recipient sucrosomes. As sucrosomes are large, dilution of lysosomal contents upon transfer into sucrosomes may have limited our ability to score early transfer events. We have reinvestigated this question using an invertase-[^{14}C]-sucrose content pairing in the donor 3T3 and recipient CHO cell lysosomes. With this pairing, content intermixing results in the invertase-dependent cleavage of sucrose to glucose and fructose, which permeate cell membranes and appear rapidly in the extracellular media. As shown in Fig. 2, almost no lag in apparent content intermixing was detectable with this assay upon fusion of CHO and 3T3 cells. Based on this more sensitive assay, we conclude that there is no significant lag period in lysosomal content exchange upon cell fusion and that any "priming" process must indeed be rapid.

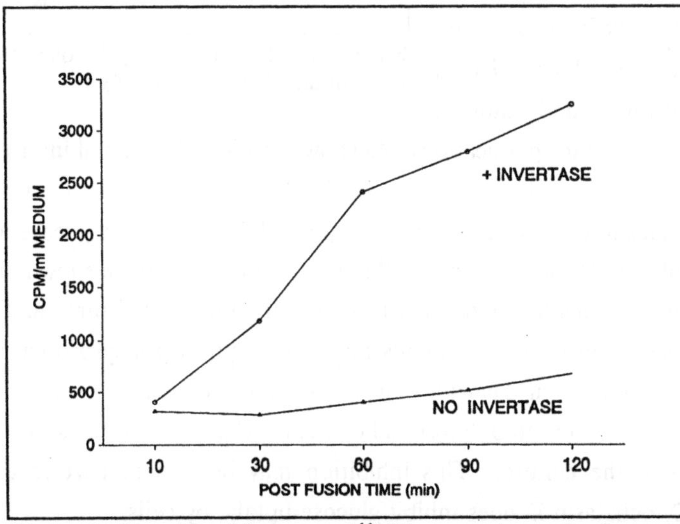

Figure 2: Results of Invertase-[^{14}C]Sucrose Assay for Lysosomal Content Intermixing

Effect of Inhibitors on Lysosome-Lysosome Exchange

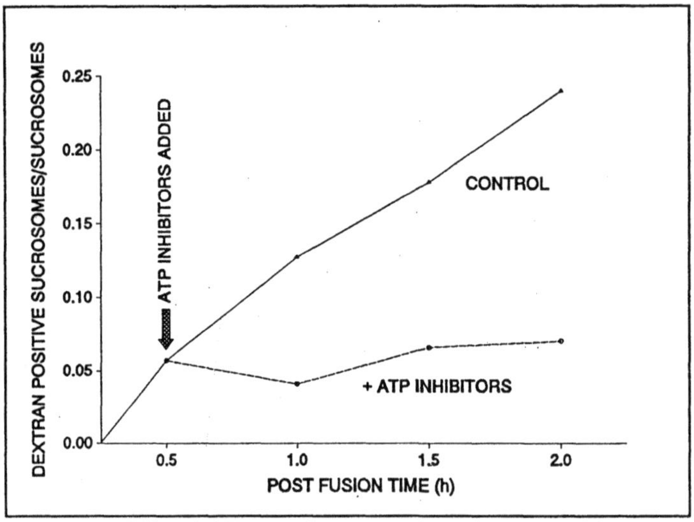

Figure 3: Effect of Deoxyglucose and Sodium Cyanide on Lysosomal Content Exchange

In previous experiments, we have shown that lysosomal intermixing is a microtubule dependent process, i.e., sensitive to nocodazole, and it exhibits a marked sensitivity to lowered temperature (Ferris el al., 1987; Deng and Storrie, 1988; Deng, Griffiths and Storrie, submitted for publication). As shown in Fig. 3, lysosomal intermixing is also strongly ATP dependent, with almost complete inhibition of transfer of Texas red dextran from 3T3 lysosomes to CHO cell sucrosomes upon addition of deoxyglucose and sodium cyanide to the culture. It should be noted that sucrosomes have been shown previously to be lysosomes, i.e., lmp-positive and MPR-negative structures (DeCourcy and Storrie, 1990). We have also asked if microfilaments might be required for lysosomal content exchange. In experiments using a pairing of invertase- and [^{14}C]sucrose-positive lysosomes, we found about a 50% inhibition of lysosomal content exchange with the addition of 5 μg/ml cytochalasin D to the culture. This inhibition may be due in part to an ATP-dependent effect. Cytochalasin D does inhibit glucose uptake by cells.

Observations on the Role of Reticular Lysosomes in Lysosome-Lysosome Exchange

A reticular nature to the vacuolar apparatus has been noted by other investigators (e.g., Swanson, Bushnell and Silverstein, 1987; Hopkins et al., 1990). Conceivably, molecular exchanges within the lysosomal population of fused cells could be the result of reticular continuities between organelles. This is a point of current research interest on which a number of initial observations can be reported. As shown previously, when CHO cells containing Texas red- and FITC-dextran marked lysosomes were fused, reticular structures were observed in live mounts and these were positive for both colored dextrans (Ferris et al., 1987). If these reticular structures are important in lysosome-lysosome exchanges, then the rate of molecular exchanges should vary with the reticularity of the donor cells. To test this, a series of cell lines differing in the frequency of reticular lysosomes were fused with sucrosome-positive CHO cells. The cell lines were NRK and HeLa cells, which contain few, if any, reticular structures, CHO cells, which were intermediate, and 3T3 cells, in which reticular lysosomes were common. In this experiment, the disappearance of sucrosomes as a result of invertase exchange was scored. No consistent correlation between the extent of reticular lysosomes in the donor cell and the rate of disappearance of sucrosomes was noted; $t_{½}$'s were: NRK cell donor, 30 min; HeLa cell donor, 1 h; CHO cell donor, 25 min; and 3T3 cell donor, 20 min. Experiments to compare rates of exchange in cases where both donor and recipient cells contain reticular structures of varying extent are planned.

DISCUSSION

The observations reported here and in previous work (Ferris et al., 1987; Deng and Storrie, 1988; Deng, Griffiths and Storrie, submitted for publication) clearly suggest that within the vacuolar apparatus of mammalian cells there is a continuous molecular exchange between members of the lysosomal population and/or segments of the PLC. This exchange, when traced by endocytized content markers or lmp distribution, is ATP dependent, temperature sensitive and microtubule dependent. In many ways, exchange within the PLC and lysosomes appear to be different processes. In the former, organelle coalescence is observed. In the latter, no change in organelle shape or size is observed. Reticular lysosomes may play a key role in this process. However, initial data suggest no correlation between the frequency of reticular lysosomes in the donor cells and the rate of lysosomal content exchange.

As summarized in Fig. 4, several pathways for the intermixing of proteins and contents within the lysosomal population can be proposed. One possibility is that the PLC may be an intermediate in this process; lysosomal membrane proteins may be in equilibrium between the PLC and lysosomes. Alternatively, intermixing within the lysosomal population may be due to direct, transient fusion within the organelle population. By time-lapse video microscopy of cells containing organelles labeled by long-term internalization of fluorescent dextran, transient collisions have been observed between lysosomes in the cell periphery (Deng and Storrie, unpublished observations). These collisions may lead to repeated rounds of exchange. Reticular connections between organelles are another possibility, although initial data do not support this suggestion. At present, a firm conclusion can not be reached as to which of the possible pathways is predominant for content and membrane exchange within the lysosomal population. These observations of rapid molecular exchange within the lysosomal population do suggest that lysosomes over time should exhibit homogeneity, although at any given instant the population may exhibit heterogeneity. Hence the balanced functionality of each organelle unit should be maintained over time.

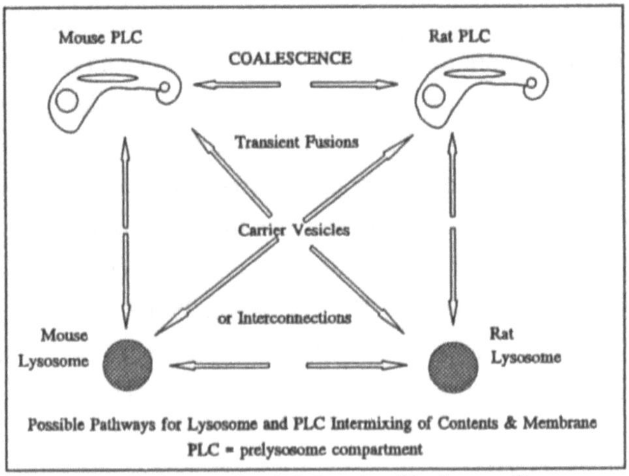

Figure 4: Pathways for Molecular Exchange Between Elements of the Lysosomal Population

LITERATURE REFERENCES

DeCourcy K, Storrie B (1990) Osmotic swelling of endocytic compartments induced by internalized sucrose is restricted to mature lysosomes in cultured mammalian cells. *Exp Cell Res*, in press

Deng Y, Storrie B (1988) Animal cell lysosomes rapidly exchange membrane proteins. *Proc Natl Acad Sci USA* 85:3860-3864

Geuze H J, Stoorvogel W, Strous J, Slot J W, Bleekmolen J E, Mellman I (1988) Sorting of mannose 6-phosphate receptor and lysosomal membrane proteins in endocytic vesicles. *J Cell Biol* 107:2491-2501

Griffiths G, Matteoni R, Back R, Hoflack B (1990) Characterization of the cation-independent mannose 6-phosphate receptor-enriched prelysosomal compartment in NRK cells. *J Cell Sci* 95:441-461

Griffiths G, Hoflack B, Simons K, Mellman I, Kornfeld S (1988) The mannose 6-phosphate receptor and the biogenesis of lysosomes. *Cell* 52:329-341

Ferris A L, Brown J C, Park R D, Storrie B (1987) Chinese hamster ovary cell lysosomes rapidly exchange contents. *J Cell Biol* 105:2703-2712

Hopkins C R, Gibson A, Shipman M, Miller K (1990) Movement of internalized ligand-receptor complexes along a continuous endosomal reticulum. *Nature* 346:335-339

ENDOSOMAL pH REGULATION AND THE MATURATION MODEL FOR LYSOSOME BIOGENESIS

Robert F. Murphy, Mario Roederer, David M. Sipe, Cynthia Corley Cain and Russell B. Wilson

Carnegie Mellon University, Department of Biological Sciences & Center for Fluorescence Research in Biomedical Sciences, 4400 Fifth Avenue, Pittsburgh, Pennsylvania 15213, U.S.A.

The mechanisms by which the recycling and degradation of endocytosed material are regulated have been the subject of intense study since the initial description of lysosomes (for reviews see Steinman et al 1983; von Figura & Hasilik 1986). While a number of questions remain, significant evidence points to the important role played by low pH in controlling and facilitating processing of endocytosed ligands. The dissociation of many ligand-receptor complexes is induced by acidification below pH 7, and many lysosomal hydrolases have low pH optima (Mellman et al 1986). Within the last five years, it has become clear that, at least in many non-erythroid cells, the pH to which material destined for lysosomes is exposed decreases continuously with time after internalization (Murphy et al 1984; Roederer & Murphy 1986; Kielian et al 1986; Roederer et al 1987; Yamashiro & Maxfield 1987).

Endosomal pH Regulation

When fluorescence methods are used to monitor the pH of compartments containing endocytosed material destined for lysosomes, a biphasic acidification pattern is often observed. A rapid acidification to near pH 6 is followed by a slower decrease to near pH 5 (Murphy et al 1984; Roederer et al 1987). For ligands which are recycled to the plasma membrane, such as transferrin, acidification is limited to pH 6 (Sipe & Murphy 1987). Comparison of the kinetics of acidification of different classes of ligands indicates that the segregation of ligands into compartments with different pH characteristics occurs within 5 min of ligand internalization (Murphy 1988). In order to explain the biphasic acidification pattern, Fuchs et al (1989) demonstrated that the acidification of isolated early endosomes, but not of late endosomes or lysosomes, could be inhibited *in vitro* by the Na^+,K^+-ATPase. These authors proposed that a membrane potential generated by the Na^+,K^+-ATPase acts to inhibit acidification by the electrogenic endosomal/lysosomal

NATO ASI Series, Vol. H 62
Endocytosis
Edited by P. J. Courtoy

proton pump. Active Na^+,K^+-ATPase was postulated to be found only in early endosomes. Evidence in support of a role for the Na^+,K^+-ATPase in regulating vesicular pH in living cells came from the analysis of a chloroquine-resistant Swiss 3T3 cell line with elevated lysosomal pH (Cain & Murphy 1988). Relative to parental cells, this cell line showed increased resistance to ouabain, an inhibitor of the Na^+,K^+-ATPase, leading to the speculation that the pH defect was due to mislocalization of Na^+,K^+-ATPase into lysosomes. Treatment with ouabain eliminated the chloroquine-resistance of the mutant, lending support to this hypothesis (Cain et al 1989). In order to directly test the effect of inhibiting Na^+,K^+-ATPase on endosomal pH, we have measured transferrin acidification in A549 cells, a human epidermoid carcinoma cell line, in the presence and absence of ouabain. Inhibition of Na^+,K^+-ATPase activity by ouabain results in acidification of transferrin to a significantly lower pH (Fig. 1). These results confirm in living cells those obtained on isolated vesicles.

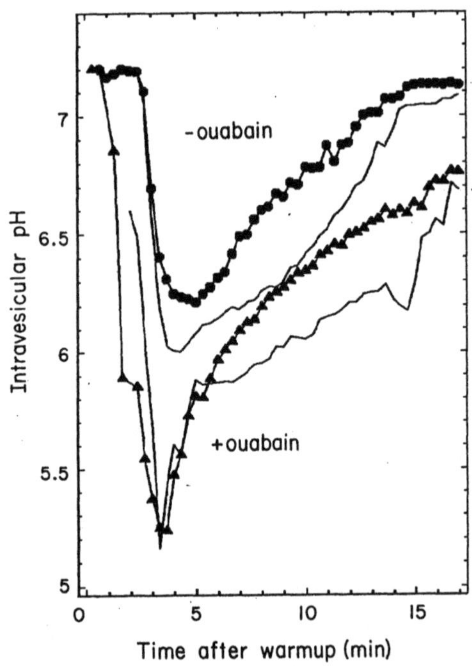

Figure 1: Effect of ouabain on endosomal pH.

A549 cells were preincubated with either 6 mM ouabain in low K^+ buffer (▲) or 10 μM ouabain in PBS (---) for 4-6 hr. Acidification kinetics were then determined (in the presence of ouabain) as described (Sipe & Murphy 1987). Cells in the absence of ouabain were similarly labeled and analyzed in low K^+ buffer (symbol(filled_circle)) or PBS (---). pH values for internalized transferrin were calculated after subtraction of the fluorescence from surface-bound transferrin. The minimum pH for the control cells was 6.0-6.2, compared to 5.2 for ouabain-treated cells. From Cain et al 1989.

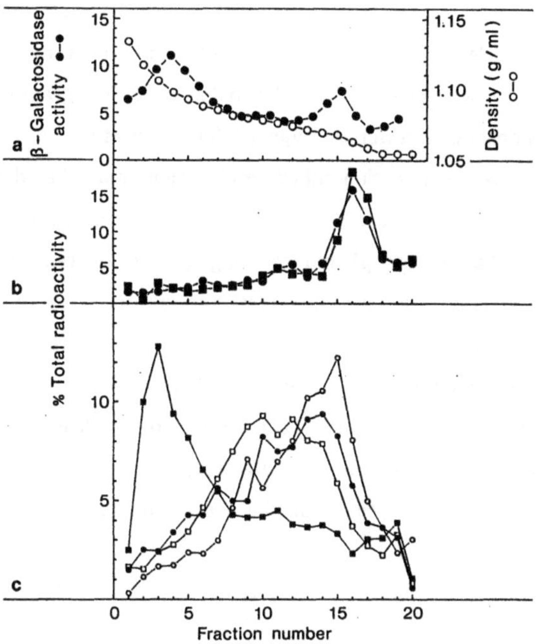

Figure 2: Endosome maturation in a cell-free system.

Postnuclear supernatants were fractionated on 27% Percoll gradients. (a) β-galactosidase activity was measured for each fraction to determine the location of dense lysosomes. (b) Endosomes were labeled by incubating cells with either ^{125}I-EGF (●) or ^{125}-transferrin (■) for 1 h at 17°C. (c) Fractions containing ^{125}-EGF were pooled and incubated at 0°C (open symbols) or 37°C (filled symbols) in the presence (□ , ■) or absence (○ , ●) of 5 mM K_2ATP for 165 min. An increase in the density of ^{125}I-EGF-containing endosomes to that of dense lysosomes is observed at 37°C in the presence of ATP. From Roederer et al 1990.

Endosome Maturation

Analysis of the kinetics of hydrolysis of endocytosed fluorogenic substrates indicates that both cathepsin B (Roederer et al 1987) and acid phosphatase (Bowser & Murphy 1990) are found in early compartments; evidence from a number of groups suggests that the initial site of exposure of endocytosed material to lysosomal enzymes is the early endosome (reviewed in Murphy 1988). In view of this, we have previously suggested that production of dense lysosomes can occur via maturation, as an alternative to or substitute for fusion with pre-existing lysosomes (Roederer et al 1987; Murphy 1988). A prediction of this model is that isolated endosomes should be capable of undergoing maturation to dense lysosomes *in vitro*. The results shown in Fig. 2 confirm this prediction.

Associated with this increase is a change in appearance of the endosomal fraction from electron-lucent to electron-dense. The density increase is inhibited by benzylamine, a membrane-soluble weak base which dissipates proton gradients, indicating that acidification of the endosome interior is required for maturation.

Two models for this acidification-linked maturation may be described. The first, based on the observation that lysosomal enzymes form a matrix at low pH (Buckmaster et al 1988), proposes that a low pH-induced aggregation of vesicle contents leads to a reduction in osmotic strength, a loss of water, and an increase in density. Membrane fission events may occur during this shrinking (e.g., to remove mannose-6-phosphate receptors and/or excess membrane). The second proposes that acidification leads to activation of lysosomal enzymes and degradation of endosomal membrane until a protective concentration of lysosomal membrane glycoproteins is reached (Kornfeld & Mellman 1989). Further analysis will be required to distinguish between these possibilities.

CONCLUSIONS

Taken together, the results described above imply a critical role for pH not only in regulating ligand-receptor sorting, but also in controlling the formation of new dense lysosomes. Further insight into this role may come from the analysis of a new class of Chinese hamster ovary cell mutants with temperature-conditional in recycling of receptors to the plasma membrane (C.C.C., R.B.W. and R.F.M., submitted). In one of these mutants, TfT1.11, a reduced ability to recycle transferrin and α_2-macroglobulin receptors is associated with a loss of dense lysosomes. The results suggest that a defect in recycling may inhibit the removal of Na^+,K^+-ATPase from vesicles destined for lysosomes, resulting in a reduced ability to form dense lysosomes.

ACKNOWLEDGEMENTS

This work was supported in part by National Institutes of Health grants GM32508, GM08067 and GM13179, and National Science Foundation Awards DCB-8351364 and DCB-8903657.

REFERENCES

Bowser R, Murphy RF (1990) Kinetics of hydrolysis of endocytosed substrates by mammalian cultured cells: Early introduction of lysosomal enzymes into the endocytic

pathway. J cell Physiol 143:110-117
Buckmaster MJ, Ferris AL, Storrie B (1988) Effects of pH, detergent and salt on aggregation of Chinese-hamster-ovary-cell lysosomal enzymes. Biochem J 249:921-923
Cain, C.C. and R.F. Murphy. 1988. A chloroquine-resistant swiss 3T3 cell line with a defect in late endocytic acidification. J. Cell Biol. 106:269-277
Cain CC, Sipe DM, Murphy RF (1989) Regulation of endocytic pH by the Na^{+},K^{+} ATPase in living cells. Proc Natl Acad Sci USA 86:544-548
Fuchs R, Schmid S, Mellman I (1989) A possible role for the Na^{+},K^{+}-ATPase in regulating ATP-dependent endosome acidification. Proc Natl Acad Sci USA 86:539-543
Kielian MC, Marsh M, Helenius A (1986) Kinetics of endosome acidification detected by mutant and wild-type Semliki Forest virus. EMBO J 5:3103-3109
Kornfeld S, Mellman I (1989) The biogenesis of lysosomes. Annu Rev Cell Biol 5:483-525
Mellman I, Fuchs R, Helenius A (1986) Acidification of the endocytic and exocytic pathways. Annu Rev Biochem 55:663-700
Murphy RF (1988) Processing of endocytosed material. Adv Cell Biol 2:159-180
Murphy RF, Powers S, Cantor CR (1984) Endosomal pH measured in single cells by dual fluorescence flow cytometry: Rapid acidification of insulin to pH 6. J Cell Biol 98:1757-1762
Roederer M, Barry JR, Wilson RB, Murphy RF ((1990) Endosomes can undergo an ATP-dependent density increase in the absence of dense lysosomes. Eur J Cell Biol 51:229-234
Roederer M, Bowser R, Murphy RF (1987) Kinetics and temperature dependence of exposure of endocytosed material toproteolytic enzymes and low pH: Evidence for a maturation model for the formation of lysosomes. J cell Physiol 131:200-209
Roederer M, Murphy RF (1986) Cell-by-cell autofluorescence correction for low signal-to-noise systems: Application to EGF endocytosis by 3T3 fibroblasts. Cytometry 7:558-565
Sipe DM, Murphy RF (1987) High resolution kinetics of transferrin acidification in BALB/c 3T3 cells: Exposure to pH 6 followed by temperature-sensitive alkalinization during recycling. Proc Natl Acad Sci USA 84:7119-7123
Steinman RM, Mellman IS, Muller WA, Cohn ZA (1983) Endocytosis and the recycling of plasma membrane. J Cell Biol 96:1-27
von Figura K, Hasilik A (1986) Lysosomal enzymes and their receptors. Annu Rev Biochem 55:167-193
Yamashiro DJ, Maxfield FR (1987) Kinetics of endosome acidification in mutant and wild-type Chinese hamster ovary cells. J Cell Biol 105:2713-2721

1.4. CELL-FREE ANALYSIS OF ENDOCYTIC TRAFFIC

(Chairperson : Jean GRUENBERG)

As discussed hereafter in the report of the roundtable, cell-free assays based on mixing of isolated subcellular fractions or the use of perforated semi-intact cells, are an essential approach to analyze the regulatory molecules involved in endocytic events. Two presentations analyzed the genesis of clathrin-coated endocytic pits in semi-intact permeabilized cells, a system allowing to easily manipulate the composition of the cytosol, especially to examine the role of coat and adaptor proteins in the regulation of the formation of endocytic coated pits. **Marc Pypaert,** from the laboratory of **Graham Warren,** reported that the invagination of coated pits is inhibited in broken mitotic HeLa cells when mitotic cytosol is added. No inhibition occurs when broken interphase cells are treated with mitotic cytosol, or when broken mitotic cells are treated with interphase cytosol. The slowly hydrolyzable analog ATPγS combined with mitotic cytosol causes a 20-fold elongation of the neck of coated pits, as compared with cells incubated with ATP and interphase cytosol. Thus, it appears that coated pit formation is regulated by phosphorylation/dephosphorylation during mitosis, and that a special "collar" protein (that could be related to the cytoskeleton) forces the membrane to invaginate.

Sandra Schmid used a glutathione-sensitive iodinated transferrin to distinguish deep invagination (monitored by the inaccessibility to anti-transferrin antibodies) from true vesiculation (resistance of transferrin release by glutathione). In HeLa cells, severe ATP-depletion abolishes the acquisition of glutathione resistance, but only partially affects antibody inaccessibility. She proposes that ATP and cytosol are required for de novo coated pit formation and budding of coated vesicles, but not for deep invagination, in agreement with Pypaert's observations.

Jean Gruenberg used cell-free mixing assays to study the regulation of the endosomal traffic. Two sequential and biochemically distinct endocytic fusion events were reconstituted in vitro, namely the lateral interaction between early endosomes and the vectorial transfer to late endosomes. Interestingly, early basolateral and apical endosomes do not fuse with each other, indicating differences in membrane composition. The most striking observation is that the fusion between early endosomes is inhibited by antibodies against the small GTP-binding protein rab5, but can be rescued by the addition of cytosol of cells overexpressing this protein. The early fusion event is also inhibited by the cdc2-protein kinase, possibly

NATO ASI Series, Vol. H 62
Endocytosis
Edited by P. J. Courtoy

reflecting the fragmentation of the early endosome during mitosis. The transfer to late endosomes is stimulated by intact microtubules and depends on both kinesin and dynein.

Finally, **Paul Luzio** studied the relation between rat liver endosomes and lysosomes by cell-free mixing assays. Asialofetuin internalized into late endosomes (essentially devoid of hexosaminidase and enriched in mannose 6-phosphate receptors) can be shifted to the equilibrium density of lysosomes, if purified lysosomal fractions are added. However, such a displacement does not occur in the absence of added lysosomes, a strong argument against the spontaneous maturation of endosomes into lysosomes. The shift requires ATP and cytosol and is blocked by N-ethylmaleimide and by the non-hydrolyzable analog GDPβS, consistent with the involvement of the fusion-promoting protein called NSF and a G protein. Early endosomes are not competent to fuse with lysosomes in this system.

EFFECT OF ATPγS ON THE FORMATION OF COATED VESICLES IN BROKEN HELA CELLS

Marc Pypaert and Graham Warren
Imperial Cancer Research Fund
PO Box 123
Lincoln's Inn Fields
London WC2A 3PX

INTRODUCTION

The mechanism of receptor-mediated endocytosis comprises a series of steps, starting with nucleation, which involves binding of triskelions to the plasma membrane. The triskelions are linked to adaptors which in turn bind to the cytoplasmic tails of receptors such as the transferrin receptor (Pearse, 1987). Nucleation is followed by invagination. The coat starts as a lattice of hexagons some of which re-arrange to pentagons causing invagination of the plasma membrane (Heuser and Evans, 1980) at the end of which the invaginated membrane is only held to the plasma membrane by a thin neck. Scission of this neck releases the coated vesicle (Smythe et al., 1989) which is then uncoated (Rothman and Schmid, 1985), releasing coat subunits for further rounds of internalisation. The uncoated vesicle is then directed to endosomes (Hopkins, 1986).

Each of these steps has been re-created in broken cells and the general requirements for each step have been determined (Rothman and Schmid, 1985; Warren et al., 1988; Gruenberg and Howell, 1989). For the purposes of this report, it is only necessary to know that the scission process requires ATP and cytosol. Invagination requires neither so incubation of broken cells in the absence of ATP and cytosol causes all coated pits to accumulate in a deeply-invaginated form with the membrane linked to the plasma membrane only by a thin neck (Smythe et al., 1989).

Endocytosis is inhibited during mitosis (Fawcett, 1965; Berlin et al., 1978) and our experiments both *in vivo* (Pypaert et al., 1987) and *in vitro* (our unpublished results) have shown that it is the invagination of coated pits which is inhibited. In the course of carrying out *in vitro* experiments using mitotic cells and cytosols we noticed that one of the analogues of ATP, namely ATPγS, had an

NATO ASI Series, Vol. H 62
Endocytosis
Edited by P. J. Courtoy

unexpected effect on the structure of the coated pits. These observations are reported here.

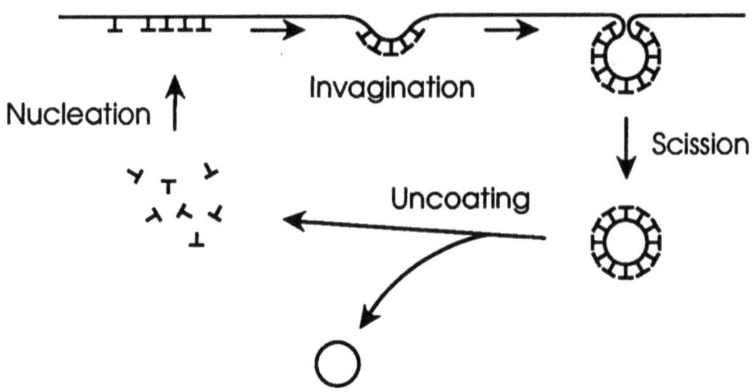

Figure 1. Schematic view of the early steps on the pathway of receptor-mediated endocytosis.

RESULTS

Mitotic HeLa cells were broken by freeze-thawing using liquid nitrogen and incubated in mitotic cytosol for 5min. at 37°C in the presence of either 2.5mM ATP or ATPγS. The samples were then diluted 20-fold into buffer containing the appropriate nucleotide and incubated for a further 5 min. at 37°C. Under these conditions we find that invagination proceeds. When examined at low power in the electron microscope (Figure 2), the cells that had been incubated with ATPγS had a striking cortical band of filaments which were probably microfilaments. This band was more than twice as thick as that observed in cells incubated with ATP (Table 1).

Table 1

Nucleotide	Width of the Cortical Band	Length of the Neck
ATP	0.64 ± 0.20μm	0.037 ± 0.061μm
ATPγS	1.42 ± 0.44μm	0.780 ± 0.410μm

Broken mitotic cells were incubated with either ATP or ATPγS and then processed for electron microscopy. Width of the cortical band represents the average distance from the plasma membrane to cytoplasmic organelles other than tubules or coated structures. Length of neck represents the average minimum distance between the plasma membrane and the base of the coated structure. Measurements were carried out using an improved stereological technique (Gundersen, personal communication).

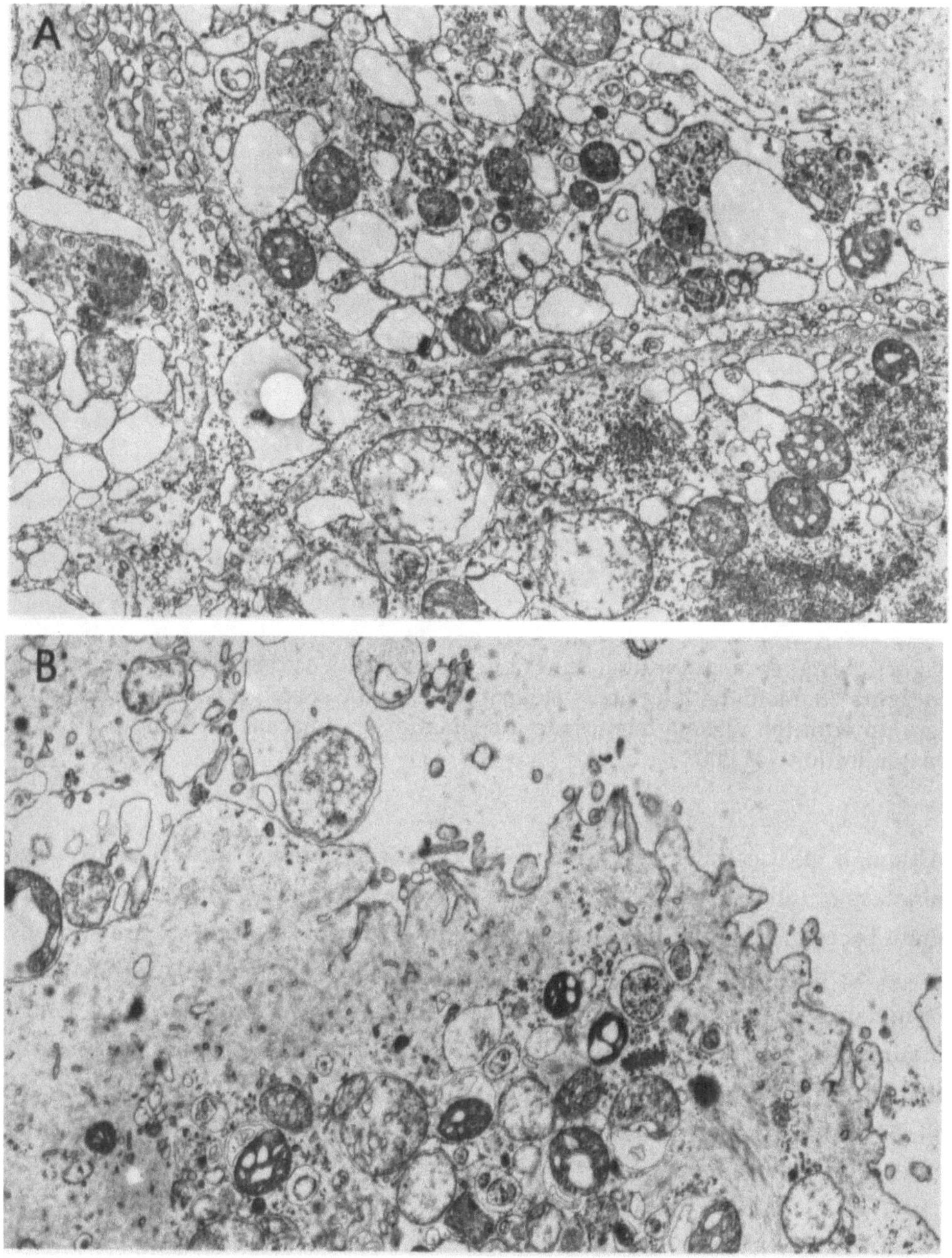

Figure 2. Mitotic HeLa cells were broken by freeze-thawing and incubated with mitotic cytosol in the presence of either ATP (A) or ATPγS (B). Cells were fixed and processed for electron microscopy in the usual way. Note the cortical band of filaments in the cells incubated with ATPγS. Magnification x15,000.

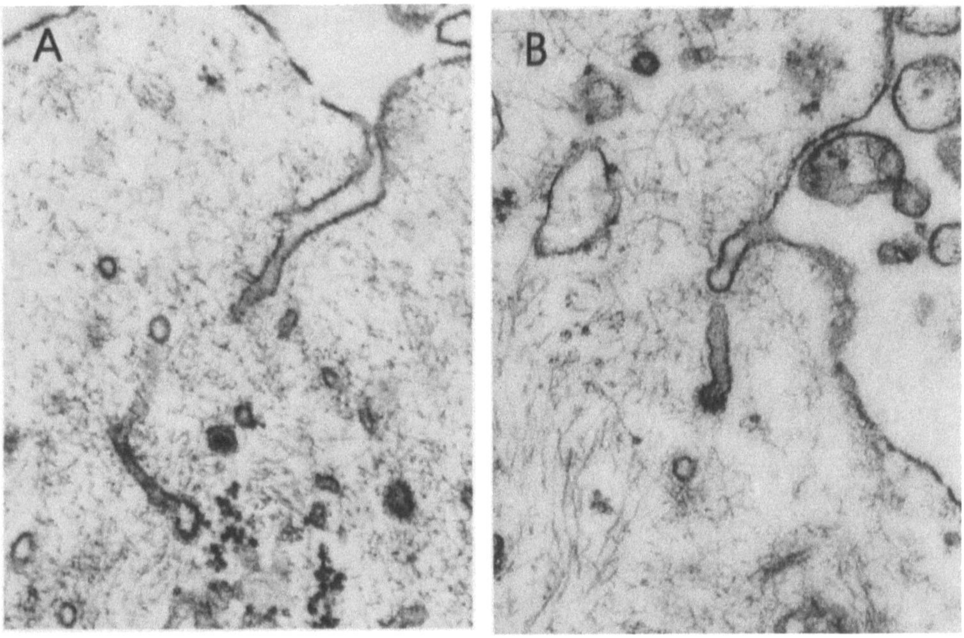

Figure 3. Higher magnification views of cells prepared as described in the legend to figure 2B. Note the long necks emanating from the coated pits. They probably join up with the plasma membrane invagination in adjacent sections. Magnification x45,000.

Although most cellular organelles were excluded from the cortical band, large numbers of tubules and coated structures remained (Figure 2B). The structures of these became clearer at higher magnification (Figure 3). Careful examination by serial section showed that the tubules originated at the cell surface and terminated in the coated structures. Measurements showed that the average minimum length of these tubules was 0.78μm, more than 20 times longer than that of coated pits in cells incubated with ATP (Table 1). The structures are similar to those reported by other groups in intact cells (Willingham and Pastan, 1983; Tooze, 1985).

DISCUSSION

Although dilution of the samples causes invagination to proceed this does not explain why the neck becomes so elongated during incubation with ATPγS. One possibility is that mitotic cytosol, together with ATPγS, encourages the formation of a cortical ring of microfilaments which pushes cytoplasmic organelles away from the plasma membrane. If polymerisation occurred between the plasma membrane and the coated pit this might force the coated pit away from the plasma membrane, deep into the cytoplasm. Another possibility is that filaments are usually involved in pulling coated vesicles into the cytoplasm (Salisbury et al., 1980). Since scission cannot occur then pulling will generate elongated necks.

Neither of these possibilities explains the shape of the neck. Irrespective of whether the coated pit is pulled or pushed, why does the plasma membrane not simply indent, yielding a conical structure, the tip of which would be a deep coated pit? The answer is probably to be found in work by Tooze (1985). He studied coated pits which were unable to undergo scission because they were taking up viruses which had not finished budding from the adjacent cell. The coated pits formed were very similar to those observed here and the simplest explanation of their shape was to suggest that the neck is normally surrounded by a collar which stays attached to the plasma membrane. Pulling the coated pit into the cytoplasm would pull the neck through this collar which would restrict its shape to that of a tubule. An obvious candidate for the collar would be the scission enzyme complex itself.

REFERENCES

Berlin, R. D., Oliver, J. M. and Walter R. J. (1978) Surface functions during mitosis I: Phagocytosis, pinocytosis and mobility of surface-bound Con A. Cell 15, 327-341.

Fawcett, D.W. (1965) Histochemical society symposium on structure and function at cell surfaces. J. Histochem. Cytochem. 13, 75-91

Gruenberg, J. & Howell, K.E. (1989) Membrane traffic in endocytosis: insights from cell-free assays. Annu. Rev. Cell Biol. 5, 453-481

Heuser, J. & Evans, L. (1980) Three-dimensional visualization of coated vesicle formation in fibroblasts. J. Cell Biol. 84, 560-583

Hopkins, C.R. (1986) Membrane boundaries involved in the uptake and intracellular processing of cell surface receptors. Trends Biochem. Sci. 11, 473-477

Pearse, B.M.F. (1987) Clathrin and coated vesicles. EMBO J. 6, 2507-2512

Pypaert, M., Lucocq, J.M. and Warren, G. (1987) Coated pits interphase and mitotic cells. Eur.J.Cell Biol. 45, 23-29

Rothman, J.E. and Schmid, S.L. (1985) Enzymatic recycling of clathrin from coated vesicles. Cell 46, 5-9

Salisbury, J.L., Condeelis, J.S. and Satir, P. (1980) Role of coated vesicles, microfilaments and calmodulin in receptor-mediated endocytosis by cultured B lymphoblastoid cells. J.Cell Biol. 87, 132-141

Smythe, E., Pypaert, M., Lucocq, J. & Warren, G. (1989) Formation of coated vesicles from coated pits in broken A431 cells. J. Cell Biol. 108, 843-853

Tooze, J. (1985) Blocked coated pits in AtT20 cells result from endocytosis of budding retrovirions. J.Cell Biol. 101, 1713-1723

Warren, G., Woodman, P., Pypaert, M. and Smythe, E. (1988) Cell-free assays and the mechanism of receptor-mediated endocytosis. Trends Biochem. Sci. 13, 462-465

Willingham, M.C. and Pastan, I. (1983) Formation of receptosomes from plasma membrane coated pits during endocytosis: Analysis by serial sections with improved membrane labelling and preservation techniques Proc.Natl.Acad.Sci. USA 80, 5617-5621

ATP IS REQUIRED FOR RECEPTOR-MEDIATED ENDOCYTOSIS BOTH *IN VIVO* AND *IN VITRO*

Sandra L. Schmid, Laura L. Carter and Elizabeth Smythe
Department of Molecular Biology
Research Institute of Scripps Clinic
10666 N. Torrey Pines Road
La Jolla, California, USA
92037

Introduction

The process of receptor-mediated endocytosis via clathrin-coated pits has been well-defined at the ultrastructural level. Receptors are concentrated in specialized "coated pit" regions of the cell surface which invaginate and pinch-off to form coated vesicles carrying receptors and bound ligands into the cell. At the molecular level, most of the constituents of the protein coat are known (although new coat constituents are still being identified, Ahle and Ungewickell, 1990) and a growing number of these have been cloned allowing primary sequence determination (for reviews see Pearse and Crowther, 1987; Brodsky, 1988; Morris et al., 1989). The major coat proteins are clathrin triskelions which consist of three 180kD clathrin heavy chains each with a tightly associated ~30kD light chain. Assembly protein (AP) complexes or 'adaptins' appear to mediate the binding of the clathrin coat to the plasma membrane perhaps through direct interaction with the cytoplasmic domains of receptors (Keen, 1987; Virshup and Bennett, 1988; Glickman et al. 1989). These complexes consist of two ~110kD , two ~50 kD and two ~20 kD subunits and are present at a 3:2 molar ratio to triskelions in the coat (reviewed by Morris et al., 1989). Two distinct protein kinases have also been identified as constituents of the coat structure, the 50kD kinase subunit of the assembly protein complex AP1 and a casein kinase II-activity. An additional light chain B kinase activity has also been identified. No function has yet been ascribed to either of these kinase activities.

Despite the relative abundance of structural and biochemical information, very little is understood about the mechanisms which drive coated pit assembly, invagination and coated vesicle budding. For example, the nature of the energy requirements for these processes has remained controversial. Since purified clathrin triskelions and AP complexes will

NATO ASI Series, Vol. H 62
Endocytosis
Edited by P. J. Courtoy

spontaneously and efficiently reassemble *in vitro* to form coats, it has been suggested that this self-assembly process drives membrane invagination and budding (see for example, McKinley, 1983; Brodsky, 1988). Support for this suggestion has come from studies in intact cells depleted of ATP by incubation with metabolic inhibitors which suggested that a 'single-round' of endocytosis could occur even at greatly reduced cellular ATP levels (Clarke and Weigel, 1985; Larkin et al., 1985). However, if coat assembly is spontaneous, what prevents this reaction from occurring in the cytosol and/or directs clathrin assembly onto the cell surface? The resolution of these issues will clearly require the establishment of functional cell-free assay systems which faithfully reconstitute these events in order to facilitate detailed biochemical analysis of the mechanisms involved.

Here we describe an assay system we have developed to study the receptor-mediated endocytosis of transferrin in semi-intact A431 cells. Transferrin internalization in this *in vitro* system is dependent on the addition of both cytosolic factors and ATP. To confirm the validity of this assay system we have re-examined the requirement for ATP *in vivo* by measuring transferrin-receptor mediated endocytosis in intact cells depleted of ATP. Our results suggest that although ligands can be efficiently sequestered into deeply invaginated pits in ATP-depleted cells, the formation of coated vesicles is severely inhibited. Results from our *in vitro* assay system have indicated at least two distinct ATP-requiring stages in the overall process of clathrin coated pit assembly and coated vesicle formation.

An Assay for Receptor-mediated Endocytosis of Transferrin in Semi-intact A431 cells

Figure 1 diagrams an assay we have developed for receptor-mediated endocytosis *in vitro* using 'semi-intact' cells: cells in which a portion of the plasma membrane has been removed so as to deplete them of cytoplasm but otherwise maintain their intracellular organelles intact and accessible to exogenously added reagents. This assay is a modification of a previously described assay (Smythe et al., 1989). Diferric transferrin which has been conjugated to biotin via a cleavable disulfide bond and iodinated to high specific activity (refered to as ^{125}I-BSST) is used as a ligand for the transferrin-receptor. Since A431 cells are tenaciously adherent, semi-intact cells can be prepared by scraping them from a tissue culture dish as previously described (Smythe et al., 1989). The cells are washed free of cytosol and reincubated at 37°C in the presence of ^{125}I-BSST, added cytosol and ATP. Internalized ^{125}I-BSST is then quantitated by one of two methods: Ab-inaccessibility, exactly as described by Smythe et al.(1989); or, GSH-resistance, where extracellular ^{125}I-BSST is reduced using membrane impermeant reducing agents such as glutathione (GSH) or ß-mercaptoethanesulfate (MesNa) and intracellular ^{125}I-BSST quantitated by adsorption to avidin-Sepharose, as described elsewhere (Schmid and Carter, 1990). The use of GSH to assess internalization had been suggested by earlier studies (Bretscher and Lutter, 1988). Internalization of ^{125}I-BSST into intact cells as assessed either by GSH-resistance or Ab-inaccessibility assays gave identical results when compared to those obtained using standard (Ciechanover et al., 1983) acid wash assay conditions (data not shown)

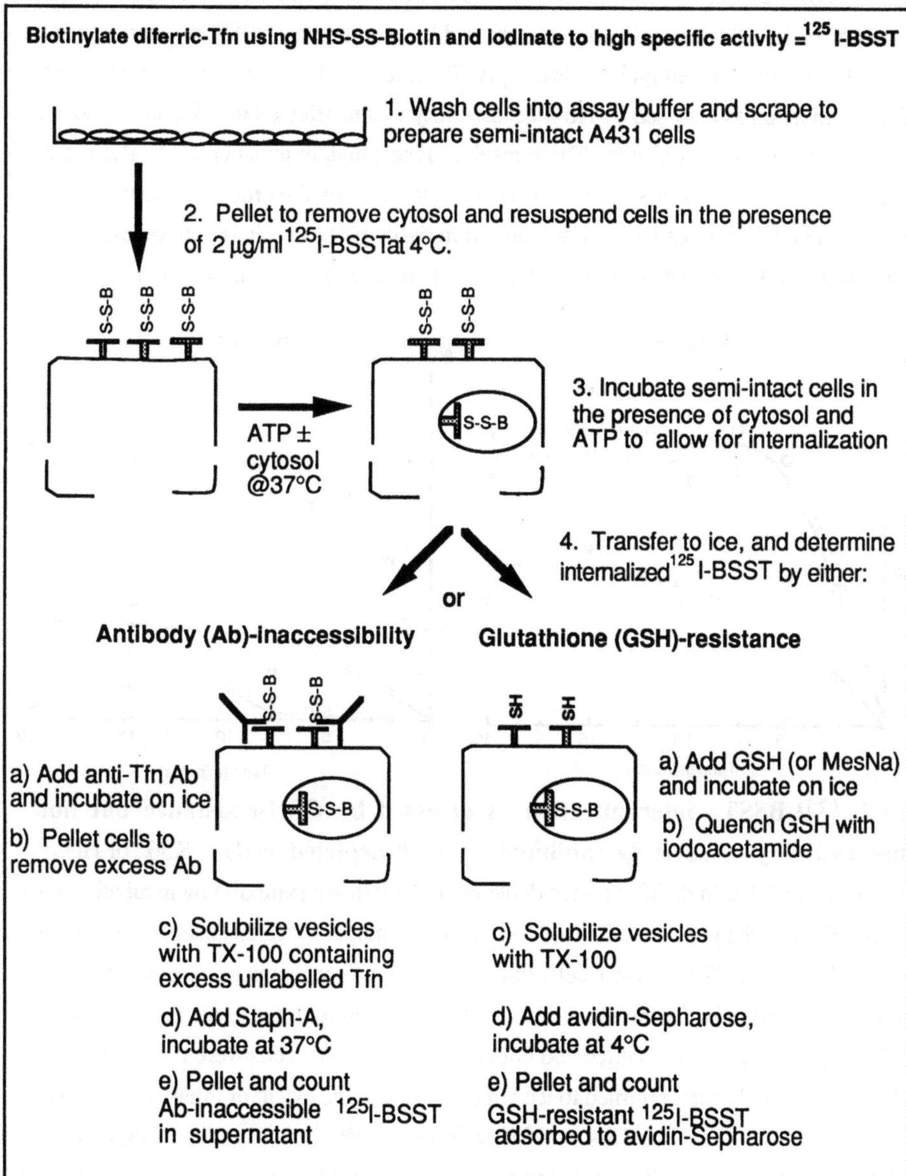

Figure 1. An assay for Receptor-mediated Endocytosis of Transferrin in Semi-intact A431 cells.

ATP is Required for Receptor-mediated Endocytosis in Intact cells

Since the question of an ATP requirement for receptor-mediated endocytosis in intact cells remains controversial, we first used these assays to assess ^{125}I-BSST internalization into intact cells depleted of ATP. Cellular ATP levels were reduced to <10% or <5% of control levels by incubation of K562 cells or HeLa cells, respectively, under N_2-atmosphere in the presence of 5 mM 2-deoxyglucose (2-dGlc). The results in Figure 2 show that both the rate and extent of ^{125}I-BSST internalization as assessed by acquisition of GSH-resistance, was severely inhibited in both cell types following ATP-depletion. In contrast, the extent of ^{125}I-BSST internalization as assessed by Ab-inaccessibility was unaffected in K562 cells and much less severely affected in HeLa cells. These results suggest that, in intact cells, ATP is required for *bona fide* endocytosis, as assessed by inaccessibility to a small membrane impermeant probe in intact cells. Sequestration of ligands from bulky probes such as antibodies, can however, occur in ATP-depleted cells. We suggest that ligands are sequestered in

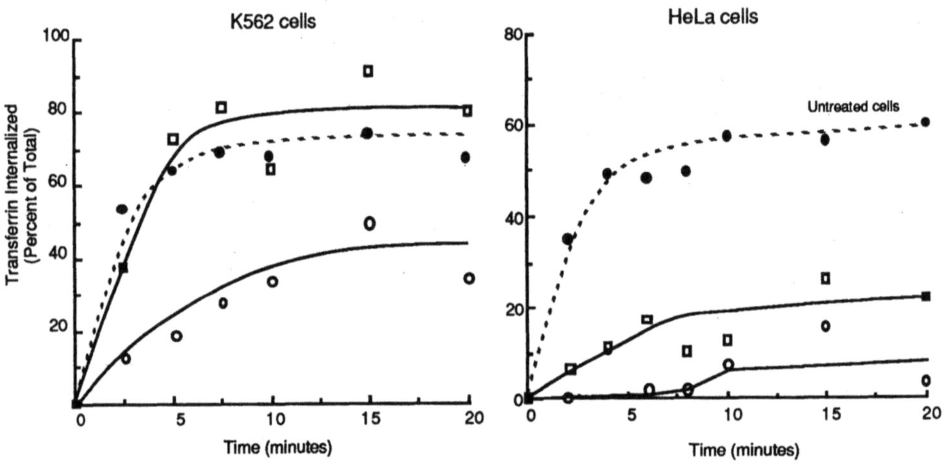

Figure 2. ^{125}I-BSST internalization as assessed by GSH-resistance but not Ab-inaccessibility is severely inhibited in ATP-depleted cells K562 or HeLa cells (~2 x 10^7 cells/ml in dPBS^{++}) were depleted of ATP (open symbols) by incubation under a N_2-atmosphere in the presence of 5 mM 2-dGlc for 15 min at 37°C, as described elsewhere (Schmid and Carter, 1990). Control cells (closed squares) were incubated exposed to air, in the presence of 5 mM Glc for the same time. Cells were then returned to ice and ^{125}I-BSST (2 µg/ml final concentration) was added and bound for 15 min. The cells were then shifted to 32°C for the indicated times, returned to ice temperature and the extent of internalization was assessed by either Ab-inaccessibility or GSH-resistance. (●) Control cells, average of results from Ab-inaccessibility and GSH-resistance assays; (❑) Ab-inaccessibility assay; (O) GSH-resistance assay.

ATP-depleted cells into deeply invaginated pits which are inhibited from budding at low cellular ATP levels. This suggestion was supported by morphological examination of ATP depleted cells which showed a two-fold increase in the number of deeply invaginated or "sealed" coated pits as compared to control cells (Schmid and Carter, 1990). We propose that the sequestration of ligands in deeply invaginated pits can explain previous findings that a "single-round" of endocytosis can occur in ATP-depleted cells (Clarke and Weigel, 19865; Larkin et al.; 1985).

ATP is Required For at Least Two Distinct Stages During ^{125}I-BSST Internalization *In Vitro*

In order to identify which stages of receptor-mediated endocytosis require ATP, we examined ATP dependence of endocytosis in our cell-free assay system. As seen for intact cells, assessment of ^{125}I-BSST internalization into semi-intact A431 cells by either Ab-inaccessibility and GSH-resistance assays provides a measure of two distinct stages of internalization. Whereas Ab-inaccessibility measures both sequestration into deeply

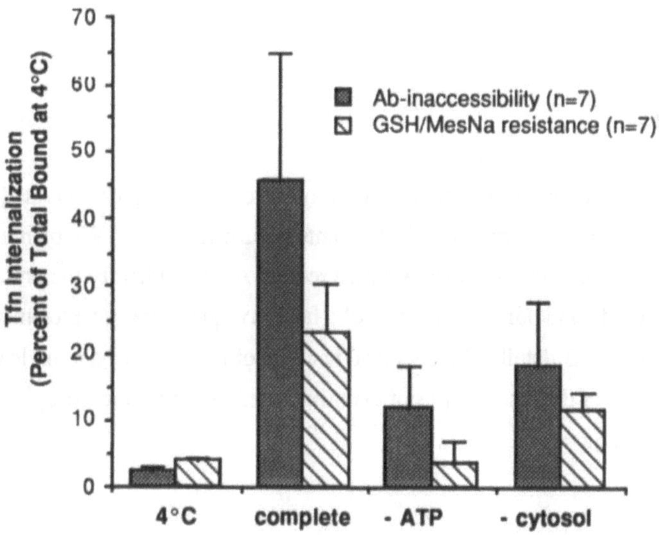

Figure 3. Cytosol and ATP are Required for ^{125}I-BSST Acquisition of Both Ab and GSH-resistance Assays were performed as described in the text. Complete incubations contained 5 mg/ml K562 cytosol, 1 mM ATP and a regenerating system of creatine/creatinephosphokinase in KSHM buffer (Smythe et al., 1989). -ATP incubations contained an ATP-depleting system of hexokinase and glucose

invaginated pits and coated vesicle formation, GSH-resistance presumably measures only coated vesicle formation.

The data in Figure 3 show that ^{125}I-BSST internalization into semi-intact A431 cells as measured by both assays was dependent on the addition of both cytosol and ATP. As observed in intact cells, ^{125}I-BSST internalization as assessed by GSH-resistance showed greater ATP dependence (essentially none over background at 4°C) than acquisition of Ab-inaccessibility. This result suggests that a late step in coated vesicle formation requires ATP. The data in Figure 3 also suggest that there exists a second ATP-dependent step which occurs early in the process of transferrin receptor-mediated endocytosis leading to the acquisition of Ab-inaccessiblity. Thus, ~35% of bound ^{125}I-BSST acquires Ab-inaccessibility in an ATP-dependent manner, whereas only ~18% of bound ^{125}I-BSST acquires GSH-resistance in an ATP-dependent manner. This additional signal (~18%) presumably results from an early ATP and cytosol-dependent step(s) which leads to sequestration of ^{125}I-BSST into deeply invaginated pits. We believe this early ATP-dependent step involves *de novo* coated pit formation. These results are consistent with previous morphological characterization of the internalization of HRP-Tfn into 'broken' A431 cells (Smythe et al. 1989) which demonstrated that preexisting shallow coated pits efficiently converted to deeply invaginated pits in an ATP and cytosol independent manner but that both *de novo* coated pit formation and coated vesicle budding required ATP.

Future Directions

These results help to validate our cell-free assay system for receptor-mediated endocytosis suggesting that it accurately reflects events occurring *in vivo*. We plan next to use this assay to examine the nature of the events which require ATP, to identify the proteins which utilize ATP and to identify and purify other cytosolic factors required for receptor-mediated endocytosis of transferrin. A detailed biochemical analysis of these events should lead eventually to elucidation of the mechanisms of clathrin coated pit formation, receptor-sorting and coated vesicle budding.

Acknowledgements

This work was supported by grants from the NIH (GM42455) and the Lucille P. Markey Charitable Trust. SLS is a Lucille P. Markey Scholar. ES was supported by a NATO/SERC Postdoctoral Fellowship.

REFERENCES

Ahle, S. and E. Ungewickell (1990) Auxilin, a newly identified clathrin-associated protein in coated vesicles from bovine brain. J. Cell Biol. **111**:19-30

Bretscher, M. and R. Lutter. (1988) A new method for detecting endocytosed proteins. EMBO J. **7**:4087-04092

Brodsky, F.M (1988) Living with clathrin: its role in intracellular membrane traffic. Science. **242**:1396-1402.

Ciechanover, A., A.L. Schwartz, A. Dautry-Varsat, and H.F. Lodish (1983) Kinetics of internalization and recycling of transferrin and the transferrin receptor in a human hepatoma cell line. J. Biol. Chem. **256**:9681-9689.

Clarke, B.L., and P.H. Weigel (1985) Recycling of the asialoglycoprotein receptor in isolated rat hepatocytes: ATP depletion blocks receptor recyling but not a single round of endocytosis J. Biol. Chem. **260**:128-133.

Glickman, J.N., E. Conibear, and B.M.F. Pearse (1989) Specificity of binding of clathrin adaptors to signals on the mannose-6-phosphate/insulin-like growth factor II receptor. EMBO J. **8**:1041-1047.

Keen, J.H (1987) Clathrin assembly proteins: affinity purification and a model for coat assembly. J. Cell Biol. **105**:1989-1998.

Larkin, J.M., W.C. Donzell, and R.G.W. Anderson (1985) Modulation of intracellular potassium and ATP: effects on coated pit function in fibroblasts and hepatocytes. J. Cell. Physiol. **124**:372-378.

McKinley, D.N. (1983) Model for transformation of the clathrin lattice in the coated vesicle pathway. J. Theor. Biol. **103**:405-419.

Pearse, B.M.F., and R.A. Crowther (1987) Structure and assembly of coated vesicles. Ann. Rev. Biophys. Biophys. Chem. **16**:49-68.

Schmid, S.L. and L.L. Carter (1990) ATP is required for receptor-mediated endocytosis in intact cells. J. Cell Biol. **111**: in press

Smythe, E., M. Pypaert, J. Lucocq, and G. Warren (1989) Formation of coated vesicles from coated pits in broken A431 cells. J. Cell Biol. 108:843-853.

Virshup, D.M., and V. Bennett (1988) Clathrin-coated vesicle assembly polypeptides: physical properties and reconstitution studies with brain membranes. J. Cell Biol. **106**:39-50.

REGULATION OF ENDOCYTIC MEMBRANE TRAFFIC

Jean Gruenberg*, Jean-Pierre Gorvel* and Morgane Bomsel+
+Dpt of Anatomy, UCSF, San Francisco, CA 94143-0452, USA.
*Europ. Mol. Biol. Lab. Postfach 10.2209, D-6900 Heidelberg, F.R.G.

1] INTRODUCTION

Markers internalized into animal cells appear sequentially in peripheral early endosomes, then in late endosomes, predominantly located in the perinuclear region, and eventually in lysosomes (review, Gruenberg and Howell, 1989; Kornfeld and Mellman, 1989; see Fig 1). Our major interest is to understand the mechanisms regulating membrane traffic in this pathway. For these studies, we use a cell-free assay measuring endocytic vesicle fusion. Avidin and biotinylated horseradish peroxidase (bHRP) are separately internalized by fluid phase endocytosis into two cell populations. After homogenization, endosomal fractions are prepared by immuno-isolation (Gruenberg and Howell, 1986, 1987; Gruenberg et al., 1989) or flotation on gradients (Tuomikoski et al., 1989; Bomsel et al., 1989; Gorvel et al., submitted). In the assay, avidin- and bHRP-labeled fractions are combined with cytosol and incubated at 37°C in the presence of ATP and biotinylated insulin, as a scavenger (Braell, 1987; Gruenberg et al., 1989). When fusion occurs, a complex is formed between avidin and bHRP, which is then immuno-precipitated after detergent extraction. Fusion is quantified by measuring the enzymatic activity of bHRP in the complex. With this assay, we have reconstituted an early (Gruenberg et al., 1989) and a late (Bomsel et al., 1990) endocytic fusion event and we have shown that these reflect lateral interaction between early endosome elements and delivery to late endosomes, respectively. Our efforts are now focused on some of the mechanisms controlling these two steps of membrane traffic. In this review, I will summarize the present status of our work.

NATO ASI Series, Vol. H 62
Endocytosis
Edited by P. J. Courtoy

2] LATERAL FUSION OF EARLY ENDOSOMES

The in vitro studies of our group and other groups show that early endosome fusion in non-polarized cells does not require the presence of polymerized microtubules, in contrast to a later endocytic fusion event (Bomsel et al., 1990) and that this fusion activity is high (Davey et al., 1985; Gruenberg and Howell, 1986, 1987; Braell, 1987; Diaz et al., 1988; Woodman and Warren, 1988). In fact, ≈ 60% of the avidin present in an early endosomal fraction is complexed to bHRP after fusion (Gruenberg et al., 1989). This high activity is specific for early endosomal elements, since in vitro fusion decreases after longer times of avidin or bHRP internalization in vivo. Our data indicate that both organellar partners of the fusion reaction correspond to early endosomal elements and that these elements undergo lateral fusion with each other in the assay. The specificity of these interactions is further demonstrated in polarized MDCK cells. In vivo observations had shown that apical and basolateral early endosomes of MDCK cells are distinct, each population lining the plasma membrane domain from where internalization occurred, whereas late endosomes are common to both pathways (Bomsel et al., 1989; Parton et al.,1989). In the cell-free assay, both the apical and the basolateral early endosomal population exhibit a high lateral fusion activity with themselves (Bomsel et al., 1990). However, apical and basolateral early endosomes do not fuse with each other. Since the cellular organization is disrupted in the assay, apical and basolateral early endosomes must be functionnally distinct. This specificity is expected to be mediated by recognition sites present on the cytoplasmic surface of apical and basolateral early endosomes, respectively.

In most cells, early endosomes exhibit a rather complex and polymorphic organization, which contains cisternal, tubular and vesicular regions (see Griffiths et al., 1989). This organization can be more developed in some cells. For exemple, apical early endosomes of rat developing intestine form an extensive tubular network (Wilson et al., 1987), and in erythroblasts, early endosomes consist of an extensive network of cisternal or tubular structures (Killisch et al., submitted). This morphological organization and the fact that individual early endosomal elements have a high and specific fusion activity with each other in vitro have lead us to propose early endosomes form a dynamic network of interacting elements in vivo (Gruenberg and Howell, 1989). As a consequence, early endososomes may, in effect, form a single compartment functionally. This organization may be, at least partially, analogous to the dynamic network observed in vivo by Hopkins et al. (1990) after transferrin internalization in Hep-2 cells. Very little is however known about the mechanisms which control fusion of early endosomes, except that this process requires a factor (Diaz et al., 1989), which is also required for transport within the Golgi (Block et al., 1988; Malhotra et al., 1988; Wilson et al., 1989) and from ER to Golgi (Beckers et al., 1989) and is affected by low concentrations of GTPgS (Mayorga et al., 1989; Tuomikoski et al., 1989). We have made use of our cell-free assay to investigate two regulatory mechanisms controlling early endosome fusion.

3] MECHANISMS REGULATING THE LATERAL FUSION OF EARLY ENDOSOMES IN VITRO

In mitotic mammalian cells, membrane traffic in the exocytic and endocytic pathways is arrested (reviews, Warren, 1985). The nuclear envelope breaks down and the Golgi complex, as well as the ER in some cells, is fragmented. Warren (1985, 1989) has proposed that both membrane traffic arrest and Golgi fragmentation may be caused by an inhibition of fusion events. We have therefore investigated the effect of mitotic cytosol on early endosome fusion in vitro (Tuomikoski et al., 1989). When the fusion assay was carried out in the presence of mitotic cytosol prepared from Xenopus eggs, we observed that fusion was inhibited when compared to interphase cytosol. This inhibition coud be reconstituted in interphase cytosol by adding affinity-purified preparations (Labbé et al., 1989) of the starfish homologue of the cell-cycle control protein kinase cdc2. The range of kinase activity required for fusion inhibition in vitro was similar to that expected during mitosis in vivo. If these observations are indicative of the in vivo situation, they may indicate that early endosome fusion is arrested during mitosis in a process regulated by the cdc2 kinase. It is tempting to postulate that this inhibition may reflect the fragmentation of the early endosomal network during mitosis. Alternatively, fusion inhibition may be indiscriminate during mitosis, possibly by cdc2-mediated phosphorylation of a factor common to more than one steps of membrane traffic.

Several lines of evidence have indicated that ras-like, small GTP-binding protein may be involved in the regulation of membrane traffic in mammalian cells, as was shown for YPT1 and SEC4 in yeast (review, Bourne, 1988). A relatively large number of mammalian genes encode ras-like GTP-binding proteins (see references in Chavrier et al., 1990), and several steps of membrane traffic are inhibited by low concentrations of GTPgS in vitro (Melançon et al., 1987; Goda and Pfeffer, 1988; Beckers and Balch, 1989; Bomsel et al., 1990), including the lateral fusion of early endosomes (Mayorga et al., 1989, Tuomikoski et al., 1989). Both the large number of genes and the GTPgS effects are consistent with the proposal (Bourne, 1988), that each step of membrane traffic may be regulated by a different GTP-binding protein. Recently, Chavrier et al. (1990) have characterized 3 GTP-binding proteins, rab2, rab5 and rab7, in BHK cells and shown that each protein has a specific localization. Rab2 was found in a compartement between ER and Golgi, whereas rab5 localized to early endosomes and the cytoplasmic face of the plasma membrane, while rab7 was restricted to late endosomes.

Since the fusion of early endosome is inhibited by GTPgS, we have tested whether rab5 was used in this process (Gorvel et al., submitted). In these experiments, we used BHK early endosomal fractions, which were separated by a flotation gradient from late endosomal fractions, containing the CI-mannose-6-phosphate receptor (MPR). Consistent with the in vivo localization studies, early and late endosomes were highly enriched in rab5 and rab7, respectively. We observed that anti-rab5 antibodies abolished the fusion reaction, whereas control antibodies or antibodies against rab2 or rab7 had no effect. After antibody-mediated inactivation, the fusion activity could however be rescued in the presence of cytosol prepared from cells overexpressing WT rab5. Recovery of fusion activity was not supported by cytosols prepared from cells overexpressing either mutant rab5Ile133, containing a single point mutation in the consensus GTP-binding domain (Pai et al., 1989), or mutant rab5ΔC, with a deletion of the 8 carboxy terminal amino acids including the 2 Cys at the consensus membrane-attachment site (Willumsen et al., 1984; Deschenes and Broach, 1987; Molenaar et al., 1988; Walworth et al., 1989). These experiments suggested that rab5 is a component of the machinery controlling early endosome recognition and/or fusion and that GTP binding and/or hydrolysis is necessary in this process, in agreement with GTPgS effects. Moreover, these observations demonstrate that the assay can be complemented by a cytosolic form of rab5. This cytosolic form may become membrane-attached during one round of fusion, since deletion of the consensus membrane attachment site abolished rab5 action.

4] TRANSFER FROM EARLY TO LATE ENDOSOMES IN VITRO

After delivery to the peripheral early endosomes, internalized markers are transported in vivo to perinuclear, late endosomes (prelysosomal compartment). In vivo, vesicles have been observed to move between these two locations and this movement requires the presence of intact microtubules (Herman and Albertini, 1984; Matteoni and Kreis, 1987; de Brabander et al., 1988). Both in non-polarized BHK (Gruenberg et al., 1989) and in polarized MDCK cells (Bomsel et al., 1990), we have observed that markers leaving the early endosomes first appear in spherical vesicles of 0.3-0.5 μm diameter with a multi-vesicular structure and then in late endosomes. In both cell-types, passage between these large vesicles and late endosomes is arrested by microtubule depolymerization, whereas this treatment has no effect on earlier endocytic events. In contrast to early endosomes, these vesicles do not fuse with each other in vitro. Neither do they fuse back with the plasma membrane. We have therefore proposed that these vesicles are endosomal "carrier" vesicles between early and late endosomes, and may correspond to the vesicles which were observed to move in vivo (see outline, Fig 1). However, the steps of membrane traffic which mediate transfer from early to late endosomes are not clear.

Fusion activity in vitro decreases with fractions prepared at stages of the pathway beyond the early endosome (review, Gruenberg and Howell, 1989). Although in vitro transfer from endosome to lysosome has been reported (Mullock et al., 1989), fusion events occuring between early and late endosomes have not been reconstituted in vitro. To study this latter stage of the pathway in vitro, we used the polarized epithelial cell--line MDCK grown on permeable filters. In vivo, the apical and basolateral endocytic pathways meet in late endosomes, which contain the bulk of the mannose-6-phosphate receptor (Parton et al., 1989). No meeting in vivo or in vitro occurred between apical and basolateral early endosomes

(Bomsel et al., 1989, 1990). Thus, MDCK cells provided an ideal system to study the late endocytic fusion events between apically- and basolaterally-labeled vesicles, which may be involved in the meeting process (Bomsel et al., 1990). Avidin and bHRP were internalized separately by fluid phase endocytosis from the apical and basolateral medium of two cell populations, respectively. The "carrier" vesicles of either pathway were labeled after 10 min internalization at 37°C in the absence of microtubules followed by a 30 min chase in marker-free medium. Cytosol-free fractions were then prepared and mixed in the assay in the presence of 5 mg/ml MDCK cytosol protein and the formation of avidin-bHRP complex during fusion was quantified after immuno-precipitation.

Under these conditions, little, if any, fusion could be detected in the assay. However, fusion was stimulated >25 fold when 5 µg taxol-stabilized microtubules prepared from purified bovine brain tubulin were added to the assay mixture. This effect of microtubules was specific, since they did not stimulate the fusion of early basolateral endosomes. Immuno-gold labeling of cryo-sections showed that meeting in vitro occurred exclusively in late endosomes, enriched in the mannose-6-phosphate receptor, as in vivo (Parton et al., 1989). Moreover, labeled structures were observed primarily in the close vicinity of microtubules, suggesting that vesicle-microtubule interactions occured in the assay. Although the precise sequence of fusion events occurring in the in vitro meeting process is not clear, our interpretation is that microtubules facilitate the fusion of apical and basolateral "carrier" vesicles with a common, MPR-positive late endosome. In any case, these studies demonstrate that both apical and basolateral "carrier" vesicles are fusogenic in our in vitro conditions.

The involvement of microtubules suggested that microtubule-directed movement and fusion may be coupled in the in vitro assay. We therefore investigated whether mechanochemical motors were also required. After immuno-depletion of kinesin with the antibody raised by Ingold et al (1988) or photocleavage of cytoplasmic dynein according to Schroer et al (1989), a 50% inhibition was observed, whereas the combination of both treatments essentially abolished fusion. These treatments had no effect on the microtubule-independent fusion of early endosomes. Recently, microtubules in MDCK cells were shown to be organized in longitudinal bundles, with their minus ends pointing towards the apical domain

(Bacallao et al., 1989). It is therefore tempting to postulate that the apical and basolateral endocytic pathway utilize cytoplasmic dynein and kinesin, respectively. However, endocytic vesicles from each pathway may also use both motors. In fact, recent studies suggest that this latter situation applies to MDCK late endosomes in vivo (Parton et al, submitted). Our future goal is to investigate the mechanisms which regulate the specificity of endocytic membrane traffic in polarized and non-polarized cells.

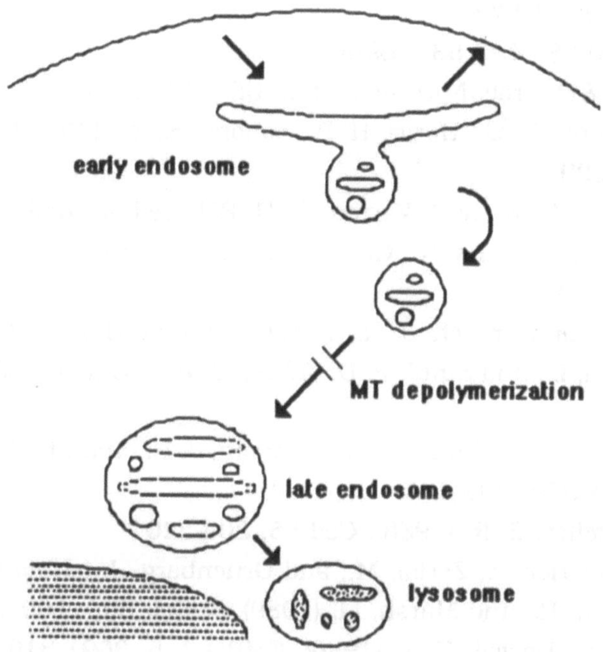

FIGURE 1. Outline of the endocytic pathway in non-polarized BHK cells. We have not indicated the organization of microtubules (MT), which radiate from the nuclear region with their plus ends pointing at the cell periphery, and the pathway followed by newly-synthesized lysosomal enzymes, which connect the trans-Golgi network with endosomes. Recycling is only indicated between early endosomes and the the plasma membrane, however other routes may also exist.

LIST OF REFERENCES

Bacallao, R. Antony, C., Dotti, C., Karsenti, E., Stelzer, E. H., and Simons, K. (1989). J. Cell Biol. 109, 2817-2832.

Beckers, C. J. M., Block, M. R., Glick, B. S., Rothman, J. E., and Balch, W. E. (1989). Nature 339, 397-398.

Block, M R., Glick, B. S., Wilcox, C. A., Wieland, F. T., and Rothman, J. (1988). Proc. Natl. Acad. Sci. USA 85, 7852-7856.

Bomsel, M., Prydz, K., Parton, R. G., Gruenberg, J., and Simons, K. (1989). J. Cell Biol. 109, 3243-3258.

Bomsel, M., Parton, R., Kuznetsov, S. A., Schroer, T. A., and Gruenberg, J. (1990). Cell 62, 719-731.

Bourne, H. R. (1988). Cell 53, 669-671.

Braell, W. A. (1987). Proc. Natl. Acad. Sci. USA. 84, 1137-1141.

Chavrier, P., Parton, R. G., Hauri, H. P., Simons, K., and Zerial, M. (1990). Cell 62, 317-329.

Davey, J., Hurtley, S. M., and Warren, G. (1985). Cell 43, 643-652.

De Brabander, M., Nuydens, R., Geerts, H., and Hopkins, C. R. (1988). Cell Mot. Cyt., 9, 30-47.

Deschenes, R. J., and Broach, J. R. (1987). Mol. Cell. Biol. 7, 2344-2351.

Diaz, R., Mayorga, L., and Stahl, P. D. (1988). J. Biol. Chem. 263, 6093-6100.

Diaz, R., Mayorga, L., Weidman, P. J., Rothman, J. E., and Stahl, P. D. (1989). Nature 339, 398-400.

Goda, Y., and Pfeffer, S. R. (1988). Cell 55, 309-320

Gorvel, J.-P., Chavrier, P., Zerial, M., and Gruenberg, J. (submitted).

Griffiths, G., Back, R., and Marsh, M. (1989). J. Cell Biol. 109, 2703-2720.

Gruenberg, J., and Howell, K. E. (1986). EMBO J. 5, 3091-3101.

Gruenberg, J., and Howell, K. E. (1987). Proc. Natl. Acad. Sci. USA 84, 5758-5762.

Gruenberg, J., and Howell, K. E. (1989). Annu. Rev. Cell Biol. 5, 453-481.

Gruenberg, J., Griffiths, G. and Howell, K. E. (1989). J. Cell Biol. 108, 1301-1316.

Herman, B. and Albertini, D. F. (1984). J. Cell Biol. 98, 565-576.

Hokins, C., Gibson, A., Shipman, M., and Miller, K. (1990). Nature, 346, 335-339.

Ingold, A. L., Cohn, S. A., and Scholey, J. M. (1988). J. Cell Biol. 107, 2657-2667.

Killisch, I., Steinlen, Römisch, K., Beug, H., and Griffiths, G. (submitted).
Kornfeld. S., and Mellman, I. (1989). Annu. Rev. Cell Biol. 5, 483-525.
Labbé, J.-C., Capony, J.-P., Caput, D., Cavadore, J.-C., Derancourt, J., Kaghad, M., Lelias, J.-M., Picard, A. & Dorée, M. (1989). EMBO J. 8, 3053-3058.
Malhotra, V., Orci, L., Glick, B. S., Block, M. R., and Rothman, J. E. (1988). Cell 54, 221-227.
Matteoni, R. and Kreis, T. E. (1987). J. Cell Biol. 105, 1253-1265.
Mayorga, L.S., Diaz, R., and Stahl, P. D. (1989). Science 244, 1475-1477.
Melançon, P., Glick, B. S., Malhotra, V., Weidman, P. J., Serafini, T., Gleason, M. L., Orci,. L., and Rothman, J. E. (1987). Cell 51, 1053-1062.
Molenaar, C. M. T., Prange, R., and Gallwitz, D. (1988). EMBO J. 7, 971-976.
Mullock, B. M., Branch, W. J., van Schaik, M., Gilbert, M., and Luzio, J. P. (1989). J. Cell Biol. 108, 2093-2100.
Pai, E. F., Kabsh, W., Krengel, U., Holmes, K. C., John, J., and Wittinghofer, A. (1989). Nature 341, 209-214.
Parton, R. G., Prydz, K., Bomsel, M., Simons, K., and Griffiths, G. (1989). J. Cell Biol. 109, 3259-3272.
Parton, R. G., Bacallao, R., and Prydz, K. (submitted).
Schroer, T. A., Steuer, E. R., Sheetz, M. P. (1989). Cell, 56, 937-946.
Tuomikoski, T., Felix, M.-A., Dorée. M., and Gruenberg, J. (1989). Nature, 342, 942-945.
Warren, G. (1985). Trends Biochem. Sci. 503, 439-443.
Warren, G. (1989). Nature, 342, 857-858.
Walworth, N. C., Goud, B., Kastan Kabcenell, A., and Novick, P. J. (1989). EMBO J. 8, 1685-1693.
Willumsen, B. M., Norris, K., Papageorge, A. G., Hubbert, N. L., and Lowy, D. R. (1984). EMBO J. 3, 2581-2585.
Wilson, J. M., Whitney, J. A., and Neutra, M. R. (1987). J. Cell Biol. 105, 691-703.
Wilson, D. W., Wilcox, C. A., Flynn, G. C., Chen, E., Kuang, W.-J., Henzel, W. J., Block, M. R., Ullrich, A., and Rothman, J. E. (1989). Nature 339, 355-359.
Woodman, P. G., and Warren, G. (1988). Europ. J. Biochem. 173, 101-108.

THE INTERACTION OF LATE ENDOSOMES WITH LYSOSOMES IN A CELL-FREE SYSTEM

J. Paul Luzio and Barbara M. Mullock
Department of Clinical Biochemistry, University of Cambridge, Addenbrooke's Hospital, Hills Road, Cambridge, CB2 2QR, U.K.

The passage of asialofetuin (ASF) through the endocytic pathway of rat hepatocytes is well described and involves binding to asialoglycoprotein receptors, internalisation via coated pits, appearance in an early, peripheral endosomal compartment(s) where the ligand becomes dissociated from receptor, appearance in a late, deep-lying endosomal compartment(s) and finally digestion in lysosomes (for references see Branch et al., 1987; Mullock et al., 1988). ^{125}I-ASF movement through the various endosomal compartments in rat hepatocytes has been demonstrated by isopycnic centrifugation on vertical density gradients of Ficoll (1 - 22% w/v) and Nycodenz (0.25M sucrose - 45% Nycodenz), which respectively separate early from late endosomal compartments and endosomes from lysosomes (Branch et al., 1987; Perez et al., 1988).

Ficoll and Nycodenz density gradient centrifugation analysis of ligand movement through endocytic compartments, after giving a single pulse of ^{125}I-ASF to an isolated perfused liver (Perez et al., 1988), suggests sequential passage through light (peaking at 1 - 2 min) and dense (peaking at 4 - 6 min) endosomes, before delivery to very dense endosomes and lysosomes. It is not clear from these experiments whether very dense endosomes are an obligatory intermediate station in the passage of ligand to lysosomes. Ligand is certainly delivered to lysosomes by 20 min after administration of the pulse dose, since soluble radioactivity derived from proteolysis of the ^{125}I-ASF may be observed on the Ficoll gradients by this time.

NATO ASI Series, Vol. H 62
Endocytosis
Edited by P. J. Courtoy

Initial development of a cell-free system

The Ficoll and Nycodenz density gradients have also been used to analyse interactions between endosomes and lysosomes in a cell-free system. Initial experiments (Mullock et al., 1989) were carried out using post-mitochondrial supernatants prepared from homogenates of rat liver that had been loaded for 10 min with ^{125}I-ASF by intravenous injection, such that >60% of the ligand was located in late endosomes. Immediate analysis of post-mitochondrial supernatants prepared from such livers on Nycodenz gradients showed that the radioactivity appeared in a broad single peak of density 1.09 - 1.13, clearly distinguished from the peak activity of β-hexosaminidase, a lysosomal marker. Density gradient analysis of the post - mitochondrial supernatant after a 30 min incubation at 37^{0}C in the presence of ATP and a regenerating system showed that >40% of the radioactivity had moved to a peak density >1.14 coincident with the peak of β-hexosaminidase activity. Further experiments using this system revealed that cell-free transfer of ^{125}I-ASF to the lysosomal position on density gradients:

(i) is complete within 30 min;

(ii) requires an ATP regenerating system, cytosolic proteins and temperatures >30°C;

(iii) requires the presence of lysosomes and late endosomes;

(iv) cannot use light (i.e. early) endosomes as a source of ^{125}I-ASF.

Cell-free interaction of endosomes and lysosomes using purified membrane fractions

Further analysis of the interaction of endosomes and lysosomes required the purification of individual endosome fractions. Endosome fractions from Ficoll gradients were particularly contaminated with smooth endoplasmic reticulum (marker glucose 6-phosphatase). Previous experiments had shown that much of this could be removed by recentrifugation on a Metrizamide gradient (Mullock et al., 1983; 1987). Both light (early) endosomes and dense endosomes were purified in this way (Table 1). By electron microscopy, both endosome fractions were vesicular, dense endosomes having a mean diameter of approx. 200nm., and light endosomes being as previously described (Mullock et al., 1987).

Table 1. Purification of Endosomes

	YIELD (% homogenate)			PURIFICATION (X) (relative to homogenate)	
	^{125}I ligand in endosomes*	Protein	G6Pase	Protein	G6Pase
Light endosomes	25	0.7	0.4	36	63
Dense endosomes	26	0.5	4	52	7

* In the purification of dense endosomes ^{125}I-ASF was used as ligand 10min after uptake into the liver. In contrast, for the purification of light endosomes, ^{125}I-pIgA was used as ligand 30min after uptake into the liver. In the whole animal a single intravenous dose of ^{125}I-pIgA continuously loads early (light) endosomes because it is not removed by a single passage through the liver (Perez et al., 1988).

Reconstitution of dense endosome - lysosome interaction in a cell free system was investigated using the purified dense endosome fraction, purified lysosomes (Maguire and Luzio, 1985), cytosol from a 1+3 liver homogenate containing 0.25M sucrose, 10mM Tes, 1mM Mg^{2+} (with other small molecules removed with Bio-Gel P6), an ATP-regenerating system, 0.5mM ATP and 1.3mM GTP. As with the crude system described above, interaction was dependent on the presence of endosomes, lysosomes, cytosol and energy and occurred only after incubation at 37^{0}C and not 4^{0}C. The use of purified membrane fractions has allowed the demonstration of inhibitory effects of N-ethylmaleimide (NEM) consistent with an involvement of NEM-sensitive factors as in fusion events on other intracellular membrane traffic pathways (Block et al., 1988) and inhibition by βSGDP (0.1 mM) suggesting an involvement of G proteins. Incubation of dense endosomes with cytosol and ATP at 37^{0}C results in an increase in their density on Nycodenz gradients which may be a necessary prelude to interaction with lysosomes.

Membrane fusion during the cell-free interaction of endosomes and lysosomes

The delivery of lumenal contents from endosomes to lysosomes requires fusion of the two membranes, and to investigate this a separate assay was developed to directly measure membrane fusion. In this assay, fusion of one membrane fraction containing self-quenching concentrations of the fluorescent probe octadecylrhodamine B-chloride (R_{18}) with an unlabelled membrane fraction results in dilution of the probe in the membrane and causes a relief of self-quenching (Hoekstra et al., 1984). The addition of lysosomes to purified dense endosomes labelled with self-quenching concentrations of R_{18}, in the presence of an energy-regenerating system, led to much more relief of self-quenching on incubation at 37°C than did addition of equivalent protein concentrations of a crude mitochondrial preparation. No increase in fluorescence occurred at 4°C.

Dense endosomes have the properties of transfer vesicles

Several properties of dense endosomes suggest that they are transfer vesicles on the endocytic pathway from early, peripheral (light) endosomes to lysosomes (see Figure 1). Such 'carrier' vesicles were previously suggested by Gruenberg and co-workers (Gruenberg et al., 1989; Bomsel et al., 1990) in the pathway from early to late endosomes. In the present experiments the dense endosomes showed the following properties of transfer vesicles :

(i) they were labelled early, although after light endosomes, and retained their label only briefly;

(ii) they contained negligible amounts of the lysosomal enzyme β-hexosaminidase;

(iii) they became specifically associated with lysosomes in vitro when incubated with an energy-regenerating system and cytosol but did not mature to lysosomal densities in the absence of added lysosomes;

(iv) experiments on NEM inhibition and on membrane fusion as measured by relief of fluorescence self-quenching indicated that some fusion of dense endosomes occurs with lysosomes.

The origin and function of very dense endosomes

Very dense endosomes were found together with lysosomes at the bottom of the 1-22%w/v Ficoll gradient but could be separated from lysosomes on the Nycodenz gradient. ^{125}I-ASF was transferred from the endosome to the lysosome position of the Nycodenz gradient by pre-incubating the Ficoll gradient bottom fraction with ATP and a regenerating system but without adding cytosol. Very dense endosomes purified on step gradients of Ficoll and Nycodenz could be R_{18} labelled and shown to fuse with lysosomes in an ATP dependent but non-cytosol dependent manner. Very dense endosome fractions were enriched in mannose 6-phosphate receptors and contained some β-hexosaminidase, consistent with their being a pre-lysosomal compartment(s). It is possible that very dense endosome fractions contain late endosomal structures that are also pre-lysosomal on the membrane traffic pathway from the trans-Golgi network (TGN) to lysosomes and also on the autophagic route of protein degradation (Figure 1).

Figure 1. Summary of rat liver endosome - lysosome interactions

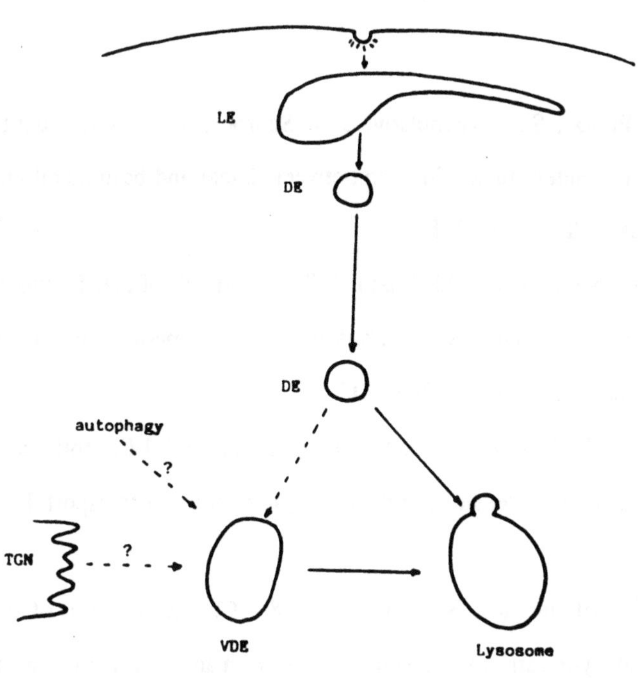

Future work - identification of endosome-specific membrane proteins and proteins involved in endosome-lysosome interaction

A general strategy, based on screening cDNA libraries in an appropriate expression vector with a polyclonal antiserum raised against Triton X-114 phase proteins of purified membrane fractions, has recently been developed to identify and characterise resident membrane proteins of intracellular organelles (Luzio et al., 1990). Experimentally, the strategy invoves the isolation of a large number of different cDNA clones with the polyclonal antiserum, followed by classification of the clones and the proteins they encode, using antibodies affinity purified on individually expressed fusion proteins. It has proven effective in cloning several Golgi membrane proteins including TGN38, a type 1 membrane protein located in the TGN. The strategy is being applied to the identification of endosomal integral membrane proteins using antisera raised against the purified fractions described in Table 1. It is also hoped to use the strategy to identify proteins involved in endosome-lysosome interaction.

Acknowledgement

This work was supported by the Medical Research Council

References

Bomsel, M., R. Parton, S.A. Kuznetsov, T.A. Schroer, and J. Gruenberg. 1990. Microtubule- and motor-dependent fusion in vitro between apical and basolateral endocytic vesicles from MDCK cells. Cell 62 : 719-731.

Branch, W.J., B.M. Mullock, and J.P.Luzio. 1987. Rapid subcellular fractionation of the rat liver endocytic compartments involved in transcytosis of polymeric immunoglobulin A and endocytosis of asialofetuin. Biochem. J. 244 : 311-315.

Block, M.R., B.S. Glick, C.A. Wilcox, F.T. Wieland, and J.E. Rothman. 1988. Purification of an N-ethylmaleimide-sensitive protein catalysing vesicular transport. Proc. Natl. Acad. Sci. USA. 85 : 7852-7856.

Gruenberg, J., G. Griffiths, and K.E. Howell. 1989. Characterization of the early endosome and putative endocytic carrier vesicles in vivo and with an an assay of vesicle fusion in vitro. J. Cell Biol. 108 : 1301- 1316.

Hoekstra, D., T. de Boer, K. Klappe, and J. Wilschut. 1984. Fluorescence method for measuring the kinetics of fusion between biological membranes. Biochemistry. 23 : 5675-5681.

Luzio, J.P., B. Brake, G. Banting, K.E. Howell, P. Braghetta, and K.K. Stanley. 1990. Identification, sequencing and expression of an integral membrane protein of the trans-Golgi network (TGN38). Biochem. J. 270 : 97-102.

Maguire, G.A., and J.P. Luzio. 1985. The presence and orientation of ecto-5'-nucleotidase in rat liver lysosomes. FEBS (Fed. Eur. Biochem.Soc.) Lett. 180 : 122-126.

Mullock, B.M., R.H. Hinton, J.V. Peppard, J.W. Slot, and J.P. Luzio. 1987. The preparative isolation of endosome fractions: a review. Cell Biochem. Function 5 : 235-243.

Mullock, B.M., W.J. Branch, M. van Schaik, L.K. Gilbert, and J.P. Luzio. 1989. Reconstitution of an endosome-lysosome interaction in a cell-free system. J. Cell Biol. 108 : 2093-2099.

Perez, J.H., W.J. Branch, L. Smith, B.M. Mullock, and J.P. Luzio. 1988. Investigation of endosomal compartments involved in endocytosis and transcytosis of polymeric IgA by subcellular fractionation of perfused rat liver. Biochem. J. 251 : 763-770.

ROUNDTABLE ON CELL-FREE SYSTEMS

(Chairperson : Kathryn Howell)

Almost every step in membrane traffic in the endocytic pathway has now been reconstituted in cell-free systems. This workshop included presentations by Mark Pypaert and Sandra Schmid on the formation of coated vesicles from coated pits, by Jean Gruenberg on the fusion of early and late endosomes, and by Paul Luzio and Robert Murphy both studying fusion of endosomes with lysosomes.

The major question all investigators working on cell-free systems need to address is how relevant their observations are to phenomena occurring in living cells. Jean Gruenberg emphasized that a safe way is to thoroughly document phenomena, e.g. by morphology, in parallel to cell-free studies. For example, they first showed that the pathways of endocytosis from the apical and basolateral domains of the MDCK cell converge at the late endosome (Bomsel et al, 1989, J. Cell Biol. 109:3243). They then were able to reconstitute the fusion of the meeting vesicles in a cell-free assay and to show requirements for polymerized microtubules and the microtubule-associated motors kinesin and dynein (Bomsel et al, 1990, Cell 62:719).

The first point of discussion was the structural relation of early tubular or cisternal endosomes with large vesicles, for each two distinctive interpretations are proposed : a continuous tubular vesicular reticulum (Hopkins et al, 1990, Nature, 346:335) vs highly dynamic early endosomal compartment wherefrom vesicular carriers are budding (Gruenberg et al, 1989, J. Cell Biol. 108:1301). Using confocal microscopy and low light video recording of Hep-2 cells loaded with transferrin labeled by Texas red, Colin Hopkins visualized an extensive tubular reticulum with expanded swellings apparently moving along its length. Gruenberg felt this is consistent with the interpretation of the early endosomes as a network of tubular or tubulovesicular compartment, if not a continuum, since isolated vesicles labeled at early stages of endocytosis rapidly fuse with each other. Hopkins agreed that a large proportion of the labeled compartment could belong to the early endosome, and that endocytic vesicles could be seen apparently independent of the reticulum. What is thus crucial is a cell-free fission assay that would indicate if the reconstituted reticulum can actually generate independent transport or recycling vesicles.

The second part of the discussion was centered on the apparent differences between the two cell-free systems dealing with the fusion of late endosomes with lysosomes. Luzio presented evidence that dense lysosomes from rat liver have to be present as acceptors for the late endosome (very

NATO ASI Series, Vol. H 62
Endocytosis
Edited by P. J. Courtoy

dense) to shift to the density of lysosomes. This shift depends on ATP and cytosol and does not occur at low temperature. Evidence for the fusion of the two compartments was obtained by a relief of self-quenching of a fluorescent membrane lipid. Murphy arrived at an apparently opposite conclusion, favoring the hypothesis that late endosomes could mature into lysosomes. In his sytem, late endosomes apparently increase in density, even in the absence of lysosomes with which to fuse. However, Murphy stressed that because the endosomal/lysosomal fractions had been isolated from different cell systems (growth-arrested liver cells versus rapidly growing cultured cells), the two experiments systems were not parallel. Since the question remains open as to whether the endocytic pathway is best described by a vesicular traffic versus a maturation model, and since it could be answered by experiments such as those used by these two groups, it becomes extremely important that reactions are further defined and that appropriate controls are carried out.

In general both the discussants and the audience felt cell-free analysis of membrane traffic in endocytosis has come a long way in the last few years. Together with the use of perforated cells, it provides the most promising approach to identify the molecules which regulate membrane traffic.

1.5. MOLECULAR PROCESSING BY ENDOSOMES

(Chairperson : Ira MELLMAN)

Besides sorting of its membranous constituents, endosomes may process their content in several ways : by acidification and possibly other alterations of the ionic composition of the endosomal fluid, by selective proteolysis, and by disulfide bridge reduction. In addition, some membranous constituents may be phosphorylated once in endosomes.

Sorting, an essential function of endosomes, generally depends on their acidification. Based on cell-free assays, **Renate Fuchs** discussed how the properties of the vacuolar H^{+}-ATPase can be regulated by other transport ATPases (e.g. Na^{+}/K^{+}-ATPase) when also present at the endosomal membrane. She further showed that, in rat liver, endocytic coated vesicles and late transcytotic endosomes are devoid of functional H^{+}-ATPase, suggesting that the proton pump is not derived in active form from endocytosis, and is sorted in endosomes from forming transcytotic vesicles.

Bernard Desbuquois reported on an endosomal protease of rat liver endosomes, capable of degrading insulin and glucagon, but not EGF, into specific pepdide fragments. Intriguingly, this protease is sensitive to o-phenanthroline, such as matrix metallo-proteinases. Using a similar in vitro assay as Fuchs, he established that proteolysis of both hormones strictly parallels ATP-dependent endosomal acidification, reflecting a genuine pH dependence of the enzyme and/or a pH-induced ligand-receptor dissociation. The endosomal protease could regulate the function of some polypeptide hormones, for example by indirectly switching off the tyrosine kinase activity of the insulin receptor.

John Bergeron demonstrated that the EGF receptor of rat liver, but not the insulin receptor, is capable of exhibiting its tyrosine kinase activity after internalization in endosomes. This suggests that the receptor may continue signalling information into the cytoplasm following internalization, or even use endocytosis to approach the crucial target. Bergeron also reported on a 55 kDa membrane protein complexed with the internalized EGF receptor, a possible candidate for the regulation of the receptor function.

NATO ASI Series, Vol. H 62
Endocytosis
Edited by P. J. Courtoy

REGULATION OF ATP-DEPENDENT ENDOSOME ACIDIFICATION

R. Fuchs[1], S. Schmid[2], I. Mellman[3], and H. Klapper[1]

[1] Institut für Allgemeine und Experimentelle Pathologie der Universität Wien, Austria;

[2] Department of Molecular Biology, Scripps Clinic, La Jolla, CA, USA, and

[3] Department of Cell Biology, Yale University, New Haven, CT, USA

INTRODUCTION

Biologically important macromolecules bind to receptors on the plasma membrane and are internalized via coated vesicles and endosomes before delivery to lysosomes. Endosomes constitute the most important branchpoint in this receptor-mediated pathway, where sorting processes occur that determine the fate of receptors and ligands. So far, the only known mechanism underlying molecular sorting in endosomes is their slightly acidic internal pH due to the activity of a proton ATPase present in the endosomal membrane.

CHARACTERISTICS OF VACUOLAR PROTON ATPases

Proton ATPases with similar characteristics have been described in organelles of the endocytic (coated vesicles, endosomes, lysosomes) and exocytic (Golgi apparatus and secretory vesicles) pathways and have therefore been classified as "vacuolar" ATPases (Mellman et al., 1986). These ion pumps mediate the electrogenic movement of protons into the vesicle lumen that is not molecularly coupled to the transport of any other ion. Consequently, this leads to the build-up of an inside-positive membrane potential, which opposes further acidification. To allow the development of a lower luminal pH this membrane potential has to be dissipated by passive influx of cations and/or efflux of anions. In contrast to other membrane ion pumps, e.g. Na^+/K^+-ATPase, F_1F_0-ATPase, K^+/H^+-ATPase, the only known "specific" inhibitor of vacuolar proton pumps is N-ethylmaleimide.

ROLE OF ENDOSOMAL pH IN MOLECULAR SORTING

Recent evidence has demonstrated that incoming receptor-ligand complexes in non-polarized cells encounter increasingly lower pH values with internalization time, suggesting a pH gradient along the endocytic pathway en route to lysosomes (Kielian et al., 1986; Yamashiro and Maxfield, 1987). Based on studies in intact cells, the pH of endosomes ranges between 6.5 and 4.6, whereas the pH of lysosomes is still lower (4.0). Many receptor-ligand complexes dissociate in the low pH environment of endosomes thus facilitating their sorting to

NATO ASI Series, Vol. H 62
Endocytosis
Edited by P. J. Courtoy

distinct cellular destinations. In general, receptors are recycled to the plasma membrane for re-use, whereas ligands are transported to lysosomes for degradation. We have isolated two distinct endosome subpopulations - early and late- from non-polarized Chinese hamster ovary (CHO) cells which differ in their function, membrane composition and internal pH (Schmid et al., 1988). Kinetically early endosomes are involved in membrane and receptor recycling and maintain pH < 6.3 in vivo, whereas "late" endosomes, which deliver content to lysosomes have a pH < 5.3 (Schmid et al., 1989). Different receptor-ligand complexes exhibit different pH tresholds for dissociation. Therefore it has been suggested that the maintainance of a fixed pH value in specific endosomal compartments plays an important role in determining where inside the cell sorting occurs. Consequently, endosomal pH has to be tightly regulated to maintain proper sorting functions.

ENDOSOMAL pH-REGULATION IN NON-POLARIZED CELLS

In order to investigate possible mechanisms of endosomal pH-regulation we have isolated endosome subpopulations selectively labeled with pH-sensitive fluorescent tracers (FITC-dextran, FITC-transferrin) from CHO cells and determined the ionic requirements for ATP-dependent endosome acidification in vitro (Schmid et al., 1989; Fuchs et al., 1989). In agreement with studies of acidification in intact cells (Murphy et al., 1984; Kielian et al., 1986; Yamashiro and Maxfield, 1988)) early endosomes acidify in vitro to a lesser extent than late endosomes. In addition, the two subpopulations could be distinguished on the basis of their differential sensitivities to inhibition by sodium: incubation in a physiological buffer containing both sodium (>20 mM) and potassium (>5mM) blocked ATP-dependent proton transport in early but not in late endosomes (Figure 1). This inhibition could be abrogated by sodium orthovanadate, an inhibitor of various ion pumps, e.g. Na^+/K^+-ATPase. Therefore the data suggest a role for Na^+/K^+-ATPase in regulating endosome acidification via modulation of the membrane potential. Na^+/K^+-ATPase mediates the inward movement of 3 sodium ions in exchange for 2 potassium ions, thereby generating an interior positive membrane potential which would limit electrogenic proton transport. Further evidence for the presence of plasma membrane derived Na^+/K^+-ATPase in early endosomes could be obtained by experiments where ouabain, a specific inhibitor of Na^+/K^+-ATPase, was co-internalized with the pH-sensitive endocytic tracer (Fuchs et al., 1989). Under this condition sodium did not inhibit endosome acidification. These findings suggest a mechanism by which pH in early endosomes could be kept less acidic than in late endosomes. Evidence that Na^+/K^+-ATPase actually regulates endosomal pH in intact cells was provided by Cain et al. (1989) using flow cytofluorometry to demonstrate that the pH of FITC-transferrin labeled endosomes is lower in cells grown in ouabain.

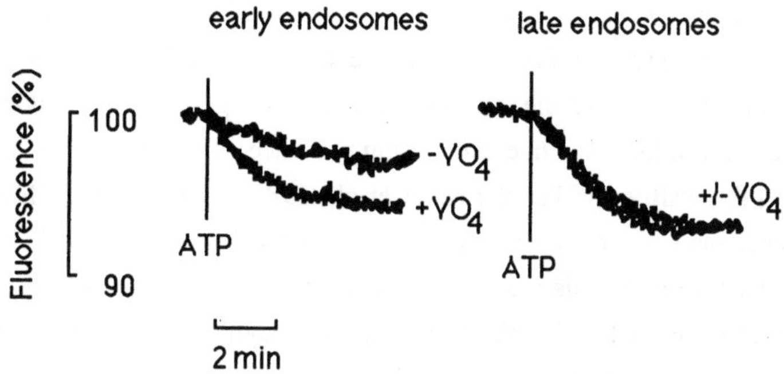

Figure 1: Na^+/K^+-ATPase inhibits acidification only in early endosomes. Kinetically early or late endosomes were selectively labeled by incubating CHO cells with FITC-dextran as described (Fuchs et al., 1989). An endosome enriched Golgi-fraction was prepared by sucrose density gradient centrifugation. Isolated endosomes were equilibrated in Na^+/K^+ buffer (145 mM NaCl, 5 mM KCl) in the presence or absence of 0.5 mM Na_3VO_4. Addition of ATP quenched the inital FITC-fluorescence, indicative of intravesicular acidification.

ENDOCYTIC PATHWAYS IN POLARIZED CELLS

The endocytic pathway in polarized cells, such as hepatocytes, is complex, involving not only constitutive recycling of receptors and plasma membrane proteins at the basolateral (sinusoidal) membrane, but also the transport of certain receptors and ligands to lysosomes or to the apical (bile canalicular) plasma membrane. Although, the final destination of receptors and ligands internalized by rat hepatocytes is different, biochemical and morphological studies have shown that e.g. asialoorosomucoid (ASOR) and polymeric immunoglobulin A (pIgA) enter a common endocytic compartment (Limet et al., 1985; Schiff et al., 1986; Branch et al., 1987). Intracellular sorting leads to selective delivery of ASOR to lysosomes and transcytosis of pIgA. It is well established that endosomes delivering ASOR to lysosomes acidify their interior (Harford et al., 1983; Sabolic et al., 1988; Fuchs et al., 1989) which leads to dissociation of ASOR from its receptor, most likely in an early endosome compartment (Mueller and Hubbard, 1986). In contrast, information about pH regulation of different organelles involved in endocytosis and transcytosis of hepatocytes is still lacking.

ENDOCYTIC COATED VESICLES DO NOT EXHIBIT ATP-DEPENDENT ACIDIFICATION

Receptors and ligands first accumulate in coated pits at the plasma membrane which then pinch off to form coated vesicles. Coated vesicle fractions isolated from a variety of cells have been found to contain a vacuolar type proton ATPase (Van Dyke et al., 1984). This has been taken as evidence that proton ATPases found in endosomes and lysosomes are derived from the plasma membrane, although the intracellular origin - Golgi derived or endocytic - of these coated vesicle preparations was unclear. Recent evidence suggests, using a cytochemically demonstrable pH-sensitive probe (Schwartz et al., 1985; Anderson and Orci, 1988), that coated vesicles found in the peripheral cytoplasm (probably endocytic coated vesicles) are not acidic. To further investigate this problem we have isolated coated vesicles (Pilch et al., 1983) from rat liver selectively labeled with pH-sensitive endocytic markers (FITC-dextran, FITC-ASOR). As shown in Figure 2A these purified coated vesicles did not exhibit ATP-dependent acidification. In contrast, using a nonselective pH-probe (acridine orange) ATP-induced proton transport could be demonstrated in this coated vesicle preparation (Figure 2B). This indicates that in polarized hepatocytes endocytic coated vesicles are not acidic and therefore may not contain a proton ATPase, whereas Golgi-derived coated vesicles are indeed acidic.

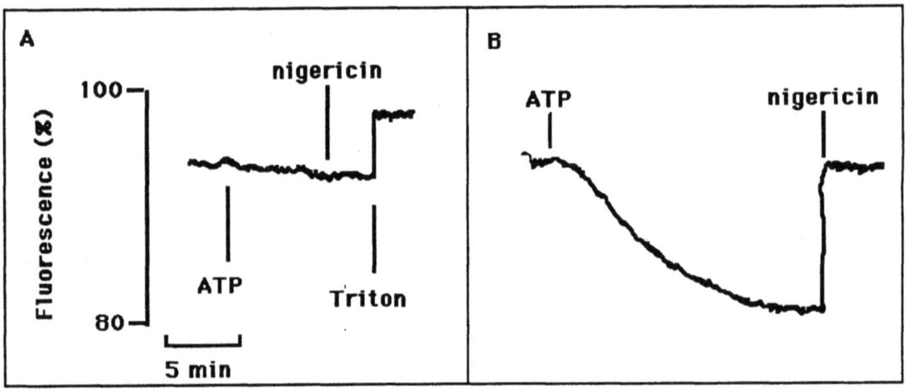

Figure 2: Endocytic coated vesicles do not exhibit ATP-dependent acidification. Rat liver endocytic vesicles were labeled by in situ perfusion with FITC-dextran and coated vesicles were prepared by equilibrium gradient cetrifugation (Pilch et al., 1983). Addition of ATP did not induce acidification of endocytic coated vesicles, that had been labeled with the pH-sensitive fluid-phase marker (2A). In contrast, ATP-dependent fluorescence quenching of the unspecific pH-probe acridine orange was observed in this coated vesicle preparation (2B).

pH-REGULATION IN ENDOSOMES INVOLVED IN TRANSPORT TO LYSOSOMES AND TRANSCYTOSIS

Next, we investigated the acidification characteristics of prelysosomal endosomes (those containing ASOR en route to lysosomes) or transcytotic endosomes (those containing pIgA en route to the bile canalicular membrane). Using free-flow electrophoresis (Fuchs et al., 1989) to confirm selective labeling, endosome subcompartments were characterized as follows: 1) an early compartment, where sorting between the lysosomal and transcytotic pathway had not yet occurred, 2) a late compartment, through which ligands pass en route to lysosomes, and 3) a late transcytotic compartment. Early endosomes, labeled with FITC-ASOR or FITC-pIgA, exhibited ATP-induced proton transport (Figure 3A,C); however, acidification of late ASOR-containing endosomes was considerably higher (Figure 3B). In contrast, late, pIgA-containing endosomes showed negligible acidification (Figure 3D).

The results indicate that ligands and receptors internalized at the basolateral apsect of polarized cells are sorted very rapidly from a common, early endosome compartment to the transcytotic and lysosomal pathway, respectively. Although endocytic coated vesicles may not contain a proton ATPase, early endosomes were found to be mildly acidic. However, transport to lysosomes involves endosomes more acidic than early endosomes, whereas transcytosis is mediated by endosomes less acidic (neutral) than early endosomes.

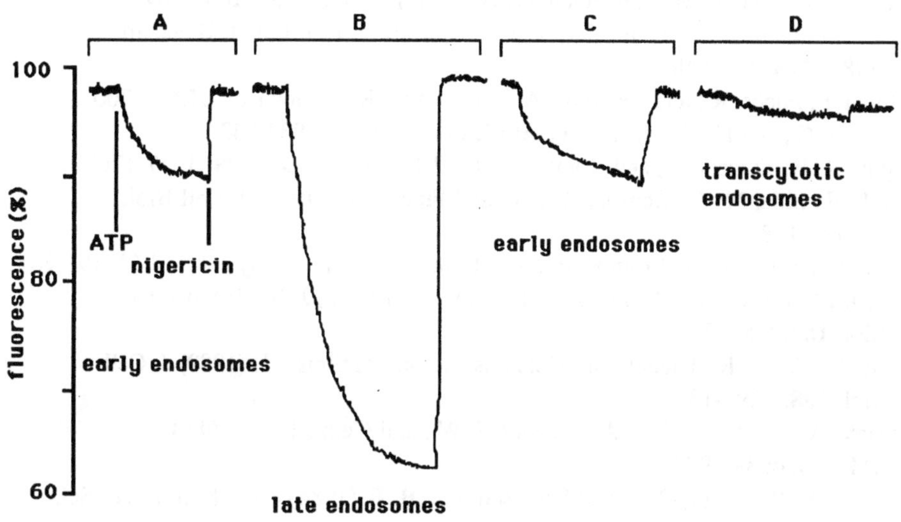

Figure 3: ATP-induced acidification of rat liver endosomes en route to lysosomes. Using an isolated liver perfusion system, kinetically early endosomes were labeled with FITC-ASOR (3A) or FITC-pIgA (3C) for one hour at 16°C. Kinetically late endosomes were labeled by a 5-min perfusion with FITC-ASOR (3B) or FITC-pIgA (3D), followed by a 8-min chase in marker-free medium at 37°C. Endosomes were isolated as described (Fuchs et al., 1989) and acidification was initiated by addition of ATP.

AKNOWLEDGEMENTS

This work was supported by the Austrian Science Research Fund, No. 6436 to R.F., by Medizinisch Wissenschaftlicher Fonds des Bürgermeisters der Bundeshauptstadt Wien to H.K, a Helen Hay Whitney Postdoctoral Fellowship to S.S., and a NIH grant to I.M. We thank Dr. J-P Vearman for providing the purified pIgA and Philippe Male and Peter Wyskovsky for excellent technical assistance with the free-flow electrophoresis apparatus.

REFERENCES

Anderson, R. G. W., and Orci, L. (1988) J. Cell Biol. **106**: 539-543

Branch, W. J., Mullock, B. M., Luzio, J. P. (1987) Biochem. J. **244**: 311-315

Cain, C. C., Sipe, D. M., and Murphy, R. F. (1989) Proc. Natl. Acad. Sci. U.S.A. **86**: 544-548

Fuchs, R., Male, P., and Mellman, I. (1989) J. Biol. Chem. **264**:2212-2220

Fuchs, R., Schmid, S., and Mellman, I. (1989) Proc. Natl. Acad. Sci. U.S.A. **86**: 539-543

Harford, J., Bridges, K., Ashwell, G., and Klausner, R. D. (1983) J. Biol. Chem. **258**: 3191-3197

Kielian, M. C., Marsh, M., and Helenius., A. (1986) EMBO J. **5**: 3103-3109.

Limet, J. N., Quintart, J., Schneider, Y. J., and Courtoy, P. J. Eur. J. Biochem. (1985) **146**: 539-548

Mellman, I., Fuchs, R., and Helenius, A. (1986) Ann. Rev. Biochem. **55**:663-700

Mueller, S. C., and Hubbard, A. L. (1986) J. Cell Biol. **102**: 932-942

Murphy, R. F., Powers, S., and Cantor, C. R. (1984) J. Cell Biol. **98**: 1757-1762.

Pilch, P. F., Shia, M. A., Benson, R. J. J., and Fine, R. E. (1983) J. Cell Biol. **93**: 133-138

Sabolic, I., Haase, W., and Burckhardt, G. (1988) Biochim. Biophys. Acta **944**: 191-201

Schiff, J. M., Fisher, M. M., Jones, A. L., and Underdown, J. B. (1986) J. Cell Biol. **102**: 920-931

Schmid, S., Fuchs, R., Kielian, M., Helenius, A., and Mellman, I. (1989) J. Cell Biol. **108**: 1291-1300

Schwartz, A. L., Strous, G. J. A. M., Slot, J. W., and Geuze H. J. (1985) EMBO J. **4**: 899-904

Van Dyke, R. W., Steer, C. J., and Scharschmidt, B. F. (1984) Proc. Natl. Acad. Sci. U.S.A. **81**: 3108-3112

Yamashiro, D. J., and Maxfield, F. R. (1987) J. Cell. Biol. **105**: 2713-2721

Yamashiro, D. J., and Maxfield, F. R. (1988) Trends Pharmacol. Sci. **9**: 190-193

DEGRADATION OF INSULIN AND GLUCAGON IN ISOLATED LIVER ENDOSOMES: FUNCTIONAL RELATIONSHIPS WITH ATP-DEPENDENT ENDOSOMAL ACIDIFICATION AND PARTIAL CHARACTERIZATION OF DEGRADATION PRODUCTS

Bernard Desbuquois, François Authier, Jean-Pierre Clot, Michel Janicot and Françoise Fouque
Unité 30 INSERM, Hôpital Necker-Enfants Malades,
149 rue de Sèvres, 75015 Paris, France

INTRODUCTION

Upon interaction with liver cells, insulin is internalized along with its receptor into nonlysosomal endocytic structures termed endosomes (Bergeron et al, 1985; Sonne et al, 1989). At least three lines of evidence suggest that the degradation of internalized insulin occurs in endosomes rather than in lysosomes. First, little association of insulin with lysosomes is demonstrable in ultrastructural and cell fractionation studies (Bergeron et al, 1985). Secondly, the insulin which is recovered from endosomal fractions isolated at various stages of endocytosis undergoes a rapid loss in integrity (Duckworth et al, 1988; Sonne et al, 1989). On HPLC, two major degradation products with an intact A chain and three products with a cleaved B chain have been identified (Hamel etal, 1988). Third, a further loss of insulin integrity occurs when endosomal fractions containing internalized insulin are incubated at 37°C in isotonic medium (Pease et al, 1985).

In the present work, the degradation of internalized insulin in cell-free endosomes has been characterized in two respects: (1) functional relationships with ATP-dependent endosomal acidification; (2) nature of the degradation products formed. Comparable studies have been carried out with glucagon, a peptide hormone which follows the same

NATO ASI Series, Vol. H 62
Endocytosis
Edited by P. J. Courtoy

intracellular pathway as insulin (Authier et al, 1990). The results presented here have been in part published previously (Desbuquois et al, 1990; Clot et al, 1990).

METHODS

A liver endosomal fraction (density, 1.05-1.16 g cm-3) was isolated in rats killed 90 sec after iv injection of ^{125}I-insulin, 20 min after injection of ^{125}I-glucagon or 10 min after injection of isothiocyanate-labeled galactosylated bovine serum-albumin (FITC-GalBSA). The fraction was incubated for 3-12 min at 30°C in 0.15 M KCl buffered with 25 mM citrate phosphate, pH 4-8, or 25 mM MOPS/KOH, pH 5-9. ^{125}I-Insulin and ^{125}I-glucagon degradation was assessed by trichloroacetic acid (TCA) precipitation, and ligand dissociation from endosomes by polyethyleneglycol (PEG) precipitation. Degradation products were isolated by HPLC and, when indicated, submitted to radiosequencing. Endosomal acidification was asssessed from the fluorescence quenching of in vitro added acridine orange and the fluorescence of in vivo injected FITC-GalBSA.

RESULTS

Biochemical characteristics of the endosomal fraction

As shown on Fig. 1, the endosomal fraction is enriched by 40-60 fold in internalized ligands, ATP-dependent acidification and galactosyltransferase, a Golgi marker. In density shift studies using diaminobenzidine cytochemistry, internalized ligands are well separated from galactosyltransferase (results not shown). Interestingly, the endosomal fraction degrades in vitro added insulin or glucagon, with a maximum at pH 4.5. This acidic degrading activity is lower than that in lysosomes (not shown) but exceeds by 4-fold that in the homogenate.

Degradation of insulin and glucagon associated with isolated endosomes.

^{125}I-insulin and ^{125}I-glucagon associated with freshly isolated endosomal fractions are, respectively, about 85 and

95% TCA-precipitable. Upon incubation, a rapid release of TCA-soluble radioactivity occurs, with a maximum at pH 5-6 for ^{125}I-insulin (Fig. 2) and 4 for ^{125}I-glucagon (results not shown). At the optimum pH, the percentage of TCA-soluble radioactivity released from ^{125}I-insulin and ^{125}I-glucagon is, respectively, 5.1 ± 0.2% and 11.6 ± 0.3% /min.

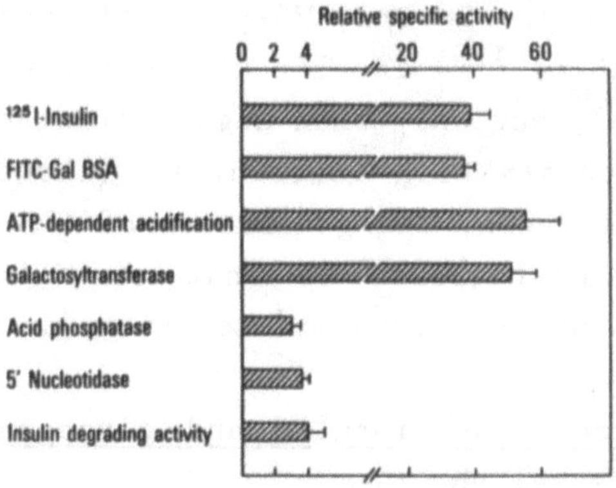

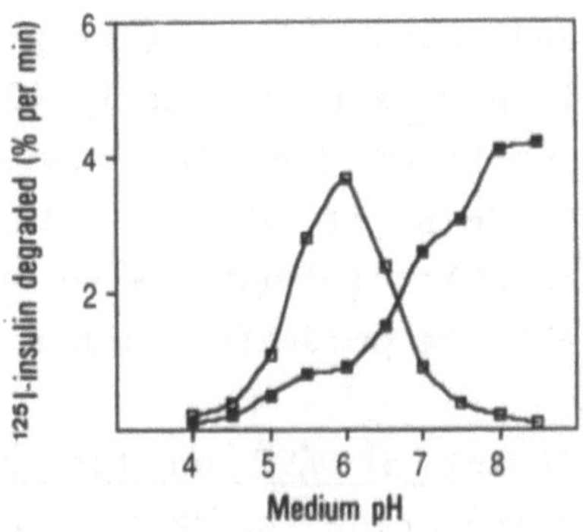

Fig. 1 (left). Specific activities of internalized ligands and enzymes associated with the endosomal fraction. Results are expressed relative to specific activities in homogenates.

Fig. 2 (right). Degradation of ^{125}I-insulin associated with endosomes as a function of medium pH. Endosomal fractions containing radiolabeled insulin were incubated for 6 min at 30°C in 125 mM KCl buffered with 25 mM citrate/phosphate in the absence (□) or presence (■) of 1 mM ATP and 5 mM Mg SO_4.

In rats injected with increasing amounts of ^{125}I-insulin, the fractional degradation of this ligand progressively decreases as more insulin is associated with endosomes, with a half-maximal effect at about 40 ng/mg protein. On a log-log plot, the absolute amount of insulin degraded is linearly related to the amount of insulin associated with endosomes, with a slope of 0.80 (results not shown).

At the optimum pH, the degradation of ^{125}I-insulin and ^{125}I-glucagon in endosomes is totally inhibited by 5 mM 1,10 phenantroline (K_i, 0.1 uM). In Brij 35-permeabilized endosomes, degradation is partially inhibited by bacitracin (5 mM), N-ethylmaleimide (3 mM) and p-mercruribenzoate (2 mM).

It is unaffected by EDTA (5 mM), glutathion (5 mM), diamide (5 mM), PMSF (1 mM) benzamidine (1 mM) and leupeptin (0.5 mg/ml).

Dissociation of insulin and glucagon from endosomes.

Upon incubation of intact endosomes containing ^{125}I-insulin, virtually all TCA-soluble products generated diffuse extraluminally as judged from the generation of a comparable amount of PEG-soluble radioactivity. In Brij 35-permeabilized endosomes, intact (TCA-precipitable) ^{125}I-insulin also diffuses in part extraluminally (5 % /min), but only when degradation is inhibited does this release increase with lowering pH (about 9% /min at pH 5). With ^{125}I-glucagon, the generation of PEG-soluble products also closely corresponds to that of TCA-soluble products, with a maximum at pH 4 (results not shown).

Effect of ATP on the pH-dependence of insulin and glucagon degradation and on internal endosomal pH

Addition of ATP (1 mM) to endosomes containing ^{125}I-insulin and ^{125}I-glucagon causes a rightward shift in the pH dependence of ligand degradation, especially marked with insulin (Fig. 2). As a result, ligand degradation is augmented in the pH interval 7-9 (insulin) and 5-8 (glucagon). Concomittantly, addition of ATP causes endosomal acidification as judged from the fluorescence quenching of acridine orange and the fluorescence of internalized FITC-GalBSA (Fig. 3). At any medium pH, the decrease in internal endosomal pH as estimated from the fluorescence of FITC-GalBSA using a standard curve is about 0.8-0.9 pH units (results not shown).

The ability of ATP to shift the pH for optimal insulin degradation is maximal at 1 mM and half-maximal at 60 uM. GTP, ITP and UTP exert comparable effects but are, respectively, only 3%, 1% and 0.1% as potent as ATP. The relative abilities of individual nucleotides to affect the pH for optimal insulin degradation correlate well with their relative abilities to acidify the internal content of the endosomes.

ATP-dependent acidification in liver endosomes is known to be optimum in the presence of permeant anions, such as Cl^-, and to be inhibited by NO_3^- ions (Fuchs et al., 1989). As shown

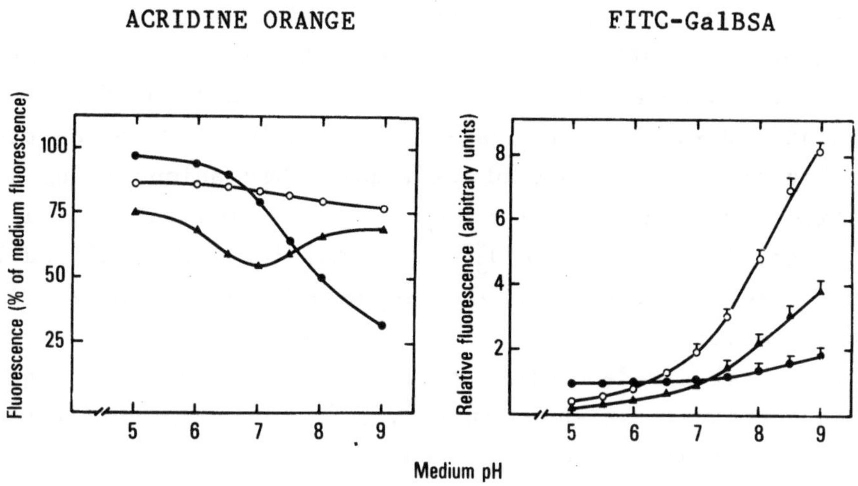

Fig. 3. Fluorescence quenching of acridine orange induced by endosomes (left) and fluorescence of FITC-GalBSA associated with endosomes (right) as a function of medium pH. Endosomal fractions from noninjected (left) or FITC-GalBSA injected (right) rats were suspended in 125 mM KCl buffered with 25 mM MOPS/KOH; with noninjected rats, acridine orange (3 uM) was included in the suspension medium. Fluorescence intensities (excitation, 493 nm; emission, 530 nm) were measured at 10-15 sec (●) and after incubating for 4 min (left) or 8 min (right) in the absence (○) or presence (▲) of 1 mM ATP and 5 mM $MgSO_4$. Adapted from Desbuquois <u>et al</u>, 1990.

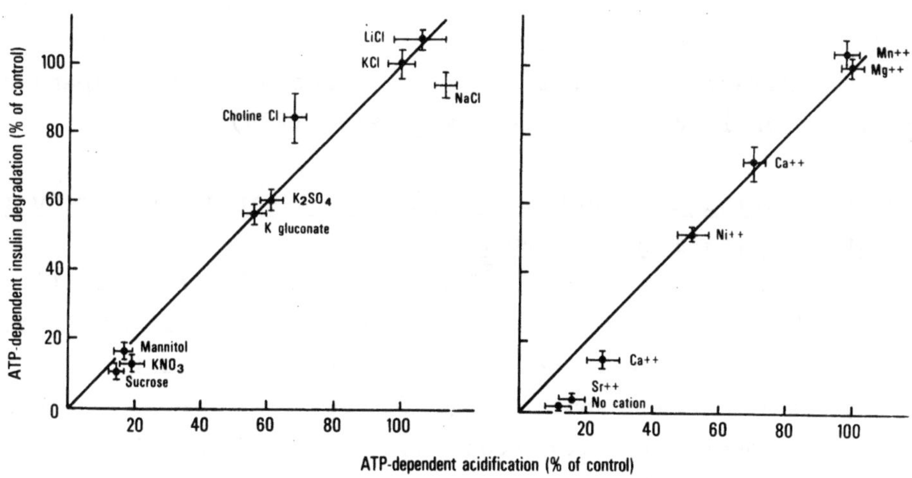

Fig. 4. Correlation between ATP-dependent acidification and ATP-dependent insulin degradation as studied in the presence of various salts (left) and divalent cations (right). Results are expressed relative to those obtained in the presence of 125 mM

on Fig. 4 (left), the abilities of various anions to support (or inhibit) the effect of ATP on the pH-dependence of insulin degradation correlate well with their abilities to affect ATP-dependent endosomal acidification. Thus, both processes occur optimally in the presence of Cl^- ions, other anions being less effective (Br^- > gluconate = SO_4^{2-} > NO_3 = sucrose = mannitol) and/or inhibitory to Cl^- (NO_3^-). In addition, Na^+, K^+ and Li^+ support more effectively insulin-dependent degradation and ATP-dependent acidification than does choline.

Endosomal ATP-dependent acidification is known to require divalent cations. As shown on Fig. 5 (right), the abilities of various divalent cations to support ATP-dependent insulin degradation correlate well with their abilities to support ATP-dependent acidification (Mg^{2+} = Mn^{2+}> Co^{2+} > Ni^{2+} = Zn^{2+}).

Endosomal ATP-driven acidification endosomes is abolished by N-ethylmaleimide (an inhibitor of the H^+ ATPase), monensin and nigericin (two proton/cation exchange ionophores), carbonyl cyanide m-chlorophenylhydrazone (a proton ionophore), and chloroquine (an acidotropic agent). As expected, these drugs also abolish the ability of ATP to affect the pH-dependence of insulin and glucagon degradation (results not shown). In contrast, neither ATP-driven acidification nor ATP-dependent acidification are inhibited by vanadate and ouabain. The lack of effect of vanadate is consistent with the finding that the Na^+, K^+ ATPase does not regulate endosomal ATP-dependent acidification in rat liver (Fuchs et al., 1989).

Characterization of insulin and glucagon degradation products generated in isolated endosomes.

Following injection of ^{125}I-insulins labeled at the A14 or B26 positions, two major degradation products slightly less hydrophobic than intact iodoinsulins are detectable by HPLC in freshly isolated endosomes (Fig. 5, upper panels). These account, respectively, for about 13% (A14 iodoinsulin) and 8% (B26 iodoinsulin) of the radioactivity in the eluate. With B26 iodoinsulin, two minor products of low hydrophobicity are also seen. Rechromatography after reduction shows that the major products from A14 iodoinsulin contain an intact A chain, whereas the major products from B26 iodoinsulin contain a broken B chain. Radiosequence analysis of the B chain products identifies cleavages at bonds B16-B17 (major products) and B23-B24 as well as B24-B25 (minor products).

Incubation of endosomes containing A14 iodoinsulin either at pH 5.5 in the absence of ATP (Fig. 5A, lower panel) or at pH 8.5 in the presence of ATP results in a time dependent decrease of intact insulin, along with an increase in intermediate products and monoiodotyrosine. With B26 iodoinsulin (Fig. 5B, lower panel) a decrease in intact insulin is also observed, but in this case monoiodotyrosine is the main product generated.

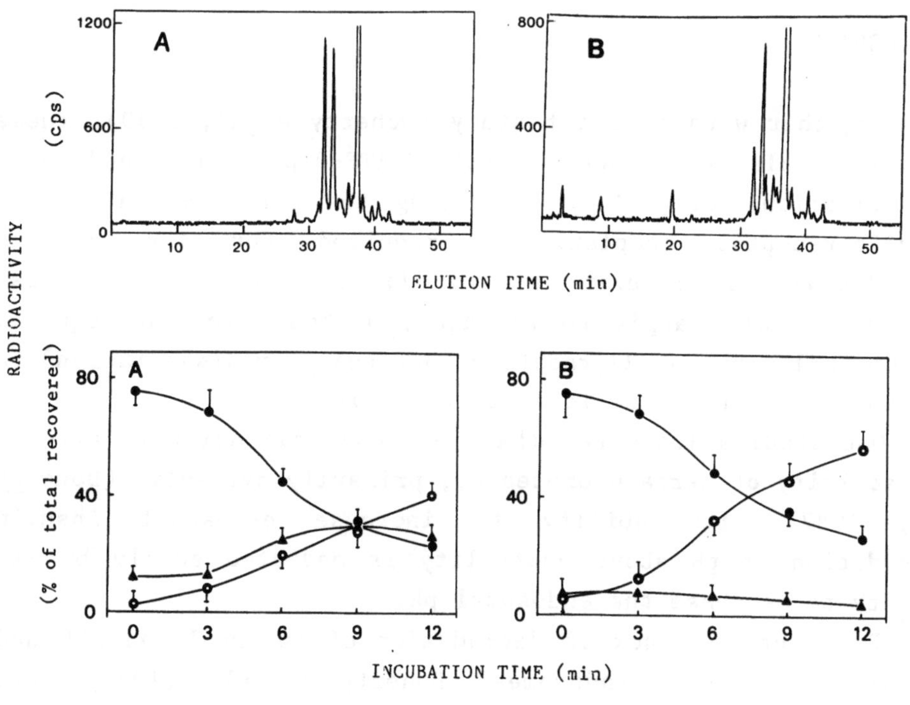

Fig. 5. HPLC analysis of the radiolabeled material generated from A14 (A) and B26 (B) radioiodinated insulins in endosomes. The upper panels show the elution profile of the radioactivity associated with freshly isolated endosomes, with intact insulin eluting at 37-38 min and iodotyrosine at 2-3 min. The lower panels show the distribution of the radioactivity between intact insulin (●), the two major degradation products (▲) and iodotyrosine (○) upon incubation of endosomes at pH 5.5 in the absence of ATP. Reproduced from Clot et al, 1990.

Following injection of ^{125}I-glucagon, > 95% of the radioactivity associated with freshly isolated endosomal fractions is recovered, on HPLC, as intact iodoglucagon. Incubation of endosomes at pH 4 results in a rapid decrease of intact iodoglucagon, along with the appearance of at least four degradation products of lesser hydrophobicity and of monoiodotyrosine (results not shown). Two of these products elute at the same position as some products previously identified in endosomal fractions of chloroquine-treated animals (Authier et al., 1990). The major degradation product contains an intact 1-13 sequence and is cleaved at an as yet undefined site beyond tyrosine 13.

DISCUSSION

Together with a recent study (Doherty et al, 1990), these studies clearly show that ATP-dependent endosomal acidification, in part by enhancing the dissociation of the insulin-receptor complex, is required for the optimum degradation of insulin within liver endosomes. These conclusions also apply to glucagon, although the pH required for optimal glucagon degradation (and receptor dissociation) in endosomes appears to be lower than for insulin.

Our studies indicate that, although ATP directly affects the activity of certain proteases, primarily cytosolic (Bond et al., 1987), its ability to increase endosomal insulin degradation at pH above neutrality is mediated solely by its ability to decrease the endosomal pH.

Based on the lack of degradation of internalized EGF and prolactin in liver endosomes, Doherty et al. (1990) have recently suggested that the degrading process is selective for insulin. According to these authors, such selectivity may be a consequence of secondary and tertiary structural differences between insulin and other peptides. The ability of isolated endosomes to degrade glucagon as rapidly as insulin indicates that the specificity of the endosomal degrading activity may be broader than initially suspected. Additional studies are needed to characterize the peptidase(s) involved and its(their) relationships with known peptidases.

AKNOWLEDGMENT

We are indebted to Dr F. Lederer for radiosequence analyses.

REFERENCES

Authier F, Janicot M, Lederer F, Desbuquois B (1990) Fate of injected glucagon taken up by rat liver in vivo. Degradation of internalized ligand in the endosomal compartment. Biochem J in the press
Bergeron JJM, Cruz J, Khan MN, Posner BI (1985) Uptake of insulin and other ligand into receptor-rich endocytic components of target cells: the endosomal apparatus Ann Rev Physiol 47: 383-403
Bond JS, Butler PE (1987) Intracellular proteases Ann Rev Biochem 56: 333-364
Clot JP, Janicot M, Fouque F, Desbuquois B, Haumont, PY, Lederer F (1990) Characterization of insulin degradation products generated in liver endosomes: in vivo and in vitro studies. Mol Cell Endocr 72: 175-185
Desbuquois B, Janicot M, Dupuis A (1990) Degradation of insulin in isolated liver endosomes is functionally linked to ATP-dependent acidification. Eur J Biochem in the press
Doherty II JJ, Kay DG, Lai WH, Posner BI, Bergeron JJM (1990) Selective degradation of insulin within rat liver endosomes J Cell Biol 110: 35-42
Duckworth WC (1988) Insulin degradation: mechanisms, products, and significance. Endocr Rev 9: 319-345
Fuchs R, Male P, Mellman I (1989) Acidification and ion permeabilities of highly purified rat liver endosomes. J Biol Chem 264: 2212-2220
Hamel FG, Posner BI, Bergeron JJM, Frank BH, Duckworth WC (1988) Isolation of insulin degradation products from endosomes derived from intact rat liver. J Biol Chem 263: 6703-6708
Pease RJ, Smith GD, Peters TJ (1985) Degradation of endocytozed insulin in rat liver is mediated by low-density vesicles. Biochem J 228: 137-146
Sonne O (1988) Receptor-mediated endocytosis and degradation of insulin. Phys Rev 68: 1129-1196

SIGNALING OF RECEPTOR TYROSINE KINASES ACROSS THE ENDOSOMAL MEMBRANE

J.J.M. Bergeron and B.I. Posner
Departments of Anatomy and Medicine
McGill University
3640 University Street
Montreal, Quebec, CANADA
H3A 2B2

The hypothesis that the ligand mediated internalization of the insulin, epidermal growth factor (EGF), and prolactin receptors regulates the physiological actions of the respective receptors was proposed approximately 11 years ago (Bergeron et al 1979; Posner et al 1980). The hypothesis was elaborated to account for our (then) surprising observations of a rapid concentrative internalization of ligand receptor complexes into novel intracellular organelles which were not lysosomes and which are now known as endosomes.

Having first documented the presence of intracellular organelles harboring receptors for peptide hormones in target cells (Bergeron et al 1973,1978; Khan et al 1981) we attempted to test if exogenously administered hormones could enter these intracellular compartments. They did (Josefsberg et al 1979; Posner et al 1980) and hence the hypothesis stated above. The in vivo radioautographic method was set up to compare our subcellular fractionation data with the corresponding in vivo compartments (Bergeron and Posner 1979; Bergeron et al 1979,1980) and also had a noteworthy ancillary use in enabling the visualization of receptor enriched target cells for a variety of peptide hormones in newly described target cells (e.g. Martineau-Doizé et al 1986; Warshawsky et al 1980; Van Houten et al 1980).

Subcellular fractionation techniques were extended (Bergeron et al 1986; Khan et al 1982,1985,1986a,1986b; Khan RJ et al 1985;) and identified distinct subcompartments of the endosomal apparatus. These compartments participated in the sequential transport of ligands and their receptors (reviewed in Bergeron et al 1985).

NATO ASI Series, Vol. H 62
Endocytosis
Edited by P. J. Courtoy

Although we initially suggested, on the basis of marker enzyme content (e.g. Josefsberg et al 1979; Khan et al 1982) and in vivo location (Bergeron et al 1979) that at least one of these components was Golgi, this issue was resolved by the use of the DAB shift protocol which completely separated cell endocytic components from the Golgi marker enzyme, galactosyl transferase (Bergeron et al 1985,1988; Kay et al 1984) as well as by detailed in vivo EM radioautography (Bergeron et al 1983; Cruz et al 1984).

The regulation of ligand receptor traffic and processing through endosomal components was assessed by means of the microtubule inhibitor colchicine and the acidotropic agent chloroquine. It was observed that both agents retarded the receptor mediated processing (degradation) of internalized insulin and prolactin in liver tissue (Posner et al 1982a, 1982b). More detailed subcellular fractionation (Bergeron et al 1986; Khan et al 1986b) and in vivo studies (Bergeron et al 1983) indicated noteworthy differences in the kinetics and extent of inhibition of degradation of the two hormones by either of the drugs used. Insulin degradation was more rapid than that for prolactin (Bergeron et al 1986,1988; Khan et al 1985,1986b; Khan RJ et al 1985). Chloroquine treatment led to the selective accumulation of insulin in endosomes defined as "early"[1]. Interestingly, negligible accumulation in secondary lysosomes was observed (Posner et al 1982b). The hypothesis that insulin degradation was occurring within 'early' endosomes was considered as a consequence of the observation of biologically inactive insulin (Posner et al 1980) and well characterized insulin fragments within endosomes (Hamel et al 1988). More recently, we have studied a cell-free endosomal system and demonstrated the presence therein of an insulinase with a low pH optimum. Degradation appeared to be selective for insulin but not prolactin nor EGF (Doherty et al 1990).

---

[1]The term 'early' and 'late' is used to signify the kinetics of ligand appearance in distinct endosomes separable by subcellular fractionation.

The effects of the alkaloid colchicine suggested an involvement of microtubules in internalization. Detailed EM radioautographic studies on intact liver demonstrated that colchicine treatment led to the accumulation of internalized prolactin in lipoprotein-filled[2] endosomes found at the periphery of the hepatocyte beneath the sinusoidal cell surface (Bergeron et al 1983). Interestingly, these endosomes became larger (Bergeron et al 1983) and on subcellular fractionation accumulated in light mitochondrial fractions (Bergeron et al 1986). Hence, we postulated that microtubules were involved in guiding the 'early' endosomes from the periphery of the hepatocyte to the Golgi region where the microbule organizing centriole had been previously observed within the region of ligand-filled endosomes (see Fig.9 of Bergeron et al 1979). However, microtubules are not required for the increase in hepatic endosomal size observed to coincide with endosomal 'maturation' (Bergeron et al 1983,1986). From our combined subcellular fractionation and in vivo studies we have therefore proposed a model for the endocytosis of insulin, prolactin, and epidermal growth factor and their receptors in liver parenchyma (Fig. 1).

Having attempted to establish the components participating in ligand receptor internalization, we set out to evaluate the hypothesis of endosomal regulation of receptor signaling. First, we documented (Kay et al 1986; Khan et al 1986,1989; Lai et al 1989) that both the insulin receptor and EGF receptor show augmented tyrosine kinase activity following internalization into endosomes. With respect to the EGF receptor, noteworthy correlations between the autophosphorylation and augmented tyrosine kinase activities have been found with both receptor internalization and receptor down-regulation (Lai et al 1989).

---

[2]As described elsewhere (Bergeron et al 1983,1985), the lipoprotein content of liver endosomes most probably represents endocytosed serum lipoprotein particles.

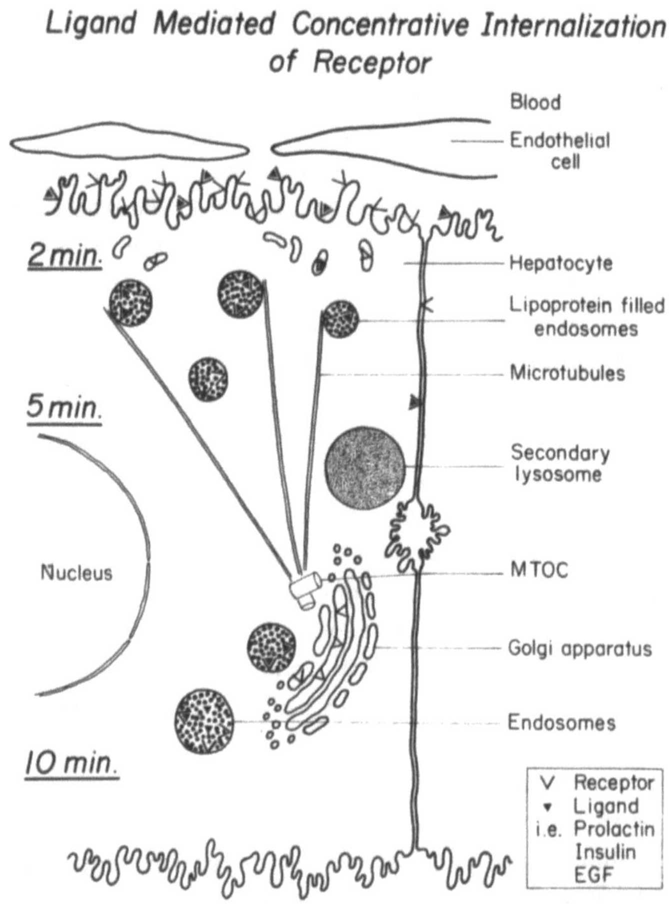

Figure 1: Receptor and ligand distributions during ligand mediated internalization of the insulin, EGF and prolactin receptors. Shortly after binding, receptors are internalized into a peripheral tubulovesicular compartment followed by their appearance in 'early' and 'late' endosomes, the latter found in the Golgi region often close to microtubule organizing centrioles (MTOC). More details are described in Fig. 1 of Bergeron et al 1988.

Hence, the transient ligand-mediated desensitization of cell surface EGF receptor autophosphorylation activity was found to correlate with the transient inhibition of receptor internalization (Lai et al 1989). Furthermore, under conditions favoring ligand-mediated down-regulation, greatly augmented autophosphorylation activity of the endosomal EGF receptor tyrosine kinase was found (Lai et al 1989). By contrast, under conditions favoring EGF receptor recycling, markedly less activity was found. Most recently, transmembrane signaling of the EGF receptor tyrosine on the endosomal membrane has been shown to extend to a novel substrate. The substrate is an integral membrane protein which forms an heterooligomer with the EGF receptor in the endosomal membrane. The substrate may be involved in either the biological response to EGF in liver parenchyma and/or the regulation of receptor distribution.

Table 1

Specific Radioactivity* of ^{32}P Tyrosine Phosphorylated EGF Receptor at the Cell Surface and in Endosomes

PM	Endosomes
10.2	37.6

*Specific radioactivity was determined from immunoprecipitated EGF receptor from PM isolated at 30 sec after the injection of EGF or in endosomes at 15 min after the injection of EGF. Phosphotyrosine content of the receptor was quantified by densitometric analysis of alkali treated gels.

Recent attempts have also focused on the determination of the phosphotyrosine content of the insulin and EGF receptors at the cell surface at peak times of activation (30 sec after hormone or growth factor administration) and in endosomes during peak times of internalization. For these experiments (Table 1) receptor content was quantified by quantitative immunoblotting with anti receptor antibodies and by Scatchard analysis. After the injection of 32Pi into the animals, EGF was injected into the hepatic portal vein, rats were sacrificed and the specific radioactivity of the respective receptors determined. The data (Table 1) show conclusively an increased specific radioactivity of the EGFR in endosomes consistent with a role for signaling therein.

References:

Bergeron JJM, Evans WH, and Geschwind II (1973) Insulin binding to rat liver Golgi fractions. J. Cell Biol. 59:771-776.

Bergeron JJM, Posner BI, Josefsberg Z, and Sikstrom R (1978) Intracellular polypeptide hormone receptors. I. The demonstration of specific binding sites for insulin and human growth hormone in Golgi fractions isolated from liver of female rats. J. Biol. Chem. 253:4058- 4066.

Bergeron JJM, Sikstrom R, Hand AR, and Posner BI (1979) Binding and uptake of ^{125}I-insulin into rat liver hepatocytes and endothelium : An in vivo radioautographic study. J. Cell Biol. 80:427-443.

Bergeron JJM and Posner BI (1979) In vivo studies on the initial localization and fate of polypeptide hormone receptors by the technique of quantitative radioautography. J. Histochem. Cytochem. 27:1512-1513.

Bergeron JJM, Rachubinski R, Searle N, Borts D, Sikstrom R, and Posner BI (1980) Polypeptide hormone receptors in vivo : Demonstration of insulin binding to adrenal gland and gastrointestinal epithelium by quantitative radioautography. J. Histochem. Cytochem. 28:824-835.

Bergeron JJM, Resch L, Rachubinski R, Patel BA, and Posner BI (1983) Effect of colchicine on internalization of prolactin in female rat liver : An in vivo radioautographic study. J. Cell Biol. 96:875-886.

Bergeron JJM, Cruz J, Khan MN, and Posner BI (1985) Uptake of insulin and other ligands into receptor-rich endocytic components of target cells: The endosomal apparatus. Annual Revs. Physiol. 47:383-403.

Bergeron JJM, Searle N, Khan MN, and Posner BI (1986) Differential and analytical subfractionation of rat liver components internalizing insulin and prolactin. Biochemistry 25:1756-1764.

Bergeron JJM, Kay DG, Lai WH, Doherty II J-J, Smith CE, Khan MN, and Posner BI (1988) Functional characteristics of the endosomal apparatus of rat liver parenchyma. IN: Cell-Free Analysis of Membrane Traffic. (Edited by D.J. Morré et al.). Alan R. Liss, New York. 391-409.

Cruz J, Posner BI, and Bergeron JJM (1984) Receptor-mediated endocytosis of ^{125}I-insulin into pancreatic acinar cells in vivo. Endocrinology 115:1996-2008.

Doherty II J-J, Kay DG, Lai WH, Posner BI, and Bergeron JJM (1990) Selective degradation of insulin within rat liver endosomes. J. Cell Biol. 110:35-42.

Hamel FG, Posner BI, Bergeron JJM, Franck BH, and Duckworth WC (1988) Isolation of insulin degradation products from endosomes derived from intact rat liver. J. Biol. Chem. 263:6703-6708.

Josefsberg Z, Posner BI, Patel B, and Bergeron JJM (1979) The uptake of prolactin into female rat liver : Concentration of intact hormone in the Golgi apparatus. J. Biol. Chem. 254:209-214.

Kay DG, Khan, MN, Posner BI, and Bergeron JJM (1984) In vivo uptake of insulin into hepatic Golgi fractions : Application of the diaminobenzidine-shift protocol. Biochem. Biophys. Res. Commun. 123:1144-1148.

Kay DG, Lai WH, Uchihashi M, Khan MN, Posner BI, and Bergeron JJM (1986) Epidermal growth factor receptor kinase translocation and activation in vivo. J. Biol. Chem. 261:8473-8480.

Khan MN, Posner BI, Verma AK, Khan RJ, and Bergeron JJM (1981) Intracellular hormone receptors : Evidence for insulin and lactogen receptors in a unique vesicle sedimenting in lysosome fractions of rat liver. Proc. Natl. Acad. Sci. USA. 78:4890-4984.

Khan MN, Posner BI, Khan RJ, and Bergeron JJM (1982) Internalization of insulin into rat liver Golgi elements : Evidence for vesicle heterogeneity and the path of intracellular processing. J. Biol. Chem. 257:5969-5976.

Khan MN, Savoie S, Khan RJ, Bergeron JJM, and Posner BI (1985) Insulin and insulin receptor uptake into rat liver : Evidence for site of chloroquine action on receptor recycling. Diabetes 34:1025-1030.

Khan MN, Savoie S, Bergeron JJM, and Posner BI (1986a) Differential kinetics and sensitivity to chloroquine of receptor-mediated insulin and prolactin endocytosis in liver parenchymal cells. Biochim. et Biophys. Acta. 888: 100-106.

Khan MN, Savoie S, Bergeron JJM, and Posner BI (1986b) Characterization of rat liver endosomal fractions : In vivo activation of insulin stimulable receptor kinase in these structures. J. Biol. Chem. 261:8462-8472.

Khan MN, Baquiran C, Brule C, Burgess J, Foster B, Bergeron JJM, and Posner BI (1989) Internalization and activation of the rat liver insulin receptor kinase in vivo. J. Biol. Chem. 264:12931-12940.

Khan RJ, Khan MN, Bergeron JJM, and Posner BI (1985) Prolactin uptake into liver endocytic components : Reduced sensitivity to chloroquine. Biochimica et Biophysica Acta. 838:77-83.

Lai WH, Cameron PH, Doherty II J-J, Posner BI, and Bergeron JJM (1989) Ligand mediated autophosphorylation activity of the EGF receptor during internalization. J. Cell Biol. 109:2751-2760.

Martineau-Doizé B, McKee MD, Warshawsky H, and Bergeron JJM (1986) In vivo demonstration by radioautography of binding sites for insulin in liver, kidney and calcified tissues of the rat. Anat. Rec. 214:130-140.

Posner BI, Patel B, Verma AK, and Bergeron JJM (1980) Uptake of insulin by plasmalemma and Golgi subcellular fractions of rat liver. J. Biol. Chem. 255:735-741.

Posner BI, Verma AK, Patel B, and Bergeron JJM (1982a) The effect of colchicine on the internalization of prolactin and insulin into Golgi fractions of rat liver. J. Cell Biol. 93:560-567.

Posner BI, Patel BA, and Bergeron JJM (1982b) Effect of chloroquine on the internalization of ^{125}I-insulin into subcellular fractions of rat liver : Evidence for an effect of chloroquine on Golgi elements. J. Biol. Chem. 257:5789-5799.

Van Houten M, Posner BI, Kopriwa BM, and Brawer JR (1980) Insulin binding sites localized to nerve terminals in rat median eminence and arcuate nucleus. Science. 207:1081-1083

Warshawsky H, Goltzman D, Rouleau MF, and Bergeron JJM (1980) Direct in vivo demonstration by radioautography of specific binding sites for calcitonin in skeletal and renal tissues of the rat. J. Cell Biol. 85:682-694.

1.6. RECYCLING PATHWAYS

(chairperson : Alice DAUTRY-VARSAT)

The major pathway followed by internalized membrane is recycling. The pathway of the bulk of internalized membrane constituents was addressed by **Lucinda Mata**, using quantitative ultrastructural autoradiography. In seminal vesicle secretory cells covalently labeled apical membrane glycoconjugates are internalized into several intracellular compartments, including the Golgi complex and secretory vesicles, and reappear at the apical plasma membrane. Data provide a direct ultrastructural evidence that the Golgi can be intensely involved in recycling of the bulk of internalized membrane. They also show a membrane shuttling between apical and basolateral domains.

The relation between the recycling (transferrin) and the degradative pathway (asialo-orosomucoid) has been studied in HepG2 cells by **Willem Stoorvogel**. He used the diaminobenzidine-induced density shift to monitor the intracellular mixing of sequentially internalized ^{125}I- and peroxidase-conjugated tracers. Data show that a pulse of internalized transferrin is largely accessible to a subsequent pulse of transferrin-peroxidase, but not at all of asialo-orosomucoid-peroxidase. In pulse-chase, asialo-orosomucoid is associated with particles recovered at gradually increasing densities, up to that of lysosomes. Since the entire pathway of asialo-orosomucoid, except lysosomes, is also accessible to subsequently internalized transferrin-peroxidase, Stoorvogel concludes that in HepG2 cells, endosomes gradually mature into lysosomes and that sorting into the recycling pathway proceeds throughout this pathway.

Two other presentations dealt with the recycling of (part of) growth factor receptors. **Alice Dautry-Varsat** studied the fate of interleukin-2 and its receptors in human T lymphocytes. The biologically active, high-affinity receptor is an $\alpha-\beta$ heterodimer which is internalized either constitutively, or at a higher rate when occupied. The β subunit is quickly degraded but the α subunit apparently parts company in endosomes and is recycled. The α subunit alone can bind but not internalize interleukin-2 and remains at the cell surface for several days. The rationale for such a complex receptor system is not clear at the moment.

Alexander Sorkin had previously demonstrated that most of EGF internalized in A431 cells is released intact in the medium. He reported the occurrence of two recycling pathways of equal importance, a rapid route which involves peripheral endosomes, and a slower route which involves

NATO ASI Series, Vol. H 62
Endocytosis
Edited by P. J. Courtoy

multivesicular endosomes and is blocked at 18°C. Experiments based on covalent labeling of the receptor by a photoactivable derivative of EGF followed by cell permeabilization suggest that EGF-receptor dissociation occurs only in the interior of lysosomes. Whether EGF-receptor complexes following the slow pathway remain at the limiting membrane or are also found in the internal vesicles awaits further studies.

Finally, the chapter of **Rudolf Tauber** and colleagues shows that plasma membrane glycoproteins of rat hepatocytes may undergo reprocessing during recycling by de- and re-glycosylation of terminal sugars, and sometimes core sugars. Removal of L-fucose presumably occurs in a prelysosomal compartment.

MEMBRANE RECYCLING IN THE EPITHELIUM OF THE SEMINAL VESICLE INTERSECTS THE SECRETORY PATHWAY

Lucinda R. Mata and Erik I. Christensen*
Department of Cell Biology, Gulbenkian Institute of Science, Apartado 14, 2781 - Oeiras - Codex, Portugal

INTRODUCTION

In the epithelial cells of the seminal vesicle, as in other exocrine gland cells , exocytosis leads to the insertion of secretory granule membrane in the cell surface (Fig.1a and b). As first pointed out by Palade (1959), a mechanism must exist to remove this excess of membrane in order to maintain a constant cell size . Whether the membrane to be removed would be degraded, dismantled into constituents to be subsequently reassembled, or reutilized as such , has become, since then, a matter for debate.

The seminal vesicle regulated secretory cells are very active in endocytosis as well as in secretion and display a short secretory cycle. The activity of these cells depends on testosterone (Mata and David-Ferreira 1982; Mata and David-Ferreira 1990), which allows their experimental manipulation. Because of these features, the seminal vesicle secretory cells have proved to be a good model to investigate endocytosis and membrane traffic.

The study of endocytosis in the hamster seminal vesicle secretory cells (Mata and David-Ferreira 1973) contributed the first evidence on the possible reutilization of endocytosed membrane in secretory vacuoles and Golgi cisternae of exocrine cells thus raising the possibility of its recycling along with secretion to the apical surface. Direct evidence demonstrating membrane recycling in this system (Mata and Christensen 1990) is briefly reviewed in this work .

INTERNALIZATION OF SOLUBLE TRACERS

The hamster seminal vesicle secretory cells endocytose

* Department of Cell Biology, Institute of Anatomy, University of Aarhus, Denmark.

NATO ASI Series, Vol. H 62
Endocytosis
Edited by P. J. Courtoy

horseradish peroxidase (HRP) as well as secretory proteins from the gland lumen (Mata and David-Ferreira 1973; Mata and

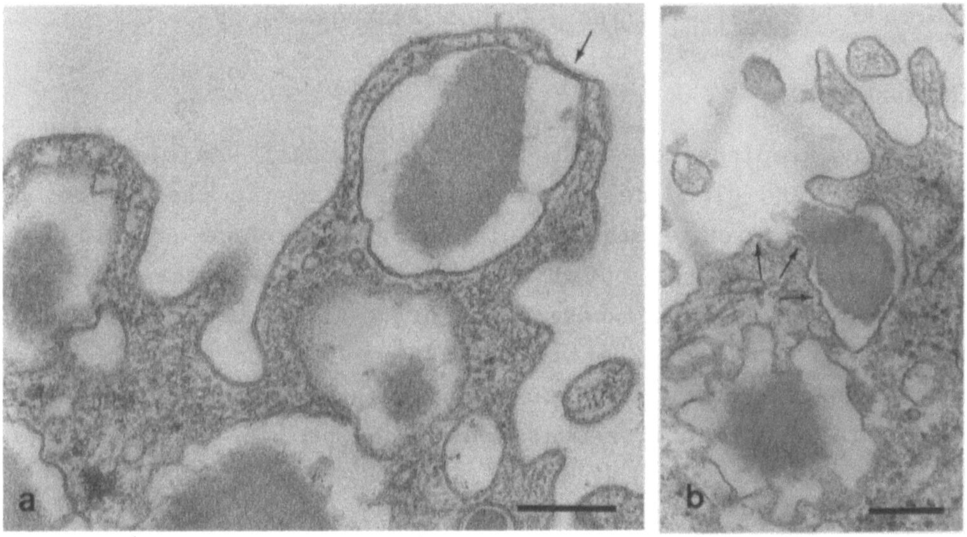

Fig. 1 - Electron micrographs from the apical region of hamster seminal vesicle secretory cells illustrating the pentalaminar membrane formed during exocytosis (a, arrow) and the release of a secretory granule leading to insertion of secretory vacuole membrane in the cell surface (b, arrows). Bars=0.2μm.

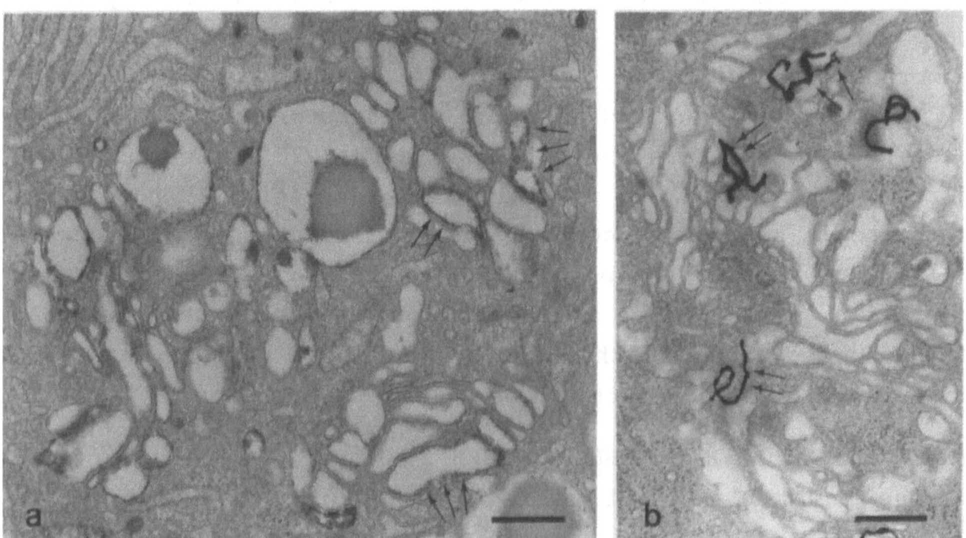

Fig. 2 - Electron micrographs illustrating that endocytosed HRP (a, arrows) as well as radioactive secretory proteins (b, arrows) label the Golgi cisternae of hamster seminal vesicle secretory cells. Bars=0.4μm.

Maunsbach 1982). The endocytic pathway includes multivesicular bodies, secondary lysosomes, secretory vacuoles and the Golgi apparatus, including its innermost and outermost cisternae (Fig. 2a and b) from where HRP is cleared later on probably along with secretion (Mata 1976). The tracers are also transcellularly tranported through the intercellular spaces bellow tight junctions.

These results are consistent with the insertion of internalized membrane in the labeled compartments and, together with biochemical evidence showing that the turn-over of membrane proteins in secretory cells is slower than required for the amount of membrane to be used in secretion (Meldolesi 1974), have further supported the hypothesis of membrane retrieval and reutilization.

The labeling of Golgi cisternae was also reported in endocrine cells (Pelletier 1973) and later extended to other secretory cells (Farquhar 1983). The notion of membrane reutilization in the exocytic pathway became thus widely accepted despite the absence of direct evidence.

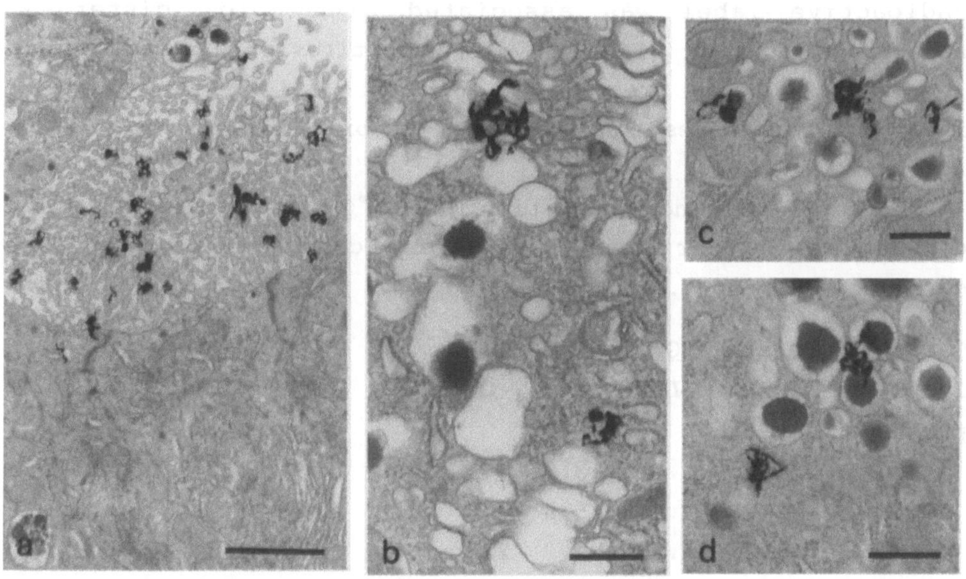

Fig. 3 - Autoradiographs showing silver grains associated with the apical cell surface immediately after labeling (a) and with Golgi cisternae (b) and secretory vacuoles (c and d) after membrane internalization. Bar in a=2μm; bars in b, c, and d=0.5μm

INTERNALIZATION OF MEMBRANE CONSTITUENTS

In this work the guinea-pig seminal vesicle was used instead of the hamster's. HRP is also endocytosed by guinea-pig seminal vesicle secretory cells and follows the same pathway as in the hamster's gland.

The apical domain of the secretory cell membrane was covalently labeled with tritiated galactose (Thilo 1983) introduced into the gland lumen (0°C, 20 min) and the epithelium incubated (37°C, 15 min) to allow endocytosis (1st incubation). Most of the label which was not internalized was removed by enzymatic hydrolysis (0°C, 30 min) and the epithelium re-incubated to allow membrane movement for 15 and 30 minutes at 37°C (2nd incubation). After each step tissue pieces were prepared for electron microscope autoradiography. The autoradiographs were quantitatively analyzed utilizing a stereological approach.

Immediately following labeling, the silver grains were almost exclusively associated with the apical surface of the secretory cells (Fig. 3a). After the 1st incubation at 37°C, 30% of the label was detected inside the cells. Internalized radioactive label was associated with Golgi cisternae, secretory vacuoles (Figs 3b-d), multivesicular bodies and lysosomes. Grains were also observed along the lateral intercellular spaces and the basal membrane of the epithelium (Fig. 4).

During the 2nd incubation at 37°C, the relative grain density was higher than 1 in every compartment except the cytoplasm (Table I).

Table I - Relative Grain Density*

Incub. time, min (37°C) post-hydrol.	AD	BLD	SV	GV	GC	MVB	LYS	CV	CYT
0	8.0	1.2	13.3	4.1	1.5	37.8	1.2	25.4	0.0
15	8.2	1.4	5.8	4.8	1.2	11.2	1.4	11.2	0.0
30	7.1	1.8	4.2	2.6	1.5	10.2	1.9	11.1	0.0

*-Relative grain density is the % of grains over the % of area. AD, apical domain of cell membrane; BLD, basolateral domain of cell membrane; SV, secretory vacuoles; GV, Golgi vesicles; GC, Golgi cisternae; MVB, multivesicular bodies; LYS, lysosomes; CV, cytoplasmic vesicles; CYT, cytoplasm.

The relative concentration of label either increased (BLD and LYS), decreased (CV, SV and MVB) or fluctuated (GC, GV and AD). These results showed that apical membrane was

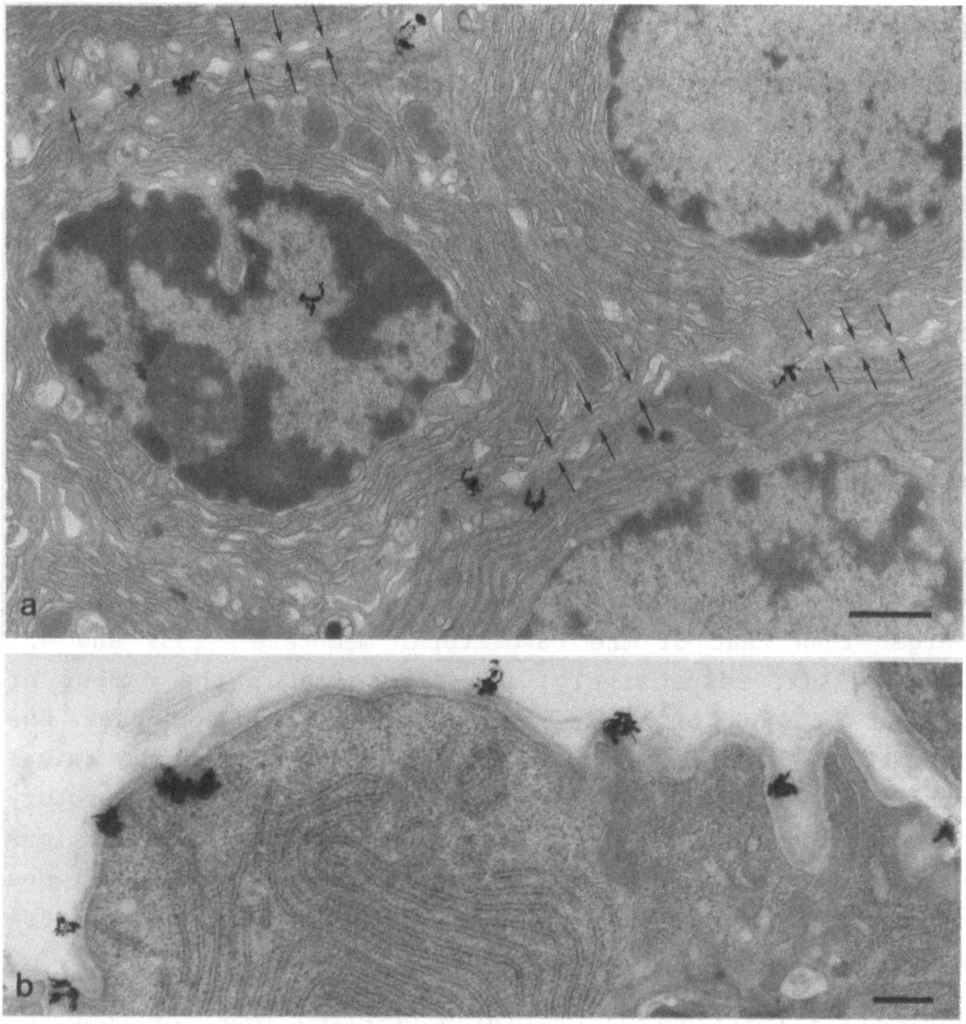

Fig. 4 - Autoradiographs illustrating the association of silver grains with lateral (a) and basal (b) cell membrane. Notice grains localized along the intercellular spaces (a, arrows). Bars=0.5µm.

internalized, transferred and redistributed through the compartments analyzed and implied that endocytosed apical membrane moved in and out of them. The way the label fluctuated in Golgi cisternae and Golgi vesicles, together with the observation of a peak in label concentration on the apical domain at 15 min, indicated that, by that time, the amount of internalized membrane that had moved to the apical cell membrane was enough for membrane recycling to be detected and suggested the involvement of the components of the exocytic pathway in the recycling pathway.

MEMBRANE RECYCLING THROUGH THE SECRETORY PATHWAY

Considering these data, the possibility of internalized label recycling back to the apical membrane during the 2nd incubation at 37°C was analyzed comparing the distribution of grains observed with that expected after 15 minutes incubation without recycling (Table II). The apical domain of the cell showed close to 50% more grains, whereas Golgi cisternae, secretory vacuoles and basolateral cell membrane showed close to 53%, 42% and 30% less grains , respectively. The decrease in the number of grains in these 3 compartments could account for 86% of the excess of grains observed on the apical membrane which was consistent with their playing an important role in the process of membrane recycling. Although the involvement of the Golgi cisternae and secretory vacuoles was relatively larger than that of the basolateral membrane (53% and 42% against 30%, respectively) the reduction in the number of grains in each of these compartments suggested that the basolateral membrane by itself could contribute to the apical membrane as much label as the Golgi cisternae and secretory vacuoles together. The difference between the two distributions was highly significant as shown by the X^2-test ($P<0.0005$) and it could therefore be concluded that internalized membrane recycled back to the cell membrane apical domain.

Table II - Distribution of grains observed (**Obs**) as compared with that expected (**Expect**) without membrane recycling after 15 min at 37°C following hydrolysis (2nd incubation).

	AD	BLD	SV	GV	GC	MVB	LYS	CV	CYT	Total
Obs(a)	277	95	26	28	17	17	14	43	22	539
Expect(b)	186	135	45	28	36	26	8	50	25	539
(a)-(b)	+91	-40	-19	0	-19	-9	+6	-7	-3	0
X^2-test		$X^2 = 83.3787$				n = 8		$P<0.0005$		

CONCLUSION

The results from the study of endocytosis in seminal vesicle secretory cells, using either HRP or radioactively labeled secretory proteins as tracers, showed that these cells display a constitutive endocytic activity and suggested that membrane recovered from the cell surface could be reutilized in the exocytic pathway, namely in the Golgi cisternae and secretory vacuoles, and recycled back to the apical cell membrane. By covalently labeling the apical membrane of seminal vesicle secretory cells, we were able to demonstrate that internalized membrane constituents are transferred and redistributed through several compartments and recycle back to the apical cell surface. Our data from these experiments confirm the involvement of secretory vacuoles and Golgi cisternae in the recycling pathway and are also consistent with membrane shuttle between the apical and basolateral domains of the cell membrane.

REFERENCES

Farquhar M G (1983) Multiple pathways of exocytosis, endocytosis, and membrane recycling: validation of a Golgi route. Federation Proc 42 : 2407-2413

Mata L R (1976) Dynamics of HRPase absorption in the epithelial cells of the hamster seminal vesicle. J Microsc Biol Cellulaire 25 : 127-132

Mata L R, Christensen E I (1990) Redistribution and recycling of internalized membrane in seminal vesicle secretory cells. Biol Cell 68 : 183-193

Mata L R, David-Ferreira J F (1973) Transport of exogenous peroxidase to Golgi cisternae in the hamster seminal vesicle. J Microscopie 17 : 103-106

Mata L R, David-Ferreira J F (1982) Testosterone interference with the intracellular transport of secretion in hamster seminal vesicle. Biol Cell 46 : 101-104

Mata L R, David-Ferreira J F (1990) Testosterone interferes with the kinetics of endocytosis in the hamster seminal vesicle. Biol Cell 68 : 195-203

Mata L R, Maunsbach A B (1982) Absorption of secretory protein by the epithelium of hamster seminal vesicle as studied by electron microscope autoradiography. Biol Cell 46 : 65-74

Meldolesi J (1974) Dynamics of cytoplasmic membranes in guinea pig pancreatic acinar cells. I. Synthesis and turnover of membrane proteins. J Cell Biol 61 : 1-13

Palade G E (1959) Functional changes in the structure of cell components. Hayashi T (ed) (1959) Subcellular particles. Ronald Press New York

Pelletier G (1973) Secretion and uptake of peroxidase by rat adenohypophyseal cells. J Ultrastruct Res 43 : 445-459

Thilo L (1983) Labeling of plasma membrane glycoconjugates by terminal glycosilation. Fleischer S (ed) (1983) Methods in enzymology 98. Academic Press London

ENDOCYTIC PATHWAYS

Willem Stoorvogel, Hans J. Geuze and Ger J. Strous

Laboratory for Cell Biology, University of Utrecht, AZU, rm H02.314, Heidelberglaan 100, 3584 CX, Utrecht, The Netherlands

Horseradish peroxidase (HRP) can catalyze the hydroxylation of a variety of aromatic compounds, including phenolic substrates such as 3,3'-diaminobenzidine (DAB), as well as tyrosine, phenylalanine, and sialic acid. These latter compounds may partly substitute for DAB as reductant. DAB cytochemistry has been well established for the localization of HRP in fixed tissue (Graham and Karnovsky, 1966). This same principle has been used by Courtoy *et al.* (1984), to distinquish non-fixed HRP-containing endosomes from cellular organelles of a similar size and equilibrium density. They made use of a conjugate of asialoglycoprotein and HRP to specifically label rat liver endosomes, which are involved in the receptor mediated uptake of asialoglycoproteins, with peroxidase activity. Later, we applied this technique after labeling tissue culture cells with asialoglycoprotein-HRP and transferrin-HRP conjugates, as well as fluid phase-endocytosed HRP (Stoorvogel *et al.* 1987; 1988; 1989; Geuze *et al.* 1988). Two major effects of DAB cytochemistry were observed. 1. Intravesicular DAB polymer is formed, and trapped within the vesicle. Due to the high density of the DAB-polymer, HRP containing vesicles are recovered at a much higher equilibrium density in a density gradient following centrifugation than non-HRP-containing microsomes. At increasing DAB concentrations this density shift becomes more pronounced (fig 1). 2. After DAB cytochemistry, proteins present within the HRP-containing compartment can no longer be extracted in a soluble form after lysis of the microsome. Encapsulation of proteins by DAB polymer as well as chemical cross-linking may play a role in this effect. The latter possibility is probably the major factor, since optimal protein cross-linking occurs at a DAB concentration (below 100 μg/ml, fig 2) at which little DAB polymer is formed as indicated by a minor density shift (see fig 1). The nature of this bond has not been defined as yet, however, aromatic residues of (glyco)proteins appear to substitute for DAB as reducing agents. Studies of receptor/ligand pathways based on these two powerful approaches are described below.

NATO ASI Series, Vol. H 62
Endocytosis
Edited by P. J. Courtoy

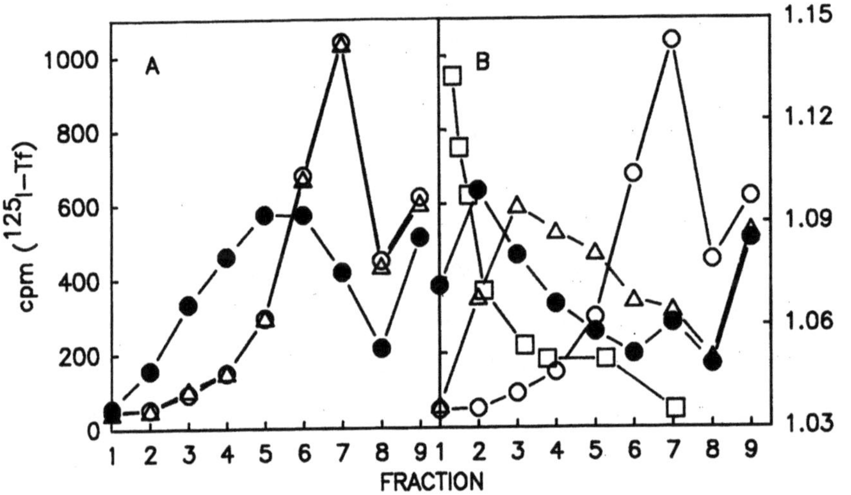

Figure 1. Density shift of HRP-containing endosomes. Human hepatoma HepG2 cells were depleted of exogenous transferrin by incubating the cells in serum free medium. Next, the cells were incubated at 37°C in medium containing both 25 μg/ml transferrin-HRP and 5 μg/ml ^{125}I-transferrin. Both ligands were endocytosed by the transferrin receptor. After uptake, and removal of cell surface-bound ligand by washing the cells sequentially at pH 5 and neutral pH, the cells were homogenized. Samples of the post-nuclear supernatant were incubated at room temperature without DAB or H_2O_2 (A and B, open circles), or with 30 μg/ml DAB only (A, triangles) or 1 mg/ml DAB only (B,triangles), or with both DAB and 0.02% H_2O_2 (A and B, closed circles). Finally the homogenates were fractionated on a Percoll gradient. At 1 mg/ml DAB a minor density shift is evident if no H_2O_2 was added, which is probably due to endogenous H_2O_2. The density shift of transferrin/HRP and ^{125}I-transferrin-containing endosomes is much greater after treatment with 1 mg/ml DAB and H_2O_2 than with 30 μg/ml DAB and H_2O_2. Density distribution in the gradient (squares).

Transferrin (Tf) is an ideal marker to study receptor recycling, since it follows the same route as its receptor during the entire recycling pathway. Tf contains two homologous domains each of which can bind a single Fe^{3+} ion with high affinity. Iron saturated Tf binds to TfR at neutral pH at the plasma membrane. After uptake via coated pits and coated vesicles, Tf loses its affinity for Fe^{3+} due to the acidic endosomal environment whereas the apotransferrin remains receptor bound because of its high affinity for the receptor at low pH (Klausner *et al.* 1983; Dautry Varsat *et al.* 1983). Apotransferrin remains TfR-bound during recycling but loses its affinity when exposed to the neutral pH of the extracellular medium and dissociates. Another ligand-receptor system in hepatic parenchymal cells is that for asialoglycoproteins (ASGP) (Ashwell and Morell, 1974). In contrast to Tf, ASGP dissociates from its receptor (ASGPR) in endosomes, and is

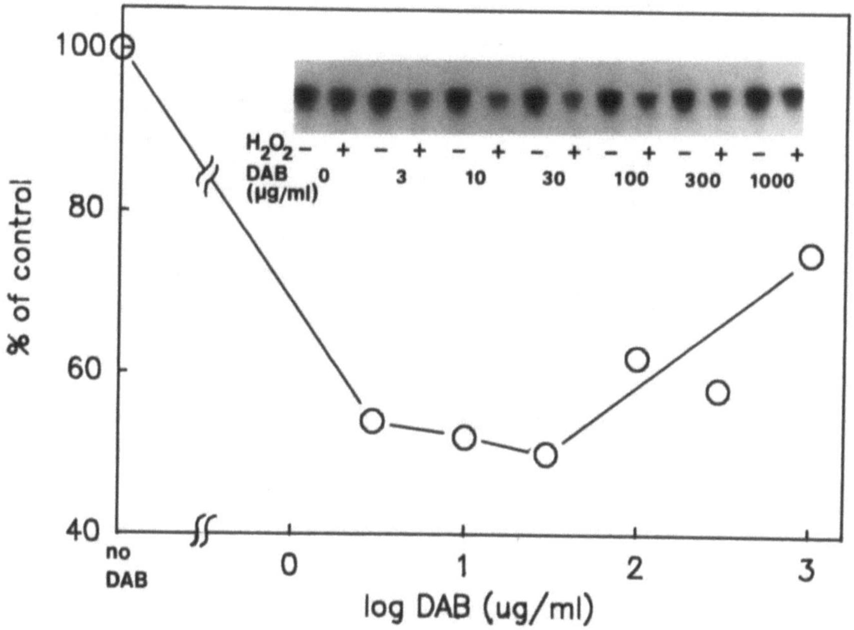

Figure 2. HRP mediated cross-linking of an endosomal protein (asialoglycoprotein receptor). HepG2 cells were depleted from exogenous transferrin by incubating the cells in serum-free medium. Next, cell surface proteins were labeled with ^{125}I by lactoperoxidase at 0°C. The cells were allowed to endocytose labeled receptors at 37°C in the presence of both 25 µg transferrin/HRP and 30 µg non-conjugated transferrin in the medium. Since transferrin is competitive with transferrin/HRP for transferrin receptor binding, suboptimal intravesicular HRP concentrations with respect to protein cross-linking were obtained. Labeled receptors which remained at the cell surface after this incubation were removed with proteinase K at 0°C. Next, the cells were homogenized. Pairs of samples of the postnuclear supernatant were incubated at room temperature in the presence of different amounts of DAB. To one sample of each pair, H_2O_2 was also added. In the presence of H_2O_2, proteins within transferrin/HRP containing endosomes were cross-linked to different extents, depending on the DAB concentration. Non cross-linked ^{125}I-ASGP receptors were visualized by immunoprecipitation and SDS-PAGE (top), and expressed as the percentage of total ^{125}I-asialoglycoprotein receptor (extracted from the sample lacking H_2O_2) (bottom). An optimal cross-linking of ^{125}I-asialoglycoprotein receptor was obtained at DAB concentrations of 30 - 300 µg/ml.

directed to lysosomes for degradation. Therefore, ASGP is a excellent marker for the degradative pathway.

We have used human hepatoma (HepG2) cells, and applied the DAB-induced density shift and cross-link principles to address questions related to the interactions between endocytic pathways. A model developed to summarize these interactions (Fig. 3) is described in the following paragraphs.

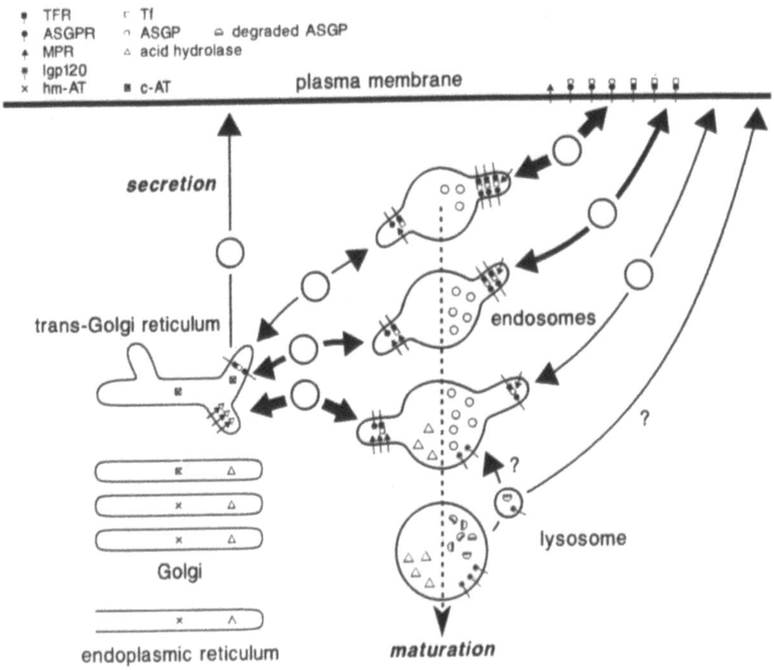

Figure 3. Interactions between endocytic pathways. The figure is explained in the text.

Simultaneously endocytosed Tf and ASGP are sorted intracellularly within minutes after uptake (Stoorvogel *et al.* 1987). This sorting process can be inhibited with primaquine, consistent with the notion that an acidic endosomal environment is a prerequisite for sorting. Endocytosed ASGPR reacycles to the plasma membrane within the same vesicles as TfR (Stoorvogel *et al.* 1989).

During intracellular trafficking, at least some of the TfR is transported to the secretory pathway (Stoorvogel *et al.* 1988). Among other secretory glycoproteins, α1-antitrypsin (AT) is converted from a high mannose (hm-AT) to a complex glycosylated form (c-AT) within the Golgi stack. Thereafter it is transported to the trans-Golgi reticulum (TGR) where it contacts recycling TfR. In the present model, however, most TfR's bypass the TGR and recycle directly to the plasma membrane (see below). Because endocytosed ASGPR is not sorted from the TfR (Stoorvogel *et al* 1989), it is conceivable that some recycling ASGPR's also pass through the TGR.

The cation-independent mannose 6-phosphate receptor (MPR) mainly shuttles directly between the TGR and late endosomes. However, a small amount of MPR is present on the plasma membrane, and in equilibrium with these major MPR pools. MPR at the plasma membrane is endocytosed via the same pathway as the TfR and ASGPR, but sorted intracellularly from these receptors (Stoorvogel *et al.* 1989). A considerable amount of the total intracellular MPR pool is accessible to both endocytosed Tf/HRP (in HepG2 cells, Stoorvogel *et al.* 1989) and fluid phase endocytosed HRP (in H4S cells, Geuze *et al.* 1988). In both cases MPR became sensitive to cross-linking with identical kinetics as compared to intracellular TfR. Optimal amounts were cross-linked between 5 and 10 min following uptake of HRP. However, within the endocytic pathway, the main localization of MPR and TfR is late endosomes (Geuze *et al.* 1988) and early endosomes, respectively. This strongly suggests that some endocytic vesicles bypass early endosomes and fuse directly with late endosomes. In such a model, plasma membrane-derived receptor-containing vesicles fuse with all, but preferably with early, endosomes. Similarly, early endosome-derived vesicles fuse preferably with the plasma membrane. Endosomes mature during this fusion and fission process, during which they gradually decrease their interaction with the plasma membrane. Simultaneously, the intensity of membrane exchange with the TGR is increased (Geuze *et al.* 1988). TfR and ASGPR preferentially exit from endosomes immediately after arrival therein (Stoorvogel *et al.* 1987), and consequently shuttle mainly between the plasma membrane and early endosomes. MPR's, however, preferentially exit from late endosomes. Thus, endocytosed MPR may remain in contact with recycling TfR by means of a "slipping coupling" mechanism: endocytosed MPR is sorted from TfR after as long as 20-45 min of uptake (Stoorvogel *et al.* 1989) by sorting from the TfR recycling pathway rather than from simultaneously endocytosed TfR.

The lysosomal glycoprotein lgp120 is mainly localized within late endosomes and lysosomes. After synthesis, lgp120 is probably transported via late endosomes to lysosomes. About half of the total lgp120 is located within endosomes (Geuze *et al.* 1988). Since lgp120 has a considerable half life, this large pool cannot be composed exclusively of newly synthesized lgp120 on transport to a final destination in lysosomes. Therefore, it seems likely that lgp120 transport to lysosomes is not unidirectional. Lgp120 transport from lysosomes to endosomes, either directly or via the plasma membrane, may induce the high lgp120 concentration in late endosomes. MPR and lgp120 are sorted within late endosomes, the former is transported to the TGR and the latter remains in the degradative pathway until late endosomes mature into lysosomes.

In summary, these studies revealed at least four distinct sites for sorting between integral membrane proteins in the endocytic pathway: 1) specific receptor uptake at the plasma membrane, 2) sorting of ASGPR and TfR from MPR during the maturation of early endosomes into late endosomes, 3) sorting between MPR and lgp120 in late endosomes, and 4) sorting of MPR from the secretory pathway in TGR. In addition, a connection between the endocytic and the secretory pathway in the TGR was shown.

REFERENCES:

Ashwell G, and Morell AG (1974). The role of surface carbohydrates in the hepatic recognition and transport of circulating glycoproteins. Adv Enzymol 41, 99-128.

Courtoy PJ, Quintart J, and Baudhuin P (1984). Shift of equilibrium density induced by 3,3'-diaminobenzidine cytochemistry: A new procedure for the analysis and purification of peroxidase-containing organelles. J Cell Biol 98, 870-876.

Dautry-Varsat A, Ciechanover A, and Lodish HF (1983). pH and the recycling of transferrin receptor during receptor mediated endocytosis. Proc Natl Acad Sci USA 80, 2258-2262.

Geuze HJ, Stoorvogel W, Strous GJ, Slot JW, Bleekemolen JE, and Mellman I (1988). Sorting of mannose 6-phosphate receptors and lysosomal membrane proteins in endocytic vesicles. J Cell Biol 107, 2491-2501.

Graham RC Jr., and Karnovsky MJ (1966). The early stages of absorbtion of injected horseradish peroxidase in the proximal tubules of mouse kidney: ultrastructural cytochemistry by a new technique. J Hist Chem 14, 291-302.

Klausner RD, Ashwell G, van Renswoude J, Harford JB, and Bridges KR (1983). Binding of apotransferrin to K562 cells: explanation of the transferrin cycle. Proc Natl Acad Sci USA 80, 2263-2266.

Stoorvogel W, Geuze HJ, and Strous GJ (1987). Sorting of endocytosed transferrin and asialoglycoprotein occurs immediately after internalization in HepG2 cells. J Cell Biol. 104, 1261-1268.

Stoorvogel W, Geuze HJ, Griffith JM, and Strous GJ (1988). The pathways of endocytosed transferrin and secretory protein are connected in the trans-Golgi reticulum. J Cell Biol 106, 1821-1829.

Stoorvogel W, Geuze HJ, Griffith JM, Schwartz AL, and Strous GJ (1989). Relations between the intracellular pathways of the receptors for transferrin, asialoglycoprotein, and mannose 6-phosphate in human hepatoma cells. J Cell Biol 108, 2137-2148.

INTRACELLULAR TRAFFIC OF THE GROWTH FACTOR INTERLEUKIN 2 AND ITS RECEPTORS IN HUMAN T LYMPHOCYTES.

M. Ferrer, A. Hémar, V. Duprez & A. Dautry-Varsat
Department of Immunology
Institut Pasteur
25, rue du Dr. Roux
75724 Paris Cedex 15 - France

INTRODUCTION

Interleukin 2 (IL2) is a polypeptide growth factor that stimulates proliferation of T lymphocytes after binding to specific glycoprotein receptors. High affinity receptors for IL2, Kd ≈ 10 to 100 pM, are made of at least two polypeptide chains α and ß of molecular weight 55kD and 70kD respectively (Sharon et al. 1986, Robb et al. 1987). Each of these polypeptides has a binding site for IL2, although they share no sequence homology. In fact, in most cases, the majority of IL2 receptors have a low affinity for IL2 (Kd ≈ 10nM) and are composed of the α chain by itself. High affinity receptors usually represent only about 10% of IL2 receptors. The ß chain alone can be found not associated to α in some specific cases, and forms a third type of receptors with an affinity of about 1nM. The high affinity receptors are responsible for transmitting the growth signal while low affinity receptors, α chain, are inactive in that respect (Smith 1988 and Smith 1989).

The expression of high affinity IL2 receptors is transient and is very tightly regulated. Resting peripheral T cells do not express IL2 receptors, and antigenic or mitogenic stimulation is required to induce receptor expression. Optimal receptor level is observed at the time the cells undergo maximum proliferation and after a few days the level of cell surface IL2 receptors declines.

Because this growth factor is quite specific for T lymphocytes, we have used it as a probe to study the receptor mediated endocytosis pathway in lymphocytes. An interesting application is that IL2 or its receptors might be modified in order to use this system as a vector to transport drugs to given intracellular compartments in T lymphocytes, which obviously might have clinical implications.

The human T lymphoma cell line IARC 301 was established in culture from the lymph nodes of a patient with a malignant T lymphoma and grows by an IL2

NATO ASI Series, Vol. H 62
Endocytosis
Edited by P. J. Courtoy

dependent autocrine mechanism (Duprez et al., 1985 ; Dautry-Varsat et al., 1988). It constitutively expresses about 3000 high affinity (αß) and 30 000 low affinity (α) IL2 receptors on its cell surface.

RESULTS AND DISCUSSION

Fate of IL2 and its high affinity receptors. After binding to its high affinity receptors, IL2 is rapidly internalized at 37°C, and it is subsequently degraded in acidic compartments (V. Duprez and A. Dautry-Varsat 1986, Fujii et al. 1986). Low affinity receptors (α chain) bind IL2 but do not allow internalization, or do so only at a very slow rate.

High affinity receptors have a short half-life on the cell surface, 1h, in the absence of IL2 (Fig. 1). This suggests that they are constantly internalized in the absence of ligand. In order to find out whether some receptors recycle to the cell membrane, high affinity receptors were stripped from the cell surface at 4°C in cells pretreated with cycloheximide, for 30 min at 37°C, to inhibit protein synthesis. Upon reincubating the cells at 37°C, no high affinity receptors reappeared on the cell surface (V. Duprez and A. Dautry-Varsat, 1986). It thus appears that in the absence of ligand, high affinity receptors are endocytosed without recycling to the cell surface.

To assess if IL2 could have an effect on receptor internalization, we incubated the cells in the presence of IL2 at concentrations such as to saturate high affinity receptors. This caused within 30 min to 1 h a down-regulation of surface receptors to about 40% of the initial value. This down-regulation is IL2 concentration dependent : it depends on the level of high affinity receptors occupancy, but not on that of low affinity receptors. It can be maintained for 48h if the high affinity receptors are occupied, but not if IL2 is removed (V. Duprez et al., 1988). This down regulation can be explained by a different rate of entry of high affinity receptors depending on whether they are occupied or not. As shown in Fig. 1, the half-life of receptors is clearly shortened from 1h to about 25 min as measured in the presence of cycloheximide. The underlying mechanism for that could be that IL2 might favor the interaction of IL2 receptors with some molecular structure, for instance coated pits, resulting in a faster rate of entry. The internalization rate measured without cycloheximide is quicker, t1/2 = 15 min. This is probably a better value than t1/2 = 25 min, because we have shown that cycloheximide slows down the internalization of IL2 by about a factor 1.5. It also slows down the endocytosis of transferrin bound to its receptors.

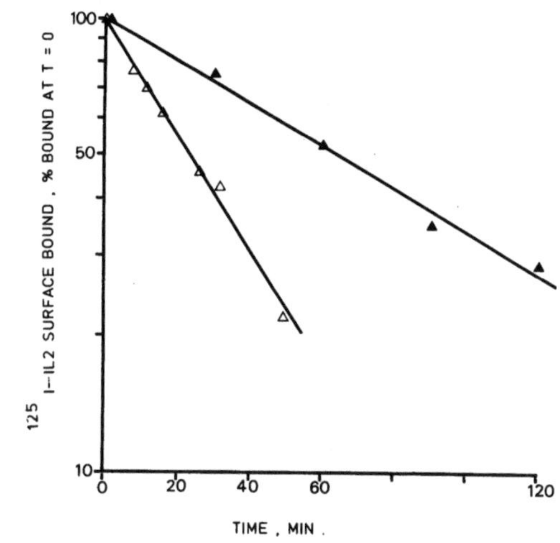

Fig. 1. Half-life of surface high-affinity receptors with or without IL2. The cells were preincubated with 0.5mM cycloheximide for 30 min at 37°C. At time t = 0, IL2 was added (△) or not (▲) and the number of surface receptors was measured at different times thereafter by ^{125}I-IL2 binding at 4°C, having removed unlabeled IL2 by acid stripping.

Fate of the IL2 receptor α chain. The stability of the receptor α chain was further studied by several methods. Under conditions of protein synthesis inhibition with cycloheximide, the half-life of the α chain was measured using a monoclonal antibody, IOT14 (Immunotech, France), by cytofluorimetry. For 6 h, no decrease of membrane α chains can be detected (Fig. 2). Other methods were also used to measure this half-life : for instance, surface iodination of proteins followed by a chase at 37°C, after which at different times the α chain was immunoprecipitated and quantified by densitometry after autoradiography (Fig. 2). For experimental reasons this value is probably underestimated and found to be about 20 h. Another approach was to make use of the immunosuppressive drug cyclosporin A which inhibits the transcription of the gene coding for α (A. Hémar and A. Dautry-Varsat, 1990). At times when transcription of this gene was inhibited and the level of its mRNA was very low, the decrease of α chain on the plasma membrane was measured by cytofluorimetry and was found to occur with a half-life of about 48 h. In a symmetrical experiment, IL2 was used to transiently increase α gene transcription which results in a rapid increase in surface α chain. After a few hours, when the gene got back to its basal level, the half-life of α surface receptors measured by cytofluorimetry was also 48 h.

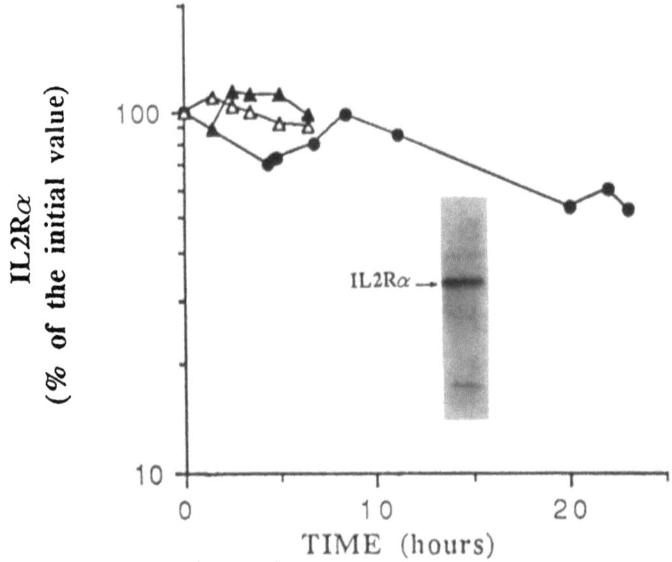

Fig. 2. Half-life of IL2Rα chain. Cell surface expression of IL2Rα on cells treated for different times with 0.5 mg/ml of the protein synthesis inhibitor cycloheximide with (▲) or without (△) 0.1 µg/ml CsA was assessed by cytofluorimetry. Determination by quantitative autoradiography of the amount of IL2Rα immunoprecipitated at different times of chase at 37°C after cell surface iodination (●). An autoradiograph is shown in the inset.

Altogether the half-life of α chains is long, about 40-48 h. 10% of the α chains are involved in interactions with ß chains to form high affinity receptors which are internalized with a half-time of 15 min in the presence of IL2 and slightly longer without IL2. If the α chain were internalized and degraded when it is part of the high affinity receptors, one would expect a half-life of about 3 to 6 h. To explain the results concerning α, there are two hypotheses : either α is not internalized with ß when involved in high affinity receptors formation, or if it is internalized, it recycles to the plasma membrane.

In order to answer this question, the intracellular localisation of α and ß was determined by confocal microscopy using anti-α and ß monoclonal antibodies. To make sure that the endocytic pathway was followed and not the secretion pathway, IL2 endocytosis was studied in the presence of cycloheximide to block protein synthesis, or of brefeldin A, a drug which disrupts the Golgi complex (Lippincott-Schwartz et al. 1990). This is an important point, first because the synthesis rate of high affinity receptors appears to be rapid and thus a large number of receptors might be located in the Golgi complex, second because in lymphocytes the cytoplasm is small and the intracellular organelles are packed together. When cycloheximide was present, the ß and α chains were concentrated in intracellular endocytic compartments after IL2 endocytosis. It is noteworthy that with cycloheximide, endocytosis was slowed down. As reported, brefeldin A disrupted the Golgi complex (Fig. 3). Under such conditions, the ß and α chains were also localized in intracellular endocytic compartments. This shows directly that indeed the α chain is internalized together with ß and IL2 in high affinity receptor complexes. Because the half-life of α is so long, it is therefore likely that α recycles after internalization.

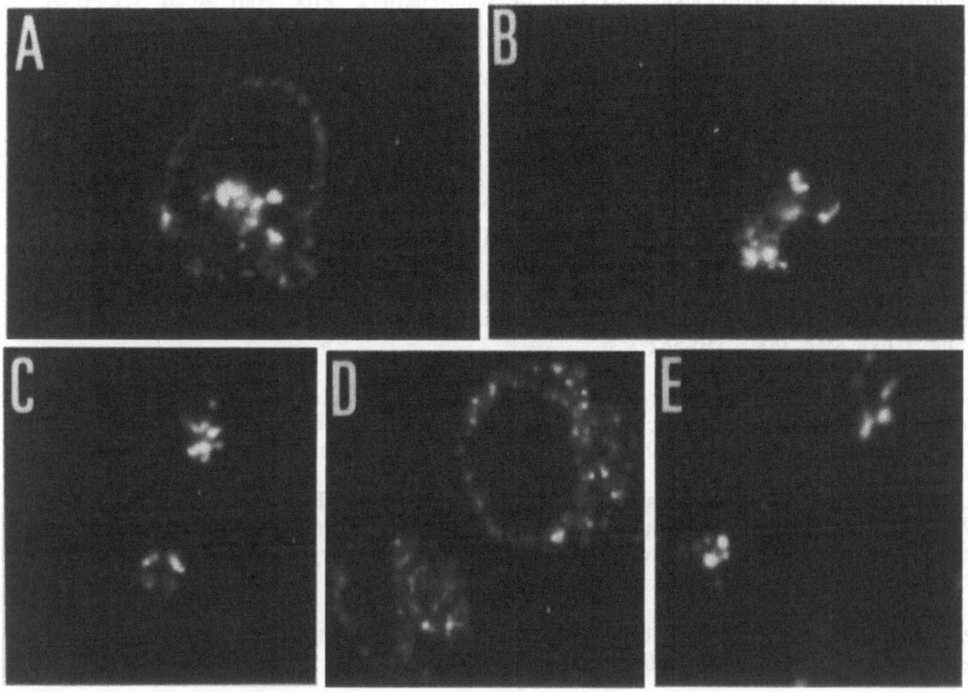

Fig. 3. Intracellular localization of α and ß chains. The cells were treated with 10μg/ml brefeldin A for 30 min at 37°C (A,B,D,E) and IL2 was added for the last 15 min (A,B). Immunofluorescence was performed on cells permeabilized with 0.05% saponin and the cells were observed in a Leitz confocal microscope. A : anti-α antibody (7G7B6, gift of Dr. Nelson). B : anti-ß antibody (Mik ß3, gift of Dr. Tsudo). C and D : anti-Golgi antibody (gift of Dr. Bornens) . E : endocytosis of fluoresceinated transferrin for 15 min at 37°C.

Finally, it can be concluded that IL2 is internalized together with the 2 components α and ß of its high affinity receptors. Sorting of these 3 molecules must occur intracellularly. IL2 probably ends up in lysosomes where it is degraded. A recent report shows that biotinylated IL2, followed by electron microscopy, seems to foolow the coated pits-coated vesicles pathway to end up in lysosomes (Peters and Norback 1990). Concerning the receptors, our data are in favor of the recycling of α to the plasma membrane and, although this has not yet been shown directly, a probable degradation of ß.

Literature references

Dautry-Varsat A, Hémar A, Cornet V & V Duprez (1988) Cyclosporin A inhibits the IL2 dependent autocrine growth of a human T cell line. Blood 72 : 588-592

Duprez V, Lenoir G & A Dautry-Varsat (1985) Autocrine growth stimulation of a tumoral human T lymphocyte cell line by Interleukin 2. Proc Natl Acad Sci. USA 82 : 6932-6936

Duprez V & A Dautry-Varsat (1986) Receptor mediated endocytosis of interleukin 2 in a human tumor T cell line : degradation of IL2 and evidence for the absence of recycling of IL2 receptors. J Biol Chem 261 : 15450-15455

Duprez V, Cornet V & A Dautry-Varsat (1988) Down-regulation of high affinity interleukin 2 receptors in a human tumor T cell line : IL2 increases the rate of surface receptor decay. J Biol Chem 263 : 12860-12865

Fujii M, Sugamura K, Sano K, Nakai M, Sugita K & Y Hinuma (1986) High-affinity receptor-mediated internalization and degradation of interleukin 2 in human T cells. J Exp Med 163 : 550-562

Hémar A & A Dautry-Varsat (1990) Cyclosporin A inhibits the interleukin 2 receptor α chain gene transcription but not its cell surface expression : the α chain stability can explain this discrepancy. Eur J Immunol (in the pres)

Lippincott-Schwartz J, Donaldson JD, Schweizer A, Berger EG, Hauri HP, Yuan L & RD Klausner (1990) Microtubule-dependent retrograde transport of proteins into the ER in the presence of brefeldin A suggests an ER recycling pathway. Cell 60 : 821-836

Peters DK & DH Norback (1990) Binding and internalization of biotinylated interleukin 2 in human lymphocytes. Blood 76 : 97-104

Robb RJ, Rusk CM, Yodoi J & WC Greene (1987) Interleukin 2 binding molecule distinct from the Tac protein : analysis of its role in formation of high affinity receptors. Proc Natl Acad Sci USA 84 : 2002

Sharon M, Klausner RD, Cullen BR, Chizzonite R & WJ Leonard (1986) Novel interleukin 2 receptor subunit detected by cross-linking under high-affinity conditions. Science 234 : 859-863

Smith KA (1988) Interleukin 2 : inception, impact, and implications. Science 240 : 1169-1176

Smith KA (1989) The interleukin 2 receptor. Annu Rev Cell Biol 5 : 397-425

TWO RECYCLING PATHWAYS OF EPIDERMAL GROWTH FACTOR-RECEPTOR COMPLEXES IN A431 CELLS

Alexander Sorkin*, Elena Kornilova and Sergey Krolenko
Institute of Cytology, Academy of Sciences of the USSR
Tichoretsky pr.4, Leningrad 194064
U.S.S.R.

We have studied the intracellular sorting of epidermal-growth-factor-receptor complexes (EGF-RC) in human epidermoid carcinoma A431 cells.

To examine the dynamics of ^{125}I-EGF recycling and degradation, A431 cells were incubated with 20-40 ng/ml of ^{125}I-EGF at 2°C for 1 h, and then for 2-15 min at 37°C without ^{125}I-EGF to allow endocytosis of the ligand. The cells were treated with 0.2 M sodium acetate buffer (pH 4.5) containing 0.5M NaCl at 2°C for 2 and 0.5 min, successively, to remove more than 90% of the surface-bound labeled EGF. Cells subjected to the above protocol are referred to as "^{125}I-EGF-loaded cells", and were used as the starting point in most experiments. These cells have a minimal amount of ^{125}I-EGF-RC on the surface and a relatively large pool of internalized ^{125}I-EGF-RC (Sorkin *et at* 1989).

The cells loaded with ^{125}I-EGF for 15 min were incubated at 37°C or 18°C in the presence of unlabeled EGF (Fig.1). At 37°C rapid accumulation of ^{125}I-EGF into the medium (Fig.1 *a*) and a corresponding decrease of the amount of intracellular ^{125}I-EGF (Fig.1 *b*) was observed. Similar rapid outflow of internalized ligand was measured after brief (2-5 min) cell loading at 37°.

If the chase of ^{125}I-EGF-loaded cells was performed at 18°C, the rate of accumulation of free ^{125}I-EGF in the medium was ≈3 times lower than at 37°C and became negligible after 2 - 3 h of continuous incubation (Fig.1 *a*). The total amount of ^{125}I-EGF recycled to the medium during 3 h at 18°C was half that at 37°C. However, when the cells exposed at 18°C for 3 h (referred to as "18°C-exposed, ^{125}I-EGF-loaded cells") were chased in fresh medium containing unlabeled EGF at 37°C, an additional outflow of intracellular ^{125}I-EGF was measured during the second chase (Fig.1 *a*). The apparent initial rate of ^{125}I-EGF outflow from 18°C-exposed, ^{125}I-EGF-loaded cells at 37°C was 2-3 times

* - Present address: Department of Biochemistry, Vanderbilt University School of Medicine, Nashville 37232-0146 Tennessee

NATO ASI Series, Vol. H 62
Endocytosis
Edited by P. J. Courtoy

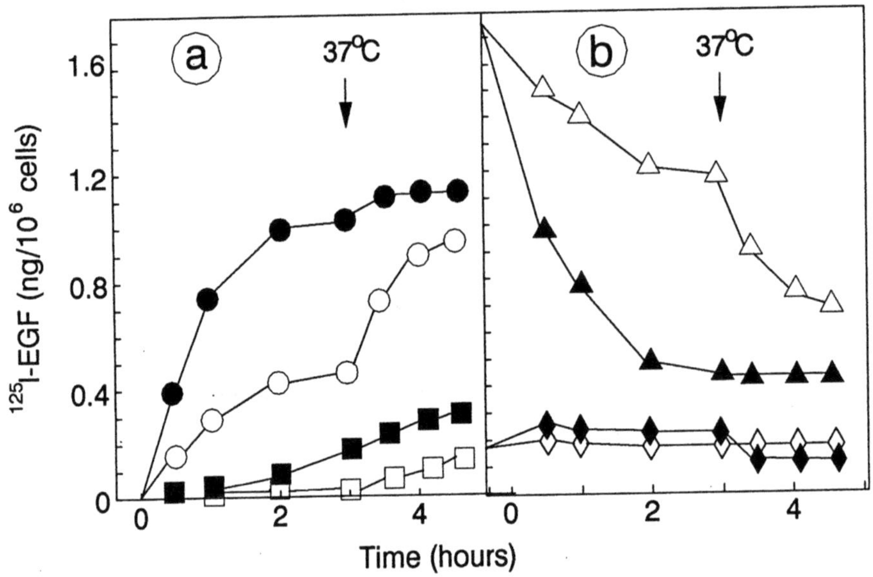

Figure 1. A431 cells were loaded with 20 ng/ml ^{125}I-EGF for 15 min and chased in the presence of excess unlabeled EGF at 37°C (closed symbols) or 18°C (open symbols). After a 3 h incubation, fresh medium was applied (arrow) and the cells were incubated for a second chase time at 37°C. At various time points, the medium (A) was assayed for intact ^{125}I-EGF (circles) and low molecular weight products of its degradation (squares), and the cells (B) were assayed for the surface-bound (rhombus) and intracellular ^{125}I-EGF (triangles).

lower than that from ^{125}I-EGF-loaded cells not-exposed to a 18°C block. These data allow us to propose that 25-30% of internalized ^{125}I-EGF is recycled via both a slower ("long-time") pathway that includes passage through a highly temperature-sensitive step and a rapid ("short-time") pathway that works at 18°C.

The $t_{1/2}$ for ^{125}I-EGF long-time recycling in 18°C-exposed, ^{125}I-EGF-loaded cells was found to be ≈20 min at 37°C. Long-time recycling would contribute only insignificantly to the initial linear rate of ^{125}I-EGF outflow measured during early stages of endocytosis. Therefore, based on the assumption that about half of the total pool of recycled EGF uses the long-time cycle (Fig.1), a $t_{1/2}$ of 5-7 min for short-time ^{125}I-EGF recycling at 37°C was calculated.

About 40-50% of internalized ^{125}I-EGF fails to be recycled and undergoes gradual degradation. The degradation of internalized ^{125}I-EGF was blocked at 18°C but was initiated with $t_{1/2}$≈2 h after the 18°C-exposed cells were warmed to 37°C (Fig.1 *a*).

The data obtained in our previous report (Sorkin *et al* 1988) by using mild treatment of the cells with Brij-58 indicated that EGF-RC do not dissociate significantly within the endosomal compartment in A431 cells. Electron microscopic studies (McKanna *et al* 1979; Carpentier *et al* 1987) also indicate that ligand remains associated with endosomal membrane in these cells. Here, using Brij treatment method, we found that at least 90-95% of ^{125}I-EGF does not dissociate from receptor intracellularly during short and long-time recycling in ^{125}I-EGF-loaded cells (data not shown).

Analysis of the dynamics of ^{125}I-EGF short-time recycling showed that the bulk of ^{125}I-EGF recycles as ^{125}I-EGF-RC and that the initial rate of outflow of ^{125}I-EGF-RC is higher than the rate of ^{125}I-EGF dissociation from the cell surface (Sorkin *et al.*,1989). The rate of long-time recycling appears to be similar or slightly slower than the rate of dissociation, since dissociation was compensated for by the insertion of recycled ^{125}I-EGF-RC into the cell surface (Fig.1 *b*). However, it was not possible to estimate from the kinetic data how much ^{125}I-EGF is recycled by long-time pathway as ^{125}I-EGF-RC.

Therefore, in order to directly demonstrate the long-time recycling of EGF-RC, EGF receptors were covalently labeled with photoreactivatable derivative of ^{125}I-EGF (ANBS-^{125}I-EGF) prepared using N-(5-azido-2-nitrobenzoyl-oxy)-succinimide ester. The 18°C-exposed, ANBS-^{125}I-EGF-loaded cells were obtained as described above for ^{125}I-EGF. UV-irradiation of these cells resulted in covalent linking of ANBS-^{125}I-EGF to receptors with an efficiency of 14-16%. All procedure steps before irradiation were carried out under weak red light.

The cells containing covalently-linked ^{125}I-EGF-RC were then incubated with unlabeled EGF for a chase time at 37°C. At the end of each chase time, the surface covalently-linked ^{125}I-EGF-RC were separated from the intracellular complexes by the cell surface immunoprecipitation method. The cells were incubated for 1 h at 2°C containing polyclonal antiserum to EGF receptor diluted to a saturating concentration, solubilized, and subjected to immunoprecipitation and electrophoresis.

As seen in Fig.2 *a* the intensity of the EGF-RC band immunoprecipitated from the cell surface was minimal in 18°C-exposed cells and increased during the chase incubation at 37°C. Intracellular receptors could be recovered from the supernatant of the cell surface immunoprecipitation. As expected, the intensity of the band corresponding to intracellular EGF-RC decreased with

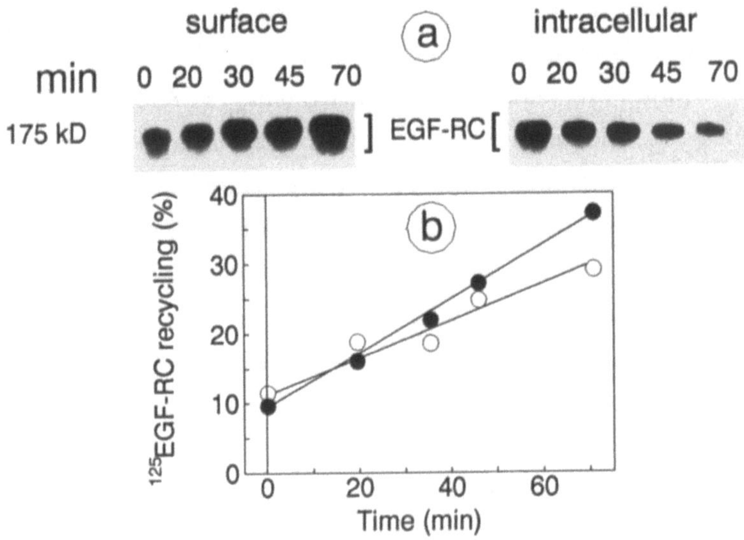

Figure 2. Cells were loaded with 40 ng/ml of ANBS-^{125}I-EGF, incubated with unlabeled EGF for 3 h at 18°C, UV irradiated, and chased at 37°C with unlabeled EGF. At the end of the chase times surface ^{125}I-EGF-RC were immunoprecipitated, and the immunoprecipitates and supernatants of the immunoprecipitates were subjected to electrophoresis. Panel A - autoradiogram. Panel B - The recycling of covalently-linked ^{125}I-EGF-RC (open circles) is expressed as a ratio of the surface to the total amount of the covalently-linked ^{125}I-EGF-RC associated with cells at each time point during a 37°C chase. The recycling of non-covalently bound ^{125}I-EGF from 18°C-exposed, ^{125}I-EGF-loaded cells (closed circles) is expressed as a ratio of the sum of free medium and surface-bound ^{125}I-EGF relative to the total amount of ^{125}I-EGF associated with cells and medium at each time point.

chase time. These data indicate that covalently-linked ^{125}I-EGF-RC were recycled via the long-time pathway.

The rate of reappearance of covalently-linked ^{125}I-EGF-RC on the cell surface was slightly lower than the rate of not covalently linked ^{125}I-EGF recycling (Fig.2 *b*). However, the data in control experiments suggest that approximately half of the ^{125}I-EGF-RC present at the cell surface during chase incubation of 18°C-exposed cells at 37°C can be internalized. Based on this estimation the rate of long-time recycling of covalently-linked ^{125}I-EGF-RC would be approximately 1.5-times higher than the apparent rate of reappearance of the complexes on the cell surface and therefore, similar to the rate of long-time recycling of not covalently linked EGF-RC that can dissociate on the cell surface.

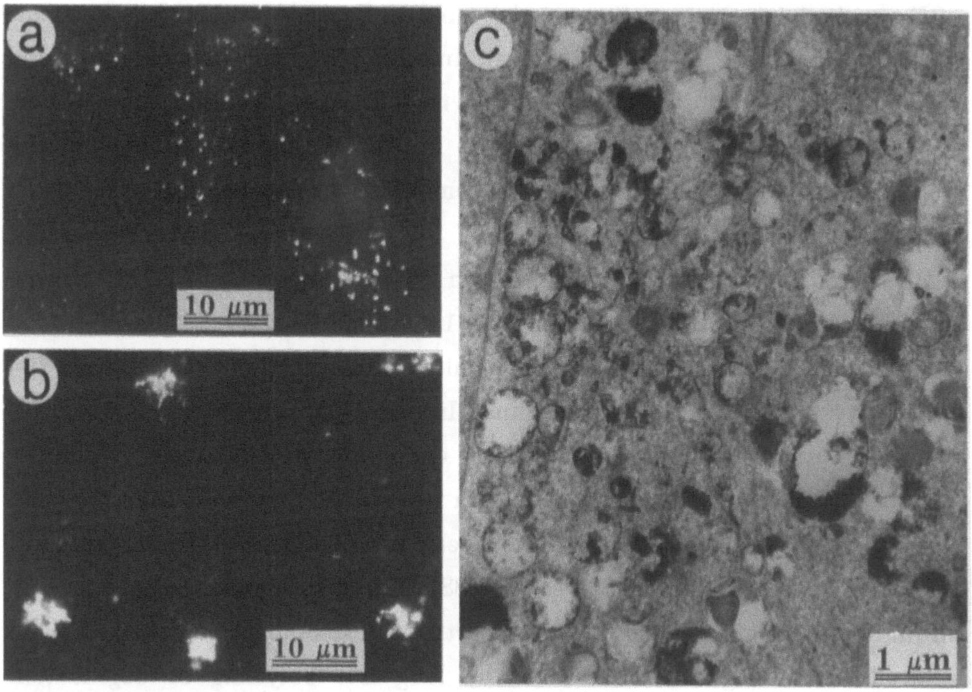

Fig.3. Cells were loaded with 50 ng/ml of PE-EGF for 5 min (a) or 15 min followed by incubation with excess unlabeled EGF at 18°C for 3 h (b). In *c*, cells were loaded with 100 ng/ml of Per-EGF and incubated with EGF at 18°C for 3h.

Taken together, the data allow us to propose that EGF remains bound to the receptor during routing to the recycling and degradative pathways. A significant dissociation of EGF-RC appears to occur in the late stages of intracellular processing, within the lysosomal compartment. In our experiments short-time recycling could be observed in cells allowed to internalize receptors during a brief (2 - 5 min) exposure to 37°C. EGF and EGF receptors have been demonstrated to localize in "early" endosomes after similar incubations (Miller *et al.* 1986). We have also observed phycoerythrin conjugate of EGF (PE-EGF) to be distributed throughout the peripheral endosomal compartment after the A431 cells were loaded for 5 min at 37°C with PE-EGF (Fig.3 *a*).

Long-time recycling of EGF-RC was highly inhibited at 18°C. Degradation of EGF (Fig.1) and EGF receptor is also blocked at 16 - 20°C (Miller *et al* 1986) in A431 cells. In our experiments a 3 h incubation at 18°C of cells loaded with PE-EGF or peroxidase conjugate of EGF (Per-EGF) in the presence of unlabeled

EGF resulted in concentration of the label exclusively within pericentriolar multivesicular endosomes (Fig. 3 *b, c*). Initiation of long-time recycling and degradation by warming 18°C-exposed cells to 37°C caused re-distribution of Per-EGF from the limited membrane to the internal vesicles and lumen of the multivesicular endosomes as well as movement of the labeled endosomes from pericentriolar complex.

By Percoll (20%) gradient centrifugation ^{125}I-EGF was seen to be distributed between two peaks with $p_{av}=1.040$ and $p_{av}=1.050$ in 18°C-exposed, ^{125}I-EGF-loaded cells. Incubation of these cells at 37°C resulted in recycling of ^{125}I-EGF from the low-density endosomes to the medium. Interestingly, recycling marker, ^{125}I-transferrin, was found only in low-density endosomes in A431 cells.

Based on morphological and kinetic data we propose that short-time recycling is due to the bi-directional movement of peripheral endosomes demonstrated in living A431 cells (De Brabander *et al* 1988) and/or to a rapid sorting process within the peripheral endosomal compartment (Stoorvogel *et al* 1988). The intracellular sorting of EGF-RC to the long-time recycling or degradation pathway appears to occur within the pericentriolar multivesicular endosomes.

REFERENCES

Carpentier, J.-L., M. F. White, L. Orci, and R. C. Kahn. 1987. Direct visualization of the phosphorylated epidermal growth factor receptor during its internalization in A-431 cells. *J. Cell Biol.* 105:2751-2762.

De Brabander, M., C. R. Hopkins, R. Nuijdens, and H. Geerts. 1988. Dynamic behaviour of the transferrin receptor followed in living epidermoid carcinoma (A431) cells with nanovid microscopy. *Cell Motil. Cytoscel.* 9: 30

McKanna J. A., H. T. Haigler, and S. Cohen. 1979. Hormone receptor topology and dynamics: morphological analysis using ferritin-labeled epidermal growth factor. *Proc. Natl. Acad. Sci. USA.* 76: 5689-5693.

Miller, K., J. Beardmore, H. Kanety, J. Schlessinger, and C. R. Hopkins. 1986. Localization of the epidermal growth factor (EGF) receptor within the endosome of EGF-stimulated epidemoid carcinoma (A431) cells. *J. Cell Biol.* 102: 500-509.

Sorkin, A., E. Kornilova, L. Teslenko, A. Sorokin, and N. Nikolsky. 1989. Recycling of epidermal growth factor-receptor complexes in A431 cells. *Bioch. Biophys. Acta.* 1011: 88-96.

Sorkin A., L. Teslenko, and N. Nikolsky. 1988. The endocytosis of epidermal growth factor in A431 cells: a pH of microenvironment and the dynamics of receptor complexes dissociation. *Exp. Cell Res.* 175: 192-205.

Stoorvogel, W., H. J. Geuze, and G. J. Strous. 1987. Sorting of endocytosed transferrin and asialoglycoprotein occurs immediately after internalization in Hep G2 cells. *J. Cell Biol.* 104: 1261-1268.

OLIGOSACCHARIDE REPROCESSING OF PLASMA MEMBRANE GLYCOPROTEINS

R. Tauber, W. Kreisel+ and W. Reutter*
Institut für Klinische Chemie und Biochemie, Freie Universität Berlin, Spandauer Damm 130, D-1000 Berlin 33

INTRODUCTION

During internalization and recycling plasma membrane glycoproteins encounter intracellular compartments containing various glycosidases and glycosyltransferases, e.g. endosomes, elements of the Golgi complex and lysosomes. Since the oligosaccharide units of plasma membrane glycoproteins are potential substrates for these enzymes, the question arises as to whether the oligosaccharide chains of the glycoproteins may undergo processing by these enzymes during recycling or whether the oligosaccharides of plasma membrane glycoproteins once formed during biosynthesis are stable unmodifiable structures. We have addressed this question in three steps, studying plasma membrane glycoproteins of rat liver and cultured rat hepatocytes as a model system:

(1) As a measure of a possible removal of sugar residues from the oligosaccharides of plasma membrane glycoproteins, half-lives of the different sugar residues were measured and compared to the half-life of the protein backbone of the glycoproteins.
(2) In order to trace a possible transfer of sugar residues to the glycoproteins during membrane recycling, incorporation of sugar precursors into plasma membrane glycoproteins was analysed.

+Medizinische Universitätsklinik Freiburg, Hugstetter Str. 55, D-7800 Freiburg i.Br., *Institut für Molekularbiologie und Biochemie, Freie Universität Berlin, Arnimallee 22, D-1000 Berlin 33. DPP IV, dipeptidylaminopeptidase IV (EC 3.4.14.5)

NATO ASI Series, Vol. H 62
Endocytosis
Edited by P. J. Courtoy

(3) Having shown that plasma membrane glycoproteins indeed undergo further processing of their oligosaccharides by partial de- and reglycosylation, the influence of cell proliferation and malignant transformation was studied in a first attempt to characterize mechanisms of regulation of this process.

PARTIAL DEGLYCOSYLATION OF PLASMA MEMBRANE GLYCOPROTEINS

The majority of plasma membrane glycoproteins of rat hepatocytes resides within the plasma membrane with comparably long half-lives ranging from approximately 20 to 90 h (for review, Hare 1990). We studied the fate of the oligosaccharide chains during this period of the glycoprotein life-span, measuring the half-lives of the different sugar residues in the glycoproteins while residing in the plasma membrane (Kreisel et al. 1980, Tauber et al. 1983, Volk et al. 1983). Half-lives were determined in pulse-chase experiments after metabolic labelling of the glycoproteins with radiolabelled sugar precursors. The different sugar residues exhibited a distinct graduation of their half-lives that was related to the position of each of the sugars within the structure of complex N-linked oligosaccharides (Tab.1) Sugar residues bound in terminal or penultimate positions, L-fucose, N-acetylneuraminic acid and D-galactose had the shortest half-lives only 1/6 to 1/3 that of the protein backbone. Conversely, the core sugar D-mannose had a half-life similar to that of the protein backbone in most of the glycoproteins (gp60, DPPIV, gp140). In some glycoproteins, however, also D-mannose had a shorter half-life either in close agreement to that of the terminal sugars (gp160) or in between that of the protein and that of the terminal sugars (gp80,gp120).

Glyco- protein	Half-live (h) of L-(3H)FUC	(3H)NeuAc	D-(3H)Gal	D-(3H)Man	Polypeptide
gp60 (1)	21	33	n.d.	58	62
gp80 (1)	17	31	n.d.	38	85
DPP IV (1)	12.5	33	20	58	70
gp120 (1)	21	27	n.d.	51	88
gp140 (1)	16	29	n.d.	66	78
gp160 (1)	16	26	n.d.	26	52
ASGP-R (2)	12	n.d.	n.d.	n.d.	24

Table 1 Half-lives of sugar residues and protein moieties of plasma membrane glycoproteins of rat liver (1) and cultured rat hepatocytes (2). ASGP-R, asialoglycoprotein receptor. Data from Kreisel et al. 1980, Tauber et al. 1983, Volk et al. 1983.

Rapid loss of terminal sugars could also be shown in plasma membrane DPP IV of cultured rat hepatocytes using a different experimental approach (Kreisel et al. 1988). Subsequent to metabolic labelling with L-(3H)fucose, D-(3H)mannose or L-(35S)methionine DPP IV molecules exposed on the cell surface were tagged with a monovalent antibody. At different times of the chase period the immune complex was isolated by use of protein A-Sepharose. During the experiment the immune complex was stable as proven by the constant specific radioactivity of the DPP IV-antibody complex after labelling of DPP IV with L-(35S)methionine (Fig.1). Whereas D-(3H)mannose remained bound to DPP IV, L-(3H)fucose was rapidly lost from the glycoprotein in accordance with the results obtained in the plasma membrane of liver.

These data show, that during the life-span of plasma membrane glycoproteins the outer sugars of their oligosaccharide units, L-fucose, N-acetylneuraminic acid and D-galactose are preferentially split off from the glycoproteins. In the majority of the membrane glycoproteins studied so far, this partial deglycosylation is restricted to the outer sugars, while the core sugars D-mannose and N-acetyl-D-glucosamine are turning over together with the protein backbone. However, the latter is not a general rule, since in distinct glycoproteins, e.g. in gp60, partial deglycosylation may also include the core region of the oligosaccharides. Since preferential removal of terminal sugars from plasma membrane glycoproteins also occurrs after transfer of the glycoproteins from the plasma membrane of rat hepatoma cells to the plasma membrane of mouse fibroblasts (Baumann et al. 1983), terminal deglycosylation is presumedly an inherent property of a given plasma membrane glycoproteins and not a function of a particular cell type.

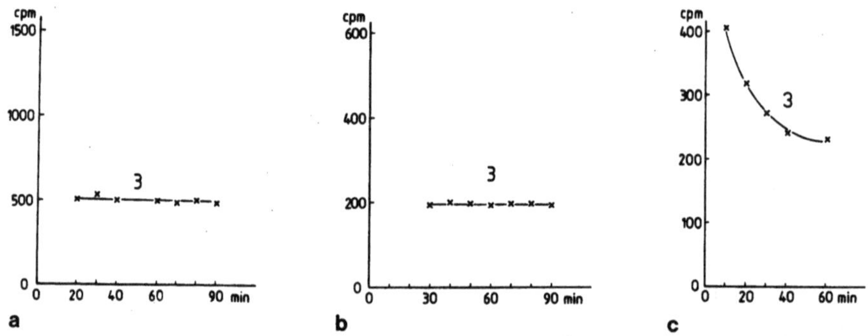

Fig. 1 Decay of specific radioactivity of plasma membrane DPP IV in primary cultures of rat heptocytes. Cells were pulse-chase labelled with L-(35S)methionine (a), D-(2-3H)mannose (b) of L-(6-3H)fucose (c) at 37° C and were then incubated with a monovalent anti-DPP IV antibody for 2 h at 4° C. After removal of unbound antibody by a washing procedure cells were rewarmed to 37° C and were further incubated for the times indicated in the presence of excess of unlabelled precursor. At the end of the incubation cells were extracted with Triton X-100/Tris buffer. The immune complex was bound to protein A-Sepharose and counted for radioactivity. Data from ref. 15.

The cellular compartments and the glycosidases involved in partial deglycosylation have not yet been identified. Chloroquine does not affect the loss of L-fucose from plasma membrane glycoproteins while significantly inhibiting degradation of their protein portion (Tauber et al. 1985). This indicates that defucosylation is an extralysosomal event. In order to estimate a possible relationship of terminal deglycosylation with membrane recycling, defucosylation was studied in the asialoglycoprotein receptor in cultured rat hepatocytes. Whereas the protein portion of the receptor was degraded with a half-life of 24 h consistent with reported data (Warren et Doyle 1981), L-fucose residues were lost from the receptor with a half-life of 12 h (Tab.1). Since the asialoglycoprotein receptor recycles with a half-life of approximately 10 min (Schwartz et al. 1984), the receptor is not routinely defucosylated during recycling.

REGLYCOSYLATION OF PLASMA MEMBRANE GLYCOPROTEINS

During recycling membrane glycoproteins may encounter sialyltransferases in the trans Golgi network as shown by use of glycoproteins experimentally rendered substrates for sialyltransferases by exogalactosylation (Duncan and Kornfeld 1988) or by desialylation (Snider and Rogers 1985) on the cell surface. However, it remained open to question if the transfer of terminal sugars to membrane glycoproteins during recycling is restricted to artificially modified glycoproteins or represents a physiologic pathway for glycosylation of plasma membrane glycoproteins subsequent to biosynthesis. Whether native cell surface glycoproteins may be reglycosylated during recycling was studied by example of DPP IV in cultured rat hepatocytes (Kreisel et al 1988). DPP IV molecules exposed on the cell surface were tagged with a monovalent antibody at 4 °C. After removal of excess antibody cells were rewarmed to 37 °C to allow DPP IV-antibody complexes to be internalized and to recycle. Thereafter de novo synthesis of N-linked oligosaccharides was blocked with tunicamycin and cells were incubated in the presence of L-(3H)fucose and of N-(3H)acetylmannosamine as a precursor of N-(3H)acetylneuraminic acid. Both L-fucose and N-acetylneuraminic acid were incorporated into DPP IV of the immune complex initially formed at the cell surface (not shown). Hence, plasma membrane glycoproteins may indeed be glycosylated also subsequent to biosynthesis at later stages of their life-cycle. If post-biosynthetic glycosylation serves as a repair mechanisms for terminally deglycosylated oligosaccharides or may also complete terminal processing of oligosaccharides initially synthesized with incomplete outer antennae is unknown. In conjunction with the turnover kinetics of the different sugar residues in the glycoproteins, the results support the concept that plasma membrane glycoproteins may undergo continuous reprocessing of the outer sugar residues of their oligosaccharide units by partial de- and reglycosylation. As shown by Snider and Rogers (1986), transferrin receptors glycosylated with N-linked oligosaccharides of the high mannose type may also recycle to cis Golgi cisternae where the high mannose oligosaccharides are trimmed by mannosidase I. However recycling of the transferrin receptor and of the two mannose-6-phosphate receptors through cis-Golgi elements and reprocessing of these glycoproteins by mannosidase I could not be shown (Neefjes et al. 1988), or was shown to occur at a very low rate (Duncan and Kornfeld 1988) by other laboratories and is a matter of controversy.

FACTORS INFLUENCING THE REPROCESSING OF PLASMA MEMBRANE GLYCOPROTEINS

In a first approach to characterize mechanisms regulating the rates and the extent of oligosaccharide reprocessing, the influence of cell proliferation and transformation was studied in regenerating liver and in hepatoma, respectively. Both the rate and the extent of partial deglycosylation differed significantly in the two tissues when compared to normal non-growing liver. In proliferating liver after partial hepatectomy the half-life of L-fucose was extended up to three-fold in DPP IV and in other plasma membrane glycoproteins, whereas half-lives of N-acetylneuraminic acid and D-mannose were affected only to a minor extent (Kreisel et al. 1984, Tauber et al. 1989). This indicates that in proliferating liver selectively the removal of L-fucose from plasma membrane glycoproteins is reduced. By contrast, in hepatoma both L-fucose and D-mannose had similar short half-lives in plasma membrane glycoproteins indicating that not only peripheral but also core sugars of the oligosaccharides are preferentially split off from the glycoproteins during the life-span of the protein backbone (Tauber et al. 1989). A number of mechanisms responsible for alterations of oligosaccharide reprocessing in proliferating and transformed cells may be proposed. These include shifts in the activity of glycosidases, different rates and routes of intracellular transport, and the formation of oligosaccharide structures with different sensitivity to specific glycosidases.

PERSPECTIVES

Many questions about oligosacchride reprocessing of plasma membrane glycoproteins are unanswered. Which glycosidases and glycosyltransferases are involved? Does reprocessing affect each of the different oligosaccharide units of a glycoprotein or is partial de- and reglycosylation restricted to distinct glycosylation sites? By analogy to oligosaccharide processing during biosynthesis, it is conceivable that the different oligosaccharide units of a glycoprotein may be reprocessed to a different extent determined by their steric accessability to glycosidases or glycosyltransferases. Is reprocessing specific for plasma membrane glycoproteins or do glycoproteins of other cellular compartments undergo similar modifications of their oligosaccharide moiety? Does oligosaccharide reprocessing represent an occasional loss of terminal sugars and subsequent repair by reglycosylation occuring during membrane recycling or does it have a biological function? Assuming that reprocessing occurs in the course of internalization and recycling, it is conceivable that de- and reglycosylation might provide signals involved in membrane trafficking. Moreover, oligosaccharide reprocessing might confer structural alterations to membrane glycoproteins rendering the protein susceptible to enter the degradative pathway.

This work was supported by the Deutsche Forschungsgemeinschaft (Sonderforschungsbereiche 154 and 312, and Re 523/3-3).

REFERENCES

Baumann, H., Hou, E. & Jahreis, G.P. (1983) J. Cell Biol. 96, 139-150.
Duncan, J.R. & Kornfeld, S. (1988) J. Cell Biol. 106, 617-628.
Hare, J.F. (1990) Biochim. Biophys. Acta 1031, 71-90.
Kreisel, W., Volk, B.A., Büchsel, R. & Reutter, W. (1980) Proc. Natl. Acad. Sci. U.S.A. 77, 1828-1831.
Kreisel, W., Reutter, W. & Gerok, W. (1984) Eur. J. Biochem. 138, 435-438.
Kreisel, W., Hanski, C., Tran-Thi, T.-A., Katz, N., Decker, K. Reutter, W. & Gerok, W. (1988) J. Biol. Chem. 263, 11736-11742.
Neefjes, J.J., Verkerk, J.M.H., Broxterman, H.J.G., van der Marel, G.A., van Boom, J.H. & Ploegh, H.L. (1988) J. Cell Biol. 107, 79-87.
Schwartz, A.L., Bolognesi, A. & Fridovich, S.E. (1984) J. Cell Biol. 98,732-738.
Snider, M.D. & Rogers, O.C. (1985) J. Cell Biol. 100, 826-834.
Snider, M.D. & Rogers, O.C. (1986) J. Cell Biol. 103, 265-275.
Tauber, R., Park, C.S. & Reutter, W. (1983) Proc. Natl. Acad. Sci. U.S.A. 80. 4026-4029.
Tauber, R., Heinze, K. & Reutter, W. (1985) Eur. J. Cell. Biol. 39, 380-385.
Tauber, R., Kronenberger, Ch. & Reutter, W. (1989) Biol. Chem. Hoppe Seyler 370, 1221-1228.
Tauber, R., Park, C.S., Becker, A., Geyer, R. & Reutter, W. (1989) Eur. J. Biochem. 186, 55-62.
Volk, B., Kreisel, W., Köttgen, E., Gerok, W. & Reutter, W. (1983) FEBS Lett. 163, 150-152.
Warren, R. & Doyle, D. (1981) J. Biol. Chem. 256, 1346-1355.

1.7. BIOGENESIS OF LYSOSOMES

(Chairperson : Brian STORRIE)

The lysosome as an end compartment of three pathways, the biosynthetic, the endocytic and the autophagic, is a challenging problem of organelle assembly and maintenance.

Bernard Hoflack focussed on the receptor for mannose 6-phosphate/insulin-like growth factor II, which delivers most of newly synthesized hydrolases to lysosomes. In NRK cells and J774 macrophages, the accessibility of a small but detectable fraction of the receptor and of phosphorylated lysosomal hydrolases to lactoperoxidase internalized at 18°C indicates that the lysosomal biosynthetic pathway may reach an early endocytic compartment of these cells. He could isolate from MDBK cells, mannose 6-phosphate receptors that are phosphorylated at two highly conserved serine residues in the cytoplasmic tail and generate a similar processing in vitro by a casein kinase II, associated with the 47 kDa subunit of the HA1 adaptor complex. Phoshorylation is believed to occur after packaging in the Golgi-derived clathrin-coated vesicles and is proposed to serve as inactivation signal, preventing the re-association of HA1 after fusion with the pre-lysosomal compartment.

The study of the so-called I-cell disease, characterized by multiple lysosomal deficiencies, has revealed that some acid hydrolases can still be delivered to lysosomes, even when the mannose 6-phosphate recognition marker cannot be generated. This second mechanism was illustrated for the lysosomal acid phosphatase (LAP) by **Christophe Peters**, from the laboratory of **Kurt von Figura**. Human LAP, transfected in BHK21 cells, is synthesized as a transmembrane protein. After insertion at the cell surface, it is repeatedly internalized and recycled, before delivery to lysosomes. Final release in the lysosomal lumen involves the sequential action of a thiol-proteinase acting on the cytoplasmic domain of LAP and an aspartyl-proteinase acting on its luminal domain. A single tyrosine at the cytoplasmic tail is essential for endocytosis and the cytoplasmic tail of LAP is apparently sufficient when attached to a non-lysosomal protein to cause its targeting to lysosomes The fate of LAP may be representative for the delivery of several lysosomal membrane glycoproteins.

Miguel Vega, from the laboratory of **Ignatio Sandoval,** reported the deduced amino-acid sequence of rat lysosomal integral membrane proteins (LIMPs). Most LIMPs are made of a luminal domain comprising two homologous sequences, each stabilized by a disulfide bridge, a

NATO ASI Series, Vol. H 62
Endocytosis
Edited by P. J. Courtoy

membrane-spanning domain and a very short cytoplasmic tail containing one tyrosine residue. In contrast, LIMP II presumably includes two transmembrane domains, which could be linked by cysteine pairing and are stabilized by palmitylation. Data suggest a direct route from Golgi to lysosomes for LIMP II and III, but a pathway similar to that of LAP for LIMP I. The lysosomal delivery of all LIMPs does not involve mannose 6-phosphate recognition as it is not affected by tunicamycine. However, N-deglycosylation dramatically increases LIMP turn-over, suggesting a protective role of carbohydrates against lysosomal degradation.

Ger Strous reported on still another mode of membrane association in HepG2 cells. By selectively extracting the soluble forms of lysosomal enzymes at low concentrations of saponin, he found that cathepsin D and glucocerebrosidase are synthesized as soluble precursors and become membrane-associated after the Golgi. This intermediate of cathepsin D can only be partially extracted when mannose 6-phosphate is added to saponin. In tunicamycin-treated cells, the transient membrane association of cathepsin D and its efficient lysosomal targetting are preserved. After a final proteolytic cleavage, cathepsin D is released in the lysosomal lumen, but glucocerebrosidase remains membrane-bound. The nature of the acquired, N-glycosylation independent membrane interaction of these enzymes remains to be discovered.

LYSOSOMAL ENZYME TRANSPORT BETWEEN GOLGI AND ENDOSOMES

Thomas Ludwig, Stéphane Méresse and Bernard Hoflack
European Molecular Biology Laboratory, Postfach 10-2209,
D-6900 Heidelberg, FRG

INTRODUCTION

Newly synthesized lysosomal enzymes are efficiently sorted and properly targeted to lysosomes by a series of interactions involving different signals and receptors (for review, see Kornfeld and Mellman, 1989). The first recognition event occurs most likely in the Cis-Golgi and leads to the synthesis of the phosphomannosyl recognition marker. In the trans-Golgi network (TGN), this marker present on every soluble lysosomal enzyme is used as a high affinity ligand by specific receptors, thus segregating this class of proteins from those destined to be secreted by a constitutive or a regulated pathway. The ligand-receptor complexes which are formed, are then delivered via clathrin-coated vesicles to a prelysosomal/endosomal compartment where the lysosomal enzymes dissociate from their receptors. While the receptors recycle back to the Golgi, the lysosomal enzymes are packaged in lysosomes by a still poorly understood process.

Two mannose 6-phosphate receptors (MPRs) are known (for review, see Dahms et al, 1989): the cation-dependent MPR (CD-MPR) and the cation-independent MPR which also binds the growth factor IGFII. Both receptors are involved in lysosomal enzyme trafficking, each being able to bind one lysosomal enzyme molecule. It is still unknown why cells need two different receptors for apparently the same biological function. It is likely that this reflects unsuspected functions of these receptors.

While many events leading to the binding of lysosomal enzymes have been studied in details at the molecular level, several aspects of lysosomal enzyme trafficking are still unclear. For example, the station of the endocytic pathway (early or late) which receives the newly synthesized hydrolases from the TGN has still to be identified since the morphological studies on the distribution of their receptors have not

NATO ASI Series, Vol. H 62
Endocytosis
Edited by P. J. Courtoy

fully resolved this issue. In addition, the molecular events leading to the packaging of the receptors into transport vesicles coated with clathrin are not completely understood.

THE BIOSYNTHETIC PATHWAY CAN MEET THE ENDOCYTIC PATHWAY IN EARLY ENDOSOMES

Many electron microscopic studies have shown that at steady state, the bulk of the MPRs are concentrated in endosomal elements, little being found over the TGN and on the plasma membrane (Geuze et al., 1985; 1988; Griffiths et al., 1988). In many cell types, the bulk (90%) of the CI-MPR is localized in late endocytic structures (see G. Griffiths'chapter in this volume). In U937 monocytes, the CD-MPR exhibits a similar distribution (Bleekemolen et al., 1989). We have investigated the distribution of the newly synthesized lysosomal enzymes along the endocytic pathway of NRK cells and J774 macrophages (Ludwig et al, submitted). Lactoperoxydase internalized by fluid phase endocytosis was first used to iodinate the intra-endosomal components. The newly synthesized lysosomal enzymes present in endosomes were purified on CI-MPR affinity columns since they should remain phosphorylated for at least a brief period of time after delivery. When LPO was internalized in early endocytic structures (internalization of 5-15 min at 37°C or 60 min at 18°C), a significant fraction (20-30%) of the phosphorylated lysosomal enzymes present in the entire pathway were labeled together with 10% of the CI-MPR. Higher amounts of phosphorylated lysosomal enzymes and the bulk of the CI-MPR became iodinated when LPO had access to late endocytic structures (30-60 min at 37°C). For subsequent electron microscopic studies, soluble fragments of the CI-MPR purified from bovine serum were conjugated to gold particles. This probe specific for the phosphomannosyl marker of newly synthesized hydrolases was then used to identify the intracellular compartments containing these molecules. In NRK cells, CI-MPR negative structures accessible to endocytic markers internalized for 5-15 min at 37°C, were found to bind significant amounts of CI-MPR gold conjugates (20%). Late CI-MPR rich structures accessible to endocytic markers internalized for 30-60 min at 37°C, contained 75% of the conjugates. As expected, lysosomes were devoid of phosphorylated lysosomal enzymes. We believe that the presence of phosphorylated lysosomal enzymes in early endocytic structures does not reflect a secretion-recapture mechanism. The same distribution was found in NRK cells grown in the presence of Man 6-P or in J774 macrophages. This macrophage-like cell line contains only the CD-MPR which does not bind the ligand at the cell surface and therefore does not internalize exogenous phosphorylated enzymes (<2% of the amount internalized by CI-MPR positive NRK cells). This data would be consistent with a model whereby newly synthesized hydrolases are delivered, at least in part, from the TGN to early endocytic elements.

PHOSPHORYLATION OF THE CI-MPR CYTOPLASMIC DOMAIN BY A HAI-ASSOCIATED KINASE

One key step in the targeting of lysosomal enzymes is the clustering of their receptors into clathrin-coated vesicles which budd from the TGN. This step involves signals that are contained in their cytoplasmic domains. It is known that a mutant CI-MPR lacking either 40 or 80 amino acids at the carboxyl-terminus of its cytoplasmic domain is unable to target efficiently its bound-ligand to the lysosomes (Lobel et al., 1989). These signals appear to interact with the Golgi-specific HAI adaptor complex (Glickman et al., 1988), thus triggering the clustering of the CI-MPR into clathrin-coated vesicles. Several studies have suggested that phosphorylation of the CI-MPR could modulate its trafficking. For example, treatment of endothelial cells with IGFII induces a down regulation of the cell surface CI-MPR and also triggers its dephosphorylation. In contrast, addition of phorbol esters to the same cells promotes the recruitment of the CI-MPR at the cell surface and its phosphorylation by Protein kinase-C (Hu et al., 1990). On the other hand, clathrin-coated vesicles are known to contain kinase activities: a casein kinase (Bar-Zvi and Banton, 1986) and a Ca++, cyclic nucleotide-independent kinase (Pauloin et al., 1982) which appears to be associated with the 50 kD subunit of the plasma membrane HAII adaptor complex (Keen et al., 1987; Manfredi and Bazari, 1987). We have investigated whether these kinases could phosphorylate the CI-MPR and therefore be involved in the modulation of its trafficking (Meresse et al, 1990).

Coat proteins extracted from purified bovine brain clathrin-coated vesicles were mixed with purified bovine CI-MPR and γ-^{32}P-ATP. Analysis of the reaction products by SDS-PAGE followed by autoradiography indicated that, in addition of coat proteins, the CI-MPR was heavily phosphorylated. During its purification, it became rapidly clear that the kinase detected in our assay, was copurifying with and therefore could be associated with the Golgi-derived HAI adaptor complex. On a Sepharose 4 B gel filtration column which allows the separation of clathrin triskelions, assembly proteins (mw 300 kD) and other low molecular weight components, the kinase behaved as a 200-300 kD protein. On hydroxylapatite column which allows the separation of HAI from the plasma membrane-derived HAII complex (Pearse and Robinson, 1984), the kinase activity also copurified with HAI. After further purification by ion exchange chromatography, the kinase containing fractions exhibited on SDS-PAGE the typical pattern of HAI formed of two 100-110 kD subunits (γ and β subunits) associated to a 47 kD polypeptide and a 20 kD polypeptide (for review see Morris et al., 1989).

The HAI associated kinase was further characterized. It uses ATP or GTP as a substrate, is stimulated by poly-L-lysine and is inhibited by heparin. This would indicate that this enzyme is a type of casein kinase II. Indeed, it phosphorylates casein (Km 500μM) but has a higher affinity for the CI-MPR (Km 50-100 nM). We took advantage of the fact that the enzyme uses GTP as a substrate to further purify the kinase. When the complex was passed over a GTP-agarose column, 40% of the activity could be retained and eluted with either Tris or GTP. Analysis of the retained fractions by SDS-PAGE showed that the kinase which phosphorylates the CI-MPR in vitro had a molecular weight of 47,000, strongly suggesting that it is the 47 kD subunit of HAI. At this step of purification, we could estimate that this enzyme was more than 200,000 fold enriched over the starting homogenate.

We used different acceptor substrates to determine which domain of the CI-MPR was phosphorylated. The first substrate was the CI-MPR purified from bovine liver. This protein contains the extracytoplasmic, transmembrane and cytoplasmic domains. Another substrate was a soluble form of the receptor purified from bovine serum which only contains the extracytoplasmic part of the receptor. Finally, we used a fusion protein from E. coli containing the 163 amino acids of the cytoplasmic domain of bovine CI-MPR. When these various proteins were used as substrate in the phosphorylation assay, it was clear that only the proteins containing the cytoplasmic domain of the receptor became phosphorylated. Analysis of the phospho-amino acids indicated that only serine residues were phosphorylated by the HAI associated kinase. This properties would also distinguish this kinase from the typical casein kinase II purified from muscle which would equally phosphorylate serine and threonine residues.

The cytoplasmic tail of the CI-MR contains several domains which are conserved among bovine, rat and human (Lobel et al, 1988; MacDonald et al, 1988; Morgan et al, 1987). The bovine CI-MPR receptor tail contains 7 serine residues, two of which are present in highly conserved domains (Ser 2421 and Ser 2492 in the full lenght bovine CI-MPR precursor) and exhibit the consensus feature for being a substrate of casein kinase II. In order to determine precisely the phosphorylation sites of the HAI associated kinase, we first synthesized a series of synthetic peptides corresponding to the various regions of the bovine receptor tail. These synthetic peptides were then assayed as potential inhibitors of the phosphorylation of the CI-MPR. Only two peptides were inhibitors. They corresponded to the stretches of amino acids 2407-2432 and 2480-2499 of the full lenght bovine CI-MPR precursor. These peptides were competitive inhibitors and therefore could also be phosphorylated by the HAI associated kinase. Since these peptides were also designed to match the peptides generated by trypsin treatment of the CI-MPR, they could be used as standards to map the serines in the CI-MPR tail which were phosphorylated by the 47 kD subunit of the HAI adaptor complex. The CI-MPR purified from liver or its cytoplasmic domain were phosphorylated by the HAI associated kinase, purified by immunoprecipitation and digested with trypsin. Analysis of the tryptic fragments indicated that indeed the kinase phosphorylates in vitro the serine residues that are contained in the stretches of amino acids 2407-2432 and 2480-2499. Additional evidence that the serines 2421 and 2492 were the phosphorylation sites came from studies using synthetic peptides in which these residues were replaced by alanine residues. These peptides were no longer phosphorylated. However, we observed that they could still inhibit the phosphorylation of the CI-MPR.

The HAI associated kinase is most likely active in vivo. The CI-MPR was immunoprecipitated from extracts of MDBK cells labelled to equilibrium with inorganic phosphate. Its tryptic phosphopeptides were analyzed. The stretches of amino acids 2407-2432 and 2480-2499 were indeed the two major phosphopeptides. It is likely that these phosphorylated peptides originate from the bulk of the CI-MPR which, at steady state, is found in endosomes (Geuze et al, 1985; 1988; Griffiths et al, 1988). We analyzed the CI-MPR purified from bovine tissue by 2 D-gel. It had a pI of 4.7 which was shifted to a pI of 4.9 upon treatment with phosphatases . We are currently investigating this aspect in more details.

We do not know yet how phosphorylation of the CI-MPR can affect its trafficking. At present, we only want to make the correlation that the 47 kD subunit of HAI complex found in the Golgi-derived clathrin-coated vesicles is potentially able to modify serine residues contained in a domain of the CI-MPR tail which is required for efficient targeting of hydrolases to lysosomes (Lobel et al., 1989). In addition these serines are present in highly conserved regions of the CI-MPR tail which are typical target sites of casein kinase II type enzymes. Our current working model is that phosphorylation occurs after packaging of non phosphorylated receptors into clathrin-coated vesicles and interaction of the HAI complex with the Golgi membrane. This would also suggest that phosphorylated receptors are delivered to endosomes. One possible fonction of the phosphorylation, either acting by itselfs of by inducing changes in the conformation of the receptor tail, could be the inactivation of a sorting signal. If in living cells, the adaptor proteins interact directly with receptor tails as proposed by Glickman and collaborators (1989), phosphorylation could modulate these interactions. For example, if phosphorylation triggers dissociation, this could explain why the HAI complex can only interact with the receptor present in the Golgi and not with the very same receptor present in endosomes.

REFERENCES

Bar-Zvi, D. and Branton, D. (1986) J. Biol. Chem. 21, 4408-4415

Bleekemolen, J., Stein, M., von Figura, K., Slot, J. and Geuze (1989) Eur. J. Cell Biol. 47, 366-372

Dahms, N., Lobel, P. and Kornfeld, S. (1989) J. Biol. Chem. 264, 12115-12118
Geuze, H., Slot, H., Strous, G., Hasilik, A. and von Figura (1985) J. Cell Biol. 101, 2253-2262
Geuze, H., Stoorvogel, G., Slot, J. Zijderhand-Bleekemolen and Mellnan, I. (1988) J. Cell Biol. 107, 2491-2502
Glickman, J., Conibear, E. and Pearse, B. (1989) EMBO J. 8, 1041-1047
Griffiths, G., Hoflack, B., Simons, K., Mellman, I. and Kornfeld, S. (1988) Cell 52, 329-341
Hu, K., Backer, J., Sahagian, G., Feener, E. and King, G. (1990) J. Biol. Chem. 265, 14864-13870
Kornfeld, S. and Mellman. I. (1989) Ann. Rev. Cell Biol. 5, 483-525
Keen, J., Chestnut, M. and Beck, K. (1987) J. Biol. Chem. 262, 3864-3871
Lobel, P., Fujimoto, K., Ye, R., Griffiths, G. and Kornfeld, S. (1989) Cell 57, 787-796
Lobel, P., Dahms, N. and Kornfeld, S (1988) J. Biol. Chem. 263, 2563-2570
Ludwig, T., Griffiths, G. and Hoflack, B. submitted
MacDonald, R., Pfeffer, S., Coussens, L., Tepper, M., Brocklebank, C., Mole, J., Anderson, J., Chen, E., Czeck, M. and Ullrich, A. (1988) Science 239, 1134-1137
Manfredi, J. and Barazi, W. (1987) J. Biol. Chem. 262, 12182-12188
Meresse, S., Ludwig, T., Frank, R. and Hoflack, B. (1990) J. Biol. Chem. in press
Morgan, D., Edman, J., Standring, D., Fried, V., Smith, M., Roth, R. and Rutter, W. (1987) Nature 329, 301-307
Morris, S., Ahle, S. and Ungewickell, E (1989) Curr. Opinion Cell Biol. 1, 684-690
Pauloin, A., Bernier, I. and Jolles, P. (1982) Nature 298, 574-576
Pearse, B. and Robinson, M. (1984) EMBO J. 3,1951-1957

The nineteen amino acid cytoplasmic tail of lysosomal acid phosphatase contains an endocytosis signal necessary and sufficient for targeting to lysosomes

Ch. Peters, M. Braun, B. Weber, M. Wendland, B. Schmidt, R. Pohlmann, A. Waheed and K. von Figura
Georg August Universität, Abteilung Biochemie II, Goßlerstraße 12D, D-3400 Göttingen

In I-cell fibroblasts deficient in the phosphotransferase activity generating mannose 6-phosphate markers of lysosomal enzymes intralysosomal activities of most lysosomal enzymes are deminished severely (Neufeld and McKusick 1983) whereas intralysosomal LAP activities are normal (Gieselmann et al 1984) suggesting that LAP is targeted to lysosomes independently of mannose 6-phosphate receptors. Furthermore segregation of LAP to lysosomes is neither affected by antibodies blocking the mannose 6-phosphate/IGF II receptor nor by NH_4Cl known to inhibit the mannose 6-phosphate receptor dependent targeting of soluble lysosomal enzymes (Gottschalk et al 1989).
Several proteins of the lysosomal membrane have been identified and characterized. Structural features common to this group of proteins are heavy N-glycosylation, a single membrane spanning domain and a short carboxy-terminal cytoplasmic domain (Granger et al and references therein). They are also routed to lysosomes independently of mannose 6-phosphate receptors. A small fraction of some of these lysosomal membrane proteins has been detected at the plasma membrane (Viitala et al 1988, Lippincott-Schwartz and Fambrough 1987).
Sorting signals have been localized in the cytoplasmic domains of receptors for LDL (Davis et al 1987), polymeric Ig (Mostov et al 1986), transferrin (Rothenberger et al 1987, Iacopetta et al 1988) and mannose 6-phosphate (Lobel et al 1989). A tyrosine residue in the cytoplasmic tail of the LDL receptor has shown to be essential for integration into coated pits and subsequent endocytosis (Davis et al 1987). An artificial endocytosis signal has been created by replacement of a cyteine by a tyrosine in the cytoplasmic tail of the influenza virus hemagglutinin (Lazarovits and Roth 1988).
Elucidation of the amino acid sequence of human LAP revealed that it resembles the structure of lysosomal membrane glycoproteins with dense glycosylation of the ectoplasmic domain, a single transmembrane domain and a short C-terminal cytoplasmic tail of 19 amino acids (Pohlmann et al 1988).

NATO ASI Series, Vol. H 62
Endocytosis
Edited by P. J. Courtoy

Synthesis and processing of LAP

For further studies human LAP was overexpressed in BHK-21 cells by stable transfection with a LAP cDNA yielding about 70-fold higher enzyme activities compared to untransfected cells. Human LAP is synthesized as a heterogeneously glycosylated precursor tightly associated with membranes. Upon transfer to the Golgi the size of a LAP monomer - LAP forms homodimers in vivo - increases from 59-61 kDa to 67 kDa due to partial processing of the N-linked oligosaccharides to complex type structures. LAP is transfered to dense lysosomes as an integral membrane protein (Waheed et al 1988). In the lysosomes it is released from the membrane by proteolytic processing involving at least two cleavages at the C-terminus of the polypeptide. The first cleavage is catalysed by a thiol proteinase yielding a membrane bound processing intermediate that lacks the major part of the cytoplasmic tail of LAP. Only after trimming of the cytoplasmic tail an aspartyl proteinase cleaves at the luminal side of the lysosomal membrane releasing mature LAP from the membrane. In turn the first proteolytic cleavage at the cytoplasmic side of the lysosomal membrane is depending upon acidification of the lysosome (Gottschalk et al 1989a).

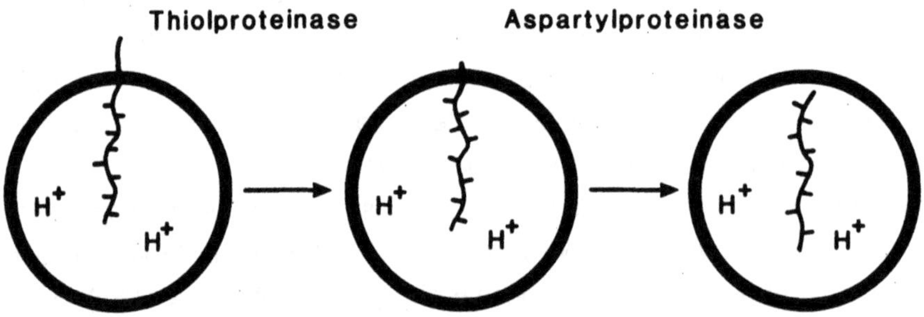

Fig. 1: Proteolytic processing of LAP in lysosomes

Transport of LAP to lysosomes

LAP has been detected in normal human fibroblasts by immunoelectron microscopy in the rough endoplasmic reticulum, lysosomes, vesicles just below the plasma membrane and at the plasma membrane (Parenti et al 1987). To quantify the fraction of LAP located at the plasma membrane overexpressing BHK cells were chilled to 4°C after metabolic labeling and cell surface glycoproteins were desialylated by incubation with neuraminidase at 4°C. The membrane bound LAP precursor was immunoprecipitated and subjected to isoelectric focussing. Untreated sialylated LAP exhibited multiple distinct forms with pIs between 4.9 and 5.6. Complete desialylation by incubation with neuraminidase converted LAP into three major forms with pls around 6.2. About 18% of LAP precursors focussed at pH 6.2 after treatment with neuraminidase at 4°C as compared to 3% in controls incubated with buffer instead of neuraminidase. These molecules represent unsialylated LAP forms that did not pass the compartment of sialytransferase. This result indicates that about 15% of the membrane bound LAP precursors are located at the plasma membrane since they were accessible to neuraminidase at 4°C.

Transport of LAP precursors from the trans Golgi to the plasma membrane requires on the average less than 10 min as determined by pulse/chase labeling, immunoprecipitation and subsequent quantitation on SDS-PAGE. Cell surface LAP is rapidly internalized and most of the molecules are transported back to the plasma membrane. To demonstrate this recycling cells were repeatedly incubated with neuraminidase at 4°C with 15 min intervals of incubation at 37°C inbetween reactivating membrane flow. 19%, 42% and 48% after the first, second and third treatment with neuraminidase were desialylated, respectively. This approach underestimated the rate of recycling due to the long incubation periods at 37°C. Incubation of cells with neuraminidase at 37°C yielded desialylation of 50% of the LAP molecules in less than 3 min. Even though some neuraminidase may enter endocytic structures at 37°C and desialylate LAP intracellularly the initial kinetics of desialylation are determined by the accessibility of LAP at the cell surface and the activity of neuraminidase. Therefore the cycle time of LAP is estimated to be around 6 min. Recycling is inhibited at temperatures below 20°C characteristic of recycling between early endosomes and plasma membrane. All LAP precursors that have passed the

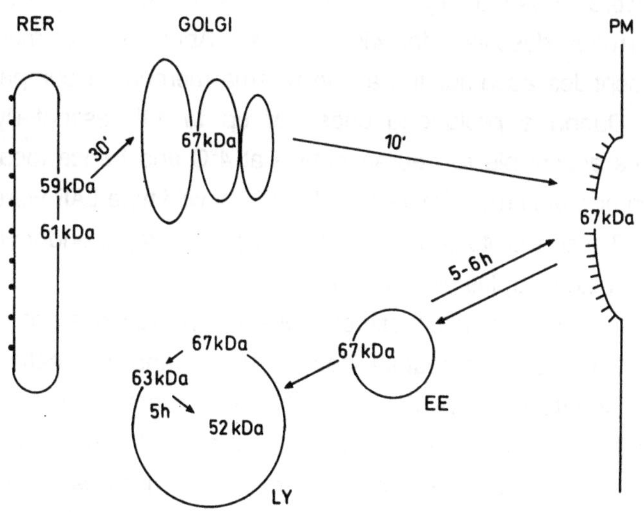

Fig. 2. Transport of LAP precursor to the lysosomes via the cell surface

trans-Golgi and not yet reached dense lysosomes are on the average for 5-6h part of the plasma membrane/endosome pool and recycle 15-50 times before they are delivered to dense lysosomes (Figure 2, Braun et al 1989). These data have been obtained with BHK cells expressing large amounts of LAP and it cannot be excluded that transport kinetics and distribution equilibria are different in cells expressing lower levels of LAP, but no obvious variations in subcellular distribution were observed between BHK cells expressing LAP levels between 10- to 600 fold above normal in immunoelectron microscopy (H. Geuze, unpublished).

Endocytosis signal in the cytoplasmic tail of LAP

As a first step to investigate the role of the cytoplasmic tail for targeting of LAP precursors to the lysosomes truncated versions of the protein were constructed by introduction of translation termination codons into the cDNA before and after the transmembrane domain. The truncated LAP lacking the transmembrane domain and the cytoplasmic tail (LAP-RP2) was secreted into the medium and exhibited acid phosphatase activity. The LAP lacking only the cytoplasmic tail (LAP-RP3), however, was retained intracellularly, transfered to dense lysosomes and converted to a soluble form indistinguishable from the mature form derived from wildtype LAP. The time required for transport of half of the precursors to dense lysosomes was extended from 6-7h for wild type LAP to about 24h for LAP-RP3 as determined by pulse/chase labeling, Percoll gradient centrifugation and subsequent quantitation on SDS-PAGE.

To determine the portion of LAP-RP3 located at the cell surface expressing cells were pulse chase labeled, subjected to neuraminidase treatment and LAP-RP3 polypeptides were subsequently immunoprecipitated and analyzed by isoelectric focussing as already described for wild-type LAP. About 80% of membrane bound LAP-RP3 polypeptides accumulated at the plasma membrane compared to 15% of wildtype LAP. During a prolonged chase of up to 14h essentially all LAP-RP3 precursors were accessible to neuraminidase at 4°C and hence located at the cell surface. In a comparable experiment only 10-13% of wild type LAP were accessible to neuraminidase. These results suggest that the signal for rapid endocytosis is located on the 19 amino acid cytoplasmic tail of LAP.

Tyrosine residues are essential constituents of endocytosis signals in the cytoplasmic tails of the LDL-receptor, a mutant hemagglutinin, and the cation independent mannose 6-phosphate receptor (Davis et al 1986, Lazarovits and Roth 1988, and Lobel et al 1989). The cytoplasmic domain of human LAP contains a single tyrosine residue at position 413. To investigate its function in endocytosis it was changed to phenylalanine by site directed mutagenesis (LAP-CP4, Figure 3). According to the desialylation/isoelectic focussing assay described above 70% of LAP-CP4 precursors

were accessible to neuraminidase at 4°C and hence located at the plasma membrane. Furthermore during a prolonged chase of 1-14h no intracellular LAP-CP4 precursors were detectable comparable to LAP-RP3, establishing Tyr_{413} as essential constituent of the endocytosis signal residing on the cytoplasmic domain of LAP.
Comparison of the amino acid sequence around this essential Tyr_{413} residue in the LAP tail with sequences around essential tyrosine residues in recycling receptors (Chen et al 1990), lysosomal membrane glycoproteins (Granger et al 1990) and the artificial endocytosis signal introduced into the cytoplasmic domain of influenza virus hemagglutinin (Lazarovits and Roth 1988) did not reveal any apparent homologies suggesting that a variety of combinations can provide the appropriate surrounding for the essential tyrosine to be functional (Table 1).

Table 1. Sequences surrounding tyrosine residues in potential endocytosis signals

MPR46	P	A	A	**Y**	R	G	V
MPR300	N	V	S	**Y**	K	Y	S
MPR300	S	Y	K	**Y**	S	K	V
h-LDL	N	P	V	**Y**	Q	K	T
HA-Tyr	S	L	Q	**Y**	R	I	C
lgpA	H	A	G	**Y**	Q	T	I
lgpB	H	A	G	**Y**	E	Q	F
h-LAP	P	P	G	**Y**	R	H	V
r-LAP	P	P	G	**Y**	H	H	V

MPR46: M_r 46 kDa mannose 6-phosphate receptor; MPR300: M_r 300 kDa mannose 6-phophate receptor; h-LDL: human LDL receptor; HA-Tyr: influenza virus hemagglutinin Cys_{543}>Tyr_{543}; lgpA: lysosomal membrane glycoprotein, type A; lgpB: lysosomal membrane glycoprotein, type B; h-LAP: human lysosomal acid phosphatase; r-LAP: rat LAP

Furthermore, phenylalanine in contrast to the LDL receptor tail (Davis et al 1987) cannot substitute for the function of the essential tyrosine residues in the tail of mutant hemagglutinin (Lazarovits and Roth, 1988) nor in the LAP tail. Recently the single tyrosine residue in the cytoplasmic tail of the human lysosomal membrane glycoprotein h-lamp-1 has shown to be necessary for efficient transport to lysosomes. Altered h-lamp-1 molecules that either did not contain the essential tyrosine residue or contained a tyrosine residue at an inappropriate position of the cytoplasmic tail were not transported to the lysosomes and accumlated at the cell surface. Williams and Fukuda (1990) conclude that lysosomal membrane glycoproteins are sorted to lysosomes by a cytoplasmic signal containing a tyrosine in a specific position, and that the sorting signal may be recognized both in the trans-Golgi network and at the cell surface.

To test whether the endocytosis in the cytoplasmic domain is not only essential but also sufficient for rapid internalization and lysosomal targeting a chimeric polypeptide consisting of the ectoplasmic and transmembrane domain of the plasma membrane sorted influenza virus hemagglutinin (HA, Kuroda et al 1986) and the cytoplasmic tail of LAP was constructed (HA-LAP, Figure 3). With the immunofluorescence internalization assay of Lazarovits and Roth (1988) it was shown that in contrast to wildtype LAP the chimera HA-LAP is rapidly internalized. Anti-HA antibodies bound to the cell surface at 0°C remained at the plasma membrane of wild type HA expressing cells whereas they were transferred to vesicular structures in HA-LAP expressing cells upon reculturing for 1h at 37°C. Substituion of the tyrosine residue in the cytoplasmic domain of HA-LAP reverted the rapid internalization almost completely.

	Transmembrane ¦ Cytoplasmic Domain
(A)	
LAP	L L T V L F ¦ R M Q A Q P P G Y R H V A D G E D H A •
LAP-CP4	L L T V L F ¦ R M Q A Q P P G F R H V A D G E D H A •
(B)	
HA	*L V F I C V* ¦ *K N G N M R C T I C I* •
HA-LAP	*L V F I C S* ¦ R M Q A Q P P G Y R H V A D G E D H A •
HA-CP4	*L V F I C S* ¦ R M Q A Q P P G F R H V A D G E D H A •

Fig. 3. Amino acid sequences of LAP mutants and hemagglutinin-LAP chimeras (HA-LAP, HA-CP4).

As a second indicator protein for construction of a chimera the M_r 46 kd mannose 6-phosphate receptor (MPR46) was used. The cytoplasmic tail and transmembrane domain of LAP were fused to the ectoplasmic domain and 6 amino acids of the transmembrane domain of MPR46. The MPR46 ectoplasmic domain of the chimera folded correctly since it bound to mannose 6-phosphate, formed dimers and its N-linked oligosaccharides were processed to complex type structures. Wild type MPR46 recycles between Golgi and endosomes and between plasma membrane and endosomes, and it is not detected in lysosomes. The receptor contains an endocytosis signal in its cytoplasmic domain (B. Weber and R. Pohlmann, unpublished results), which is lacking in the deletion mutant MPR46-M6 containing only 6 amino acids of the MPR46 cytoplasmic domain. In the immunofluorescence/endocytosis assay (Lazarovits and Roth 1988) wildtype MPR46 and the chimera MPR46-LAP were readily endocytosed, whereas the deletion mutant MPR46-M6 remained at the plasma membrane. The half life of the MPR46-LAP chimera is about 2h. In contrast to the luminal domain of LAP the ectoplasmic domain of MPR46 is likely to be sensitive to degradation by lysosomal proteinases since wildtype MPR46 does not enter this compartment. To test this lysosomal proteolysis was inhibited by lysosomal proteinase inhibitors leupeptin and pepstatin and the chimera was immunoprecipitated after pulse chase labeling and subsequently analyzed by SDS-PAGE. The results showed that the chimera was stabilized in the presence of the inhibitors whereas they had no effect on the detectable level of wild type MPR46. Under the protection of these proteinase inhibitors it was possible to show in a Percoll density gradient experiment that the chimera MPR46-LAP is delivered to dense lysosomes whereas wildtype MPR46 and MPR46-M6 do not enter this compartment (Peters et al 1990). From these results we conclude that the 19 amino acid cytoplasmic tail of LAP is not only necessary for rapid endocytosis and targeting to lysosomes but also sufficient for this function.

Targeting of lysosomal membrane glycoproteins

LAP and a group of lysosomal membrane proteins (Granger et al 1990) share structural homologies; they all contain a heavily glycosylated luminal domain a single transmembrane domain and a short cytoplasmic domain with a single tyrosine residue. For h-lamp-1 (Williams and Fukuda 1990) and LAP it has been shown that this tyrosine residue is essential for rapid endocytosis and that mutants of these proteins devoid of this signal accumulate at the plasma membrane.

Therefore passage of the cell surface and rapid endocytosis may be a general mechanism for efficient targeting of a group of membrane bound glycoproteins to the lysosomes.

References

Chen, W.-J., Goldstein, J.L., and Brown, M.S. (1990) J.Biol. Chem. 265, 3116-3123.

Braun, M., Waheed, A., and von Figura, K., (1989) EMBO J. 8, 3633-3640.

Davis, G.G., Lehmann, M.A.,Russell, D.W., Anderson R.G.W., Brown, M.S. and Goldstein, J.L. (1986) Cell, 45, 15-24.

Davis, G.G., van Driel, I.R., Russell, D.W., Brown, M.S., and Goldstein, J.L. (1987) J. Biol. Chem. 262, 4075-4082.

Gieselmann, V., Hasilik, A., and von Figura, K. (1984) Biol. Chem. Hoppe-Seyler 365, 651-660.

Gottschalk, S., Waheed, A., and von Figura, K. (1989) Biol. Chem. Hoppe-Seyler 370, 75-80.

Gottschalk, S., Waheed, A., Schmidt, B., Laidler, P., and von Figura, K. (1989a) EMBO J. 8, 3215-3219.

Granger, B.L., Green, S.A., Gabel, C.A., Howe, C.L., Mellmann, I., and Helenius, A. (1990) J. Biol. Chem. 265, 12036-12043.

Iacopetta, B.J., Rothenberger, S., and Kühn, L.C. (1988) Cell 54, 485-489

Kuroda, K., Hauser, C., Rott, R., Klenk, H.-D., and Doerfler, W. (1986) EMBO J. 5, 1359-1365.

Lazarovits, J., and Roth, M. (1988) Cell 53, 743-752.

Lippincott-Schwartz, J. and Fambrough D.M. (1987) Cell 49, 669-677.

Lobel, P., Fujimoto, K., Ye, R.D., Griffiths, G., and Kornfeld, S. (1989) Cell 57, 787-796.

Mostov, K.E., de Bruyn Kops, A., and Deitcher, D.L. (1986) Cell 47, 359-364.

Neufeld, E.F., and McKusick, V.A. in: The Metabolic Basis of Inherited Disease, eds. Stanbury, J.B., Wyngaarden, J.B., Fredrickson, D.S., Goldstein, J.L. and Brown, M.S. (McGraw-Hill, New York 1983), pp. 778-787.

Parenti, G., Willemsen, R., Hoogeveen, A.T., Verleun-Mooyman, M., van Dongen, J.M., and Galjaard, H. (1987) Eur. J. Cell Biol. 43 121-127.

Peters, C., Braun, M., Weber, B., Wendland, M., Schmidt, B., Pohlmann, R., Waheed, A., and von Figura, K. (1990) EMBO J., in press.

Pohlmann, R., Krentler, C., Schmidt, B., Schröder, W., Lorkowski, G., Culley, J., Mersmann, G., Geier, C., Waheed, A., Gottschalk, S., Grzeschik, K.-H., Hasilik, A. and von Figura, K. (1988) EMBO J. 7, 2343-2350.

Rothenberger, S., Jacopetta, B.J., and Kühn, L.C. (1987) Cell 49, 423-431.

Tanaka, Y., Himeno, M., and Kato, K. (1990) J. Biochem. 108, 287-291.

Viitala, J., Carlsson, S.R., Siebert, P.D., and Fukuda, M. (1988) Proc. Natl. Acad. Sci. U.S.A. 85, 3743-3747.

Waheed, A., Gottschalk, S., Hille, A., Krentler, C., Pohlmann, R., Braulke, T., Hauser, H., Geuze, H., and von Figura, K. (1988) EMBO J. 7, 2351-2358.

Williams, M.A., and Fukuda, M. (1990) J. Cell Biol. 111, 955 -966.

BIOCHEMICAL CHARACTERIZATION, STRUCTURE AND TRANSPORT OF LYSOSOMAL INTEGRAL MEMBRANE PROTEINS

B. Seguí-Real, M.A. Vega, J.G. Barriocanal[1], L. Yuan[1], J. Alcalde and I.V. Sandoval

Centro de Biología Molecular, Cantoblanco, 28049 Madrid, Spain

INTRODUCTION.

Lysosomes are terminal degradative compartments involved in the degradation of extracellular and intracellular macromolecules (Kornfeld and Mellman, 1989). Degradation is performed by a large variety of soluble acidic hydrolases, many of which are well characterized. The functions of the lysosomal membrane proteins are less well understood, but they are believed to be involved in such diverse roles as, the maintenance of the membrane structure, the transport of ions and small molecules through the membrane, and in the movement and fusion of lysosomes with other organelles. Only recently, the first membrane proteins have been characterized (Pohlmann et al., 1988; Chen et al., 1988; Howe et al., 1988; Viitala et al., 1988; Fukuda et al., 1988; Noguchi et al., 1989; Cha et al., 1990).

The biogenesis of lysosomes involves the selective fusion of Golgi-derived vesicles loaded with newly synthesized lysosomal proteins, and endocytic elements transporting extracellular macromolecules and plasma membrane components. In mammalian cells the transport of newly synthesized soluble lysosomal proteins is dependent on their recognition by receptors for mannose 6-phosphate (Man-6-P). The Man-6-P signal is acquired during transport of the soluble enzymes from the endoplasmic reticulum to the trans-Golgi network where it is recognized by specific receptors (Kornfeld and Mellman, 1989). The recognition results in transport of the enzymes by coated vesicles to a post-Golgi compartment (Brown et al., 1986). Recent results indicating that this compartment is endocytic

[1] National Institutes of Health, Bethesda, Maryland 20892, USA

NATO ASI Series, Vol. H 62
Endocytosis
Edited by P. J. Courtoy

in nature and evolves to lysosomes (Geuze et al., 1988), suggest that further transport of the lysosomal proteins is not required. In contrast, the mechanisms involved in the sorting and transport of lysosomal membrane proteins from the Golgi complex to lysosomes are poorly understood. Yet, it is clear that they are not dependent on Man-6-P (Barriocanal et al., 1986).

Here we report our studies on the characterization of lysosomal membrane proteins, their structure, transport and cellular distribution.

RAT LYSOSOMAL INTEGRAL MEMBRANE PROTEINS.

I. Biochemical characterization.

The use of monoclonal antibodies have been a powerful tool for the characterization of lysosomal integral membrane proteins (LIMPs). By using this approach, we have identified five classes of LIMPs in rat (Barriocanal et al., 1986, Bonifacino et al., 1990). They cover, with the exception of acid phosphatase , all the lysosomal membrane proteins characterized in other animal species.

Table I shows some of the more relevant biochemical properties of LIMPs. The molecular weights of the polypeptide chains range from 20 to 50 KDa. All contain a large number of N-linked oligosaccharide chains, and with the exception of LIMP II, are highly syalilated. LIMPs I, II and III are insensitive to O-glycanase, suggesting the absence of O-linked carbohydrates. Particularly important is that in contrast to soluble lysosomal proteins, LIMPs are not phosphorylated (Barriocanal et al., 1986).

TABLE I.

	LIMP I	LIMP II	LIMP III	LIMP IV	LIMP V
N-deglycos. (KDa)	20	50	37	40-45	38
Precursor (KDa)	28	72	86	80-90	55
Mature form (KDa)	35-50	74	90-100	110	80

II. Structure.

Study of the structural organization of LIMPs is of great importance to advance our knowledge of their transport, localization and function. For this purpose, LIMP II, III and IV have been purified by affinity chromatography and microsequenced (Vega et al., 1990). Analysis of their N-terminal sequences reveals that LIMPs III and IV correspond, respectively, to the A and B classes of lysosomal membrane proteins recently described. Both classes of proteins possess between 380-396 amino acids and 16-20 asparagine sites for glycosilation. They have two homologous luminal domains, each with four bridged cysteine residues, separated by a segment 20-35 amino acids long, rich in proline and serine (class A), or proline and threonine (class B). Furthermore, they have short, 10-11 amino acids long cytoplasmic tails displaying 19% identity and conserved Gly-Tyr pairs. Class A proteins are rat lgp 120 (which is the same as rat LGP 107), mouse and human LAMP-1, and chicken LEP-100. Rat lgp 110 (equivalent to rat LGP 96), mouse lgp 110 and human LAMP-2 constitute class B (Pohlmann et al., 1988; Chen et al., 1988; Howe et al., 1988; Viitala et al., 1988; Fukuda et al., 1988; Noguchi et al., 1989; Cha et al., 1990).

The primary sequence of LIMP II has been deduced from a cDNA clone (Vega et al., 1990). Transfection of COS cells with the cDNA of LIMP II resulted in expression of the protein and its transport to lysosomes. The cDNA clone encodes a protein 478 amino acids long. All the amino acids with the exception of the first methionine residue are present in the mature form. The first two amino acids (Ala-Arg) of the polypeptide chain are followed by a hydrophobic stretch of 23 amino acids. This stretch is likely to constitute the hydrophobic core of an uncleaved signal peptide. A second hydrophobic region is located between residues 433-458, near the carboxyl terminus of the chain. This region probably acts as a stop transfer signal that anchors the protein to the membrane. Between the signal peptide and the stop transfer signal lies a segment 406 amino acids long. Spreaded throughout this segment are 11 potential sites for N-glycosilation, and five cysteine residues are clustered in its C-terminal. Its high content in N-linked carbohydrates, and resistance to proteinase K incubated with intact lysosomes makes, this region a strong candidate for the luminal domain.

The inner location of the two segments that anchor the protein to the membrane and their separation by a large luminal domain predicts the existence of two short cytoplasmic tails, one in the amino terminus and the other in the carboxyl end (Figure 1).

Cysteines residues at positions 4, 5 and 458, could be important in determining the boundaries of the tail. Pairing between cysteines 458 and one of the two cysteines located at the N-terminal segment would bring the two cytoplasmic tails together. Furthermore, since paired cysteines are known to become extremely hydrophobic, after formation of a disulfide bond they are likely to be embedded into the membrane. It is also noteworthy that LIMP II is palmitylated in vivo (unpublished results). Palmitic acid is frequently post-translationaly incorporated into cysteine residues and plays an important role in anchoring proteins to membranes. In either case, cysteine pairing, palmitylation, or both, would increase the stability of the protein anchor to the membrane.

A search into the EMBL database revealed an extensive homology between LIMP II and the human cell surface protein CD36. The two proteins share 34% of the amino acids. The identity was primarily restricted to the extracytoplasmic domains. Like LIMP II, CD36 also contains an uncleaved signal peptide and a stop transfer signal. CD36 displays 10 potential sites for N-glycosilation, 3 of them conserved in LIMP II, and the rest in close proximity. Seven of the eight cysteines found in LIMP II are conserved in CD36. In contrast to LIMP II, CD36 contains O-linked carbohydrates and is highly syalilated (Oquendo et al., 1990). All these data suggest that LIMP II and CD36 display identical topologies and similar structures. It is relevant that in comparison to the ubiquitous presence of LIMP II, the expression of CD36 appears to be restricted to endothelial cells, platelets, monocytes, and melanoma cells. Since CD36 is a receptor for collagen type I (Tandon et al., 1989) and for an unidentified component of the membranes of red blood cells parasitized by Plasmodium falciparum (Oquendo et al., 1990), it would be interesting to study whether LIMP II has conserved some of these functional properties.

No major similarities can be drawn between LIMP II and LIMPs III and IV, except the high content of carbohydrates, and short cytoplasmic tails.

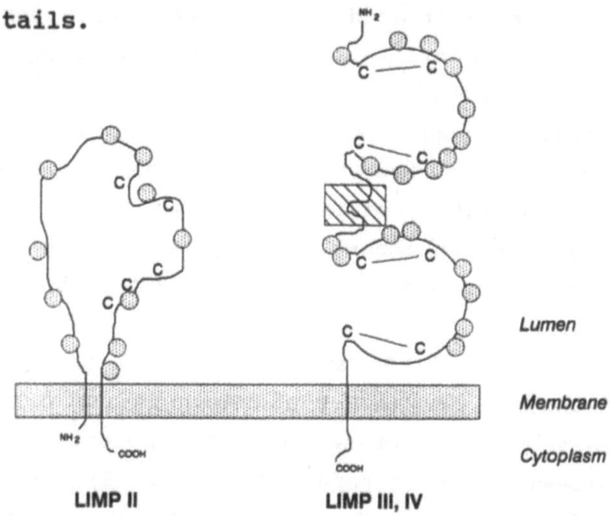

Figure 1. Topologies of LIMPs II, III, and IV. The cartoon shows the identical structural organization of the luminal portions of LIMPs III and IV in two homologous domains separated by a hinge region. Note the different structural organization of the luminal portion of LIMP II. It can be observed that whereas LIMPs III and IV are anchored to the membrane through a single transmembrane region, LIMP II is fastened by two distant transmembrane domains. Circles represent the potential sites for N-glycosilation found in LIMPs II and III. The positions of cysteine residues contained in the luminal domains are marked by letter C.

III. Pathways of transport.

Transport of newly synthesized LIMPs I, II, and III from the endoplasmic reticulum to Golgi proceeds rapidly and at similar rates, 30-60 min, as determined by the acquisition of resistance to endoglycosidase H (Barriocanal et al., 1986). Sorting of LIMPs to lysosomes, and to secretory granules loaded with lysosomal enzymes, occurs in the trans-Golgi network. Two hours after their synthesis the transport of LIMPs II and III to lysosomes is virtually completed. In contrast, only 40% of LIMP I is found in lysosomes after the same period of time (Barriocanal et al., 1989).

The slow transport of LIMP I to lysosomes might be related with the relative high expression of the protein on the cell surface (5%) as compared to that of LIMPs II and III (0.2%) (unpublished results). With respect to this, it is interesting that the lysosomal membrane protein acid phosphatase is transported from the Golgi to the plasma membrane before being delivered to lysosomes at a slow rate (5-6 h) (Braun et al., 1989). Therefore, the possibility should be considered that whereas some proteins are transported directly from the Golgi to lysosomes (i.e. LIMPs II and III) others could be delivered to lysosomes via the plasma membrane (i.e. LIMP I, acid phosphatase). The speed of transport of LIMPs from the plasma membrane to lysosomes is probably determined by their rate of internalization and recycling to the plasma membrane (unpublished results). LIMPs are transported to lysosomes by a mechanism independent of Man-6-P as shown by the normal rates of transport observed in the presence of the potent N-glycosylation inhibitor tunicamycin (Barricanal et al., 1986), and the massive accumulation of lysosomal membranes in cells with a defect in the incorporation of Man-6-P into lysosomal hydrolases (Sandoval et al., 1989). Double immunofluorescence microscopy experiments indicate that different LIMPs colocalize to the same lysosomes. Study of the half lives show significant differences between LIMPS (LIMP I, 8 h; LIMP II, 20 h; LIMP III, 10 h) and a dramatic increase in the turnover of the N-deglycosilated forms (Barriocanal et al., 1986). These results point to a different turnover of the components of the lysosomal membrane and to a fundamental role of the carbohydrates in protecting LIMPs from degradation by lysosomal hydrolases.

The simultaneous presence of LIMPs in lysosomes and secretory granules of some cells of the immune system, raises important questions about their sorting (Bonifacino et al., 1990). First, it is not known if LIMPs are delivered to the two organelles by different mechanisms and pathways. Second, it is also not known whether the incorporation of LIMPs into the plasma membrane upon stimulation of secretion, subsequent internalization (10 min) and recycling to newly formed secretory granules involves or not additional mechanisms of sorting not found in other cells.

IV. Mechanisms of transport.

As described above, Man-6-P plays no role in the transport of LIMPs from the Golgi complex to lysosomes. All the four lysosomal membrane proteins so far characterized (classes A and B, acid phosphatase, LIMP II) have in common the display of short carboxyl cytoplasmic tails containing a tyrosine residue, and can be localized in coated vesicles budding from the trans Golgi network and the plasma membrane. Tyrosines contained in cytoplasmic tails have been shown to play an important role in protein endocytosis mediated by clathrin coated vesicles (Davis et al., 1987, Lazarovits and Roth, 1988, Lobel et al., 1989, Vega and Strominger, 1989, Jing et al., 1990). In line with these observations, Williams and Fukuda have recently demonstrated that the tyrosine contained in the cytoplasmic tail of lysosomal membrane protein human LAMP-1 (equivalent to LIMP III) plays a major role in the delivery of the protein from the Golgi complex to lysosomes, and internalization from the plasma membrane (Williams and Fukuda, 1990). By contrast, LIMP II neither has tyrosine residues at the cytoplasmic tails nor shows any homology to the cytoplasmic tails of LIMPs III and IV (see figure 2). These features suggest the existance of more than one structural motive for the targeting of LIMPs to lysosomes.

Figure 2. Carboxyl cytoplasmic tails of LIMPs.

LIMP II	R G Q G S T D E G T A D E R A P L I R T
LIMP III	R K R S H A G Y Q T I
LIMP IV	K R H H T G Y E Q F

Comparison of the amino acid sequences of the carboxyl cytoplasmic tails of new LIMPs, as well as detailed site-directed mutagenesis studies should help to determine the structural nature of the sorting signal(s).

References.

- Barriocanal, J., Bonifacino, J., Yuan, L. and Sandoval, I.V. (1986) J. Biol. Chem. 261, 16755-16763.
- Bonifacino, J.S., Yuan, L., and Sandoval, I.V. (1990) J. Cell Sci. 92, 701-712.
- Braun, M., Waheed, A. and von Figura, K. (1989) EMBO J., 8, 3633-3640.
- Brown, W.J., Goodhouse, J., and Farquhar, M.G. (1986) J. Cell Biol. 103, 1235-1247.
- Cha, Y., Holland, S.M. and August, T. (1990) J. Biol. Chem. 265, 5008-5013.
- Chen, J.W., Cha, Y., Yuksel, K.U., Gracy, R.W. and August, T. (1988) J. Biol. Chem. 263, 8754-8758.
- Davis, C.G., van Driel, I.R., Russell, S.W., Brown, M.S. and Goldstein J.L. (1987) J. Biol. Chem. 262, 4075-4082.
- Fukuda, M., Viitala, J., Matteson, J. and Carlsson, S.R. (1988) J. Biol. Chem. 263, 18920-18928.
- Geuze, H.J., Stoorvogel, W., Strous, G.J., Slot, J.W., Bleekemolen, J.E. and Mellman, I. (1988) J. Cell Biol. 107, 2491-2501.
- Howe, C.L., Granger, B.L., Hull, M., Green, S.A., Gabel, C.A., Helenius, H. and Mellman, I. (1988) Proc. Natl. Acad. Sci. USA. 85, 7577-7581.
- Jing, S., Spencer, T., Miller, K., Hopkins, C. and Trowbridge, I.S. (1990) J. Cell Biol. 110, 283-294.
- Kornfeld, S. and Mellman, I. (1989) Ann. Rev. Cell Biol., 5, 483-525.
- Lazarovits, J. and Roth, M. (1988) Cell 53, 743-752.
- Lobel, P.K., Fujimoto, K., Ye, R.D., Griffiths, G. and Kornfeld, S. (1989) Cell 57, 787-796.
- Noguchi, Y., Himeno, M., Sasaki, H., Tanka, Y., Kono, A., Sakaki, Y. and Kato, K. (1989) Biochem. Biophys. Res. Commun. 164, 1113-1120.
- Oquendo, P., Hundt, E., Lawler, J., Seed, B. (1990) Cell 58, 95-101.
- Pohlmann, R., Krentler, C., Schmidt, B., Schroeder, W., Lorkowski, G., Culley, J., Mersmann, G., Geier, C., Waheed, A., Gottschalk, S., Grzeschik, Hasilik, A., and von Figura, K. (1988) EMBO J. 7, 2343-2350.
- Sandoval, I.V., Chen, J.W., Yuan, L. and August, T. (1989) Arch. Biochem. Biophys. 271, 157-167.
- Tandon, N.N., Kralisz, U. and Jamieson, G.A. (1989) J. Biol. Chem. 264, 7576-7583.
- Vega, M.A. and Strominger, J.L. (1989) Proc. Natl. Acad. Sci. USA 86, 2688-2692.
- Vega, M.A., Seguí-Real, B., Alcalde, J., Calés, C., Vanderkerckhove, J., and Sandoval, I. (1990). Submitted for publication.
- Viitala, J., Carlsson, S.R., Siebert, P.D. and Fukuda, M. (1988) Proc. Natl. Acad. Sci. USA 85, 3743-3747.
- Williams, M.A. and Fukuda, M. (1990) J. Cell Biol. 111, 955-966.

MANNOSE PHOSPHATE-INDEPENDENT MEMBRANE ASSOCIATION OF LYSOSOMAL ENZYMES OCCURS AFTER PASSAGE OF THE GOLGI COMPLEX

Ger J. Strous, Hans J. Geuze, *Hans M.F.G. Aerts, *Joseph M. Tager and Simon Rijnboutt

Laboratory of Cell Biology, University of Utrecht, Heidelberglaan 100, 3584 CX Utrecht, and *E.C. Slater institute for Biochemical Research, University of Amsterdam, AMC, Amsterdam, The Netherlands

INTRODUCTION

The degradative action of lysosomes depends primarily on the hydrolytic enzymes present within their lumen. However, the number of lysosomes per cell and their level of activity is regulated by events controlled by constituents of their limiting membrane. Therefore, study of formation of new lysosomes, the composition of lysosomal membranes and their maintenance is the basis for understanding the biology of lysosomes. Much is known about the biosynthesis, sorting and turnover of soluble lysosomal enzymes (Von Figura and Hasilik, 1986; Kornfeld and Mellman, 1989). They are targeted through endosomes to lysosomes via the mannose phosphate receptor (MPR). In the endosomes MPR separates from the ligand (the soluble lysosomal enzyme) and returns to the Golgi complex or the cell surface, while the enzyme is transported to the lysosomes. Other mechanisms to direct proteins to the lysosomes must exist (Neufeld and McKusik, 1983). The glycoproteins of the limiting lysosomal membrane have no mannose 6-phosphate and yet are very efficiently transported to the lysosomes. In addition, some enzymes such as glucocerebrosidase (deficient in Gaucher disease) and acid phosphatase have their own unique pathways to reach the lysosomes (Erickson, *et al.*, 1985; Braun *et al.*, 1989). We have studied the intracellular transport between the Golgi complex and the lysosomes of two lysosomal glycoproteins cathepsin D, a MPR-dependent soluble enzyme and glucocerebrosidase, a MPR-independent enzyme after their synthesis in the rough endoplasmic reti-

NATO ASI Series, Vol. H 62
Endocytosis
Edited by P. J. Courtoy

culum (RER).

Recently, Diment *et al.* (1988) found that in rabbit macrophages 90% of the 53-kD precursor of cathepsin D was membrane-associated. In this paper we present data which support and extend these observations. Using the human liver hepatoma cell line HepG2, we combined metabolic labeling with differential permeabilization at low concentrations of saponin. The plasma membrane and endomembranes can be permeabilized in the presence of saponin (Strous and van Kerkhof, 1989) which is primarily due to complex formation with cholesterol (Bangham and Horne, 1962). Differential permeabilization with saponin can induce release of secretory proteins but not of integral membrane proteins (Wassler *et al.*, 1987; Strous and van Kerkhof, 1989). This feature is used to demonstrate transient membrane association of cathepsin D. We also were able to establish the localization of membrane attachment of glucocerebrosidase following its biosynthesis. (Rijnboutt *et al.*, 1990)

METHODS

Nearly confluent HepG2 cells grown on 35 mm petri dishes were labeled for 15 min with [^{35}S]methionine (60 C/ml) (800-1200 Ci/mmol, The Radiochemical Centre, Amersham). After metabolic labeling the cells were washed in phosphate buffered saline without Ca^{2+} (PBS) at 0°C and then placed on a rocking platform in 1 ml of PBS containing various concentrations of saponin for 30 min. Thereafter, the permeabilization medium was removed and the cells were washed with PBS and lysed in 1% Triton X-100, 1 mM phenylmethylsulfonyl fluoride in PBS. Aliquots of the cell extracts and the permeabilization media were used for immunoprecipitation. Immunoprecipitations were carried out as previously described (Strous and Lodish, 1980). Aliquots of the Triton X-100 soluble material were immunoprecipitated with either normal rabbit anti-human asialoglycoprotein receptor (a kind gift of Dr. Alan Schwartz, St.Louis, MO), anti-cathepsin D or anti-glucocerebrosidase and the immunoprecipitates were analyzed in 10% polyacrylamide gels in the presence of sodium dodecylsulphate (SDS-PAGE).

RESULTS

Soluble proteins transported through cells via the vacuolar system can be released using low concentrations of saponin (Strous and Van Kerkhof, 1989). To test this we have in-

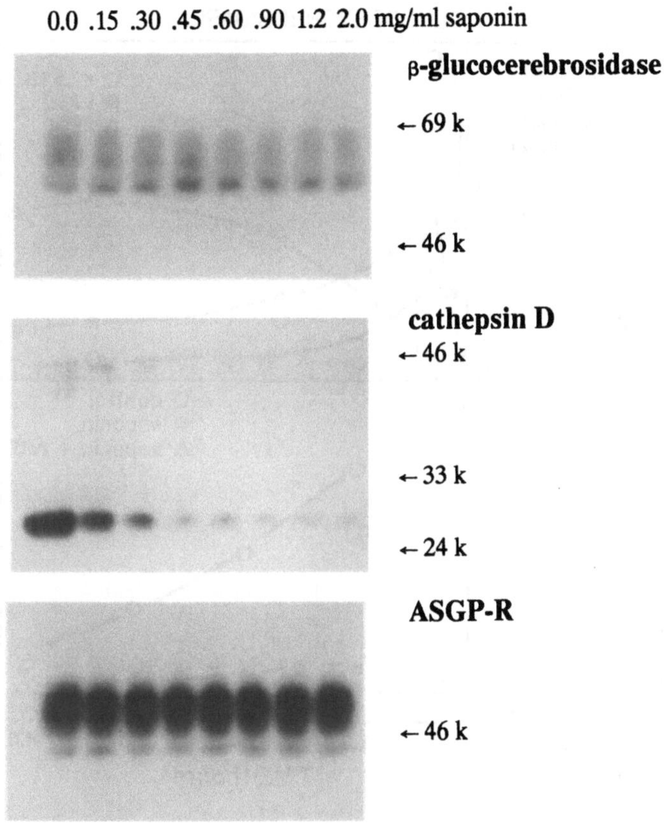

Fig. 1. *Saponin titration of protein release from various intracellular compartment.* HepG2 cells were incubated with saponin at indicated concentrations. The cells were solubilized and the proteins remaining with the cells were separated by SDS-PAGE and blotted onto nitrocellulose. The blots were incubated with antisera against β-glucocerebrosidase, cathepsin D or the receptor for asialoglycoproteins. The IgG binding was visualized after incubation with ^{125}I-protein A and autoradiography.

cubated HepG2 cells in the presence of increasing concentrations of this reagent. After 30 min the permeabilized cells were washed and the proteins present in these cells were separated by SDS-PAGE, transferred to nitrocellulose and immunostained for cathepsin D, glucocerebrosidase and the asialoglycoprotein receptor. As seen in Fig. 1 the 2 lysosomal enzymes behave different. While the mature protein cathepsin D (31 kDa) can be completely washed out at 2 mg/ml saponin, the major portion of glucocerebrosidase remains associated with the cells at this saponin concentration. This is in agreement with previous observations that glucocerebrosidase is a membrane-associated lysosomal enzyme. To test whether the effect of saponin is restricted to soluble proteins without affecting integral membrane proteins or even whole cells, we quantitated the effect of this reagent on the asialoglycoprotein receptor. No decrease of the amount of this protein was measured in the range of saponin concentrations tested.

The experiment depicted in Fig. 1 also shows that not all immunoreaction of cathepsin D had disappeared at the highest concentration used: a 53 kDa band remained associated

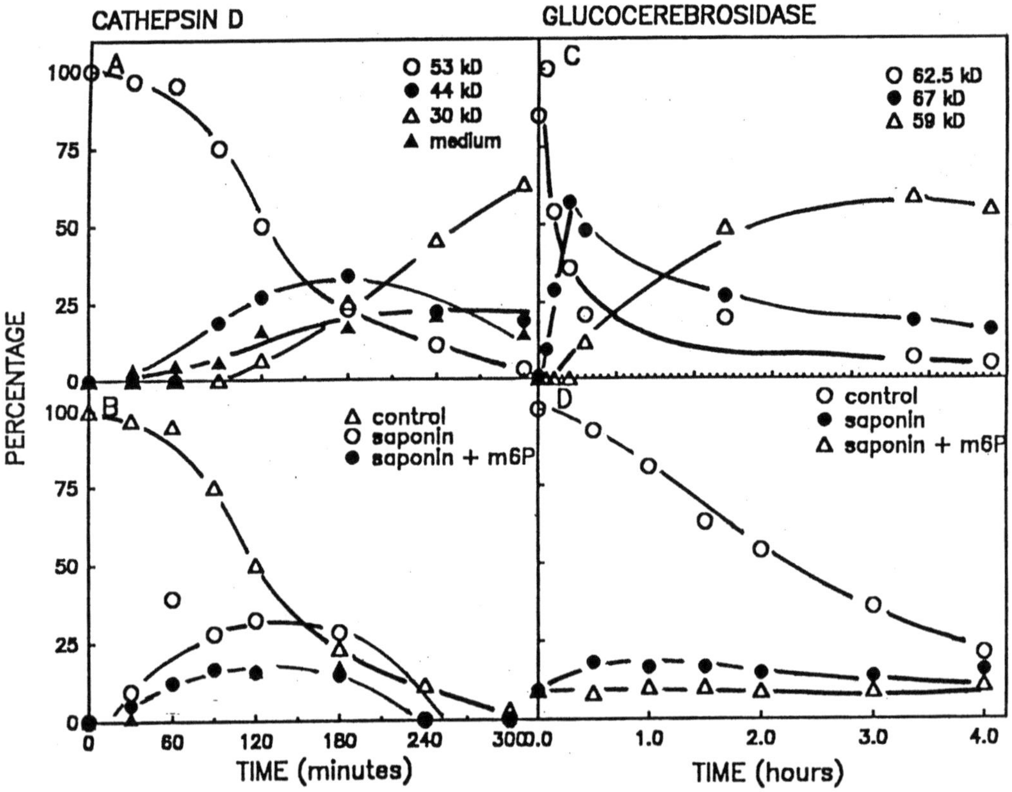

Fig. 2. *Release of cathepsin D and glucocerebrosidase from pulse-chase labeled HepG2 cells.* In A and B time courses of the various species of cathepin D and glucocerebrosidase, respectively are shown in percentages of label present in the initially synthesized enzyme. In B and D percentages of 51-53-kDa of cathepsin D and the 62.5-kDa precursor of cerebrosidase, resepectively are plotted. In case of cathepsin D saponin wash removes the 51 kDa precursor, but not the 53-kDa intermediate. Note, the difference in time scale between C and D.

with the cells. This band originates from the pro-enzyme which is synthesized in the rough ER and transported to the Golgi, where it is provided with mannose phosphate residues. To examine this further HepG2 cells were pulse-labeled in the presence of [^{35}S]methionine and chased for various periods of time. The cells were treated with saponin, the cell-associated cathepsin D was immunoprecipitated, and the radioactivity present in the pro-enzyme as wll as the mature protease was quantitated after SDS-PAGE. Fig 2A shows the biosynthesis of cathepsin D: it is synthesized as a 51-kDa polypeptide and within 60 min converted to a 53-kDa protein, carrying phosphorylated mannose residues. It is this latter species which becomes partly associated (Fig. 2B). All other species (51, 44, 31 and 16 kDa) can be completely removed by saponin. Next, we tested

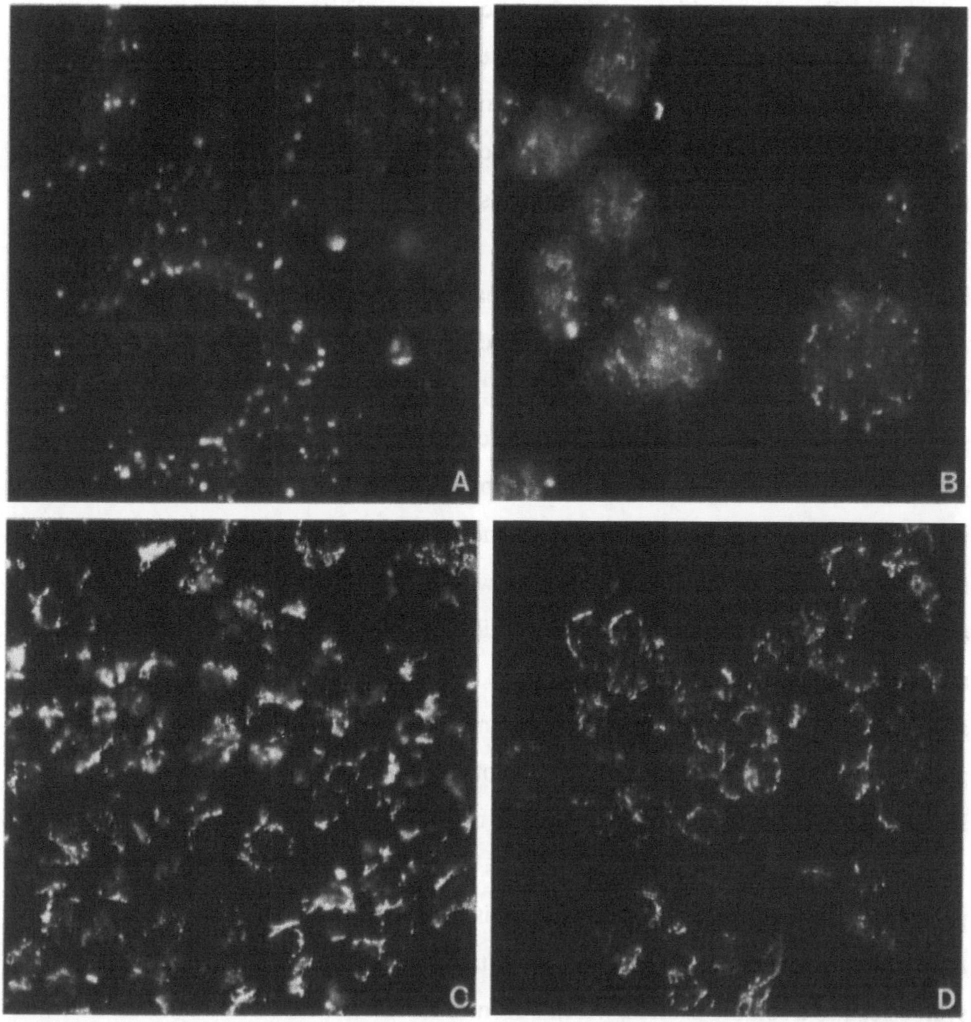

Fig. 3. ***Immunofluorescent localization of cathepsin D after saponin treatment.*** Cells were fixed and incubated with antibodies against cathepsin D (A and B) or against galactosyltransferase (C and D). Prior to fixation the cells were treated with 2 mg/ml saponin (B and D). A and C represent control cells.

the involvement of the mannose phosphate receptor in the membrane association of the 53 kDa band using a combination of saponin and 10 mM mannose 6-phosphate. As shown in Fig. 2B only 50% of the membrane associated cathepsin D was released by this competitive ligand for the receptor. Thus, cathepsin D can associate with the membranes of some intermediate compartments on its way to the lysosome, probably of the trans Golgi reticulum or of endosomes. A similar phenomenon is observed for glucocerebrosidase. The immunoblot in figure 1A shows that one of the bands, intermediate in molecular mass, disappears at higher saponin concentration. This was further investigated in a pulse-chase experiment (Fig. 2D). The initial precursor of this enzyme (62 kDa) is completely soluble as is apparent from the experiment with saponin. Concomittantly with conversion to the 68-kDa intermediate glucocerebrosidase gets membrane-associated; this is unchanged for all later processed forms including the mature enzyme (59 kDa after 15 h of chase).

Finally, we asked the question in which compartment the transient membrane association of the 53-kDa form occurs. To answer this, cells grown on cover slips were treated with saponin and then fixed in methanol/acetone, immunoreacted with anti-cathepsin D and stained with fluorescent second antibody. In control cells, cathepsin D-positive staining is mostly present in lysosomes. Some background is visible near the nucleus. In cells treated with saponin only faint labeling in vesicles mostly near the nucleus is visible. To exclude the possiblity that this labeling is associated with the Golgi complex we reacted cells with antibodies against galactosyltransferase. While the integrity of the Golgi complex is not affected by the saponin treatment, it is clear that cathepsin D remaining in the cell after saponin treatment is not associated with membranes of the Golgi complex. The labeling pattern suggests that membrane association of cathepsin D is in endosomes.

In summary, we have shown that in HepG2 cells cathepsin D transiently associates with the membrane of one of the compartments between the rough ER and the lysosome, probably of the endosome. This association is for at least 50% independent on the mannose phosphate receptor. In addition, we established that glucocerebrosidase is initially synthesized as a soluble protein, which becomes membrane-associated later in the transport pathway.

REFERENCES

Bangham, A.D., and Horne, R.W. (1962) Action of saponin on biological membranes. Nature 196, 952-953.

Braun, M., Waheed, A., and Von Figura, K. (1989) Lysosomal acid phosphatase is transported to lysosomes via the cell surface. EMBO J. 8, 3633-3640.

Diment, S., Leech, M.S., and Stahl, P.D. (1988) Cathepsin D is membrane-associated in macrophage endosomes. J. Biol. Chem. 263, 6901-6907.

Kornfeld, S., and Mellman, I. (1989) The biogenesis of lysosomes. Annu. Rev. Cell Biol. 5, 483-525.

Neufeld, E.F., and McKusik, V.A. (1983) Disorders of lysosomal enzyme synthesis and localization: I-cell disease and pseudo-Hurler polydistrophy. eds. JB Stanburry, JB Wyngaarden, DS Frederickson, JL Goldstein, MS Brown) In: The Metabolic Basis in inherited disease. pp. 778-787.

Prieels, J.-P., Pizzo, S.V., Glasgow, L.R., Paulson, J.C., and Hill, R.L. (1978) Hepatic receptor that specifically binds oligosaccharides containing fucosyl α1-3 N-acetylglucosamine linkages. Proc. Natl. Acad. Sci. USA 75, 2215-2219.

Rijnboutt, S., Aerts, H.M.F.G., Geuze, H.J., Tager, J.M., and Strous, G.J. (1990) Mannose 6-phosphate-independent membrane association of cathepsin D, glucocerebrosidase and sphingolipid-activating protein in HepG2 cells. J. Biol. Chem. *in press*.

Strous, G.J., and Van Kerkhof, P. (1989) Release of soluble resident as well as secretory proteins from HepG2 cells by partial permeabilization of rough endoplasmic reticulum membranes. Biochem. J. 257, 159-163.

Von Figura, K., and Hasilik, A. (1986) Lysosomal enzymes and their receptors. Annu. Rev. Biochem. 55, 167-193.

Wassler, M., Jonasson, I., Persson, R., and Fries, E. (1987) Differential permeabilization of membranes by saponin treatment of isolated rat hepatocytes. Release of secretory proteins. Biochem J. 247, 407-415.

1.8. HETEROGENEITY OF LYSOSOMES AND AUTOPHAGY

(Chairperson : Trond BERG)

Whereas lysosomes as a whole can be functionally defined as the organelles containing the bulk of acid hydrolases and primarily responsible for hetero- and autophagic degradation, their heterogeneity is puzzling. **Simone Wattiaux-De Coninck** described the distribution and possible function of a 94 kDa rat liver lysosomal membrane glycoprotein, identified by a monoclonal antibody. This antigen is found in hepatocytes but predominates in sinusoidal cells. In homogenates, it is associated with three separable populations of particles containing different amounts of lysosomal enzymes. The antibody immunoprecipitates a complex resembling that described by Galjaard and colleagues, made of β-galactosidase, neuraminidase and a protective protein.

Trond Berg further emphasized that the biochemical heterogeneity of lysosomes in rat liver homogenates reflects not only differences between cell types, but also within a given cell. Distributions obtained by a "trap-label" (a technique allowing retention of the degradation product) indicate that only a subgroup of lysosomes is actively involved in degradation, both in the heterophagic and the autophagic pathways. The remainder, defined as residual bodies, are gradually recruited. He proposes that a two-step degradation pathway may provide adaptation to changing loads of substrate. By discharging their content to residual bodies, active lysosomes remain fully available for new substrates coming from endosomes or amphisomes.

Another important contribution of lysosomes is the degradation of cellular material by autophagy, a nonselective process amounting to up to 4 % of protein in isolated rat hepatocytes. **Per Seglen** has carried out a detailed dissection of this pathway and identified useful inhibitors. The first step, sequestration in an autophagosome, is best inhibited by 3-methyladenine, and is sensitive to hormones affecting the level of c-AMP and protein phosphorylation. This is followed by a vinblastine-sensitive merging with a late prelysosomal stage of the heterophagic pathway, producing a mixed organelle, called amphisome. The final transport from amphisomes to lysosomes is blocked by asparagine. These pharmacological tools should lead to the isolation and characterization of the various intermediates.

NATO ASI Series, Vol. H 62
Endocytosis
Edited by P. J. Courtoy

LGP10D10, A LYSOSOMAL MEMBRANE PROTEIN

S. Wattiaux-De Coninck[1], MM Gonze[1], L. De Waele[1], F. Mainferme[1], P. Van Der Smissen[2], P.J. Courtoy[2], J. Thirion[1], JJ Letesson[3] and R. Wattiaux[1]
[1]Laboratoire de Chimie Physiologique, Facultés Universitaires ND de la Paix, Namur, Belgium, [2] Cell Biology UCL-ICP, Brussels, Belgium and [3]Laboratoire d'Immunologie, Facultés Universitaires ND de la Paix, Namur, Belgium

Introduction

LGP10D10 is a 94 Kd glycoprotein recognized by a mouse monoclonal antibody directed against rat liver lysosomal membrane (Gonze et al., 1987). In rat liver and other tissues, lysosomes are heterogeneous. That results from the existence of probably different kinds of lysosomes in the same cell (intracellular heterogeneity) and from the presence of different types of cells in the organ (intercellular heterogeneity). Therefore we believe that when a lysosomal membrane protein has been identified in a tissue, it is interesting to see if its distribution amongst lysosomes is relatively homogeneous or is heterogeneous. To answer such questions, there are two experimental approaches. One is morphological and is based on the immunocytochemical detection of the membrane protein and of some lysosomal enzymes. A second consists in establishing the distribution by centrifugation of the membrane protein and to compare such distribution with that of lysosomal enzymes. This is what has been done in the work reported here. In addition, the biogenesis of that protein has been investigated. Results suggest that LGP10D10 could be related to a lysosomal complex containing ß galactosidase, neuraminidase and a protective protein, described by Verheijen et al. (1982)

Materials and Methods

Isopycnic centrifugation was performed on an ML fraction originating from normal rat liver, from a rat injected with 0.9 mg invertase/100 g body weight for 3 hours, from a rat injected 4 days before with 85 mg Triton WR-1339/100 g body weight. Rats were fasted overnight. Purification of rat liver lysosomes was performed as described previously except that Nycodenz (Nyegaard, Oslo, Norway) was used instead of Metrizamide (Wattiaux et al. 1978)). Lysosomal matrix and membranes were separated by centrifugation at 105000 g for 40 min after an hypotonic shock; the remaining pellet was washed successively with 0.15 M NaCl, 1.5 M NaCl and 0.15 M NaCl. Lysosomes, lysosomal matrix and lysosomal membrane preparations were labelled in vitro with ^{125} I as described by Greenwood et al. (1963)

Enzymes and LGP10D10 assays: Cathepsin C and arylsulfatase activities were determined as described by Jadot et al. (1986). LGP10D10 content in isolated fractions was

NATO ASI Series, Vol. H 62
Endocytosis
Edited by P. J. Courtoy

measured by a competition assay using ^{35}S methionine labelled antibody. Dilutions of the fractions to be analyzed were preincubated with labelled antibody for 4 hours at 4°C and incubated overnight at 4°C in 96-well plates coated with purified rat liver lysosomal membranes; after washing, wells were counted for radioactivity. Results were expressed as units of inhibition i.e. the fraction dilution corresponding to 50% binding inhibition of the antibody at the bottom of the wells, as compared with a control.

Antibodies : 10D10 is a monoclonal antibody raised against purified rat-liver lysosomal membrane (Gonze et al., 1987)). 7C7, a mouse monoclonal antibody used as control, was raised against bovine Major Histocompatibility Complex of class II. Mouse monoclonal and rabbit polyclonal antibodies raised against ß-galactosidase of E.Coli were obtained respectively from Promega (Madison, WI-USA) and Cappel (Cochranville, PA-USA). An antiserum against neuraminidase from Vibrio Cholerae (Sigma, St Louis, USA) was obtained by immunization of Balb/C mice. Rabbit anti-cathepsin C was obtained as described by Mainferme et al.(1985)

Immunoprecipitation : ^{125}I labelled lysosomes, lysosomal membrane and lysosomal matrix were solubilized in RIPA medium (100 mM Tris/HCL pH 7.4, 0.1% SDS, 1% Triton X100, 1% Na deoxycholate, 1 mM PMSF, 0.02% azide, 1 μM iodoacetate, 1 μM aprotinin, 10 mg/ml leupeptin), or in NP4O medium (100 mM phosphate pH 7.4, 150 mM NaCl, 5 mM EDTA, 2% NP40, 1 mM PMSF, 0.02% azide, 1 μM iodoacetate, 1 μM aprotinin, 10 mg/ml leupeptin). ^{35}S methionine pulse-chase labelled hepatocytes were solubilized in NP40 medium.

Immunocytochemistry: Control male Wistar rats or rats injected intraperitoneally 16 hours before sacrifice with 4 ml of 50% (w/v) sucrose were fasted overnight. The liver was perfusion-fixed with 4% formaldehyde, plus 0.1% glutaraldehyde in 100 mM phosphate buffer pH 7.4. The liver was removed and postfixed by immersion in formaldehyde alone at 4°C. Frozen sections (0.5 μm nominal thickness) were incubated with a mixture of mouse monoclonal 10D10 and rabbit polyclonal anti-cathepsin C Ig. Fluorescein-conjugated goat anti-rabbit IgG (Cappel) and rhodamine-conjugated sheep anti-mouse IgG were added as secondary antibodies.

Results and Discussion.

1. Molecular weight of the mature antigen

Immunofixation, immunoprecipitation and immunoblotting of purified lysosomal membranes (labelled or not with ^{125}I) and solubilized in RIPA buffer, allow to detect a 94 kDa polypeptide as the unique antigen.

2. Intracellular and intercellular localization of LGP10D10 in rat liver

Differential centrifugation of a rat liver homogenate shows that 40-45% of LGP 10D10 is associated with the total mitochondrial fractions ML. Such a fraction contains 75-80% of acid hydrolases.

1) Isopycnic centrifugation of an ML fraction of normal rat liver layered at the top of a sucrose gradient (density limits 1.09-1.26 g/ml) shows a distribution quite similar for LGP 10D10, arylsulfatase and cathepsin C (Fig.1a). The median equilibrium density of these

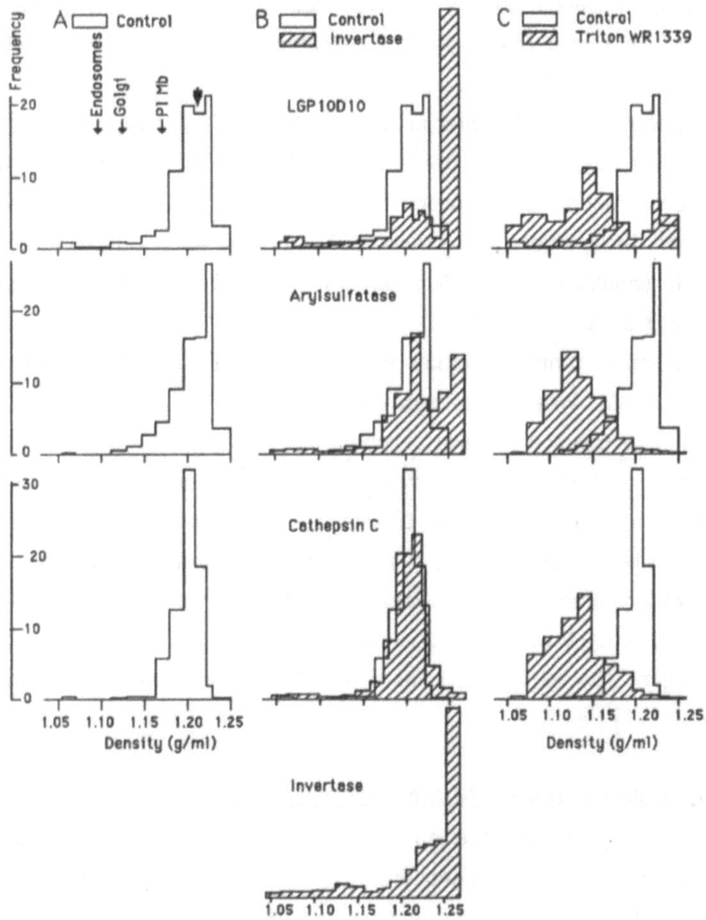

Fig.1. Distribution of LGP10D10, arylsulfatase, cathepsin c and invertase after isopycnic centrifugation in a sucrose gradient of liver mitochondrial fractions (ML) from a control rat (A), a rat injected with invertase (B) and a rat injected with Triton WR1339 (C). *Ordinate* : frequency i.e. the % of activity found in the fraction divided by the limits of density of the fraction. *Abscissa*: density in g/ml.

three components is between 1.2O4-1.209 g/ml. Distribution profiles and median equilibrium densities of LGP 10D10, arylsulfatase and cathepsin C were definitely different from those of alkaline phosphodiesterase (1.17 g/ml, plasma membrane), galactosyltransferase (1.127 g/ml, Golgi apparatus) and asialofetuin, 10min after its injection (1.104 g/ml, hepatocytes endosomes). These results confirm that in rat-liver LGP 10D10 is associated with lysosomes.

2) Injection of invertase to rats allows to establish the repartition of a lysosomal component between endothelial cells and hepatocytes (Jadot et al., 1985), owing the specific shift of density that affects endothelial cell lysosomes. If the rat was injected with invertase for 3 hours, (Fig. 1b), about 55% of invertase was recovered in the high density regions of the gradient (1.255 g/ml). The distribution of cathepsin C was not modified at all by the injection of invertase. Cathepsin C is a lysosomal enzyme mainly located in parenchymal cells as its distribution in such a gradient is not influenced by the accumulation of invertase. Arylsulfatase distribution is only partly shifted towards high densities. This enzyme is shared between parenchymal and endothelial cells. 65 % of LGP 10D10 is recovered in the last fraction of the gradient along with 55% of invertase. LGP 10D10 is thus most concentrated in endothelial cells. About 35 % of LGP10D10 of the liver is localized in the lysosomes of hepatocytes.

3) Injection of Triton WR1339 to rats decreases the density of all lysosomes (Wattiaux et al.1963). When a rat was injected with Triton WR 1339 (Fig.1c) lysosomal marker enzymes distribution remains unimodal and is shifted towards low densities (1,130 g/ml instead of 1.206 g/ml). However, the distribution of LGP10D10 becomes heterogeneous. About 25% of LGP10D10 remains at the bottom of the gradient demonstrating that these organelles containing that protein did not endocytose Triton WR1339. These denser LGP10D10 bearing structures are devoid of cathepsin C and arylsulfatase activities. The remainder of LGP10D10 was divided in two populations equilibrating at densities of 1.11 g/ml and 1.16 g/ml. The first one contains a lower level of the two enzymes, the second one, a higher level.

3. Immunolocalization of LGP10D10 and cathepsin C in rat liver.

After swelling by a sucrose load, in double labelling experiments, LGP10D10 signal can be clearly attributed to the periphery in a large number of sectioned particles, while cathepsin C is diffusely found in the entire granule.(Fig.2)

The similarity between the two labelling patterns gives evidence for the lysosomal localization of LGP10D10. Whereas all structures containing cathepsin C are also labeled for LGP10D10, a small amount of structures found in the same region of hepatocytes are labelled only for LGP10D10.

Immunocytochemistry on the liver of a rat injected with invertase, using a polyclonal antibody against invertase and 10D10 (results not shown) confirms the sinusoïdal localization of invertase and of the major part of LGP10D10. Hepatocytes contain LGP10D10 in a less significant proportion.

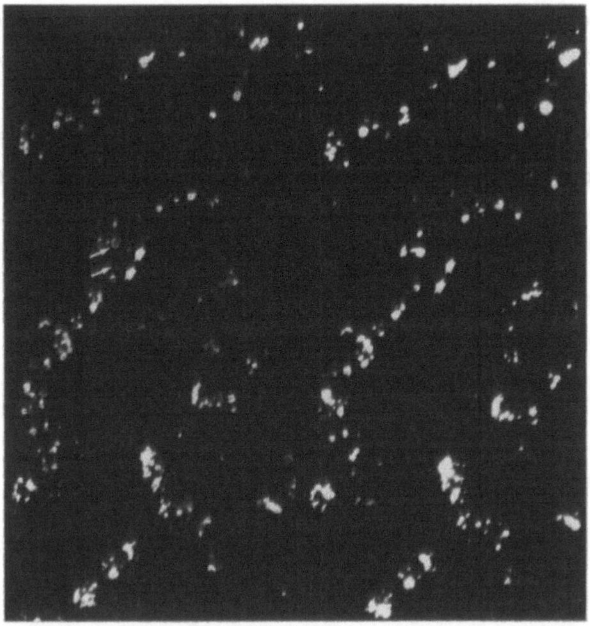

Fig.2. Immunolocalization by double labeling of LGP10D10 and cathepsin C in a rat liver injected with sucrose. Left panel:10D10. Right panel: Anti-cathepsin C. Magnification: 700 X

4. Biogenesis of LGP 10D10

Non Idet P40 medium was used for extraction of proteins. This detergent is known to respect multiprotein complexes (Labeta et al.,1988); this is not the case for detergents like sodium deoxycholate or sodium dodecylsulfate (RIPA medium)

Immunoprecipitation with 10D10 of extracts of hepatocytes pulse-labelled for 5 min and chased for *30 min* shows a family of polypeptides of MW : >200 KD, 200 KD, 125 KD, 94KD, 85 KD, 66 KD, 64 KD, 54 KD and 48 KD. However when chase lasted for *4 hours*, the number of polypeptides of the immunoprecipitate decreases (Fig 3). Finally after 14 hours of chase, only LGP 10D10 (94KD) subsisted.

As it was demonstrated that NP40 is a non dissociating detergent for multiprotein complexes and as, on the other hand a complex containing ß galactosidase (64KD: monomer, 125 KD: dimer and 200 KD: tetramer), soluble neuraminidase (66KD) and a protective protein (54KD) has been described (Verheijen et al.), and that after *30 min* of chase bands of the same MW are immunoprecipitated by 10D10, we wondered if LGP10D10 could be related to this complex.

As we had no rat liver anti-ß galactosidase, we used a monoclonal antibody against ß galactosidase of E.Coli. Comparison of immunoprecipitates with anti-ß galactosidase and 10D10 after 5 min pulse labelling with ^{35}S methionine and *30 min* of chase of hepatocytes (Fig 3) demonstrates that all bands except those of 94 KD and 85 KD are present in both lanes. The protein of 85 KD is probably directly linked to the 94 KD LGP10D10. Indeed, both are not immunoprecipitated by an anti ß galactosidase antibody. Furthermore the 85 and the 94 KD products can be detected 30 min after the pulse labelling; after 4 hours of chase the 94 KD polypeptide remains apparent and the labelling of the 85 KD totally vanishes. Therefore the 85 KD polypeptide is probably the precursor of the mature 94 KD product. Treatment with endoglycosidase F shows that the polypeptidic core of LGP10D10 has an apparent molecular weight of 48 KD .

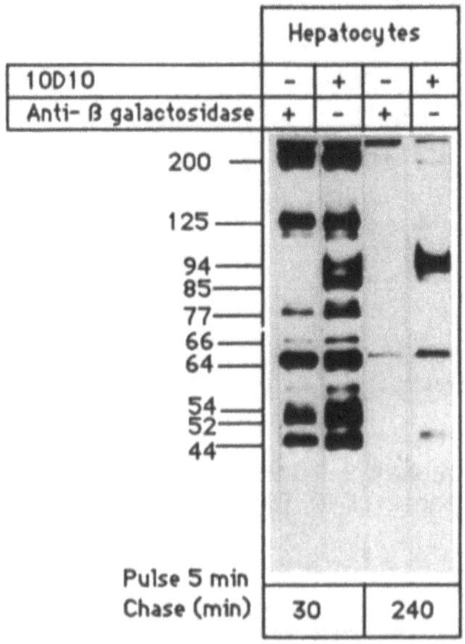

Fig. 3. Biosynthesis of LGP10D10 by isolated hepatocytes. Hepatocytes were pulse-labelled and chased as indicated. Cells were extracted with a solubilization medium containing NP40, and immunoprecipitates were obtained with 10D10 and anti-ß galactosidase (E. Coli).

Why, after a chase of *4 hours*, 10D10 does not precipitate the complex from pulse labeled hepatocytes. To clear up this question, ^{125}I labelled lysosomes, lysosomal membrane and matrix were used, in which all protein synthetized, recently or formerly, can be detected. Immunoprecipitates with 10D10 of total lysosomes, of lysosomal membrane but not of lysosomal matrix extracts contain the proteins of 125 KD, 94 KD, 66 KD, 64 KD, 54 KD and 48 KD (Fig 4). Thus the components of the complex can be exchanged with preexisting non labeled components as it is the case in pulse-chase labelled hepatocytes

The next question was to know why the monoclonal antibody against ß galactosidase of E. Coli doesn't precipitate anything after *4 hours* of chase? Fig 4 shows that this monoclonal antibody is unable to immunoprecipitate any protein either from in vitro labelled total lysosomes or lysosomal membrane or lysosomal matrix. A polyclonal antibody against E. Coli ß galactosidase, but not the monoclonal antibody, recognizes the mature form of the 64 KD protein of lysosomes. These data suggest that in isolated hepatocytes and rat liver lysosomes, the proteins of 64 KD, 125 KD and 200 KD could be ß galactosidase. What could be the significance of LGP10D10? Verheijen et al. described a soluble neuraminidase of 66 KD in

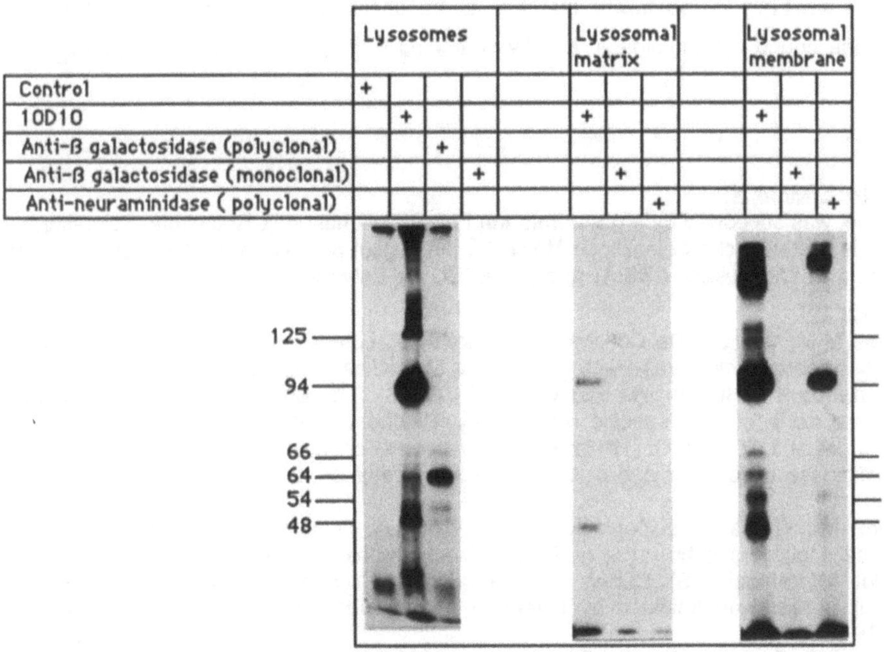

Fig. 4. Immunoprecipitation of peptides from purified in vitro labelled lysosomes, lysosomal membrane and lysosomal matrix. Antibodies were 10D10, monoclonal and polyclonal anti-ß galactosidase (E. Coli) and anti-neuraminidase (V. cholerae).

the lysosomal complex containing also ß galactosidase and a protective protein. However, Horvat and Touster (1968) described that lysosomal neuraminidase is a lysosomal membrane enzyme. LGP10D10 could be this membrane associated form of neuraminidase, whose MW has not been ascribed so far. For lack of purified rat-liver lysosomal membrane neuraminidase and the corresponding antibody, we used a polyclonal antibody against V. Cholerae neuraminidase. This antibody immunoprecipitates from an extract of lysosomal membrane but not from lysosomal matrix a protein of 94 KD.

These results suggest the following hypothesis.

As LGP10D10 is a lysosomal membrane protein of 94 KD, immunoprecipitated by 10D10 monoclonal antibody,

as LGP10D10 is immunoprecipitated, if NP40 detergent is used, along with proteins of MW of 200 KD, 125KD, 66 KD, 64 KD, 54 KD,

as these MW correspond to those described for ß-galactosidase (64KD, 125KD, 200 KD), for soluble neuraminidase (66 KD) and protective protein (54 KD),

as a protein of 94 KD is immunoprecipitated by Vibrio cholerae anti-neuraminidase, and finally

as lysosomal neuraminidase activity has been described as a lysosomal membrane enzyme, we suggest that LGP10D10 could be the lysosomal membrane neuraminidase.

This is a **working hypothesis** that has to be checked by purifying rat-liver lysosomal membrane bound neuraminidase, and by searching if this enzyme is immunoprecipitated by 10D10.

ACKNOWLEDGEMENT.
This work was supported by a grant from the Fonds National de la Recherche Scientifique, the Fonds de la Recherche Scientifique Médicale, the Institut pour la Recherche Scientifique dans l'Industrie et l'Agriculture (IRSIA) and the ASBL Air Escargot.

REFERENCES

- Gonze M M, Wattiaux-De Coninck S, Mainferme F, Letesson JJ and Wattiaux R (1987): A study of lysosomal membrane with monoclonal antibodies. Bioch.Soc.Trans. 15:436
- Greenwood FC, Hunter WM and Glover JS (1963): The preparation of I-labelled human growth hormone of high specific radioactivity. Biochem.J., 89: 114-123.
- Horvat, A and Touster O.,(1968): On the lysosomal occurrence and the properties of the neuraminidase of rat liver and of Ehrlich Ascites tumor cells. J. of Biol. Chem. 243: 4380-4390.
- Jadot, M., Wattiaux-De Coninck, S. and Wattiaux, R., (1985) Effect on lysosomes of invertase endocytosed by rat liver. Eur. J. Biochem. 151: 485-488.
- Jadot M, Misquith S, Dubois F, Wattiaux-De Coninck S and Wattiaux R (1986) Intracellular pathway followed by invertase endocytised by rat liver.Eur.J.Biochem. 161: 695-700.
- Labeta MO, Fernandez N and Festenstein H (1988): Solubilisation effect of Nonidet P-40, Triton X-100 and CHAPS in the detection of MHC-like glycoproteins. J. of Immunol. Meth, 112: 133-138.
- Mainferme,F., Wattiaux,R. and von Figura,K. (1985):Synthesis, transport and processing of cathepsin C in Morris hepatoma 7777 cells and rat hepatocytes. Eur.J. Biochem. 153: 211-216
- Verheijen,F., Brossmer,R. and Galjaard,H. (1982): Purification of acid ß galactosidase and acid neuraminidase from bovine testis: evidence for an enzyme complex. Biochem.Biophys.Res.Comm. 108: 868-875.
- van der Horst GTJ, Galjart NJ, d'Azzo A, Galjaard H and Verheijen FW (1989): Identification and in vitro reconstitution of lysosomal neuraminidase of human placenta. J. of Biol. Chem., 264: 1317-1322.
- Wattiaux,R., Wattiaux-De Coninck,S., Ronveaux-Dupal,M.F. and Dubois,F. (1978): Isolation of rat-liver lysosomes by isopycnic centrifugation in a metrizamide gradient. J.Cell Biol. 78: 349-367.
- Wattiaux R. Wibo M, Baudhuin P: Influence of the injection of Triton WR1339 on the properties of rat liver lysosomes. In de Reuck AV, Cameron MP (eds). Ciba Symposium on Lysosomes. J.&A. Churchill Ltd London 1963: 176-200.

RECEPTOR-MEDIATED ENDOCYTOSIS IN LIVER ENDOTHELIAL CELLS. EVIDENCE OF LYSOSOMAL HETEROGENEITY

T. Berg, S. Magnusson, E. Stang, N. Roos
Department of Biology, University of Oslo
P.O.Box 1050, Blindern 0316 Oslo 3, Norway

Hepatic lysosomal heterogeneity is partly due to differences between lysosomes in the different types of liver cells. It has been shown, for instance, that the composition and contents of various lysosomal enzyme activities are distinctly different in the parenchymal cells (PC) and the nonparenchymal liver cells (Berg and Boman 1973). The biochemical and morphological differences are also reflected in the finding that lysosomes of PC, Kupffer cells (KC) , and endothelial cells (EC) show different density distributions after isopycnic centrifugation in sucrose and Nycodenz gradients (Kindberg et al 1990). The density distribution of lysosomes of individual cells can be followed after injecting intravenously 125I-tyramine-cellobiose-labelled ligands that are selectively taken up by receptor-mediated endocytosis in only one cell type (Berg et al 1985). Following degradation the labelled degradation products are trapped in the lysosomes of the cells in which uptake took place and may therefore serve as markers for these organelles. Lysosomal heterogeneity may also be seen in the individual cells. By means of subcellular fractionation techniques combined with the "trap-label" method it has been possible to observe subpopulations of lysosomes in PC and EC that may be involved in degradation of substrates brought into the lysosomes by heterophagy and autophagy. Nutritional step-down conditions induce autophagy in surface-cultures of rat hepatocytes. During a transition period two populations of lysosomes, differing in buoyant density in Nycodenz gradients, can be observed (Berg et al 1987). Evidently, the autophagosomes fuse with only a subgroup of active lysosomes, rendering these organelles more buoyant. With

NATO ASI Series, Vol. H 62
Endocytosis
Edited by P. J. Courtoy

time of incubation gradually more "resting lysosomes" are recruited and eventually all lysosomes are involved in autophagy. A dual distribution of lysosomes may also be observed in connection with receptor-mediated endocytosis in both PC and EC. Using 14C-sucrose-labelled asialo-fetuin it was found some years ago that the first degradation products appeared in a buoyant group of lysosomes (in hepatocytes in suspension cultures), banding in sucrose gradients at about 1.12 g/ml. Subsequently degradation products accumulated at the same density as the lysosomal enzymes, 1.19 g/ml (Tolleshaug et al 1983). The purpose of the present report is to summarize our studies of receptor-mediated endocytosis in rat liver endothelial cells and discuss the possible involvement of two groups of lysosomes in these cells.

The liver endothelial cells have an essential homeostatic role by removing from blood denatured proteins and macromolecules that have leaked out of damaged cells and tissues (Review: Smedsrød et al 1990). We have particularly studied uptake mediated by the mannose receptor and the scavenger receptor, and the ligands used include ovalbumin (OA), mannosylated serum albumin (MSA), formylated serum albumin (FSA), tissue plasminogen activator (tPA), and ricin.

It is typical of receptor-mediated endocytosis in liver endothelial cells that internalization of receptor-ligand complex is exceedingly rapid. The T1/2 for internalization of both OA (Magnusson and Berg 1989) and FSA (Eskild et al 1989) is about 15 sec, and this rate is at least 10 times higher than that of internalization of asialoglycoproteins in PC (Tolleshaug 1981). The reason for the relatively rapid internalization may in part be that the receptors are concentrated in coated pits. Immunogold-labelling of OA and tPA in cryosections from liver following intravenous injection of the ligands have demonstrated label exclusively in coated pits (Stang et al 1990). These data indicate that the receptors are reinserted very rapidly into coated pits after recycling back to the plasma membrane, the transport in the plasma membrane to the coated pits will therefore not be a rate-limiting step (Ward and Kaplan 1990). The role of clatrin is nicely demonstrated in the study of receptor-mediated uptake of ricin in EC. Ricin is internalized both via the mannose receptor in coated pits and via galactose-terminated glycoproteins and glycolipids mainly outside coated pits. There are at least 100 times more lectin binding sites than mannose receptors. Nevertheless, the mannose-receptor mediated uptake is more efficient than that taking place outside coated pits (Magnusson and Frenoy, unpublished observations).

The intracellular transport of ligands in EC has been followed both by means of biochemical methods (subcellular fractionation) and by electron microscopy of ultrathin cryosections in which the ligands were identified by immunogold labelling. The electronmicroscopic investigations showed that tPA and OA were, following uptake in coated pits, in coated vesicles and then in small tubular and cisternal (cup-shaped) structures reminiscent of those described by Griffiths and coworkers (1989). After 6 min the ligand had entered endosomes which contained Lgp120. These "late" endosomes did not contain cathepsin D that could be identified with the colloidal-gold method. After 12 min and 24 min the ligand could also be seen in lysosomes that contained both Lgp120 and cathepsin D (Fig 1).

Subcellular fractionation by means of sucrose density gradients following intravenous injection of 125I-TC-OA, 125I-TC-MSA or 125I-TC-FSA showed that the ligands were sequentially in three groups of organelles of increasing density (Eskild et al 1989, Kindberg et al 1990). During the first min after injection the radioactivity was at about 1.14g/ml, then from 1-6 min a peak appeared at 1.17 g/ml, and after 6 min the ligand accumulated at 1.22 g/ml. With time after injection acid soluble radioactivity accumulated at this density. Invertase is selectively taken up in liver EC after intravenous injection (Wattiaux et al 1986) and renders the EC lysosomes selectively denser in a sucrose gradient. When invertase was injected into rats 24 hours prior to injection of 125I-TC-OA and subcellular fractionation in sucrose gradients was performed at various times after the injection it was observed that labelled degradation products appeared sequentially at two

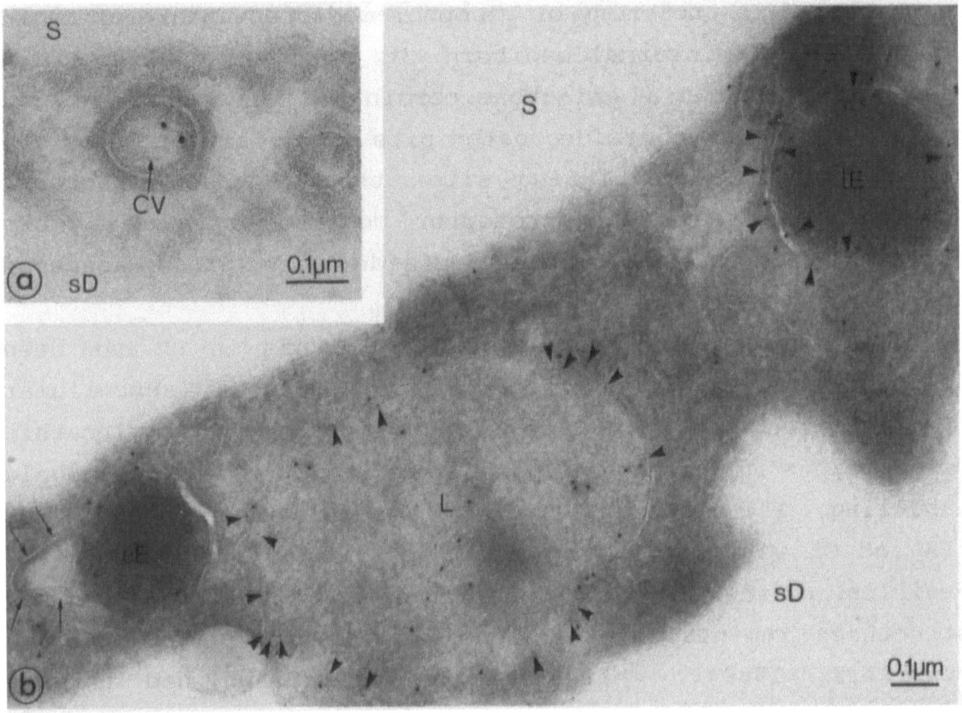

Fig. 1. Electron micrographs showing immunocytochemical localization of tissue-type plasminogen activator (t-PA) endocytosed in EC. tPA is taken up via coated pits and coated vesicles (CV) shown in 1a. Fig 1b shows the distribution of t-PA (large gold particles) and lgp120 (small gold particles, arrow heads) 12 min after intravenous injection. t-PA is in early endosomes (eE) consisting of cisternae (arrows) with vesicular dilatations, in late endosomes (lE) or carrier vesicles which differ from eE by the presence of lgp-120 in their surrounding membrane, and in large electron lucent lysosomes (L).

density regions of the gradient.

These results suggested that degradation of 125I-TC-OA was initiated in a more buoyant organelle than that containing the bulk of the invertase. Wattiaux and coworkers at the same time observed that labelled degradation products from 125I-TC-FSA (injected intravenously into rats) appeared sequentially in two groups of lysosomes of increasing density in Percoll gradients

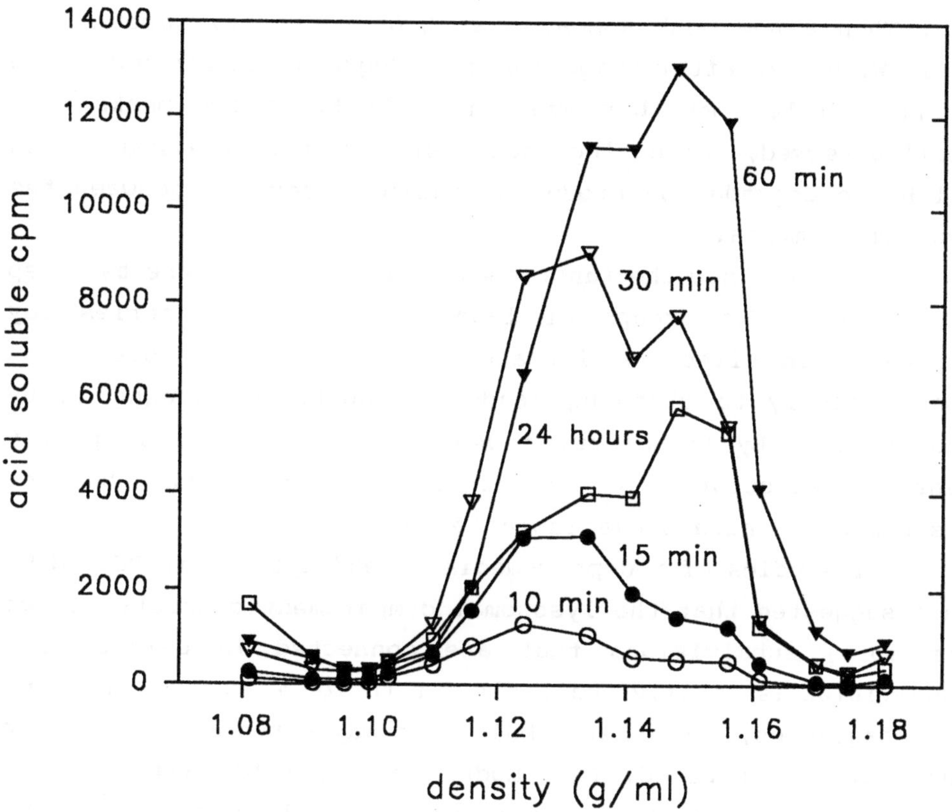

Fig. 2. Distribution in Nycodenz gradients of labelled degradation products formed from 125I-TC-MBSA in liver EC. Postnuclear fractions were prepared at the indicated times from livers of rats which had been injected intravenously with 125I-TC-MSA (0.8 nmol). Degraded ligand in each fraction was determined as radioactivity soluble in 10% trichloroacetic acid.

(Misquith et al 1988). Their data, combined with our findings, suggested that isotonic gradients (e.g. Nycodenz gradients) might separate lysosomes that coincided in the hypertonic sucrose gradients. We found, in accordance with this notion, that isopycnic centrifugation in Nycodenz gradients revealed two groups of lysosomes involved in the degradation of 125I-TC-MSA. In a typical experiment (Fig. 2) the degradation products were first seen at about 1.12 g/ml after 6-10 min. The degradation products accumulated at this density for about 10 min. Then a new peak appeared at higher density, about 1.14 g/ml. With time after injection more degradation products were found at 1.14 g/ml, but even after 24 hours two peaks were still observed, indicating that a steady state was reached in which the degradation products possibly recycled between the two compartments.

We do not know the functional significance of the two step degradation. Two groups of lysosomes acting in series may represent an effective intracellular digestive system that adapts easily to changing loads of substrate brought in by endocytosis. By transferring degradation products or partly degraded substrate to the denser lysosomes, the buoyant lysosomes are always available for new substrate.

Our studies of receptor-mediated endocytosis in PC and EC have suggested that the lysosomal compartment consists of two (or more) subdivisions that are connected in series. One subdivision is readily available for fusion with endosomes and autophagosomes, the other subdivision may serve as a reservoir that may be recruited. This model is compatible with earlier observations suggesting that only a subdivision of the lysosomes ("active lysosomes") is readily available for fusion with endosomes and phagosomes. The remainder of the lysosomes ("residual bodies") are recruited more slowly (Griffiths et al 1989, Seglen and Solheim 1980, Wattiaux et al 1989).

References

Berg T, Boman D (1973) Distribution of lysosomal enzymes between parenchymal and Kupffer cells of rat liver.Biochim.biophys Acta 321:585-596

Berg T, Kindberg, GM, Ford T, Blomhoff R (1985) Intracellular transport of asialoglycoproteins in rat hepatocytes. Evidence for two subpopulations of lysosomes. Exp Cell Res 161:285-296

Eskild W, Kindberg GM Smedsrød B, Blomhoff R, Norum KR, Berg T (1989) Intracellular transport of formaldehyde- treated albumin in liver endothelial cells after uptake via scavenger receptors. Biochem J 258:511-520

Griffiths G, Back R, Marsh M (1989) A quantitative analysis of the endocytic pathway in baby hamster kidney cells. J Cell Biol 109:2703-2720

Kindberg GM, Magnusson S, Berg T, Smedsrød B (1990) Receptor-mediated endocytosis of ovalbumin by two carbohydrate specific receptors in rat liver cells. Biochem J 270:197-

203
Kindberg GM, Refsnes M, Christoffersen T, Norum KR, Berg T (1987) The relationship between autophagy and the intracellular degradation of asialoglycoproteins in cultured rat hepatocytes. J Biol Chem 262:7066-7071
Kindberg GM, Stang E, Andersen KJ, Roos N, Berg T (1990) Intracellular transport of endocytosed proteins in rat liver endothelial cells. Biochem J 270:205-211
Magnusson S, Berg T (1989) Extremely rapid endocytosis via the mannose receptor of sinusoidal endothelial rat liver cells. Biochem J 257:651-656
Misquith S, Wattiaux-De Coninck, Wattiaux R (1988) Uptake and intracellular transport in rat liver of formaldehyde-treated bovine serum albumin labelled with 125I-tyramine cellobiose. Eur J Biochem 174:691-697
Seglen PO, Solheim AE (1985) Conversion of dense lysosomes into light lysosomes during hepatocytic autophagy. Exp Cell Res 157:550-555
Smedsrød B, Pertoft H, Gustafson S, Laurent TC (1990) Scavenger functions of the liver endothelial cell. Biochem J 266:313-327
Stang E, Kindberg GM, Berg T, Roos N (1990) Endocytosis mediated by the mannose receptor in liver endothelial cells. An immunocytochemical study. Eur J Cell Biol 52:67-76
Tolleshaug H (1981) Binding and internalization of asialo-glycoproteins by isolated rat hepatocytes. Int J Biochem 13:45-51
Tolleshaug H, Blomhoff R, Berg T (1983) Intracellular transport of endocytosed glycoproteins in rat hepatocytes. In: Protein Synthesis pp 267-285. Abraham KA, Eikholm TS, Pryme IF (eds) The Humana Press
Ward DM, Kaplan J (1990) The rate of internalization of different receptor-ligand complexes in alveolar macrophages is receptor-specific. Biochem J 270:369-374
Wattiaux R, Jadot M, Misquith S, Dubois F, Wattiaux-De Coninck S (1986) Differences in the cellular location of substances endocytosed by rat liver as observed from the distribution patterns obtained after isopycnic centrifugation in a sucrose gradient. Biochem Biophys Res Comm 136:504-509
Wattiaux R, Misquith S, Wattiaux-De Coninck S, Dubois F (1989) Fate of asialofetuin endocytosed by rat liver. Biochem Biophys Res Comm 158:313-318

AUTOPHAGIC-ENDOCYTIC INTERACTIONS IN HEPATOCYTES

P.O. Seglen, P.B. Gordon, I. Holen, H. Høyvik, A.L. Kovács, P.E. Strømhaug and T.O. Berg
Department of Tissue Culture, Institute for Cancer Research, The Norwegian Radium Hospital, Montebello, 0310 Oslo 3, Norway

Autophagy: a multistep pathway

Autophagy is the major mechanism by which cells degrade their cytoplasmic macromolecules to provide amino acids and other nutrients in response to the needs of the organism. In contrast to the non-autophagic mechanisms for protein degradation, which in hepatocytes collecytively account for a more or less constant degradation rate of 1.0-1.5 %/h (Seglen et al., 1979), the rate of autophagic protein degradation may vary from zero to 4 %/h depending on the ambient levels of autophagy-suppressive amino acids (Seglen, 1983).

Ultrastructural and biochemical studies have resolved a number of individual steps in the autophagic pathway. (1) *Autophagic sequestration*, the initial step, is performed by a characteristic organelle to which the name *phagophore* has been assigned (Seglen, 1987). The phagophore has a multilamellar structure staining heavily with lead citrate (Fig. 1). From the condensed phagophore in the cytoplasm a thick membrane spreads out to envelop a region of the cytoplasm, eventually forming a closed vacuole called an *autophagosome* (Fig. 1).

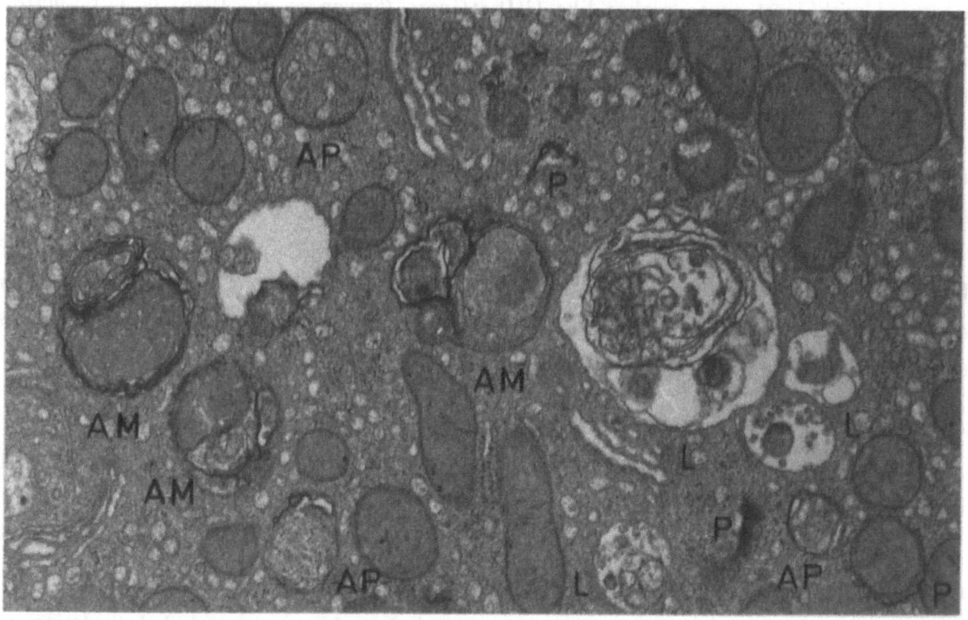

Figure 1. *Organelles of the autophagic pathway.* P, phagophore; AP, autophagosome; AM, amphisome; L, lysosome. Hepatocyte incubated 1h at 37°C with asparagine (10 mM). Fixed with glutaraldehyde (primary) and OsO_4; stained with lead citrate. x 20,000.

NATO ASI Series, Vol. H 62
Endocytosis
Edited by P. J. Courtoy

(2) *Amphisome loading*: Several phagophores may deliver their contents to the same prelysosomal collecting vacuole, probably by penetration through the vacuole wall (Tooze et al., 1990). This prelysosomal autophagic vacuole can also receive endocytic material, and has therefore been given the name *amphisome* (Gordon and Seglen, 1988). Whereas the simpler amphisomes may be difficult to distinguish from autophagosomes without immunocytochemical criteria, complex amphisomes (Tooze et al., 1990), accumulating e.g. in asparagine-treated hepatocytes (Gordon and Seglen, 1988), are easily recognized (Fig. 1). (3) *Fusion* between lysosomes and prelysosomal autophagic vacuoles brings the autophaged cytoplasm in contact with lysosomal hydrolases. (4) *Degradation* (intralysosomal hydrolysis) of proteins and other macromolecules, visible as structural deterioration and the formation of clear zones inside the vacuole now defined as a secondary, active lysosome (Fig. 1) is the last step in the autophagic pathway.

Biochemical approaches to autophagy

Autophagy can be studied biochemically by taking advantage of the fact that the phagophore generates a closed vacuole, thereby rendering soluble cytosol components sedimentable. Electrodisruption of isolated hepatocytes followed by centrifugation of the leaky "cell corpses" through a density cushion effectively separates sedimentable from soluble components (Gordon and Seglen, 1982). The autophagic sequestration step can be measured with inert sugar probes like [^{3}H]raffinose (Seglen et al., 1986) or [^{14}C]sucrose (Gordon et al., 1985), the sugars being loaded into the cytosol by reversible electropermeabilization of the plasma membrane (Gordon and Seglen, 1982; Seglen and Gordon, 1984). Cytosolic enzymes, e.g. LDH, can also be used as sequestration probes, provided the cells are incubated in the presence of leupeptin to prevent intralysosomal proteolysis of the enzymes (Kopitz et al., 1990). In the absence of leupeptin, autophagically sequestered enzymes accumulate only if post-sequestrational steps (fusion or degradation) are blocked (Kopitz et al., 1990); this is also the case with electroinjected [^{14}C]lactose, which is degraded by the lysosomal ß-galactosidase (Høyvik et al., 1986). Both lactose and LDH can therefore be used to probe these later steps in the autophagic pathway (Plomp et al., 1989).

Autophagic sequestration: a nonselective bulk process

Whether autophagy contributes to differential protein turnover or not is a long-standing issue. Ultrastructural studies indicate that autophagy is basically a bulk process, but there is some qualitative evidence for autophagic degradation of specific organelles during tissue remodeling or involution (Holtzman, 1989) which suggests that a kind of crude selectivity may exist. The relationship between autophagy and the selective lysosomal degradation of cytosolic proteins carrying a certain signal domain (Dice and Chiang, 1989) has not been clarified.

To address the question of autophagic selectivity, the sequestration of seven different cytosolic enzymes with half-lives ranging from 1 to 19 h was measured in hepatocytes

Table 1. *Cytosolic liver enzymes with different half-lives are autophagically sequestered at the same rate.*

Enzyme	Half-life (h)	Autophagic sequestration (%/h)	Enzyme degradation (%/h) -Leupeptin	Enzyme degradation (%/h) +Leupeptin	Calculated autophagic degradation (%/h)
Aldolase	18.9	3.6 ± 0.2 (13)	3.6 ± 0.1	0.3 ± 0.2	3.3 ± 0.1
Lactate dehydrogenase	18.9	3.6 ± 0.2 (2)	3.6 ± 0.2	0.1 ± 0.1	3.5 ± 0.2
Glucokinase	12.5	3.4 ± 0.5 (4)	5.4 ± 0.4	2.1 ± 0.6	3.3 ± 0.2
Serine dehydratase	10.7	3.1 ± 0.2 (5)	6.3 ± 0.2	2.8 ± 0.3	3.5 ± 0.1
Tryptophan oxygenase	3.6	3.7 ± 0.1 (2)	17.6 ± 0.1	12.4 ± 1.0	5.2 ± 1.0
Tyrosine aminotransferase	3.4	3.5 ± 0.1 (4)	18.6 ± 0.4	14.8 ± 0.6	3.8 ± 1.0
Ornithine decarboxylase	1.1	3.3 ± 0.1 (3)	47.4 ± 0.5	42.1 ± 1.7	5.3 ± 1.2

Autophagic sequestration (in the presence of 0.3 mM leupeptin) or degradation (in the presence of 1 mM cycloheximide, ± leupeptin) of cytosolic enzymes was measured in hepatocytes incubated at 37°C. Each value is the mean ± S.E./range of two experiments or of the number of experiments given in parentheses. Modified from Kopitz et al., 1990.

incubated in the presence of leupeptin under conditions of maximal autophagy (amino acid-free medium). The enzymes were all sequestered at the same rate of 3.5-4.0 %/h (Table 1), supporting the concept of autophagy as a non-selective bulk process (Kopitz et al., 1990), and suggesting that additional non-autophagic degradation mechanisms must be responsible for differential enzyme turnover.

Regulation of autophagy at the sequestration step

Biological regulation of autophagy is exerted at the sequestration step. The major physiological regulators, the amino acids, were found to suppress autophagic sequestration both individually (leucine and tryptophan being particularly active) and in combination (Seglen, and Gordon, 1984) (Table 2). Adenine, adenosine and several of their analogues were also effective autophagy inhibitors. 3-methyladenine (3MA), which is routinely used to shut down hepatocytic autophagy (Seglen and Gordon, 1982; Plomp et al., 1989) prevented autophagic sequestration virtually completely.

Polypeptide hormones like glucagon, bradykinin, bombesin, angiotensin, vasopressin and insulin had little effect on their own, but insulin potentiated the autophagy-suppressive effect of amino acids (Table 2), thus providing a mechanism for the synergism previously seen in studies on autophagic-lysosomal protein degradation (Poli et al., 1981). Epinephrine also suppressed autophagic sequestration, apparently through an α_1-adrenergic mechanism as suggested by phenylephrine agonism and prazosin antagonism (Tab. 2). The β_2-agonist salbutamol was inactive; indeed β-adrenergic agonists failed to elevate cyclic AMP in our

Table 2. *Effect of amino acids, purines, hormones, cyclic nucleotides and phosphorylation on autophagic sequestration in hepatocytes*

	% Inhibition of autophagic sequestration
Amino acids	
Amino acid mixture (25 mM)	84 ± 7 (6)
Tryptophan (10 mM)	39 ± 2 (3)
Leucine (10 mM)	18 ± 4 (4)
Asparagine (10 mM)	1 ± 10 (5)
Purines	
Adenosine (0.3 mM)	55 ± 3 (3)
N^6-Dimethyladenosine (0.2 mM)	85 (1)
Adenine (10 mM)	81 ± 2 (2)
3-Methyladenine (10 mM)	93 ± 2 (3)
Hormones	
Glucagon (10^{-6} M)	-1 ± 3 (11)
Bradykinin (10^{-6} M)	-1 (1)
Bombesin (10^{-6} M)	2 ± 3 (3)
Angiotensin (I or II, 10^{-6} M)	4 ± 2 (2)
Vasopressin (10^{-6} M)	18 ± 3 (2)
Insulin ($3x10^{-7}$ M)	13 ± 3 (6)
Amino acids (7.5 mM mixture)	46 ± 4 (11)
Amino acids (7.5 mM) + Insulin	70 ± 7 (5)
Epinephrine (100 µM)	42 ± 4 (7)
Epinephrine + prazosin (10^{-7} M)	19 ± 2 (2)
Phenylephrine (α_1-agonist; 100 µM)	35 ± 3 (6)
Salbutamol (β_2-agonist; 100 µM)	-8 ± 6 (3)
Cyclic nucleotides/phosphorylation	
Dibutyryl-cyclic AMP (300 µM)	47 ± 7 (8)
Deacetyl-forskolin (100 µM)	42 ± 18 (3)
Theophylline (10 mM)	91 ± 5 (5)
Isobutyl-methylxanthine (1 mM)	81 ± 10 (2)
Okadaic acid (10^{-6} M)	76 ± 18 (3)

Hepatocytes loaded with [^{3}H]raffinose were incubated for 1-3h at 37°C in the presence of the agents indicated, and the transfer of soluble (cytosolic) radioactivity to sedimentable cell corpses was measured. Per cent inhibition relative to control incubations is expressed as the mean ± S.E. of the number of experiments given in parentheses.

experiments. Dibutyryl-cyclic AMP nevertheless inhibited autophagic sequestration, as did the adenylate cyclase activator deacetyl-forskolin (Table 2). Sequestration was also suppressed by phosphodiesterase inhibitors (theophylline, isobutyl-methylxanthine and many others), suggesting that cyclic AMP may play a role in the regulation of autophagy. Whether the effect of cyclic AMP is mediated by a protein kinase is not clear, but the strong inhibition of sequestration by the protein phosphatase inhibitor okadaic acid (Table 2) makes it likely that protein phosphorylation is involved in autophagy control.

Asparagine: a specific inhibitor of autophagic-lysosomal fusion

As indicated in Table 2 asparagine (Asn) had very little effect at the autophagic sequestration step, yet high concentrations of this amino acid are able to inhibit autophagic protein degradation almost completely (Seglen et al., 1980; Seglen, 1987). A post-sequestrational site of action would therefore seem likely, and is supported by the accumulation of autophagic vacuoles and autophaged [^{14}C]lactose in Asn-treated hepatocytes (Seglen, 1987). The inhibition by Asn appears to be specific for the autophagic pathway, as no effect on degradation of endocytosed protein was observed (Seglen et al., 1980).

As shown in Fig. 2, the degradation of autophaged [^{14}C]lactose was strongly suppressed by Asn, and both lactose and autophaged LDH accumulated in the presence of the amino acid (Fig. 3). To determine whether Asn affected a penultimate fusion step or the final degradation step, the lysosomes were loaded with autophaged [^{14}C]lactose in the presence of the lysosome inhibitor propylamine (Seglen, 1983). Propylamine was then washed out and the cells incubated in the presence of 3MA to allow the detection of intralysosomal lactolysis without any disturbance from additional lactose autophagy (Plomp et al., 1989). Table 3 shows that this intralysosomal lactolysis was inhibited by propylamine, but unaffected by Asn, indicating that the amino acid did not affect the final degradation step in the autophagic pathway. In contrast, the hydrolysis of [^{14}C]lactose accumulated in the presence of Asn could be prevented both by Asn and by propylamine, suggesting that Asn acts at a prelysosomal step to prevent the transfer of autophaged material to the lysosomes. Vinblastine, a fusion inhibitor effective in both the autophagic and the endocytic pathway (Kovács et al., 1982), likewise protected the prelysosomal lactose (accumulated under Asn) but not the lysosomal lactose (accumulated under propylamine), suggesting that, like Asn, it suppressed fusion but not intralysosomal degradation.

The amphisome: a prelysosomal unloading station

Having established that Asn inhibits autophagic-lysosomal fusion but not intralysosomal degradation, it follows that the autophaged lactose accumulating in the presence of Asn must reside in prelysosomal autophagic vacuoles. We have previously shown that exogenously added ß-galactosidase, taken up by hepatocytic endocytosis, can degrade this prelysosomal lactose, indicating autophagic-endocytic convergence in the prelysosomal vacuoles called amphisomes (Gordon and Seglen, 1988). To see if hepatocytic amphisomes might be accessible

Table 3. *Effects of asparagine and propylamine on degradation of [^{14}C]lactose accumulated in lysosomes or in prelysosomal vacuoles*

Additions during final 1-h incubation with 3MA	Degradation of autophaged [^{14}C]lactose (% of total cellular radioactivity /h)	
	Preincubated 2h with propylamine	Preincubated 2h with asparagine
None	2.8 ± 0.2 (8)	2.3 ± 0.2 (6)
Propylamine	0.9 ± 0.3 (7)[a]	1.1 ± 0.5 (4)[c]
Asparagine	2.3 ± 0.4 (5)	1.2 ± 0.3 (3)[b]
Vinblastine	2.7 ± 0.2 (3)	1.3 ± 0.3 (3)[b]

[a]*P<0.001;* [b]*P<0.02;* [c]*P<0.05* vs. the respective control.
Hepatocytes, electroloaded with [^{14}C]lactose, were preincubated 2h with Asn (10 mM) or propylamine (10 mM), then washed and incubated for another hour with 3MA (10 mM) and Asn or propylamine as indicated. Net hydrolysis of [^{14}C]lactose during the latter period was measured with HPLC and expressed as % of total cellular radioactivity. Each value is the mean ± S.E. of the number of experiments given in parentheses.

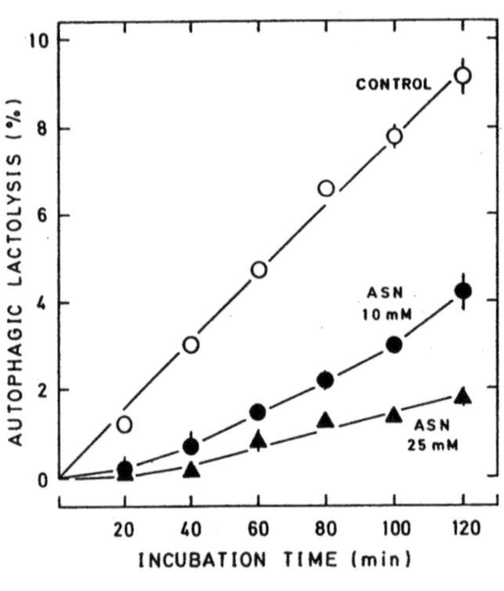

Fig. 2. *Effect of asparagine on autophagic lactolysis.* Hepatocytes loaded with [^{14}C]lactose were incubated at 37°C without additions (O), or with 10 (●) or 25 (▲) mM Asn. The degradation of lactose was measured by HPLC and expressed as % of total cellular radioactivity. Each value is the mean ± S.E. of three experiments.

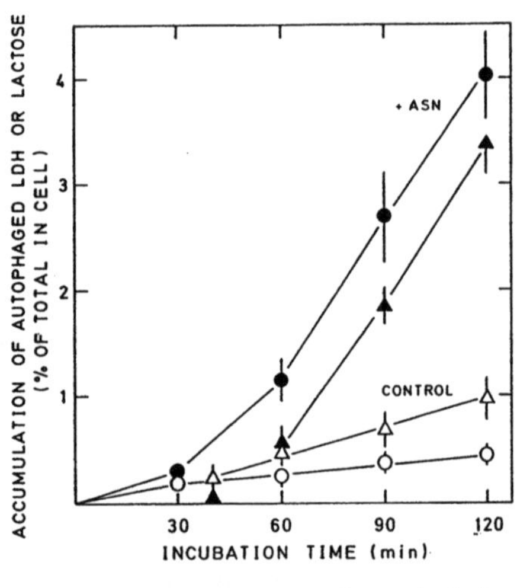

Fig. 3. *Asparagine-induced accumulation of autophaged lactate dehydrogenase and lactose in the presence of asparagine.* Hepatocytes loaded with [^{14}C]lactose were incubated in the absence (open symbols) or presence (closed symbols) of 20 mM Asn, and the accumulation of autophaged LDH (circles) or lactose (triangles) was measured in sedimentable cell corpses and expressed as % of total cellular amount. Each value is the mean ± S.E. of 6-11 experiments.

Table 4. *Protection of autophaged sucrose from endocytosed invertase by vinblastine but not by asparagine*

	- Invertase	+ Invertase
Control	5.8 ± 0.3 (4)	0.8 ± 0.4 (8)
Asparagine	4.3 ± 0.4 (4)	0.2 ± 0.2 (5)
Vinblastine	4.0 ± 0.3 (2)	2.5 ± 0.2 (3)[a]
Asparagine + Vinblastine	4.5 ± 0.6 (4)	1.8 ± 0.3 (6)[b]

[a]*$P<0.02$ vs.* invertase alone; [b]*$P<0.005$ vs.* Asn + invertase. Hepatocytes electroloaded with [^{14}C]sucrose were incubated for 1h with or without added invertase (4,000 units/ml) in the presence or absence of asparagine (10 mM) or vinblastine (50 μM) as indicated. The accumulation of autophaged sucrose is expressed as % of the total cellular radioactivity. Each value is the mean ± S.E. of the number of experiments given in parentheses.

to endocytosed enzymes other than ß-galactosidase, a series of convergence experiments were performed using sucrose and invertase (a sucrose-degrading enzyme) in place of lactose and ß-galactosidase. In contrast to lactose, sucrose is not degraded intralysosomally, and therefore accumulates continuously in the absence of added enzyme. As shown in Table 4, added invertase prevented the accumulation of autophaged [^{14}C]sucrose regardless of whether the hepatocytes were incubated in the absence or presence of Asn, i.e. both lysosomal and prelysosomal sucrose was apparently accessible to endocytosed enzyme. On the other hand, vinblastine protected lysosomal as well as prelysosomal sucrose against invertase, presumably by blocking endocytic influx into both compartments.

Both ß-galactosidase and invertase would thus seem capable of entering amphisomes. In addition a direct (autophagy-independent) entry of invertase into the lysosomes has been demonstrated (Høyvik et al., 1987). The rules which determine whether endosomes take the amphisomal or the direct route to the lysosomes are not known; presumably both the type of ligand and the autophagic flux rate are important factors. Endocytosed horseradish peroxidase has been shown to reach the amphisomes (Tooze et al., 1990), whereas in hepatocytes neither EGF (Dunn, 1990) nor asialoglycoproteins (Seglen and Gordon, 1982; Schwarze et al., 1985) would appear to take the amphisomal (3MA-sensitive) route. In fibroblasts, the starvation-induced degradation of endocytosed α_2-macroglobulin (and to a lesser extent that of EGF) was found to be 3MA-sensitive, implicating a coupling to the autophagic pathway (Hendil et al., 1990). The starvation-induced degradation of endogenous fibroblast protein was likewise 3MA-sensitive (Hendil et al., 1990), suggesting that this signal domain-dependent, selective protein degradation (Dice and Chiang, 1989) may similarly be coupled to autophagy at the amphisome junction. A picture is thus emerging of the amphisome as a central unloading station for several intracellular vacuolar pathways.

References

Dice, J.F. & H.-L. Chiang (1989) Peptide signals for protein degradation within lysosomes. Biochem. Soc. Symp. 55:45-55

Dunn, W.A (1990) Studies on the mechanisms of autophagy: Formation of the autophagic vacuole. J. Cell Biol. 110:1923-1933

Gordon, P.B. & P.O. Seglen (1982) Autophagic sequestration of [^{14}C]sucrose, introduced into isolated rat hepatocytes by electropermeabilization. Exp. Cell Res. 142:1-14

Gordon, P.B. & P.O. Seglen (1988) Prelysosomal convergence of autophagic and endocytic pathways. Biochem. Biophys. Res. Commun. 151:40-47

Gordon, P.B., H. Tolleshaug & P.O. Seglen (1985) Use of digitonin extraction to distinguish between autophagic-lysosomal sequestration and mitochondrial uptake of [^{14}C]sucrose in hepatocytes. Biochem. J. 232:773-780

Hendil, K.B., A.-M.B. Lauridsen & P.O. Seglen (1990) Both endocytic and endogenous protein degradation in fibroblasts is stimulated by serum/amino acid deprivation and inhibited by 3-methyladenine. Biochem. J., in press.

Holtzman, E. (1989) Lysosomes. Plenum Press, New York and London, pp. 1-439.

Høyvik, H., P.B. Gordon & P.O. Seglen (1986) Use of a hydrolysable probe, [^{14}C]lactose, to distinguish between pre-lysosomal and lysosomal steps in the autophagic pathway. Exp. Cell Res. 166:1-14

Høyvik, H., P.B. Gordon & P.O. Seglen (1987) Convergence of autophagic and endocytic pathways at the level of the lysosome. Biochem. Soc. Transact. 15:964-965

Kopitz, J., G.Ø. Kisen, P.B. Gordon, P. Bohley & P.O. Seglen (1990) Non-selective autophagy of cytosolic enzymes in isolated rat hepatocytes. J. Cell Biol. 111:941-953

Kovács, A.L., A. Reith & P.O. Seglen (1982) Accumulation of autophagosomes after inhibition of hepatocytic protein degradation by vinblastine, leupeptin or a lysosomotropic amine. Exp. Cell Res. 137:191-201

Plomp, P.J.A.M., P.B. Gordon, A.J. Meijer, H. Høyvik & P.O. Seglen (1989) Energy dependence of different steps in the autophagic-lysosomal pathway. J. Biol. Chem. 264:6699-6704

Poli, A., P.B. Gordon, P.E. Schwarze, B. Grinde & P.O. Seglen (1981) Effects of insulin and anchorage on hepatocytic protein metabolism and amino acid transport. J. Cell Sci. 48:1-18

Seglen, P.O., B. Grinde & A.E. Solheim (1979) Inhibition of the lysosomal pathway of protein degradation in isolated rat hepatocytes by ammonia, methylamine, chloroquine and leupeptin. Eur. J. Biochem. 95:215-225

Seglen, P.O., P.B. Gordon & A. Poli (1980) Amino acid inhibition of the autophagic/lysosomal pathway of protein degradation in isolated rat hepatocytes. Biochim. Biophys. Acta 630:103-118

Seglen, P.O (1983) Inhibitors of lysosomal function. Meth. Enzymol. 96:737-764

Seglen, P.O., P.B. Gordon, H. Tolleshaug & H. Høyvik (1986) Use of [^{3}H]raffinose as a specific probe of autophagic sequestration. Exp. Cell Res. 162:273-277

Seglen, P.O. & P.B. Gordon (1984) Amino acid control of autophagic sequestration and protein degradation in isolated rat hepatocytes. J. Cell Biol. 99:435-444

Seglen, P.O. & P.B. Gordon (1982) 3-Methyladenine: a specific inhibitor of autophagic/lysosomal protein degradation in isolated rat hepatocytes. Proc. Natl. Acad. Sci. USA 79:1889-1892

Seglen, P.O (1987) Regulation of autophagic protein degradation in isolated liver cells. In Lysosomes: Their Role in Protein Breakdown. H. Glaumann & F.J. Ballard (eds), Academic Press, London, p. 369-414

Tooze, J., M. Hollinshead, T. Ludwig, K. Howell, B. Hoflack & H. Kern (1990) In exocrine pancreas, the basolateral endocytic pathway converges with the autophagic pathway immediately after the early endosome. J. Cell Biol. 111:329-345.

1.9. THE BIOGENESIS OF EPITHELIAL POLARITY

(Chairperson : Ann HUBBARD)

The understanding of transcytosis requires first to elucidate how polarity is established and how it is maintained despite, or due to, transcytosis.

Ann Hubbard used two perturbations to clarify the role of transcytosis in hepatocyte polarity. She showed that bile duct ligation causes the accumulation of newly synthesized apical membrane proteins in the pericanalicular region. Interestingly, a low dose of colchicine leads to the retention of apical membrane markers at the basolateral domain, and largely reroutes into bile newly synthesized proteins destined to plasma. **Michèle Maurice**, who studied the repolarization of isolated rat hepatocytes, also found that apical relocalization is inhibited by colchicine. Hence, microtubules are important in the guidance to the appropriate membrane domains of hepatocytes.

Using Caco-2 cells as a model, **Leo Ginsel** showed by immunolocalization that newly synthesized apical proteins can follow different routes in the same cell. Sucrase-isomaltase is directly transported to the apical membrane domain, whereas a significant fraction of dipeptidylpeptidase IV is first routed to the basolateral membrane and reaches its final destination by transcytosis. A detectable amount of this enzyme is also directly transported to lysosomes and degraded without ever reaching the cell surface. Part of newly synthesized lysosomal enzymes are secreted at the basolateral surface, except acid α-glucosidase, which is apically secreted. These observations raise the question of different versus unequally efficient polarity signals.

In addition to differences in protein composition, **Gerrit van Meer** emphasized that the polarized MDCK cells show differences in the lipids of the outer leaflet at the apical plasma membrane domain (enriched in glycolipids) and at the basolateral domain (enriched in phosphatidylcholine). He especially discussed the model that sorting occurs in the luminal side of the trans-Golgi network, where glycolipid self-association in patches is the positive signal for insertion into vesicles destined to the apical domain. The destination of other lipids would depend whether they are excluded from (phosphatidylcholine) or partially included in glycolipid patches (sphingomychine). The basolateral route is here viewed as a "default" pathway. Two predictions were confirmed, namely the Golgi-luminal localization of biosynthetic enzymes, and the vesicular

NATO ASI Series, Vol. H 62
Endocytosis
Edited by P. J. Courtoy

traffic between Golgi and plasma membrane. How this model applies to cells where the apical addressing involves a preliminary basolateral stop-over remains to be examined.

PLASMA MEMBRANE BIOGENESIS AND VESICLE TRAFFIC IN POLARIZED HEPATOCYTES: WHAT CAN PERTURBATIONS TELL US?

Ann L. Hubbard and Valarie A. Barr
Department of Cell Biology and Anatomy
The Johns Hopkins University School of Medicine
725 North Wolfe Street
Baltimore, Maryland 21205

Plasma membrane (PM) biogenesis in polarized epithelial cells involves the generation of specialized PM domains, each facing a different environment and containing a distinct set of proteins. Many domain specific integral membrane proteins have complex type glycosylation and are cotranslationally inserted into the rough endoplasmic reticulum and pass through the Golgi complex together. Afterwards, they must be sorted to the correct membrane domain. We are studying PM biogenesis in the rat hepatocyte, a polarized epithelial cell with multiple basal (sinusoidal) and apical (bile canalicular) domains.

All studies of polarized epithelial cells have found that newly synthesized basolateral (BL) membrane proteins are sorted in the trans Golgi network and are delivered directly to the correct domain. In contrast, apical membrane proteins appear to be sorted at different sites in different types of epithelial cells. Our 1987 kinetic analysis of 5 domain specific membrane proteins arriving at the PM, indicated that in hepatocytes apical membrane proteins are sorted after delivery to the BL surface (Bartles et al 1987). Studies done on MDCK cells have shown that viral glycoproteins and several endogenous apical membrane proteins are sorted in the trans Golgi network followed by delivery to the apical surface (Misek et al 1984; Simons and Fuller 1985; Matlin 1986; Lisanti et al 1989). It is not clear how apical membrane proteins are sorted in intestinal cells. There is evidence that sorting can occur after delivery of these proteins to the BL surface in rat enterocytes (Massey et al 1987) and in Caco 2 cells (Matter et al 1990). However, other reports show that apical proteins move directly to the apical membrane from the trans Golgi network (Ahnen et al 1982; Fransen et al 1985; Danielson and Cowell 1985).

NATO ASI Series, Vol. H 62
Endocytosis
Edited by P. J. Courtoy

To study the sorting of apical membrane proteins in hepatocytes, we wanted to selectively inhibit protein delivery to the apical surface in vivo, thereby accumulating intermediates which could be identified and studied. Bile duct ligation (BDL) and administration of the microtubule-depolymerizing drug colchicine have been reported to affect plasma membrane protein distributions and vesicle traffic in hepatocytes (Rank and Wilson 1983; Durand-Schneider 1987; Reaven and Reaven 1980; Barnwell and Coleman 1983; Mullock et al 1980).

BDL: The relative amounts of seven hepatocyte proteins were quantitated by immunoblot analysis of liver homogenates from BDL (6-72 hrs) or sham operated rats. There was a decrease in the amounts of 3 BL membrane proteins: CE 9, the asialoglycoprotein receptor (ASGP-R) and the epidermal growth factor receptor (EGF-R). In contrast, the amounts of albumin (Alb) and pIgA-R, as well as the amounts of 2 apical membrane proteins HA 4/ecto ATPase and dipeptidylpeptidase IV (DPP IV) increased. These changes occurred within 6 hrs, so ligation probably directly affected the rates of synthesis and/or degradation of these proteins, but the selective accumulation of apical membrane proteins and pIgA-R is consistent with a specific effect on transcytotic protein traffic.

After BDL, the locations of several domain specific antigens were determined by immunofluorescence (IMF). In sham operated animals, the fluorescence staining of DPP IV, a representative apical antigen, was restricted to the apical PM (Fig. 1). Punctate fluorescence became visible around the BC membrane within 10 hrs after BDL and continued to increase with time after surgery until about 48 hrs post BDL. The IMF patterns of pIgA-R and 2 other apical membrane proteins, HA 4/ecto-ATPase and aminopeptidase N (APN), also showed an increase in punctate fluorescence near the apical membrane. This fluorescence resembled accumulated vesicles surrounding the bile front. These putative vesicles could transport proteins to the apical membrane and should contain newly synthesized proteins, or they could be formed by the internalization of the BC membrane and should contain older proteins. Cycloheximide (CHX) treatment of ligated animals for 6 hrs was used to distinguish between these two possibilities. This greatly reduced the punctate fluorescence in the IMF pattern of DPP IV, although the staining of the BC membrane remained. CHX treatment of ligated animals had the same effect on the IMF of APN and HA 4.

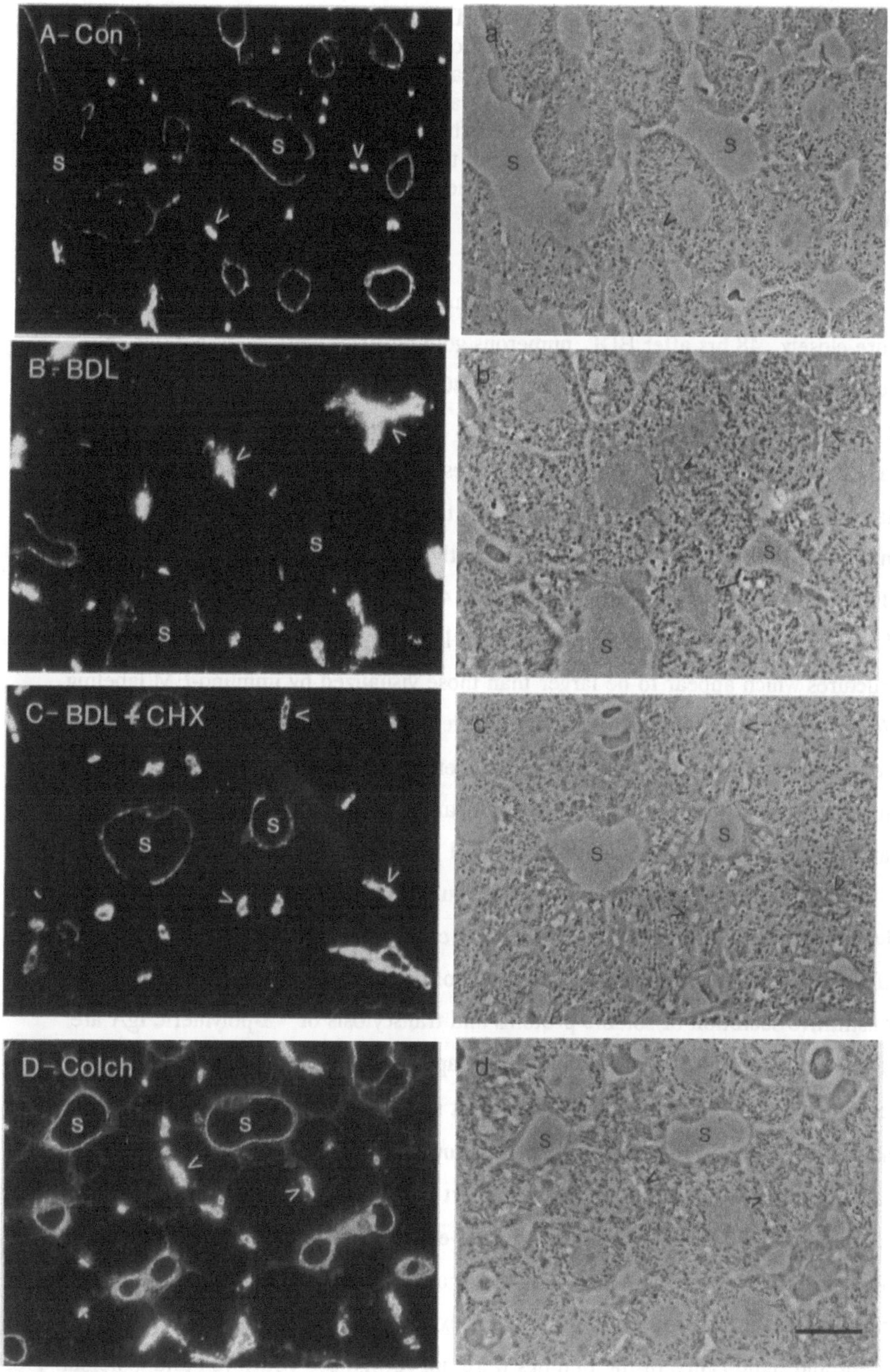

FIGURE 1

Fig. 1. Immunofluorescence localization of DPPIV in hepatocytes. Sinusoids (S) and bile canaliculi (>) are identified on all photographs. (A and a): In the hepatocytes of sham operated animals, DPPIV is localized at the canalicular membrane. The fluorescence at the sinusoids is due to the staining of endothelial cells. (B and b): 48 hr after BDL, bright punctate staining is visible around the bile canaliculi. (C and c): 48 hrs after BDL plus 6 hr CHX treatment, the punctate IMF signal is greatly reduced. (D and d): 12 hr colchicine treatment, DPPIV can be seen faintly at the basolateral surface. Bar, 15 μm. x800.

Immuno-EM labeling was used to examine the accumulation of apical antigens more closely. 48 hrs after BDL, numerous gold particles were seen marking the location of APN in the cytoplasm near the BC, although the number of gold particles at the canalicular membrane decreased (Fig. 2). The immunolabel in the cytoplasm was often within membrane limited structures which were usually vesicles with diameters less than 100 nm and tubules less than 300 nm in length. The immunolabeling of these structures was greatly reduced by treating BDL animals with CHX for 6 hrs. Immuno-EM studies of HA 4 and DPP IV gave similar results. Horseradish peroxidase (HRP) injected into ligated animals accumulated near BC in structures which appear to be larger than those visualized by immuno-EM labeling. We do not yet know if these represent different compartments or if several molecules are in the same structures. Further experiments are needed to determine if the putative block in transport caused by BDL can be reversed and to determine the nature of the structures containing the apical proteins.

Microtubule Disruption: We have found that microtubule disruption by in vivo administration of low doses of the drug colchicine (0.2 mg/100 gm body weight) affects three vesicle-mediated processes in polarized hepatocytes. That is, both constitutive secretion of soluble proteins and transcytosis of ^{125}I-polymeric IgA are slowed in the absence of microtubules, and apical plasma membrane proteins accumulate in the basolateral domain. Most surprisingly, in light of recent reports that apical secretion of soluble proteins is **slowed** in MDCK (Parczyk et al 1989) and Caco-2 cells (Paschal et al 1987) (and in part redirected to the basolateral surface), we found that apical secretion, virtually non-existent in normal liver, is dramatically **increased** after microtubule disruption. Although others earlier had reported secretion of plasma proteins into bile after exposure of livers to colchicine, the full extent and significance of the findings were not recognized (Barnwell and Coleman 1983).

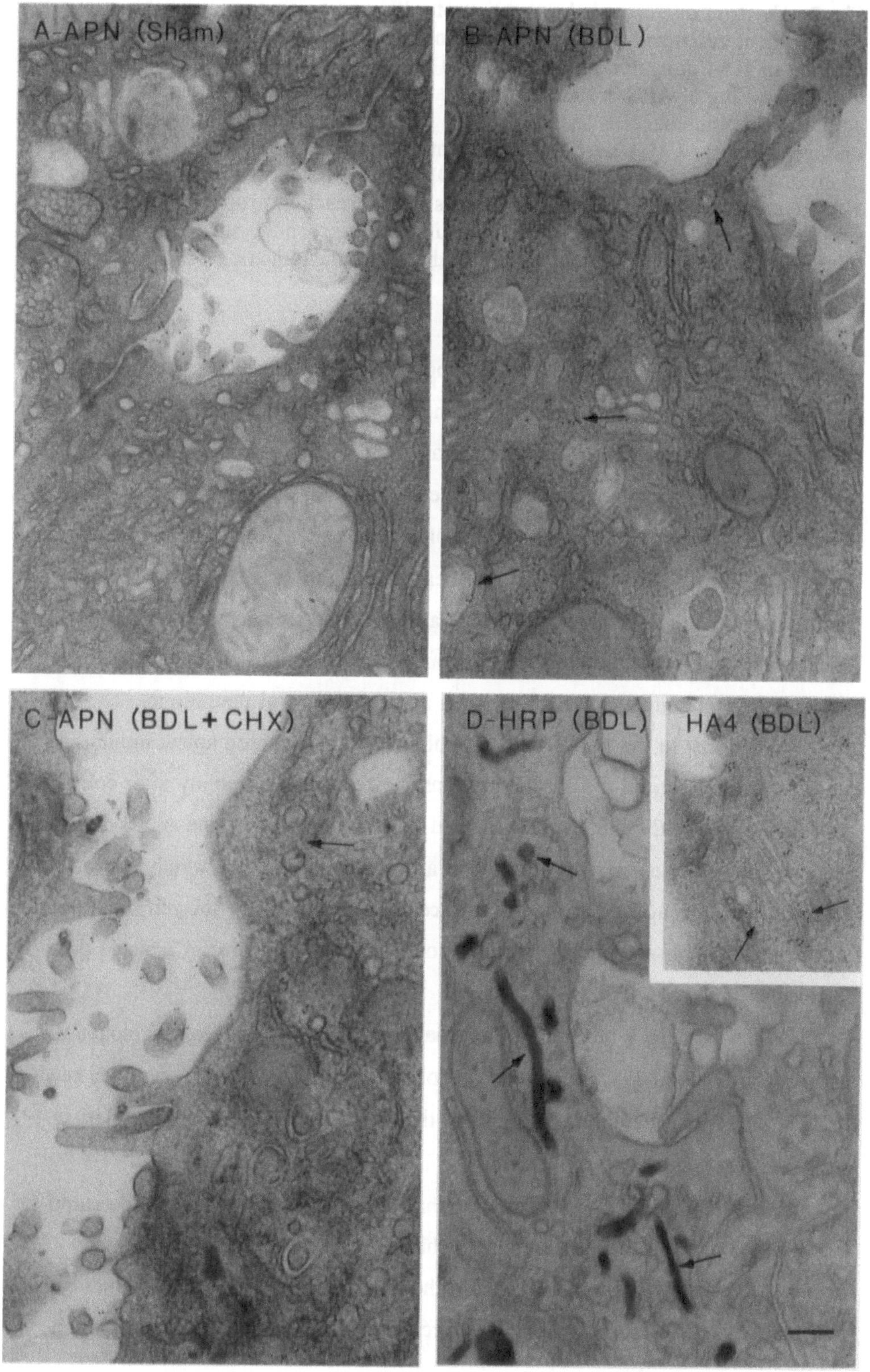

FIGURE 2

Fig. 2. Electron microscopic localization of APN and a comparison of the compartment visualized by HRP injection of ligated animals and by immuno-EM labeling of HA4 in ligated animals. A) In the livers of sham operated animals immunolabeling of APN is seen primarily at the canalicular membrane. B) 48 hr after BDL, immunolabel is seen within the hepatocyte, often within small vesicles and tubules (arrows). C) After CHX treatment of ligated animals, the immunolabeling of internal structures is greatly reduced. D) HRP reaction product in comparison to HA4 immunolocalization in ligated animals. HRP reaction product is present in vesicles and tubules near the bile canaliculi. The insert shows smaller vesicles and tubules containing αHA4 immunolabel at the same magnification. Bar, 0.5 μm. x43000.

We used IMF and metabolic labeling followed by subcellular fractionation to study the effect of in vivo colchicine treatment on the steady state distributions of seven hepatocyte proteins. The three apical proteins (represented in Figure 1 by DPPIV) showed strong labeling around bile canaliculi at all times after microtubule depolymerization and increasingly detectable staining along the basolateral surfaces up to 12 hr after in vivo colchicine treatment, although it was still much less intense than that in the apical membrane. 24 hrs after a single colchicine injection the three apical membrane proteins exhibited distributions close to that at steady state; however, trace amounts could still be seen at the basolateral membrane.

In contrast to the parallel changes observed for the three apical membrane proteins, the basolateral membrane proteins showed more variability after colchicine treatment. Two (CE9 and HA321) showed no change, another (the ASGP-R) shifted to more apical intracellular locations, and a fourth (the EGF-R), which in controls is detectable by IMF at the basolateral surface (Dunn et al 1986), showed redistribution to the apical membrane and intracellular locations 8 and 12 hr after a single dose of colchicine.

When we used a biochemical approach adapted from our earlier biogenetic studies (Bartles et al 1987) to determine colchicine's effect on the delivery of newly-synthesized plasma membrane proteins to their final destinations, we obtained inconsistent results.

We next examined the secretion behavior of ^{35}S-albumin in rats pretreated for 4 hr with 0.2 mg/100 gm body weight colchicine before pulse-chase metabolic labeling. It is important to reiterate that the only secretory pathway in hepatocytes is the one carrying plasma proteins to the circulation (Miller et al 1951). Although proteins are present in bile, they gain access to this space via the blood or basolateral surface.

Figure 3 shows that the secretion of ^{35}S-albumin into the circulation was slowed after colchicine pretreatment. More striking however, was the effect of microtubule disruption on the **polarity** of constitutive secretion in rat hepatocytes. Secretory vesicles normally discharge their contents in the circulation that bathes the

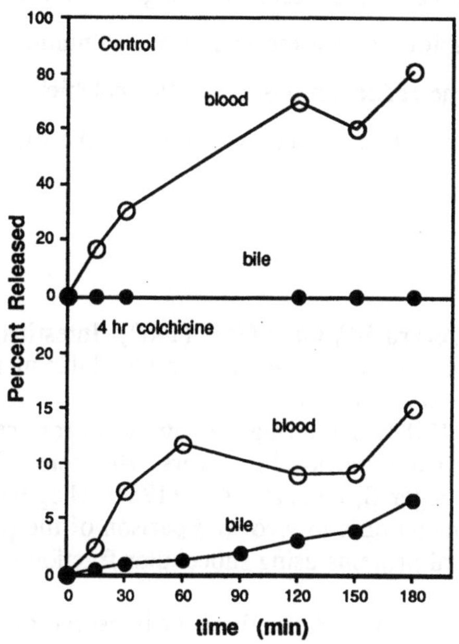

Fig. 3. Microtubule disruption slows secretion of ^{35}S-albumin into the circulation but enhances its secretion into bile. Rats (pretreated or not with colchicine) were injected iv with ^{35}S-trans-label (pulse) followed 10 min later by unlabeled methionine (chase). Albumin was immunoprecipitated from blood and bile collected during the chase and from final liver homogenates. After SDS-PAGE and autoradiography, the developed bands were quantified and expressed as % of total in bile + blood + liver.

basolateral surface with extremely high fidelity (> 99/1). After colchicine treatment vesicles containing ^{35}S-albumin fused inappropriately with the apical membrane and released substantial amounts of newly-synthesized plasma proteins directly into the bile. The fidelity of secretion into the blood dropped to 2.5/1. When we analyzed total ^{35}S-proteins in bile and serum from control and colchicine-treated animals by SDS-PAGE and autoradiography, it was evident that many plasma proteins in addition to albumin appear in bile after microtubule disruption. This result--the secretion of substantial amounts of newly-synthesized plasma proteins at both poles of the cell--has implications for molecular sorting at the level of the trans-Golgi, particularly in the formation of vesicles and inclusion of content in them.

We found that the blood clearance of ^{125}I-pIgA injected intravenously and its accumulation in liver was not affected by colchicine but that its appearance in bile was slowed. There was an increasing effect with longer colchicine pretreatments, up to a maximum of 75% inhibition 6 hr after colchicine. A number of control experiments indicate that the reduction observed after colchicine reflects primarily decreased vesicle traffic and not decreased availability of the pIgA-receptor at the basolateral front.

References

Ahnen DJ, Santiago NA, Cezard J-P, Gray GM (1982) Intestinal aminooligopeptidase. In vivo synthesis on intracellular membranes of rat jejunum. J Biol Chem 257:12129-12135

Barnwell SG, Coleman R (1983) Abnormal secretion of proteins into bile from colchicine-treated isolated perfused rat livers. Biochem J 216:409-414

Bartles JR, Feracci HM, Stieger B, Hubbard AL (1987) Biogenesis of the rat hepatocyte plasma membrane in vivo: comparison of the pathways taken by apical and basolateral proteins using subcellular fractionation. J Cell Biol 105:1241-1251

Danielsen EM, Cowell GM (1985) Biosynthesis of intestinal microvillar proteins. Evidence for an intracellular sorting taking place in, or shortly after, exit from the Golgi complex. Eur J Biochem 152:493-499

Dunn WA, Connolly TP, Hubbard AL (1986) Receptor-mediated endocytosis of epidermal growth factor by rat hepatocytes: receptor pathway. J Cell Biol 102:24-36

Durand-Schneider A-M, Maurice M, Dumont M, Feldmann G (1987) Effect of colchicine and phalloidin on the distribution of three plasma membranes in rat hepatocytes: comparison with bile duct ligation. Hepatology 7:1239-1248

Fransen JAM, Ginsel LA, Hauri H-P, Sterchi E, Blok J (1985) Immunoelectron microscopical localization of a microvillus membrane disaccharidase in the human small intestinal epithelium with monoclonal antibodies. Eur J Cell Biol 38:6-15

Lisanti MP, LeBivic A, Sargiacomo M, Rodriguez-Boulan E (1989) Steady-state distribution and biogenesis of endogenous Madin-Darby canine kidney glycoproteins: evidence for intracellular sorting and polarized cell surface delivery. J Cell Biol 109:2117-2127

Massey D, Feracci H, Gorvel J-P, Rigal A, Soulie JM, Maroux S (1987) Evidence for the transit of aminopeptidase N through the basolateral membrane before it reaches the brush border of enterocytes. J Membrane Biol 96:19-25

Matlin KS (1986) The sorting of proteins to the plasma membrane in epithelial cells. J Cell Biol 103:2565-2568
Matter K, Brauchbar M, Bucher K, Hauri H-P (1990) Sorting of endogenous plasma membrane proteins occurs from two sites in cultured human intestinal epithelial cells (Caco-2). Cell 60:429-437
Miller LL, Bly CG, Watson ML, Bale WF (1951) The dominant role of the liver in plasma protein synthesis. A direct study of the isolated perfused rat liver with the aid of lysine-e-C14. J Exp Med 94:431-453
Misek DE, Bard E, Rodriguez-Boulan E (1984) Biogenesis of epithelial cell polarity: intracellular sorting and vectorial exocytosis of an apical plasma membrane glycoprotein. Cell 39:537-546
Mullock BM, Jones RS, Peppard J, Hinton RH (1980) Effect of colchicine on the transfer of IgA across hepatocytes into bile in isolated perfused rat livers. FEBS Lett 120:278-282
Parczyk K, Haase W, Kondor-Koch C (1989) Microtubules are involved in the secretion of proteins at the apical cell surface of the polarized epithelial cells. J Biol Chem 264:16837-16846
Paschal BM, Shpetner HS, Vallee RB (1987) MAP 1C is a microtubule-activated ATPase which translocates microtubules in vitro and has dynein-like properties. J Cell Biol 105:1273-1282
Rank J, Wilson ID (1983) Changes in IgA following varying degrees of biliary obstruction in the rat. Hepatology 3:241-247
Reaven EP, Reaven GM (1980) Evidence that microtubules play a permissive role in heaptocyte very low density lipoprotein secretion. J Cell Biol 84:28-39
Simons K, Fuller SD (1985) Cell surface polarity in epithelia. Annu Rev Cell Biol 1:243-288

PLASMA MEMBRANE POLARITY OF RAT HEPATOCYTE IN PRIMARY CULTURE

M. Maurice, A.-M. Durand-Schneider, J.-C. Bouanga and G. Feldmann
INSERM U327, Laboratoire de Biologie Cellulaire, Faculté de Médecine Xavier-Bichat, 16 rue Henri Huchard, 75018 Paris, France.

INTRODUCTION

In vivo, hepatocytes are polarized epithelial cells. The bile canaliculi constitute the apical pole which is separated by tight junctions from the basal-lateral pole facing the blood. Specific proteins are asymmetrically distributed in the plasma membrane, thus defining the apical, basal, and lateral domains (Evans, 1980; Simons and Fuller, 1985). Such a polarized organization of the plasma membrane is the result of several processes including initial targeting, specific-domain recycling, and inter-domain exchanges of membrane constituents. Therefore, intracellular membrane traffic in polarized epithelial cells cannot be dissociated from cell polarity, a fact which must be taken into account especially when studies are made on *in vitro* models.

Hepatocytes cultured as primary monolayers have been shown to reconstitute apical poles resembling bile canaliculi *in vivo* (Wanson et al., 1977). In order to gain information on the organization of hepatocytes in this model, we examined the formation of the apical pole in primary cultures of rat hepatocytes, and the fate of three plasma membrane proteins which are preferentially associated with one domain in the rat liver *in vivo* (Maurice et al., 1988). We also studied the effect of microtubule disruption on the development of cell polarity, since microtubules have been implicated in intracellular membrane traffic, especially in the transport of apical membrane proteins (Hubbard and Stieger, 1988).

EXPERIMENTAL PROCEDURES

Hepatocyte isolation and culture : Hepatocytes were isolated from adult rat liver by collagenase perfusion (Seglen, 1975; Maurice et al., 1988). Primary cultures were performed in 35 mm dishes either uncoated or coated with type I collagen, in MEM medium supplemented with 25 mM Hepes, 10% fetal calf serum, 1% albumin, 2mM glutamine, 20 μg/ml insulin, 70 μM hydrocortisone, penicillin, streptomycin, and 2% dimetylsulfoxide.

NATO ASI Series, Vol. H 62
Endocytosis
Edited by P. J. Courtoy

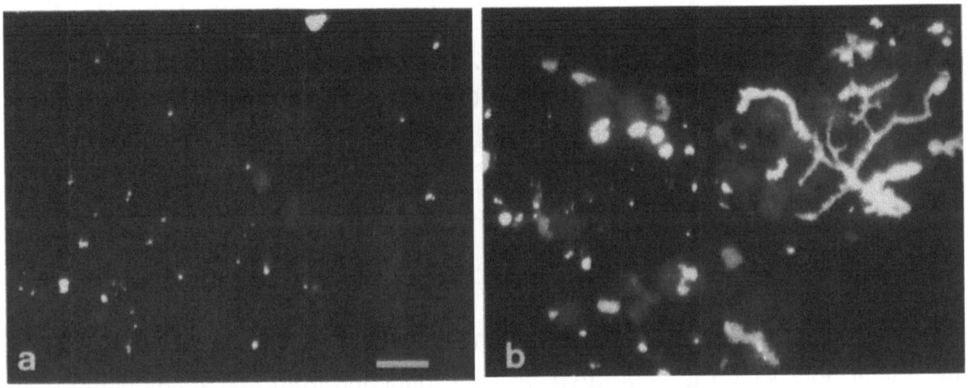

Fig. 1. Fluorescence micrographs of hepatocytes cultured for 1 day (a) or 3 days (b). The cultures were incubated with fluorescein diacetate. Accumulation of the fluorescent compound indicates the formation of functional sealed apical poles. Bar, 25 μm.

Characterization of the plasma membrane proteins : The three plasma membrane proteins studied here are defined by monoclonal antibodies. All three have been characterized as integral membrane glycoproteins with M_r of 30-36,000 (A39), 100,000 (B1) and 150,000 (B10) (Maurice et al., 1988; Scoazec et al., 1989). A39 is mainly associated with the basal domain, B1 with the lateral domain and B10 with the apical domain (Maurice et al., 1985).

Transport of fluorescein diacetate : Hepatocytes in culture were incubated for 45 min with Na fluorescein diacetate (Barth and Schwarz, 1982), washed and immediately examined under epifluorescence.

Immunocytochemistry : Immunolocalization of the membrane glycoproteins was performed by an indirect immunoperoxidase technique using the monoclonal antibodies, after fixation of the cells with paraformaldehyde and permeabilization with saponin (Maurice et al., 1988).

Drug treatment : Colchicine (10^{-5}M) or nocodazole (3.3×10^{-5}M) were added to the culture medium either 2-3 h after seeding the cells or after 1 or 2 days of culture. They were maintained for 1 to 24 h before fixing the cells for immunocytochemistry.

RESULTS AND DISCUSSION

Development of morphological and functional polarity

Immediately after isolation, hepatocytes were round-shaped and had no apparent polarity. During the first hours of culture, they plated and began to establish cell-cell contacts. After 1 day of culture, small dilations between adjacent hepatocytes could be observed by phase-contrast microscopy. Hepatocytes were able to capture fluorescein diacetate, and to transport and accumulate the dye within those dilations. This functional test made it possible to visualize the dilations and identify them as functional sealed bile canalicular structures. They appeared as fluorescent spots after 1 day of culture, enlarged and formed branched channels during the following days (Fig.1).

Repolarization of the glycoprotein B1 on the lateral domain

The glycoprotein B1, which was distributed over the entire cell surface of isolated hepatocytes rapidly regained a localization at the level of cell-cell contacts (Fig.2a). This phenomenon was observed 4 h after seeding the cells, before bile canaliculus-like structures had formed. Thus, cell-cell contacts appear to be sufficient to induce heterogeneous distribution of this protein in the plasma membrane and differentiation of the lateral domain. The redistribution is likely to occur by movement in the plane of the membrane because the phenomenon was rapid and no intracellular labelling for B1 was observed (Maurice et al., 1988).

Repolarization of the glycoprotein B10 on the apical domain

After 1 day of culture, the membrane of the newly-formed bile canaliculi was already enriched with the apical protein B10 (Fig.2c), and was devoid of any labelling for B1 (Fig.2b). After 2 and 3 days, the apical membrane still enriched with B10 as the bile canalicular structures developped and enlarged. Moreover, the rest of the plasma membrane became less labelled for B10 and the apical protein was almost restricted to the bile canaliculi after 3 days (Fig.2d). Polarization of the glycoprotein B10 is unlikely to occur by lateral diffusion as membrane proteins are not allowed to cross tight junctions (Dragsten et al., 1981). The numerous vesicles carrying the glycoprotein B10 which were observed inside the cytoplasm on the first day of culture (Fig.2c) are likely to be involved in the repolarization process. Since it has been shown that apical membrane proteins are first routed to the basolateral domain and then transcytosed to the apical domain (Bartles et al., 1987), B10 might be transported from incorrect insertion sites toward the membrane of the newly-formed bile canaliculi by the same intracellular route.

Relocalization of the glycoprotein A39

The protein A39 did not fully regain its *in situ* localisation which is mainly on the basal domain. It was always detectable on the whole membrane. However, the apical membrane which was strongly labelled after 1 day of culture became less

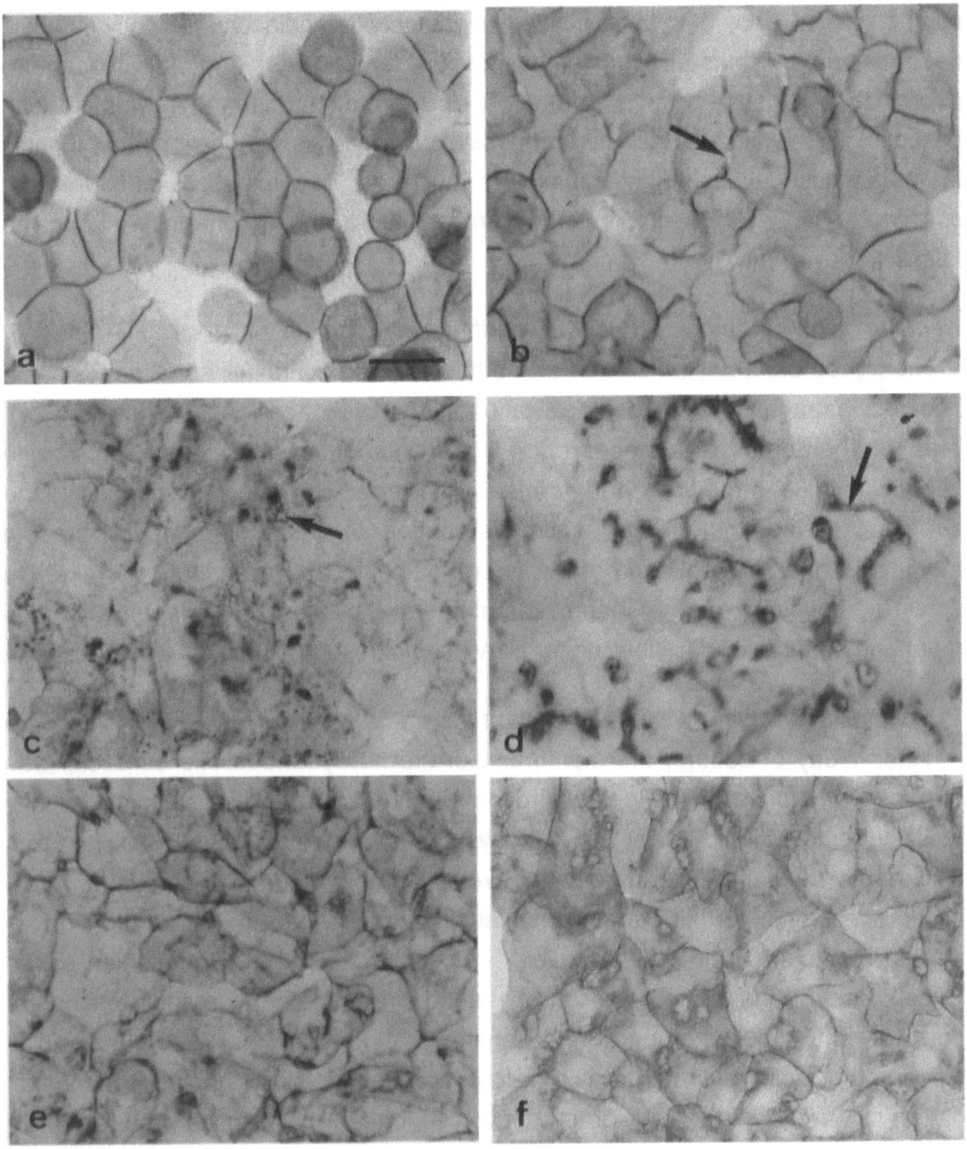

Fig.2. Repolarization of the plasma membrane in primary cultures of rat hepatocytes. Indirect immunoperoxidase. (a) 4 h after seeding the cells, B1 localizes at the level of cell-cell contacts ; (b) after 1 day of culture, the membrane of the newly-formed bile canaliculi is unlabelled (arrow). (c) In 1-day cultures, B10 begins to repolarize on the membrane of the bile canaliculi (arrow) ; intracytoplasmic vesicles are seen ; (d) After 3 days, B10 is almost exclusively present on the apical membrane (arrow). (e) A39 is present on the bile canalicular structures as well as on the rest of the membrane after 1 day; some intracytoplasmic vesicles are also seen ; (f) after 3 days, the apical membrane is slightly less labelled. Bar, 25 µm.

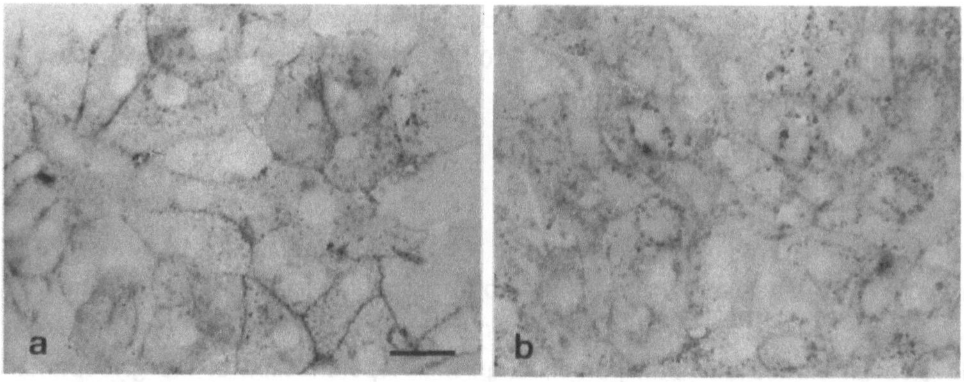

Fig.3. Effect of microtubule disruption on the polarization of the apical glycoprotein B10. Indirect immunoperoxidase. (a) Colchicine was added after 1 day of culture and maintained for 4 h. Note the accumulation of small vesicles, especially just beneath the plasma membrane; compare to control cultures on fig.2c. (b) Colchicine was added after 1day of culture and maintained for 24 h. B10 accumulates in larger vesicles resembling lysosomal structures. Bile canaliculi are not visible. Bar, 25µm.

labelled after 3 days (Fig.2e,f). A labelling of the bile canaliculi for A39 has also been observed in the fetal liver (Moreau et al., 1988), and might be ascribed to differences in the expression of this protein during development. As in the case of B10, several vesicles were observed inside the cytoplasm during the first days of culture (Fig.2e), suggesting that intracellular traffic of this protein did occur.

Effect of microtubule disruption

Colchicine and nocodazole, two drugs known to disrupt the microtubular network, inhibited the formation of the apical pole. Addition of the drugs 2-3 h after plating greatly reduced the appearance of the structures able to concentrate fluorescein that were observed in control cultures after 1 day. When the drugs were added after the formation of the canaliculi, these structures progressively disappear within 4-18h, and the few remaining fluorescent spots appeared to be inside the cytoplasm (not shown).

Microtubule disruption did not prevent the redistribution of B1 on the lateral membrane. However, it inhibited the repolarization of the apical glycoprotein B10 in specialized areas of the plasma membrane (the apical domain), and induced intracellular accumulation of vesicles carrying A39 and B10. When the drugs were applied for 1-4 h, the vesicles were small and many were located just beneath the plasma membrane (Fig.3a). After 24 h, larger vesicles were distributed throughout the cytoplasm and resembled lysosomal structures (Fig.3b). These observations suggest that colchicine and nocodazole block the migration of vesicles involved in the intracellular transport of A39 and B10. In these conditions, the vesicles would be diverted to the lysosomal system.

CONCLUSION

Hepatocytes in primary culture regain a polarized organization and are able to relocalize at least some plasma membrane proteins in their correct domains. Repolarization of the glycoprotein B1 on the lateral domain does not appear to involve an intracellular transport mechanism. Repolarization of the glycoprotein B10 on the membrane of the bile canaliculi involves intracellular membrane vesicles whose transport is inhibited by microtubule disruption. Thus, hepatocytes in primary culture appear to have retained or recovered specialized membrane transport pathways that regulate polarized insertion of membrane components. They provide an interesting *in vitro* model in which to study the development of the apical pole, the sorting of membrane proteins, and the generation of plasma membrane domains.

REFERENCES

Barth , CA and Schwarz, LR (1982). Proc. Natl. Acad. Sci. 79 : 4985-4985.

Bartles, JR, Feracci, HM, Stieger, B and Hubbard, AL (1987). J. Cell Biol. 105 : 1241-1251.

Dragsten, PR, Blumenthal, R and Handler, JS (1981). Nature, Lond. 294 : 718-722.

Evans, WH (1980). Biochim. Biophys. Acta 604 : 27-64.

Hubbard, AL and Stieger, B (1988) J. Cell Biol. 107 : 447a.

Maurice, M, Durand-Schneider, AM, Garbarz, M and Feldmann, G (1985). Eur. J. Cell Biol. 39 : 122-129.

Maurice, M, Rogier, E, Cassio, D and Feldmann, G (1988). J. Cell Sci. 90 : 79-92.

Moreau, A, Maurice, M and Feldmann, G (1988). J. Histochem. Cytochem. 36 : 87-94.

Scoazec, JY, Maurice, M, Moreau, A and Feldmann, G (1988). Cancer Res. 48 : 6882-6890.

Seglen, PO (1975). Meth. Cell Biol. 13 : 29-83.

Simons, K. and Fuller, SD (1985). Ann. Rev. Cell Biol. 1 : 243-288.

Wanson, JC, Drochmans, P, Mosselmans, R and Ronveaux, MF (1977). J. Cell Biol. 74 : 858-877.

SORTING OF ENDOGENOUS PROTEINS IN INTESTINAL EPITHELIAL CELLS

Leo A. Ginsel, Judith Klumperman, Joseph M. Tager and Jack A.M. Fransen
c/o Laboratory for Electron Microscopy, University of Leiden, Rijnsburgerweg 10, 2333 AA Leiden
The Netherlands

I. Introduction

The intestinal epithelial cell (enterocyte) forms an interesting type of polarized cell to study the sorting of endogenous membrane proteins (for reviews see Ginsel et al., 1988; Fransen et al., 1988; and Hauri, 1988). It is characterized by a typical brush border (apical or microvillar) membrane containing hydrolases involved in the terminal digestion of dietary carbohydrates and peptides, and a basolateral membrane with different structural and functional properties. Studies on the pathways and mechanisms of protein sorting in this type of cell have been going on for a relatively long period. The data on the transport pathways were however, controversial. It was unclear whether the brush-border hydrolases are directly transported from the Golgi apparatus to the apical membrane (Danielsen and Cowell, 1985; Fransen et al., 1985) or, indirectly, via the basolateral membrane (Hauri et al., 1979; Massey et al., 1987).
In recent years we have used the differentiated intestinal epithelial cell line Caco-2 (Pinto et al., 1983) as a model to study the intracellular transport of apical and basolateral membrane glycoproteins, as well as the transport of lysosomal enzymes. These cells form tight monolayers, and when grown on filters the apical and basolateral sides of the cells can be approached separately. Using this attractive model and a number of monoclonal antibodies to surface domain-specific membrane proteins and to lysosomal enzymes, we came to the following conclusions on the targetting of apical, basolateral and lysosomal proteins.

II. Sorting of brush border enzymes (Fig.1)

(1) Newly synthesized brush border enzymes are transported to the apical plasma membrane by two different pathways, a direct intracellular route and an indirect route via the basolateral plasma membrane.
Evidence for this conclusion was provided by the experiments of Matter et al. (1990). After metabolic labeling of Caco-2 cells,

NATO ASI Series, Vol. H 62
Endocytosis
Edited by P. J. Courtoy

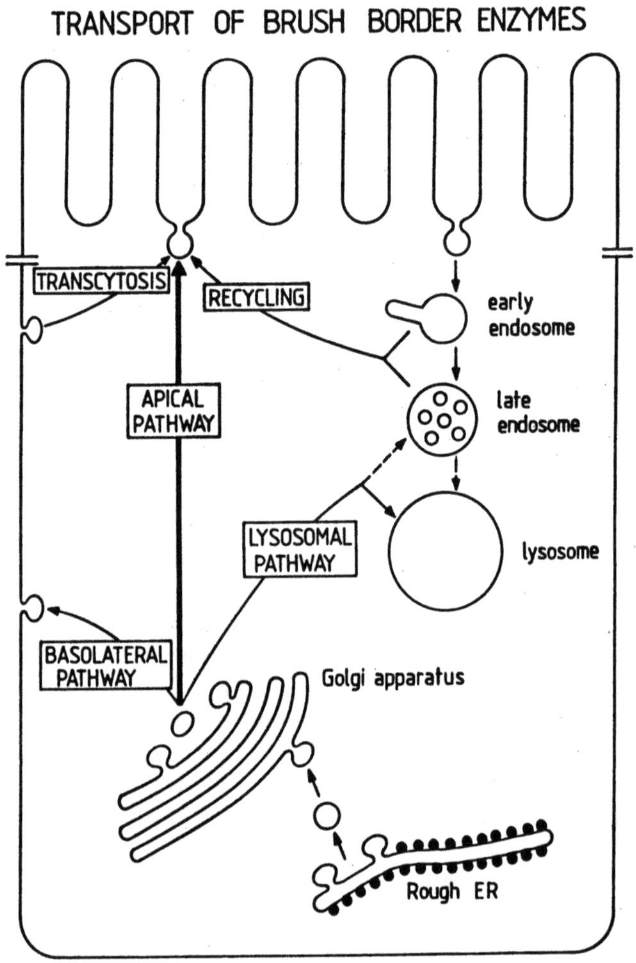

Fig.1 Model for the transport of brush border enzymes in intestinal epithelial cells (Caco-2).

*After synthesis and further processing of brush border enzymes in rough endoplasmic reticulum and Golgi apparatus, they are transported from the trans-Golgi reticulum into several directions by small vesicles. The main route is to the apical plasma membrane (**APICAL PATHWAY**), which is considered to be signal mediated and excludes basolateral proteins. Part of the brush border enzymes escape the direct pathway and are, together with basolateral proteins, transported to the basolateral plasma membrane (**BASOLATERAL PATHWAY**). These missorted brush border enzymes can be redirected to the apical plasma membrane (**TRANSCYTOSIS**). Brush border enzymes present in the apical plasma membrane can be endocytosed and transported to early and late endosomes. From there brush border enzymes can recycle to the apical plasma membrane (**RECYCLING**). An important fraction of newly synthesized brush border enzymes is directly transported from the trans-Golgi reticulum to the lysosomes (**LYSOSOMAL PATHWAY**). This pathway may also include late endosomes.*

different biochemical assays were used to study the cell surface appearance of the apical hydrolases sucrase-isomaltase (SI), dipeptidylpeptidase IV (DPPIV), and aminopeptidase N (ApN).
SI could be detected in the brush-border membrane with kinetics similar to those previously determined (Stieger et al., 1988). However, a small fraction of SI was also detected in the basolateral membrane . With immunoelectron microscopy SI was not observed at the basolateral membrane, neither in enterocytes (Fransen et al., 1985), nor in Caco-2 cells (Klumperman, 1990). The data suggest that, although a small amount of SI reaches the basolateral membrane, the enzyme is predominantly sorted intracellularly and delivered directly (without involvement of the basolateral membrane) to the apical plasma membrane.
DPPIV showed a more complex pattern of surface appearance. By immunoelectron microscopy it has been localized both in the apical and basolateral plasma membrane , as well as in intracellular compartments (Klumperman, 1990). The biochemical data indicated that DPPIV is simultaneously transported to both membrane domains (Matter et al., 1990a). However, insertion into the basolateral membrane was only transient and after about 3hr of chase the basolaterally labeled DPPIV started to appear in the brush-border membrane. This suggests that basolaterally located DPPIV is transcytosed to the apical plasma membrane. In addition, similar observations have also been made for ApN (Matter et al., 1990a), although the rate of transcytosis from basolateral to the apical plasma membrane domain was much slower for DPPIV than for ApN. A possible reason for this difference might be that transcytosed DPPIV is first delivered to an intracellular pool. Interestingly, EM-immunocytochemical studies have localized DPPIV in numerous small vesicles (Klumperman, 1990), which may indeed function as a storage compartment for DPPIV.
In view of the above data obtained from Caco-2 cells, the question arises whether this model also holds for other enterocyte-like cells. For instance, in both jejunal enterocytes (Fransen, unpublished) as well as in optimally differentiated HT 29 cells (Klumperman, 1990), no label for DPPIV was found on the basolateral membrane by means of immunoelectron microscopy. Although with this technique we might be unable to detect low concentrations of DPPIV, it may also point to a different time course of transport or indicate that the transport via the basolateral membrane is cell-type specific.

(2) The transport of newly synthesized brush border enzymes to the lysosomes of Caco-2 cells is mainly by a direct intracellular pathway and not by endocytosis. Apically endocytosed brush border enzymes are mainly recycled to the apical plasma membrane domain.
Immunoelectron microscopical studies on human jejunal enterocytes

have shown that brush border enzymes are present in the lysosomes (Fransen et al., 1985). Similar observations were made by Klumperman (1990) in Caco-2 cells. To investigate the pathways along which SI and DPPIV are transported to the lysosomes, Matter et al. (1990b) applied new subcellular fractionation procedures. About 8% of metabolically labeled SI and of DPPIV were present in the lysosomes after about 7-8h of chase. Appearance of these enzymes in the lysosomes was biphasic. Endocytosis of radioiodinated anti-(enzyme) Fab fragments and iodinated proteins revealed that the transport of the enzymes to lysosomes is only slow and at a low level. However, both enzymes were internalized with different efficiencies and recycled to the cell surface. In view of these findings it is of interest that internalized SI and DPPIV were absent from the trans-Golgi reticulum, while a constant amount of label was found in early and late endosomes (multivesicular bodies) as detected by immunoelectron microscopy. This distribution suggests that the latter organelles may be involved in membrane recycling (Klumperman, 1990).
The results indicate that in Caco-2 cells a significant amount of newly synthesized SI and DPPIV is directly imported into lysosomes bypassing the brush-border membrane (Matter et al., 1990a; cf.Ginsel et al., 1973). A direct intracellular pathway, (i.e. crinophagic pathway, Farquhar, 1969) to the lysosomes, may point to a regulatory function of these organelles in the cell surface expression of brush-border membrane proteins (Blok et al., 1984), or may be related to a late "product control" mechanism which would remove from the biosynthetic pathway proteins that, are not for example correctly processed and have no proper signal for their transport to the microvillar membrane.

(3) Mechanisms of brush-border enzyme transport

Based on the above data, the following hypothesis for the sorting of brush-border enzymes is proposed. Brush-border enzymes destined for the apical membrane can be sorted prior to cell surface expression in a signal-dependent manner. This sorting most likely occurs in the trans-Golgi reticulum. The mechanisms for this sorting process and possible variations between different enzymes are far from clear. However, recent observations of Fransen et al. (submitted), in patients with a sucrase-isomaltase deficiency suggest that SI carries a signal recognition element that resides in the isomaltase subunit. A fraction of the brush border enzymes can be transported first to the basolateral membrane, and then by a transcytotic mechanism to the apical membrane. This second sorting mechanism probably functions in the retrieval of apical brush-border enzymes which have first been (mis)sorted to the basolateral plasma membrane.

The mechanisms by which this Golgi to basolateral and basolateral to apical transport occur, are as yet unclear.

III.Sorting of lysosomal enzymes (Fig.2)

(1) Newly synthesized lysosomal enzymes are transported to the lysosomes, or secreted via the basolateral plasma membrane. A precursor form of one particular lysosomal enzyme, acid α-glucosidase, is secreted mainly from the apical plasma membrane. Evidence for this conclusion was found in studies on the biosynthesis and transport of lysosomal acid α-glucosidase, cathepsin D, ß-hexosaminidase and ß-glucuronidase (Klumperman, 1990). Metabolic labeling revealed that in Caco-2 cells α-glucosidase is synthesized as a precursor form of 110 kDa, which is converted into a precursor of slightly higher Mr (112 kDa) by the addition of complex oligosaccharide chains. A combination of metabolic labeling studies with subcellular fractionation showed that the 112 kDa precursor of α-glucosidase is transported to the lysosomes. However, the same form is secreted into the culture medium (20% of newly synthesized enzyme after 4h of chase). Immuno-precipitation of α-glucosidase from the culture medium derived from the apical and basolateral sides of radiolabeled cells showed that 70-80% of the total amount of secreted α-glucosidase precursor is secreted from the apical membrane, which was also observed by measuring enzyme activities. This finding is in accordance with immunoelectron-microscopical observations which showed that the precursor of α-glucosidase is present on the apical but not the basolateral membrane of Caco-2 cells (Klumperman, 1990), jejunal enterocytes (Fransen et al., 1988) and also human proximal tubule epithelial cells (Oude Elferink et al., 1989).

The secretion of lysosomal enzymes by Caco-2 cells has been reported earlier by Rindler and Traber (1988). Their radiolabeling studies showed that there was secretion of ß-hexosaminidase, 85% of which was found in the basolateral medium, which is compatible with our results on cathepsin D, ß-hexosaminidase, and ß-glucoronidase. It was postulated by these authors that in cultured intestinal epithelial cells: "the basolateral pathway represents a default pathway for exocytosis, not requiring a specific signal".

Other data (Eilers et al., 1989) strongly suggest that the majority of newly synthesized α-glucosidase is, unlike aminopeptidase N (Matter et al, 1990b), not delivered to the basolateral membrane prior to the transport to the microvilli. The preferential apical secretion of α-glucosidase suggests that this enzyme possesses a specific signal that leads to its segregation into a signal-dependent apical pathway. The question

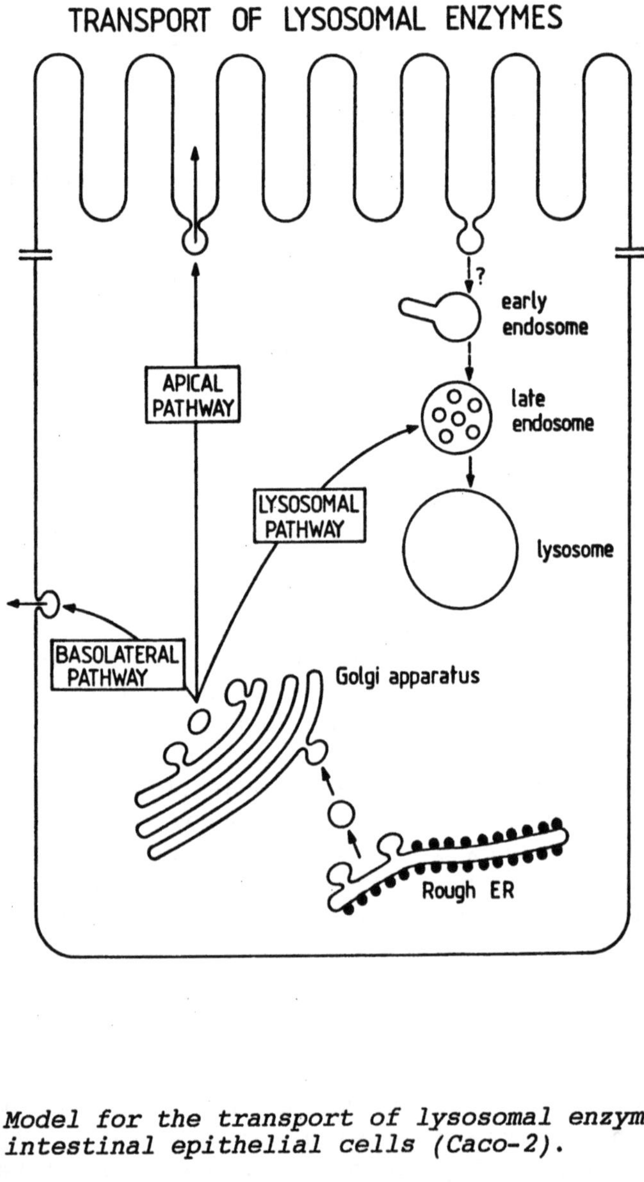

Fig. 2 *Model for the transport of lysosomal enzymes in intestinal epithelial cells (Caco-2).*

*Newly synthesized lysosomal enzymes are targeted by small vesicles most likely from the trans-Golgi reticulum to several directions. They are transported by mannose 6-phosphate receptors, via late endosomes, to lysosomes (**LYSOSOMAL PATHWAY**). A fraction of lysosomal enzymes is transported in a signal independent manner to the basolateral membrane (**BASOLATERAL PATHWAY**) and secreted. A precursor form of one particular lysosomal enzyme, acid α-glucosidase, is transported to the apical plasma membrane, and mainly secreted (**APICAL PATHWAY**). The signal for this apical transport is not yet known. Furthermore, it is not known whether (apical) endocytosis of lysosomal enzymes occurs.*

thus arises whether the transport of α-glucosidase to the apical membrane is mediated by the same mechanism as that underlying the transport of brush-border enzymes like sucrase-isomaltase.

(2) Lysosomal enzymes are most likely transported to the lysosomes by a direct intracellular pathway.

Similar to the findings on fibroblasts and certain other cell types (for a review see Kornfeld and Mellman, 1989), we suppose that lysosomal enzymes are transported to the lysosomes via the cation independent mannose 6-phosphate (M6P) receptor. This receptor was localized in a prelysosomal compartment, (multivesicular body) which also participates in the endocytic pathway (Klumperman, 1990). Although the secretion of the precursor of α-glucosidase might be indicative for an alternative transport route of this enzyme to the lysosomes, i.e. by secretion and subsequent endocytosis, we have as yet no evidence for the occurence of such mechanism in Caco-2 cells. In addition, although a M6P receptor related secretion-endocytosis pathway is certainly present in a number of cell types, it is generally viewed as a salvage mechanism of quantitatively less importance, the major portion of the enzymes being transported by a completely intracellular pathway.

(3) Mechanisms of lysosomal enzyme transport

In general it is thought that most soluble lysosomal enzymes are transported to their final destination by M6P receptors. However, we have observed in Caco-2 cells that a substantial amount of a precursor form of the lysosomal enzyme α-glucosidase is not segregated into lysosomes, but instead secreted from the apical membrane. Therefore the question arised whether this process is mediated by M6P receptors. We localized the cation independent M6P receptor both on the plasma membrane and in intracellular compartments by immunoelectron microscopy (Klumperman, 1990) However, we obtained no evidence for the actual involvement of this receptor in the transport of α-glucosidase: incubation with the weak base ammoniumchloride had little or no effect on the apical or basolateral secretion of α-glucosidase. Furthermore, when cells were incubated with buffer of low pH or medium supplemented with mannose 6-phosphate, no effect was found on the amount of α-glucosidase in the microvilli. Similar investigations for cathepsin D, ß-hexosaminidase and ß-glucuronidase have shown that incubation with ammoniumchloride leads to an enhanced secretion from the basolateral side. We thus conclude that although M6P receptor mediated transport is functionally active in Caco-2 cells, an additional or alternative mechanism must contribute to the transport of α-glucosidase to the apical membrane and probably also to the lysosomes.

IV. Perspectives for future research

One of the main questions arising from our work, is how brush-border and lysosomal enzymes are associated with the (apical) plasma membrane and which sorting signals direct these enzymes to their final destination in the cell and regulate their secretion. We plan to start studying the association of α-glucosidase with the apical (microvillar) membrane and the mechanism which underlies the secretion of this enzyme. An additional approach to study the nature of the signal which directs α-glucosidase to the apical membrane will be the transfection of polarized cells with the gene encoding human α-glucosidase. A study of mutant proteins by both immunoelectron microscopical and biochemical methods may provide new insights in the nature of the sorting signal and the mode of membrane association. Similar studies might be carried out with the brush-border enzyme sucrase-isomaltase, which is 40% identical to α-glucosidase at the protein level (Hoefsloot et al., 1988). It will be interesting to compare these two enzymes, because they are transported to the same destinations in the cell, i.e. apical membrane and lysosomes, but with different efficiencies.

REFERENCES

Blok J, Fransen JAM, Ginsel LA (1984) Cell Biol Int Rep 8:993-1014
Danielsen EM, Cowell GM (1985) Dig Dis Sci 30:1079-84
Eilers U, Klumperman J, Hauri HP (1989) J Cell Biol 108:13-22
Farquhar MG, (1969) In: Dingle JT, Fells HB (eds): Lysosomes in Biology and Pathology 2:462-482 North Holland Publ Comp Amsterdam
Fransen JAM, Ginsel LA, Hauri HP, Blok, J (1985) Eur J Cell Biol 38:6-15
Fransen JAM, Ginsel LA, Cambier PH, Klumperman J, Tager JM (1988) Eur J Cell Biol 47:72-80
Fransen JAM, Klumperman J, Oude Elferink RPJ, Hauri HP, Tager JM, Ginsel LA (1988) In: Mammalian brush border membrane proteins (Eds Lentze HJ, Sterchi JJ) 12-28 Thieme New York
Ginsel LA, Fransen JAM, Klumperman J, Hauri HP, Hilgers J, Oude Elferink RPJ, Tager JM (1988) In: Benga G, Tager JM (eds) Biomembranes. Basic and Medical Research 61-80 Springer Heidelberg
Hauri HP, Quaroni A, Isselbacher KJ (1979) Proc Natl Acad Sci USA 76:5183-5186
Hauri HP (1988) Subcell Biochem 12:155-197
Hoefsloot LH, Hoogeveen-Westerveld M, Kroos MA, Van Beeumen J, Oostra BA (1988) EMBO J 7:1697-1704
Klumperman, J (1990) Transport of lysosomal and brush-border enzymes in polarized colon carcinoma cell lines. Thesis University of Leiden
Kornfeld S, Mellman I (1989) Ann Rev Cell Biol 5:483-525
Massey D, Feracci H, Gorvel JP, Rigal A, Maroux S (1987) J Membr Biol 96:19-25
Matter K, Brauchbar M, Bucher K, Hauri HP (1990a) Cell 60:429-437
Matter K, Stieger B, Klumperman J, Ginsel LA, Hauri HP (1990b) J Biol Chem 265:3503-3512
Oude Elferink RPJ, Fransen JAM, Klumperman J. Ginsel LA, Tager JM (1989) Eur J Cell Biol 50:299-303
Pinto C. Robine-Leon S, Appay MD, Kedinger M, Triadou N, Dassaulx E, Lacroix B, Simon-Assmann D, Haffen K, Fogh J, Zweibaum A (1983) Biol Cell 47:323-330
Rindler MJ, Traber MG (1988) J Cell Biol 107:471-479
Stieger B, Matter K, Baur B, Bucher K, Hoechli M, Hauri HP (1988) J Cell Biol 106:1853-1861

LIPID SORTING BY THE FORMATION OF MICRODOMAINS OF SPHINGOLIPIDS; A MODEL AND ITS CONSEQUENCES.

Gerrit van Meer, Wouter van 't Hof, Ida van Genderen and Alex Sandra
Department of Cell Biology, Medical School
AZU H02.314 University of Utrecht
Heidelberglaan 100
3584 CX Utrecht, The Netherlands.

1. MODEL FOR EPITHELIAL LIPID SORTING.

Epithelial cells display (protein and) lipid polarity. The apical plasma membrane domain on the side of the external milieu and the basolateral domain facing the neighbor cells and the underlying tissue possess unique lipid compositions. The tight junction acts as a barrier to lipid diffusion in the outer, exoplasmic leaflet of the plasma membrane, and thereby maintains the enrichment of glycosphingolipids in the outer leaflet of the apical domain, and of phosphatidylcholine (PC) in the outer leaflet of the basolateral domain. Some years ago, we postulated that the lipid polarity is generated by a lipid sorting event in the *trans* Golgi network (TGN). In the luminal leaflet of the TGN membrane glycolipids would aggregate into a microdomain that would vesiculate into a transport vesicle destined for the apical plasma membrane domain. PC on the other hand, would be excluded from the glycolipid microdomain and would leave the TGN in the luminal leaflet of vesicles traveling to the basolateral surface. The third lipid that is abundant in the exoplasmic leaflet of plasma membranes, sphingomyelin (SPH), would display no preference for apical or basolateral transport (Simons and van Meer, 1988; van Meer, 1989b).

2. DIFFERENT EPITHELIA DISPLAY SIMILAR BIOSYNTHETIC LIPID SORTING

The model of epithelial lipid sorting is based on experimental work using the exchangeable lipid N-6[7-nitro-2,1,3-benzoxadiazol-4-yl] aminocaproyl-sphingosine (C6-NBD-ceramide) in Madin-Darby canine kidney (MDCK) cells (van Meer et al., 1987). When inserted into the plasma membrane from liposomes, this lipid is converted to C6-NBD-glucosylceramide (GlcCer) and C6-NBD-SPH, the former being a fluorescent analog of a lipid that is enriched on the apical surface of epithelial cells whereas SPH is essentially non-polarized (van Meer, 1988). At temperatures below 20°C, each fluorescent product accumulated in the Golgi area. Transport to the plasma membrane could subsequently be monitored at 37°C by trapping the fluorescent molecules immediately upon arrival at the cell surface by selective extraction onto BSA in the

NATO ASI Series, Vol. H 62
Endocytosis
Edited by P. J. Courtoy

medium ("back-exchange"). Transport to the two surfaces could be assayed separately, as the junctional complex acts as a diffusion barrier for both NBD-lipids and BSA. The seminal observation leading to the model was that whereas equal amounts of C6-NBD-SPH reached the two plasma membrane domains, nearly three-fold more C6-NBD-GlcCer was delivered to the apical than to the basolateral surface. Thus, newly synthesized sphingolipids had been sorted before reaching the cell surface.

Generation of epithelial lipid polarity was also studied in the human intestinal Caco-2 cells. It has been reported that in Caco-2 some apical proteins are incompletely sorted in the biosynthetic pathway and must be transcytosed to establish the steady state polarized situation. This result has been interpreted to indicate a difference in the Golgi protein sorting mechanism between MDCK and Caco-2. Moreover, in Caco-2 the direct pathway from the Golgi to the apical cell surface is thought to be a minor pathway (discussed in van 't Hof and van Meer, 1990). In these cells, where C6-NBD-ceramide was converted equally to C6-NBD-GlcCer and C6-NBD-SPH, C6-NBD-GlcCer was enriched 3-fold apically, whereas once again C6-NBD-SPH was equally distributed over both domains. Within 10 min from the start of synthesis, both products displayed their typical surface polarity. Furthermore, when we measured lipid transcytosis, it displayed a $t_{1/2}$ of hours. Therefore, the latter process was not required for generating the *in vivo* lipid polarity. In conclusion, in Caco-2 cells newly synthesized sphingolipids were sorted before reaching the cell surface just like in MDCK. The sorting in the biosynthetic pathway was sufficient to generate the *in vivo* lipid polarity of small intestinal cells.

Since not only lipid sorting, which we define as the difference between the polarity of delivery of C6-NBD-GlcCer and that of C6-NBD-SPH, was identical between Caco-2 and MDCK but also the individual polarities of delivery, we conclude that there is no reduction in the apical pathway size in Caco-2 relative to MDCK (van 't Hof and van Meer, 1990).

3. DEMANDS ON LIPID TRANSPORT MADE BY THE MODEL

The model requires that the glycosphingolipids, SPH and PC reaching the outer leaflet of the epithelial plasma membrane have actually undergone the sorting event in the TGN. Therefore, these lipids must have been initially present in the lumen of the TGN, and should be unable to bypass the Golgi by aqueous exchange followed by transbilayer translocation from the cytoplasmic into the outer leaflet of the plasma membrane. The clearest case for this can be made for SPH.

3.1. SPH synthesis and transport.

It is now generally accepted that SPH, the only major phospholipid in animal cells with a sphingosine backbone, is synthesized from PC by the enzyme PC:ceramide cholinephosphotransferase (see Eppler et al., 1987; Futerman et al., 1990). Less appreciated is the fact that the Golgi complex has, in the recent years, been established as the primary site of SPH biosynthesis. While early data located SPH biosynthesis partially or exclusively at the plasma membrane (Marggraf et al., 1981; 1982; Nelson and Murray, 1982; Voelker and Kennedy, 1982; Marggraf and Kanfer, 1984; 1987; van den Hill et al., 1985), other findings, especially those of Lipsky and Pagano (1983; 1985a,b), suggest that the reported activity in plasma membrane fractions may have been due to contamination with Golgi elements. On a sucrose gradient the SPH biosynthetic activity was shown to peak at the location of a Golgi marker and not of the plasma membrane or lysosomes. In agreement with this finding, during biosynthetic labeling of SPH with radioactive choline the specific activity of plasma membrane SPH notably lagged behind that of SPH in a Golgi/ER fraction (Cook et al., 1988). Finally, it has been demonstrated by a variety of techniques that C6-NBD-SPH is synthesized in the Golgi complex (Lipsky and Pagano, 1983; 1985a,b; van Meer et al., 1987; Kobayashi and Pagano, 1989; van 't Hof and van Meer, 1990). On sucrose gradients, like for natural SPH, the maximum of the biosynthetic activity of the fluorescent analog colocalized with a Golgi marker and not the plasma membrane or lysosomes, while the bulk of ER on such gradients bands at a distinctly higher density. Furthermore, the fluorescence of newly synthesized C6-NBD-SPH and C6-NBD-GlcCer initially localized by microscopy to the perinuclear region in fibroblasts and epithelial cells. In fibroblasts the fluorescence colocalized with the Golgi markers (1) thiamine pyrophosphatase, (2) a Golgi specific antibody, and (3) rhodamine-conjugated wheat germ agglutinin (Lipsky and Pagano, 1983; 1985a,b). Finally, NBD-fluorescence was dispersed over the cytoplasm in mitotic cells (Kobayashi and Pagano, 1989), and cells treated with nocodazole (van Meer, unpublished) or colcemid, while monensin led to swelling of the C6-NBD-lipid containing compartment (Lipsky and Pagano, 1985b), all consistent with a Golgi location.

Compelling evidence has been presented for the localization of most SPH to the exoplasmic, luminal bilayer leaflet of various biological membranes. Data on the relative availability of SPH to hydrolysis by sphingomyelinase from either side of the membrane support this conclusion for the plasma membrane of erythrocytes in a variety of animal species and human blood platelets (reviewed in Zachowski and Devaux, 1989), erythroleukemic cells (Rawyler et al., 1983; 1985), epithelial cells (Venien and Le Grimellec, 1988a,b), and muscle cells (Post et al., 1988), and also for a number of intracellular organelles (Nilsson and Dallner, 1977a,b; Vale, 1980). It has been further proposed that this method underestimates the exoplasmic pool (Allan and Walklin, 1988), and that, in fact, *all* SPH is situated in the exoplasmic leaflet. This has been demonstrated for C6-NBD-SPH, that was introduced into the cell biosynthetically using C6-NBD-ceramide as the precursor. C6-NBD-SPH in the plasma membrane could be quantitatively extracted from the exoplasmic surface by back-exchange to liposomes or BSA (Lipsky and Pagano, 1983; 1985a,b; van Meer et al., 1987; van't Hof and van Meer, 1990). It is unlikely that this accessibility was due to rapid flip-flop across the plasma membrane. If C6-NBD-SPH would have had access to the cytoplasmic leaflet of the plasma membrane, it would have been expected to redistribute over all organelles: C6-NBD-lipids rapidly exchange from all surfaces studied through the aqueous phase due to their non-physiological C6-NBD-chain. Such equilibration over the intracellular organelles is not observed. C6-NBD-SPH was confined to the luminal leaflet of intracellular membranes, since BSA in the cytoplasm could not extract C6-NBD-SPH from the Golgi of perforated cells (Simons and Virta, 1987), nor from the cytoplasmic side of intracellular vesicles (Bennett et al., 1988) within the time-scale of these experiments, which was typically 1 h at 37°C.

A low rate for SPH flip-flop is also suggested by the fact that flip-flop of spin-labeled SPH was too slow to be measured in the human erythrocyte plasma membrane (Zachowski et al., 1985; Calvez et al., 1988; Rosso et al., 1988) or platelets (Sune et al., 1987; Sune and Bienvenue, 1988), while a difference in fatty-acyl composition between the sphingomyelinase-susceptible and -resistant pools in human erythrocytes has led to the same conclusion (Boegheim et al., 1983). A reliable interpretation of the much faster reduction of spin-labeled SPH measured after insertion into the plasma membrane of nucleated cells (Zachowski et al., 1987; Sune et al., 1988) must await experiments by independent techniques.

The exoplasmic, luminal location of SPH in cellular membranes in combination with a slow flip-flop implies that the rapid transport of C6-NBD-SPH between the Golgi and the plasma membrane, $t_{1/2}$= 20-30 min (Lipsky and Pagano, 1985a; van Meer et al., 1987; van 't Hof and van Meer, 1990), must be mediated by carrier vesicles. In favor of this notion, it has been observed that transport of newly synthesized C6-NBD-SPH is partially inhibited by monensin (Lipsky and Pagano, 1985a), and is completely interrupted in mitotic cells (Kobayashi and Pagano, 1989). Putative exocytic transport vesicles containing C6-NBD-SPH in a luminal orientation have recently been isolated from perforated cells (Bennett et al., 1988). Moreover, it has been observed that C6-NBD-SPH, after insertion into the exoplasmic plasma membrane leaflet, is endocytosed and recycles to the plasma membrane via the same intracellular organelles as transferrin (Koval and Pagano, 1989). All these observations support a vesicular route as the major intracellular transport mechanism for SPH, which is consistent with the confinement of intracellular SPH to organelles that are connected by vesicular traffic (van Meer, 1989a).

The latest evidence has assigned the SPH biosynthesis to the *cis*- and *medial*-Golgi by subfractionation of rat liver, with <13% of the activity present at the plasma membrane (Futerman et al., 1990; Jeckel et al., 1990). Evidence for a luminal orientation of SPH-synthase was provided by controlled protease digestion of intact and permeabilized Golgi vesicles. This location is very exciting in the context that there may be an SPH-sorting event taking place in the early Golgi, by which SPH is prevented from traveling backwards into the ER (van Meer, 1989a). SPH is, therefore, shuttled outwards and may play a role in concentrating cholesterol in the plasma membrane (van Blitterswijk et al., 1987; van Meer, 1989a).

3.2. How does PC reach the outer leaflet of the plasma membrane.

In a study on transport of radiolabeled PC from its intracellular site of synthesis, ER and Golgi, to the plasma membrane, Kaplan & Simoni (1985a) have described how newly synthesized PC rapidly equilibrated with that in the plasma membrane [$t_{1/2}$(37°C)« 2 min]. However, since PC flip-flop in plasma membranes is very slow (Pagano and Sleight, 1985; Zachowski and Devaux, 1989, 1990), the problem of PC transport to the plasma membrane has to be approached from the point of view that, on time-scales of up to a couple of hours, the plasma membrane PC consists of two independent pools. When looking for two-pool kinetics, it is evident from the data that a significant fraction of the newly synthesized PC reached the plasma membrane with a $t_{1/2}$ of 30-60 min, a process that was influenced by energy poisons and inhibited at 15°C, all of which are characteristics of vesicular transport. At this moment we are trying to test the obvious prediction that this PC fraction represents the exoplasmic PC pool by a direct experiment.

3.3. Do glycolipids pass the luminal leaflet of the TGN.

Galactosylceramide and the complex glycosphingolipids based on it are synthesized in the luminal leaflet of a Golgi membrane. This is also the case for the complex glycolipids based on glucosylceramide (see Thompson and Tillack, 1985). Since flip-flop of these lipids is slow they necessarily pass through the luminal leaflet of the TGN.

However, the situation is far more complex for the simple glycosphingolipid glucosylceramide. On the one hand, there are a series of observations predicting that this lipid reaches the plasma membrane by vesicular traffic from the TGN. On the first place, it seemed to accumulate in the Golgi of Caco-2 cells at low temperature (van 't Hof and van Meer, 1990). Secondly, the temperature dependence of its transport is identical to that of SPH (van 't Hof and van Meer, unpublished). Thirdly, glucosylceramide transport to the plasma membrane was inhibited in mitotic cells (Kobayashi and Pagano, 1989). Finally, transport to the apical plasma membrane of MDCK cells was inhibited to the same extent as that of SPH (van Meer, unpublished). On the other hand, the synthesis of the lipid was found to be located at the cytoplasmic side of a Golgi membrane, and the lipid product was first located in the cytoplasmic leaflet (Coste et al., 1986; Sasaki, 1990). We are now trying to confirm these results by independent techniques, and if true, the site of glucosylceramide translocation will have to be established.

Acknowledgements:

The present work was made possible by (senior) fellowships from the Royal Netherlands Academy of Arts and Sciences (G.v.M.), the Netherlands Heart Foundation (grant 88.045, I.v.G.), and the Fogarty International Fellowship Program (A.S.).

References:

Allan, D., and C. M. Walklin. 1988. Endovesiculation of human erythrocytes exposed to sphingomyelinase C: a possible explanation for the enzyme-resistant pool of sphingomyelin. Biochim. Biophys. Acta 938: 403-410.

Bennett, M. K., A. Wandinger-Ness, and K. Simons. 1988. Release of putative exocytic transport vesicles from perforated MDCK cells. EMBO J. 7: 4075-4085.

Boegheim Jr., J. P. J., M. van Linde, J. A. F. Op den Kamp, and B. Roelofsen. 1983. The sphingomyelin pools in the outer and inner layer of the human erythrocyte membrane are composed of different molecular species. Biochim. Biophys. Acta 735: 438-442.

Calvez, J.-Y., A. Zachowski, A. Herrmann, G. Morrot, and P. F. Devaux. 1988. Asymmetric distribution of phospholipids in spectrin-poor erythrocyte vesicles. Biochemistry 27: 5666-5670.

Cook, H. W., F. B. St. Palmer, D. M. Byers, and M. W. Spence. 1988. Isolation of plasma membranes from cultured glioma cells and application to evaluation of membrane sphingomyelin turnover. Anal. Biochem. 174: 552-560.

Coste, H., M. B. Martel, and R. Got. 1986. Topology of glucosylceramide synthesis in Golgi membranes from porcine submaxillary glands. Biochim. Biophys. Acta 858: 6-12.

Eppler, C. M., B. Malewicz, H. M. Jenkin, and W. J. Baumann. 1987. Phosphatidylcholine as the choline donor in sphingomyelin synthesis. Lipids 22: 351-357.

Futerman, A. H., B. Stieger, A. L. Hubbard, and R. E. Pagano. 1990. Sphingomyelin synthesis in rat liver occurs predominantly at the cis and medial cisternae of the Golgi apparatus. J. Biol. Chem. 265: 8650-8657.

Jeckel, D., A. Karrenbauer, R. Birk, R. R. Schmidt, and F. Wieland. 1990. Sphingomyelin is synthesized in the cis Golgi. FEBS Lett. 261: 155-157.

Kaplan, M. R., and R. D. Simoni. 1985. Intracellular transport of phosphatidylcholine to the plasma membrane. J. Cell Biol. 101: 441-445.

Kobayashi, T., and R. E. Pagano. 1989. Lipid transport during mitosis. Alternative pathways for delivery of newly synthesized lipids to the cell surface. J. Biol. Chem. 264: 5966-5973.

Koval, M., and R. E. Pagano. 1989. Lipid recycling between the plasma membrane and intracellular compartments: Transport and metabolism of fluorescent sphingomyelin analogues in cultured fibroblasts. J. Cell Biol. 108: 2169-2181.

Lipsky, N. G., and R. E. Pagano. 1983. Sphingolipid metabolism in cultured fibroblasts: Microscopic and biochemical studies employing a fluorescent ceramide analogue. Proc. Natl. Acad. Sci. USA 80: 2608-2612.

Lipsky, N. G., and R. E. Pagano. 1985a. Intracellular translocation of fluorescent sphingolipids in cultured fibroblasts: Endogenously synthesized sphingomyelin and glucocerebroside analogues pass through the Golgi apparatus en route to the plasma membrane. J. Cell Biol. 100: 27-34.

Lipsky, N. G., and R. E. Pagano. 1985b. A vital stain for the Golgi apparatus. Science 228: 745-747.

Marggraf, W.-D., F. A. Anderer, and J. N. Kanfer. 1981. The formation of sphingomyelin from phosphatidylcholine in plasma membrane preparations from mouse fibroblasts. Biochim. Biophys. Acta 664: 61-73.

Marggraf, W.-D., and J. N. Kanfer. 1987. Kinetic and topographical studies of the phosphatidylcholine:ceramide cholinephosphotransferase in plasma membrane particles from mouse ascites cells. Biochim. Biophys. Acta 897: 57-68.

Marggraf, W.-D., and J. N. Kanfer. 1984. The phosphorylcholine receptor in the phosphatidylcholine:ceramide cholinephosphotransferase reaction. Is the enzyme a transferase or a hydrolase? Biochim. Biophys. Acta 793: 346-353.

Marggraf, W.-D., R. Zertani, F. A. Anderer, and J. N. Kanfer. 1982. The role of endogenousphosphatidylcholine and ceramide in the biosynthesis of sphingomyelin in mouse fibroblasts.Biochim. Biophys. Acta 710: 314-323.

Nelson, D. H., and D. K. Murray. 1982. Dexamethasone increases the synthesis of sphingomyelin in 3T3-L1 cell membranes. Proc. Natl. Acad. Sci. USA 79: 6690-6692.

Nilsson, O. S., and G. Dallner. 1977a. Enzyme and phospholipid asymmetry in liver microsomal membranes. J. Cell Biol. 72: 568-583.

Nilsson, O. S., and G. Dallner. 1977b. Transverse asymmetry of phospholipids in subcellular membranes of rat liver. Biochim. Biophys. Acta 464: 453-458.

Pagano, R. E., and R. G. Sleight. 1985. Defining lipid transport pathways in animal cells. Science 229: 1051-1057.

Post, J. A., G. A. Langer, J. A. F. Op den Kamp, and A. J. Verkley. 1988. Phospholipid asymmetry in cardiac sarcolemma. Analysis of intact cells and 'gas-dissected' membranes. Biochim. Biophys. Acta 943: 256-266.

Rawyler, A. J., B. Roelofsen, J. A. F. Op den Kamp, and L. L. M. van Deenen. 1983. Isolation and characterization of plasma membranes from Friend erythroleukaemic cells. A study with sphingomyelinase C. Biochim. Biophys. Acta 730: 130-138.

Rawyler, A., P. H. van der Schaft, B. Roelofsen, and J. A. F. Op den Kamp. 1985. Phospholipid localization in the plasma membrane of Friend erythroleukemic cells and mouse erythrocytes. Biochemistry 24: 1777-1783.

Rosso, J., A. Zachowski, and P. F. Devaux. 1988. Influence of chlorpromazine on the transverse mobility of phospholipids in the human erythrocyte membrane: relation to shape changes. Biochim. Biophys. Acta 942: 271-279.

Sasaki, T. 1990. Glycolipid transfer protein and intracellular traffic of glucosylceramide. Experientia 46: 611-616.

Simons, K., and G. van Meer. 1988. Lipid sorting in epithelial cells. Biochemistry 27: 6197-6202.

Simons, K., and H. Virta. 1987. Perforated MDCK cells support intracellular transport. EMBO J. 6: 2241-2247.

Sune, A., P. Bette-Bobillo, A. Bienvenüe, P. Fellman, and P. F. Devaux. 1987. Selective outside-inside translocation of aminophospholipids in human platelets. Biochemistry 26: 2972-2978.

Sune, A., and A. Bienvenue. 1988. Relationship between the transverse distribution of phospholipids in plasma membrane and shape change of human platelets. Biochemistry 27: 6794-6800.

Sune, A., M. Vidal, P. Morin, J. Sainte-Marie, and A. Bienvenue. 1988. Evidence for bidirectional transverse diffusion of spin-labeled phospholipids in the plasma membrane of guinea pig blood cells. Biochim. Biophys. Acta 946: 315-327.

Thompson, T. E., and T. W. Tillack. 1985. Organization of glycosphingolipids in bilayers and plasma membranes of mammalian cells. Annu. Rev. Biophys. Biophys. Chem. 14:361-86

Vale, R. D. 1987. Intracellular transport using microtubule-based motors. Ann. Rev. Cell Biol. 3: 347-378.

van Blitterswijk, W. J., B. W. van der Meer, and H. Hilkmann. 1987. Quantitative contributions of cholesterol and the individual classes of phospholipids and their degrees of fatty acyl (un)saturation to membrane fluidity measured by fluorescence polarization. Biochemistry 26: 1746-1756.

van den Hill, A., G. P. H. van Heusden, and K. W. A. Wirtz. 1985. The synthesis of sphingomyelin in the Morris hepatomas 7777 and 5123D is restricted to the plasma membrane. Biochim. Biophys. Acta 833: 354-357.
van Meer, G. 1988. How epithelial cells grease their microvilli. TIBS 13: 242-243.
van Meer, G. 1989a. Lipid traffic in animal cells. Annu. Rev. Cell Biol. 5: 247-275.
van Meer, G. 1989b. Polarity and polarized transport of membrane lipids in a cultured epithelium. Modern Cell Biology (Satir, B.H., ed.), Functional Epithelial Cells in Culture. (Matlin, K.S., Valentich, J.D., eds.), Alan R. Liss, Inc., New York 8: 43-69.
van Meer, G., B. Gumbiner, and K. Simons. 1986. The tight junction does not allow lipid molecules to diffuse from one epithelial cell to the next. Nature 322: 639-641.
van Meer, G., and K. Simons. 1986. The function of tight junctions in maintaining differences in lipid composition between the apical and the basolateral cell surface domains of MDCK cells. EMBO J. 5: 1455-1464.
van Meer, G., E. H. K. Stelzer, R. W. Wijnaendts-van-Resandt, and K. Simons. 1987. Sorting of sphingolipids in epithelial (Madin-Darby canine kidney) cells. J. Cell Biol. 105: 1623-1635.
van 't Hof, W., and G. van Meer. 1990. Generation of lipid polarity in intestinal epithelial (Caco-2) cells: Sphingolipid synthesis in the Golgi complex and sorting before vesicular traffic to the plasma membrane. J. Cell Biol. 111: 977-986.
Vénien, C., and C. Le Grimellec. 1988. The involvement of cytoskeletal proteins in the maintenance of phospholipid topology in renal brush-border membranes. Biochim. Biophys. Acta 946: 307-314.
Venien, C., and C. Le Grimellec. 1988. Phospholipid asymmetry in renal brush-border membranes. Biochim. Biophys. Acta 942: 159-168.
Voelker, D. R., and E. P. Kennedy. 1982. Cellular and enzymic synthesis of sphingomyelin. Biochemistry 21: 2753-2759.
Zachowski, A., and P. F. Devaux. 1989. Bilayer asymmetry and lipid transport across biomembranes. Comments Mol. Cell. Biophys. 6: 63-90.
Zachowski, A., and P. F. Devaux. 1990. Tansmembrane movements of lipids. Experientia 46: 644-656.
Zachowski, A., P. Fellman, and P. F. Devaux. 1985. Absence of transbilayer diffusion of spin-labeled sphingomyelin on human erythrocytes. Comparison with the diffusion of several spin-labeled glycerophospholipids. Biochim. Biophys. Acta 815: 510-514.
Zachowski, A., A. Herrmann, A. Paraf, and P. F. Devaux. 1987. Phospholipid outside-inside translocation in lymphocyte plasma membranes is a protein-mediated phenomenon. Biochim. Biophys. Acta 897: 197-200.

1.10. TRANSCYTOSIS

(Chairperson : Hubert REGGIO)

Transcytosis refers to the vesicular transfer of soluble or membranous constituents across polarized cells. So far, transcytosis has been studied in epithelial cells, where it occurs in the two opposite directions : from basolateral-to-apical, and from apical-to-basolateral.

Pierre Courtoy summarized detailed analytical subcellular fractionation studies of endosomes in rat hepatocytes *in situ*, by proposing a branching-maturation model. This model distinguishes an early endosomal compartment including most recycling membrane (possibly equivalent to the endosomal reticulum of Hopkins), and distinct carrier vesicles which mature and so become competent to fuse either with lysosomes or with the apical (i.e. canalicular) membrane. By the combination of a refined differential sedimentation protocol, isopycnic centrifugation and diaminobenzidine-induced density shift, he could isolate early and late endosomes involved in basolateral-to-apical transcytosis and show differences in membrane composition. Along this route, $F(ab')_2$ fragments of antibodies against the receptor for polymeric immunoglobulins are partially reduced into Fab' in the endosomes. Reduction of disulfide bridges could be important in the physiology of endosomes.

Renate Fuchs has applied a low temperature block on the perfused liver system to dissect the transcellular pathway. She found unexpectedly that low temperature arrests transcytosis of the high-molecular weight fluid-phase tracer sucrose. The regurgitation of sucrose into the perfusate is increased after rewarming, excluding a paracellular pathway. It remains to elucidate the significance of these two putative transyctotic routes and the mechanism whereby the molecular weight of a fluid-phase tracer can affect its vesicular transport.

The opposite apical-to-basolateral route permits the receptor-mediated transport of IgG across the syncytiotrophoblast of the neonatal enterocytes. To compare the two opposite transcytotic pathways in the same cell, **Ira Mellman** doubly transfected the polarized MDCK cells with the receptor for IgG (Fc receptor) and the receptor for polymeric immunoglobulins. Interestingly, the transfected isoform FcRII-B2 concentrates in clathrin-coated pits while the FcRII-B1 does not. Hybrid constructs of the two isoforms identify domains of their cytoplasmic tail acting as positive and negative signals for coated pit localization. In the transfected cells, the microtubule-depolymerizing agent nocodazole

NATO ASI Series, Vol. H 62
Endocytosis
Edited by P. J. Courtoy

completely blocks the basolateral-to-apical transcytosis of pIgA, but not the apical-to-basolateral transcytosis of IgG.

The amount of fluid and membrane transcytozed in both directions by MDCK cells is impressive. **Robert Parton** reported that, although the rate of endocytosis at the basolateral domain is fourfold higher than at the apical surface, about equal amounts of fluid-phase tracers internalized from the apical and basolateral surfaces of MDCK cells become regurgitated from the same surface or are transcytozed. In terms of membrane flow, > 10 % of biotinylable membrane constituents from each membrane domain are found at the opposite domain after 3 hours. This underscores the magnitude of the correcting traffic, necessary to maintain the well-differentiated membrane domains of these cells.

As pointed out by **Erik Christensen**, these values should not be lightly extrapolated to the corresponding non-transformed cells in the living organism. He studied protein handling by apical and basolateral surfaces in micropunctured-perfused proximal tubules. The apical uptake generally involves a poorly characterized "high capacity-low affinity" adsorptive endocytosis, which results in complete uptake of insulin from the ultrafiltrate, but only 5 % uptake of EGF. Most internalized apical membrane constituents are recycled, with < 5 % transcytosis. Basolateral uptake is relatively small and mixing of apical and basolateral tracers is very limited.

ANALYTICAL SUBCELLULAR FRACTIONATION OF RECEPTOR-MEDIATED TRANSCYTOSIS IN RAT HEPATOCYTES

Pierre-J. Courtoy[1], Michèle Leruth-Deridder[1], Jean-Pierre Vaerman[2] and Pierre Baudhuin[1]
University of Louvain Medical School & International Institute of Cellular and Molecular Pathology, (1) Cell Biology Unit & (2) Experimental Medicine Unit, Avenue Hippocrate 75, 1200 Brussels, Belgium

INTRODUCTION

Transcytosis is defined as the vesicular transfer of soluble or membranous constituents across polarized cells. Transcytosis can occur from basolateral-to-apical, or apical-to-basolateral in the same cell (Bomsel et al 1989). The three classical modes of endocytosis, namely fluid-phase, adsorptive and receptor-mediated, also apply to transcytosis. Receptor-mediated transcytosis makes the transcellular transfer of solutes both specific and efficient, by the selective uptake of ligands and their subsequent guidance along the complex transcellular route. This process is particularly important for the transfer of antibodies across epithelial barriers. Receptor-mediated transcytosis ensures the selective passage of IgG from the maternal blood across the placenta or, in some species, its intestinal absorption from mother's milk, resulting in neonatal humoral immunity. Receptor-mediated transcytosis is also responsible for the transfer of J-chain containing polymeric immunoglobulins (pIg) A and M into secretions, generating mucosal immunity, conferring protection to suckling infants and allowing for the disposal of circulating pIg-immune complexes.

RAT LIVER AS A MODEL OF TRANSCYTOSIS

We shall here review our subcellular fractionation analysis of receptor-mediated transcytosis in rat hepatocytes *in situ*. It is implicit that this study is limited to basolateral(sinusoidal)-to-apical(canalicular) transcytosis. The rat liver offers several advantages for a study of transcytosis : (1) in this species, the liver is the major site of clearance for polymeric immunoglobulin A (pIgA) (Delacroix et al 1983) ; this is due the high number of receptors for pIg exposed at the hepatocyte sinusoidal surface (estimated at 300,000 exposed copies per cell), combined with its major blood supply ; (2) the pIg receptor and its bound ligands are constitutively internalized at the sinusoidal membrane and transferred to the canalicular membrane (Mullock

NATO ASI Series, Vol. H 62
Endocytosis
Edited by P. J. Courtoy

et al 1980) the transfer is rapid (peaking within 15 to 30 min after uptake), specific (up to 80 % of injected ligands) and depends on the phosphorylation of a serine residue at the cytoplasmic domain of the pIg receptor (Casanova et al 1990) ; (3) at the canalicular membrane, the ectodomain of the receptor is cleaved by a leupeptin-sensitive endoprotease (Musil and Baenziger 1988) and is released into bile, where it is known as the secretory component (SC) ; this also accounts for the discharge of bound ligands and facilitates the monitoring of their transcytosis, using bile-duct cannulation ; (4) SC is a major protein of bile, from which it can be easily purified and it is a strong immunogen ; appropriate antibody fragments bound to SC faithfully follow its route through hepatocytes and can thus substitute to pIg as ligands ; (5) the liver is an ideal object for the subcellular fractionation of the endocytic apparatus (reviewed in Courtoy 1991) ; (6) transcytosis can be compared with the degradative pathway.

POLYMERIC IMMUNOGLOBULIN A AND ANTI-SECRETORY COMPONENT $F(ab')_2$ ARE EQUIVALENT TRACERS OF BASOLATERAL-TO-APICAL TRANSCYTOSIS IN RAT HEPATOCYTES

We have labeled the transcytotic apparatus of rat hepatocytes by three probes : radiolabeled pIgA purified from either rat or human myelomas, its conjugates with horseradish peroxidase (HRP), and iodinated $F(ab')_2$ fragments of affinity-purified rabbit antibodies against free SC purified from rat bile. On the average, 55 ± 7 % of injected pIgA and 62 ± 14 % of injected $F(ab')_2$, are recovered in bile. Compared to the peak of liver uptake, this corresponds to ≥ 80 % efficiency of transfer into bile. The rate of uptake and the efficiency of transfer are somewhat reduced when pIgA is conjugated to HRP (Quintart et al 1989). In our hands, the clearance of the anti-SC $F(ab')_2$ appears to be faster than for the natural ligand, resulting in an almost complete uptake within 2 min. With such a probe, a pulse-chase is thus spontaneously achieved, without the need of a perfused liver system (see Perez et al 1988).

Using cultured rat hepatocytes, we have observed that anti-SC Fab' conjugated to HRP and pIgA adsorbed to colloidal gold particles largely co-distribute in coated pits and endosomes (Courtoy 1985), while other investigators have reported that anti-SC IgG follow a quite different intracellular route than pIgA, namely uptake by a different type of endocytic pit, followed by preferential discharge into lysosomes (Kim et al 1985). This discrepancy is probably related to the much lower efficiency of transfer of anti-SC IgG into bile, compared to its Fab' and especially F(ab')2 fragments (Fig 1).

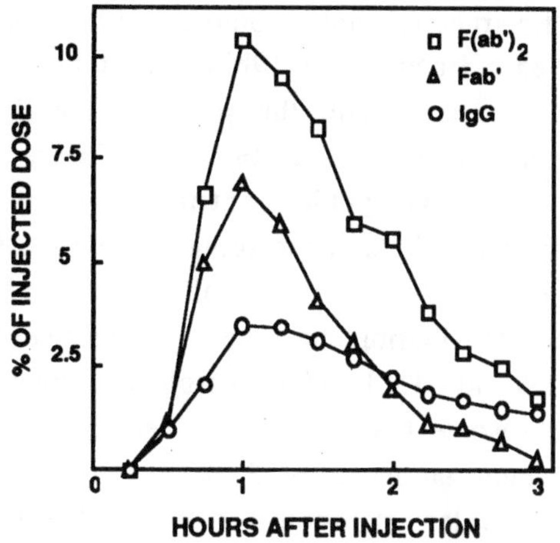

Fig 1. Secretion into bile of anti-SC IgG, F(ab')2 and Fab'

(Reproduced with permission from Lemaître-Coelho et al 1982).

ALL TRANSCYTOTIC ENDOSOMES HAVE SIMILAR PHYSICAL PROPERTIES

By differential sedimentation, the bulk of injected pIgA (Limet et al 1985), pIgA-peroxidase (Quintart et al 1989) and F(ab')2 anti-SC (Fig 2) remain associated throughout transcytosis with the classical microsomes (particles sedimenting between 275,000 and 3 x 10^6 *g* x min.

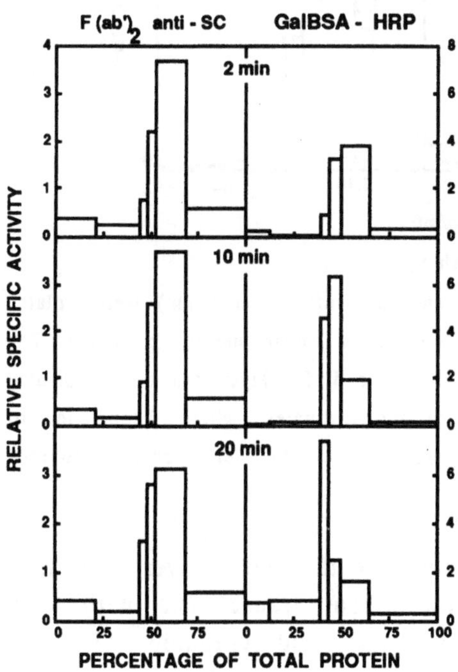

Fig 2. Differential sedimentation of endosomes

At the indicated time before sacrifice, the rats were injected with either ^{125}I-F(ab')2 anti-SC as a tracer of trancytosis, or ^{3}H-GalBSA-HRP as a tracer of the degradation pathway. The homogenate was resolved by differential sedimentation into, from left to right, a N pellet (6,000 *g* x min), a M pellet (33,000 *g* x min), a L pellet (275,000 *g* x min), a F fraction (fluffy layer corresponding to particles loosely sedimenting over the packed L pellet) a P' pellet (microsomes, 3 x 10^6 *g* x min) and a final supernatant (S). Distributions are averages of 2 to 6 experiments.

For spherical particles, this can be translated into an minimal average diameter of 130 nm (Beaufay et al 1976), in good agreement with the ultrastructural data obtained with pIgA-peroxidase conjugates (Courtoy et al 1985 ; Hoppe et al 1985). About one tenth of the F(ab')2 sediments faster and is recovered with the particles loosely sedimenting on the 275,000 *g* x min pellet, referred to as the fluffy layer.

By isopycnic centrifugation in linear sucrose gradients, the labeled particles of the microsomal fraction essentially equilibrate as a single peak at 1.13 g/ml in density, at all intervals after injection (Fig 3). In the fluffy layer, the distribution becomes progressively bimodal, with a similar low-density peak at 1.13 g/ml (probably particles similar to those recovered in microsomes) and a higher density peak at 1.18 g/ml. This second component accounts at most for 3 % of liver-associated ligand and co-distributes with the canalicular membrane marker alkaline phosphodiesterase I, suggesting the transient association of the antibodies with bile canaliculi.

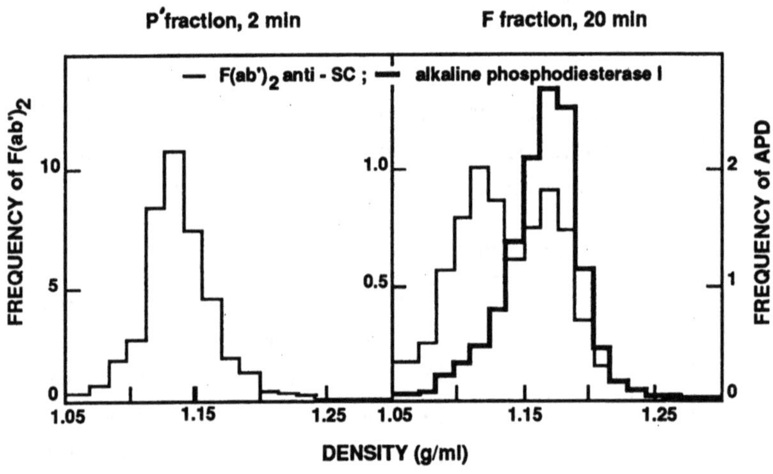

Fig 3. Density distribution of anti-SC F(ab')2

Particles of the microsomal fraction (left) and the fluffy layer (right) were isolated at the indicated times after injection and equilibrated in linear sucrose gradients. The distribution of alkaline phosphodiesterase I (APD) in the fluffy layer has been indicated for reference. Notice the difference in scale of the ordinates, where frequencies of constituents are normalized according to the percentage of the homogenate recovered in the fraction.

Thus, except for the minor contribution of the fluffy layer, the three transcytotic tracers remain associated throughout transcytosis with very small particles equilibrating at low densities and coincide with the distribution of none of the marker enzymes studied so far. This monotony is in marked contrast with the degradation pathway, along which tracers are successively associated with small particles equilibrating around 1.13 g/ml, larger particles with first a lower and then a much higher equilibrium density (around 1.11 g/ml, followed by 1.18 g/ml) and finally very dense particles (1.21 g/ml). The latter correspond to secondary lysosomes (Limet et al 1985).

EARLY ENDOSOMES ARE COMMON FOR TRANSCYTOSIS AND THE DEGRADATION PATHWAY

Conventional fractionation also suggests that the early stages of the degradation and of the transcytotic pathways are common, since both involve endosomes of similar physical properties. A direct evidence of identity has been obtained using the diaminobenzidine-induced density shift, a procedure capable of selectively retrieving a particular endocytic compartment that has been selectively labeled by peroxidase (Courtoy et al 1984). In this case, rats were injected with radiolabeled pIgA together with galactosylated bovine serum albumin conjugated to HRP (GalBSA-HRP), which serves as a tracer of the degradation pathway. At 5 min after injection, about 70 % of pIgA of a comprehensive low-density fraction was displaced concomitantly with the conjugate, indicating physical association with the same endosomes. This proportion declined markedly at later time intervals, demonstrating sorting and segregation into distinct endosomes (Fig 4).

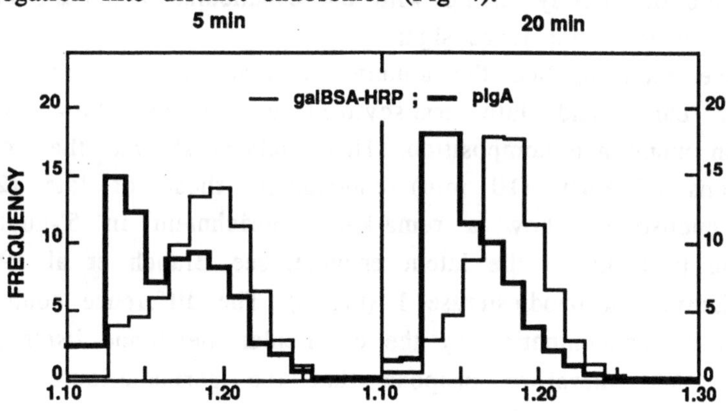

Fig 4. Identity of the early, but not the late, endosomes constituting the transcytotic and the degradative pathways

Low-density preparations were isolated at 5 or 20 min after a combined injection of GalBSA-HRP and pIgA, incubated in diaminobenzidine and H_2O_2 and equilibrated again in linear sucrose gradients. GalBSA-HRP is shifted to heavy densities, together with the fraction of pIgA present in the same particles (reproduced from Courtoy et al 1988).

The low-density fraction has been further resolved into common, sorting and segregated endosomes (Kennedy and Cooper 1988). These biochemical results are in good agreement with the ultrastructural localization of the ligands and their receptors using double immunolabeling on ultrathin frozen sections, showing sorting in tubulospherical endosomes at the periphery of hepatocytes (Geuze et al 1984).

LATE TRANSCYTOTIC ENDOSOMES HAVE A DISTINCT MEMBRANE COMPOSITION

A fascinating question in this system is the final subcellular addressing. Indeed, late endosomes committed either to the degradation or the transcytotic pathway show a low level of misaddressing, despite their close vicinity with each other and with their respective target in the pericanalicular region. This implies the existence of specific recognition systems. Selective fusion of the endosomes of the degradative pathway with lysosomes has been recently reconstituted in a promising cell-free system (Mullock et al 1989). We have explored the complementary aspect, namely the recognition between transcytotic endosomes and the canalicular membrane. The ultrastructural studies using pIgA-HRP indicate a lag between pericanalicular clustering and discharge into bile (Courtoy et al 1985 ; Hoppe et al 1985). This suggests the possibility that transcytotic endosomes are not capable of directly fusing with the canalicular membrane and must first acquire a fusion competence signal.

We have again applied the diaminobenzidine-induced density shift to ask whether early and late transcytotic endosomes show detectable differences in membrane composition. This analysis showed that, compared to preparations of early (10 min) endosomes, those of late (20 min) transcytotic endosomes show a remarkable enrichment in 5'-nucleotidase (corresponding probably to the latent enzyme, see Branch et al 1987) and especially alkaline phosphodiesterase 1 (Fig 5). The difference could not be accounted for by contamination by the canalicular membrane itself (Quintart et al 1989). We therefore propose that the pericanalicular delay is concomitant with an enrichment in canalicular membrane markers and corresponds to a rate-limiting fusion with a pool of apical recycling vesicles. The occurrence of this pathway has recently been documented in MDCK cells (Bomsel et al 1989).

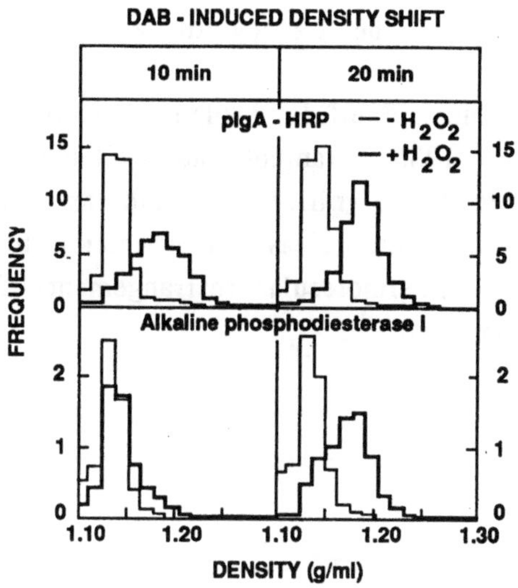

Fig 5. Association of transcytotic endosomes with an endogenous plasma membrane marker

Low-density preparations enriched in endosomes were isolated at 10 or 20 min after injection of pIgA-HRP, incubated with diaminobenzidine in the presence (thick line) or absence of H_2O_2 (thin line) and equilibrated again in linear sucrose gradients. The concomitant displacement of alkaline phosphodiesterase I is used as a criterion of its association with the transcytotic endosomes. Frequencies are normalized as at Fig 3. (Adapted from Quintart et al 1989).

TRANSCYTOTIC ENDOSOMES CAN REDUCE DISULFIDE BRIDGES

When injected F(ab')2 anti-SC is recovered in bile, its electrophoretic mobility in SDS-PAGE corresponds to that of Fab' produced by mild *in vitro* reduction, but it remains as F(ab')2 when incubated in freshly collected bile (Limet et al 1985). During transcytosis, an increasing proportion of the tracer isolated with the low-density microsomes corresponds to Fab', demonstrating that reduction can occur in the transcytotic endosomes. However, reduction does not exceed 50 % in those preparations, whereas it is virtually complete in the particles of the fluffy layer equilibrating at the density of alkaline phosphodiesterase, indicating that reduction is completed at the canalicular membrane (Fig 6). Interestingly, the disulfide bond between the two chains of Fab' is largely unaffected in both fractions, showing their limited reduction potential.

This observation could explain how pIgA becomes covalently linked to its receptor by a disulfide bridge during transcytosis. Because the rat (like the human) SC have an even number of cysteine residues (Banting et al 1989), all of which could be engaged in disulfide bridges (as in the case of human SC ; Eiffert et al 1984), a prerequisite for the generation of the intermolecular bond would be the opening of an intramolecular bridge in the receptor molecule (Lindh and Björk 1977). The formation of this bond is promoted *in vitro* by a microsomal disulfide interchange enzyme (Murkofsky and Lamm 1979). Whether this is artificially due to the enzyme protein disulfide isomerase contributed by vesicles derived from the endoplasmic reticulum, or reflects an intrinsic activity of the transcytotic endosomes remains to be established.

The capacity of endosomes to reduce disulfide bridges has been demonstrated in other studies (Shen et al 1985 ; Janicot and Desbuquois 1987). This activity could have a broader significance, such as the dissociation of the A and B chain of insulin, the release of hormonogenic peptides from the thyroglobulin termini, or intermolecular rearrangement of various recycling multimeric membrane proteins.

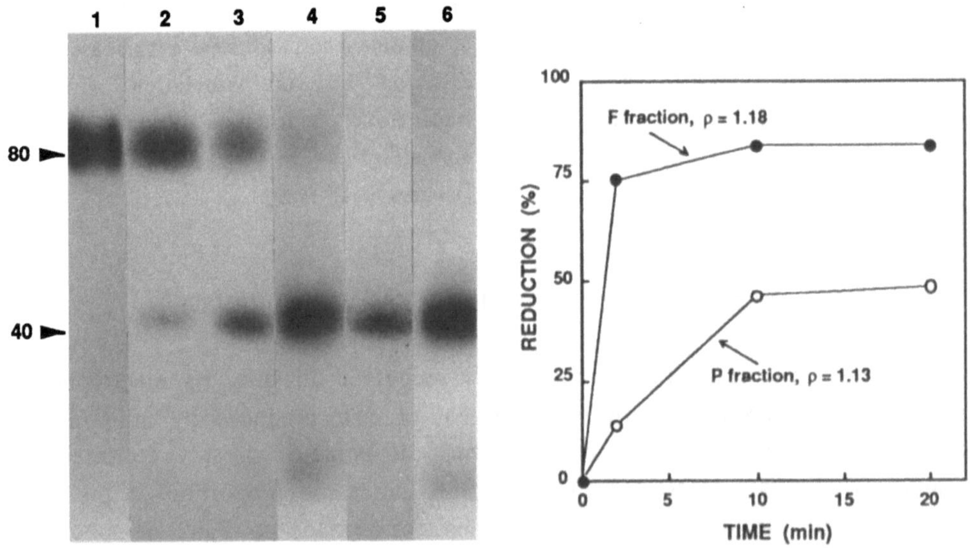

Fig 6. Reduction of disulfide bridges in transcytotic endosomes

The left panel shows autoradiographs after SDS-PAGE (non reducing conditions) of (1) injected F(ab')2 ; (2, 3) tracer recovered in microsomes at 2 and 20 min ; 4) tracer recovered in the particles of the fluffy layer equilibrating at 1.18 g/ml ; 5) tracer recovered in bile ; 6) tracer reduced *in vitro* by 5 mM cysteamine The right panel quantifies the kinetics of reduction in low-density microsomes (open circles) and the high-density peak of the fluffy layer (closed circles), expressed as a percentage of the tracer recovered in the indicated fraction.

REFERENCES

Banting G, Brake B, Braghetta P, Luzio JP, Stanley KK (1989) Intracellular targetting signals of polymeric immunoglobulin receptors are highly conserved between species. FEBS Lett 254:177-183

Beaufay H, Amar-Costesec A (1976) Cell fractionation techniques, in "Methods in membrane biology" (ED Korn ed) vol 6 Plenum press, pp 1-100

Bomsel M, Prydz K, Parton RG, Gruenberg J, Simons K (1989) Endocytosis in filter-grown Madin-Darby canine kidney cells. J Cell Biol 109:3243-3258

Branch WJ, Mullock BM, Luzio JP (1987) Rapid subcellular fractionation of the rat liver endocytic compartments involved in transcytosis of polymeric immunoglobulin A and endocytosis of asialofetuin. Biochem J 244:311-315

Casanova JE, Breitfeld PP, Ross SA, Mostov KE (1990) Phosphorylation of the polymeric immunoglobulin receptor required for its efficient transcytosis. Science 248:742-745

Courtoy P.J., Quintart J., Baudhuin P (1984) Shift of equilibrium density induced by 3,3'-diaminobenzidine cytochemistry : a new procedure for the analysis and purification of peroxidase-containing organelles. J. Cell Biol. 98:870-876.

Courtoy PJ (1985) Receptor-mediated endocytosis in rat hepatocytes, with special reference to the pathways of polymeric IgA and galactose-exposing proteins. Thesis (University of Louvain) pp 1-156

Courtoy PJ, Quintart J, Limet JN, De Roe C, Baudhuin P (1985) Polymeric IgA and galactose-specific pathways in rat hepatocytes : evidence for intracellular ligand sorting, in "Endocytosis" (I Pastan and MC Willingham, eds), Plenum Press, New York pp 163-194

Courtoy PJ, Quintart J, Draye JP, Baudhuin P (1988) The DAB-induced density shift : principle, validity and application to endosomes, in "Cell-free analysis of membrane traffic", Progress in Clinical and Biological Research, vol. 270 (DJ Morré, KE Howel, GMW Cook and WH Evans, eds)Alan Liss, New York pp 169-183

Courtoy PJ (1991) Analytical subcellular fractionation of endosomal compartments in rat hepatocytes, in "Identification and characterization of endocytic components" (Bergeron JJM ed), Subcellular Biochemistry, Plenum press, "in press"

Delacroix DL, Furtado-Barreira G, De Hemptinne B, Goudswaard J, Dive C, Vaerman JP (1983) The liver in the IgA secretory immune system : dogs but not rats and rabbits are suitable model for human studies. Hepatology 3:980-988

Eiffert H, Quentin E, Decker J, Hillemeir S, Hufschmidt M, Klingmüller D, Weber MH, Hilschmann N (1984) Die Primärstruktur der menschlichen freien Sekretkomponente und die Anordnung der Disulfidbrücken. Hoppe-Seyler's Z Physiol Chem 365:1489-1495

Geuze HJ, Slot JW, Strous GJAM, Peppard J, Von Figura J, Hasilik A,Schwartz AL (1984) Intracellular receptor sorting during endocytosis : comparative immunoelectron microscopy of multiple receptors in rat liver. Cell 37:195-204

Hoppe CA, Connolly TP, Hubbard AL (1985) Transcellular transport of polymeric IgA in the rat hepatocyte : biochemical and morphological characterization of the transport pathway. J Cell Biol 101:2113-2123

Janicot M, Desbuquois B (1987) Fate of injected ^{125}I-labeled cholera toxin taken up by rat liver *in vivo :* generation of the active A_1 peptide in the endosomal compartment. Eur J Biochem 163:433-442

Kennedy J, Cooper C (1988) The time-dependent distribution of ^{125}I-asialo-orosomucoid-horseradish peroxidase and ^{131}I-immuno-globulin A among three endosomal subfractions isolated from rat liver. Biochem J 252:739-752

Kim E, Hradek GT, Jones AL (1985) Degradative intracellular transport of anti-secretory component in cultured rat hepatocytes : an alternate pathway for the immunoglobulin A receptor. Gastroenterology 88:1791-1798

Lemaître-Coelho I, Meykens R, Vaerman JP (1982) Anti-receptor antibodies : a comparison in the rat of the plasma-to-bile transfer of purified IgG, $F(ab')_2$ and Fab' antibodies against rat secretory component. Protides Biol Fluids, 29th Colloquium (Peeters H ed), Pergamon press, Oxford, pp 419-422

Limet JN, Quintart J, Schneider YJ, Courtoy PJ (1985) Receptor-mediated endocytosis of polymeric IgA and galactosylated serum albumin in rat liver : evidence for intracellular ligand sorting and identification of distinct endosomal compartments. Eur J Biochem 146:539-548

Lindh E, Björk I (1976) Binding of secretory component to dimers of immunoglobulin A *in vitro* : mechanism of the covalent bond formation. Eur J Biochem 62:263-270

Mullock BM, Jones RS, Hinton RH (1980) Movement of endocytic shuttle vesicles from the sinusoidal to the bile canalicular face of hepatocytes does not depend on occupation of receptor sites. FEBS Lett 113:201-205

Mullock B.M., Branch W.J., Van Schaik M., Gilbert L.K., Luzio J.P. (1989) Reconstitution of an endosome-lysosome interaction in a cell-free system, J Cell Biol 108:2093-2099.

Murkofsky NA, Lamm ME (1979) Effect of a disulfide interchange enzyme on the assembly of human secretory immunoglobulin A from immunoglobulin A and free secretory component. J Biol Chem 254:12181-12184

Musil LS, Baenziger JU (1988) Proteolytic processing of rat liver membrane secretory component : cleavage activity is localized to bile canalicular membranes. J Biol Chem 263:15799-15808

Perez JH, Branch WJ, Smith B, Mullock BM, Luzio JP (1988) Investigation of endosomal compartments involved in endocytosis and transcytosis of polymeric immunoglobulin A by subcellular fractionation of perfused rat liver. Biochem J 251:763-770

Quintart J, Baudhuin P, Courtoy PJ (1989) Marker enzymes in rat liver vesicles involved in transcellular transport. Eur J Biochem 184:567-574

Shen WC, Ryser HJP, LaManna L (1985) Disulfide spacer between methotrexate and poly(D)-lysine. J Biol Chem 260:10905-10908

TEMPERATURE DEPENDENCE OF TRANSCYTOTIC PATHWAYS IN RAT LIVER

H. Klapper, J. Graf and R. Fuchs
Institut für allgemeine und experimentelle Pathologie
der Universität, Vienna, Austria

INTRODUCTION

The liver is a transporting epithelium, composed of sheets of hepatocytes that separate the canalicular lumen from the perisinusoidal space. Transport of macromolecules from blood to bile is thought to occur by a transcellular vesicular pathway (Brown and Kloppel, 1989; Lake et al. 1985), where hepatocytes internalize extracellular material either by specific receptors or by fluidphase endocytosis. Ligands taken up by receptor-mediated endocytosis are selectively delivered to lysosomes (e.g. asialoglycoproteins) or to bile (e.g. transcytosis of polymeric immunoglobulin A). Transcellular vesicular transport of the majority of other blood derived biliary proteins (Reuben, 1984) does not involve specific receptor binding.

Pathways for biliary secretion of fluid phase markers have been investigated using the isolated perfused rat liver (Lake et al., 1985; Hardisson et al., 1989; Hayakawa et al., 1990). Appearance of inert molecules in the bile exhibits different kinetics dependent on the molecular weight of these substances, however, there is no clear-cut correlation (Lake et al., 1985). This has been taken as evidence that small molecules (e.g. erythritol, sucrose) enter bile by a paracellular route (Hardisson et al., 1989) in addition to a vesicular transcellular pathway (Graf, 1983; Lake et al, 1985). Paracellular and transcellular pathways may exert charge selectivity for fluid phase markers (Bradley and Herz, 1978; Hardisson et al., 1989), although these data are also contradictory (Lake et al., 1985). Recently, differences in the endocytic uptake rates of chemically different markers have been described in non-polarized tissue culture cells: specifically a faster initial rate of uptake of sucrose than that of horseradish peroxidase and Lucifer Yellow (Griffith et al., 1989).

To gain more insight into the mechanism of intracellular trafficking of fluid phase markers, this study addresses the question whether multiple transcytotic pathways for fluid phase markers with different molecular weight (^{14}C-sucrose, FITC-dextran) and receptor mediated transcytosis of ^{125}I-ASOR can be distinguished on the basis of different temperature sensitivities of their transport kinetics.

NATO ASI Series, Vol. H 62
Endocytosis
Edited by P. J. Courtoy

MATERIALS AND METHODS

Materials: FITC-Dextran (Mw=70000) was purchased from Sigma and dialysed extensively before use. Asialoorosomucoid (ASOR) was prepared by enzymatic desialylation of α1-acid glycoprotein (Dunn et al., 1979) and iodinated in the presence of iodogen (Sigma) to a specific activity of 3,3 x 10^6 cpm / µg. ^{14}C-Sucrose was obtained from Amersham.

Animals: Male Louvain rats (200-250g) were used in all experiments.

Perfusion protocols: Isolated rat livers were perfused in a non recirculating system with Krebs-Henseleit bicarbonate buffer (KHB) as described (Graf et al., 1983). Livers were equilibrated 20 minutes at 16°C and then perfused with ^{125}I-ASOR (5µg) for two minutes. After additional 55 minutes the temperature was shifted from 16°C to 37°C. Trichloracetic acid precipitable and soluble radioactivity was determined as intact ASOR and ASOR degradation products, respectively in single bile drops. Fluid phase markers (2 mM ^{14}C-sucrose (200µCi) or FITC-dextran (3mg/ml) were administered as a 5 minutes pulse at 16°C. Concentration of ^{14}C-sucrose in bile and perfusate was determined by liquid scintillation counting. Fluorescence intensity of FITC-dextran was measured on a Hitachi spectrofluorometer (excitation: 485 nm, emission: 515nm). Concentration of FITC-dextran in bile samples was corrected for autofluorescence of bile using excitation of 485 nm, and emission of 570 nm.

Data presentation: Biliary secretion of fluid phase markers was calculated as the product of biliary concentration and bile flow and was corrected for biliary deadspace (20 µl).

Mathematical procedures: Efflux of the respective markers into the effluent perfusate was analyzed by assuming three independent compartments that release their contents with first order kinetics. The sizes of the compartments and rate constants of release were obtained by a least square fit using the computer program BMDP 3R. All results are expressed as mean values ±S.D.

RESULTS AND DISCUSSION

1) Transport of ASOR: Although, the vast majority of internalized ASOR is targeted to lysosomes for degradation, a small amount (3%) is missorted to the transcytotic pathway (Schiff et al., 1984). Since it has been demonstrated that delivery of ligands and fluid phase markers from endosomes to lysosomes can be blocked at temperatures below 20°C (Dunn et al., 1980) we first investigated the influence of low temperature perfusion on transcytosis of intact ASOR. As shown in Figure 1, neither TCA-soluble nor TCA-precipitable radioactivity could be detected in bile during low temperature perfusion. Shifting the perfusate temperature to 37°C resulted in the appearance of intact and degraded ASOR in bile.

These results clearly show that receptor mediated transcytosis can also be completely blocked at low temperatures. These observations are also consistent with the assumption that at 16°C ligand accumulates in a kinetically early endosome compartment (Mueller et al., 1988) where sorting between both routes has not yet occurred (Fuchs et al., 1988).

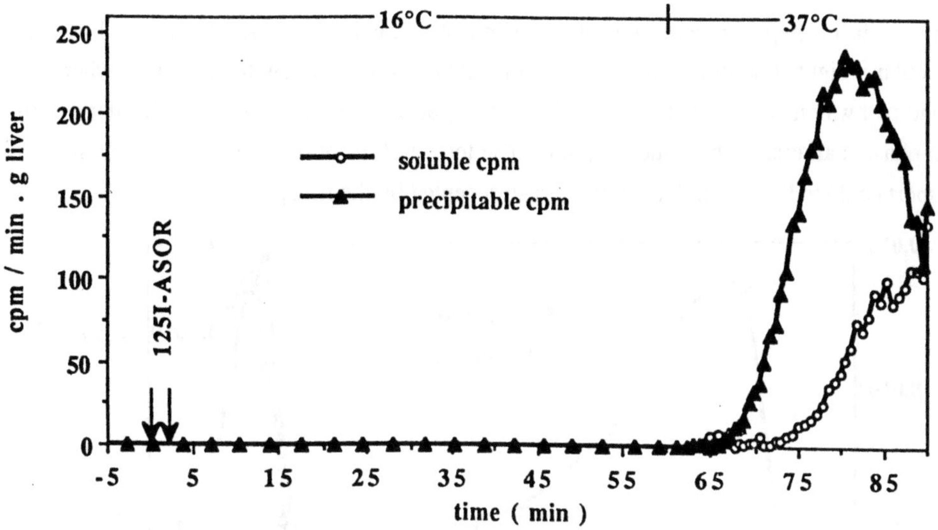

Figure 1: Effect of low temperature on degradation and transcytosis of ASOR. The liver was loaded with a 2-min pulse of ^{125}I-ASOR at 16°C. 60 minutes afterwards the perfusate temperature was shifted to 37°C. Intact ASOR and ASOR degradation products were determined in single bile drops.

2) Transport of fluid phase markers: Next, we determined whether transport of fluid phase markers into bile is also blocked at 16°C. Rat livers were perfused with FITC-dextran for 5 minutes (Figure 2) followed by perfusion with marker-free buffer. During continous perfusion at 16°C negligible amounts of FITC-dextran were secreted into bile. Warming the liver to 37°C rapidly initiated secretion. These data demonstrate that transport of FITC-dextran into bile can be accounted for entirely by a transcellular pathway. In addition, cell fractionation studies have shown that at low temperature FITC-dextran and ^{125}I-ASOR can be colocalized in an early endosome compartment (sorting compartment, see above), indicative of identical or similar transcytotic pathways taken by ligands and high molecular weight fluid phase markers.

In contrast, sucrose was secreted into bile even at 16°C (Figure 2). Interestingly, biliary sucrose secretion continued to increase after sucrose had been omitted from the perfusate, and reached a peak concentration after 10 minutes. Raising the perfusate temperature to 37°C resulted in a brisk, but transient increase of biliary sucrose secretion. These data do not support the notion that paracellular diffusion of sucrose occurrs to any significant extent at 16°C. If so, removal of sucrose from the perfusate would have resulted in diluting biliary sucrose through back diffusion. Instead, a bile to perfusate concentration ratio of 300:1 was maintained throughout the period of perfusion at 16°C. Furthermore, sucrose enters bile via a transcellular pathway that is not blocked by low temperature.

Taken together, the data suggest the existence of at least two different transcytotic pathways in rat hepatocytes: transcytosis of receptor-mediated molecules as well as high molecular weight fluid phase markers is completely blocked by low temperature, whereas a second pathway is less temperature sensitive and appears accessible to low molecular weight fluid phase markers, such as sucrose, only. The temperature sensitive component of sucrose transport could indicate that this marker is also accepted by the first pathway.

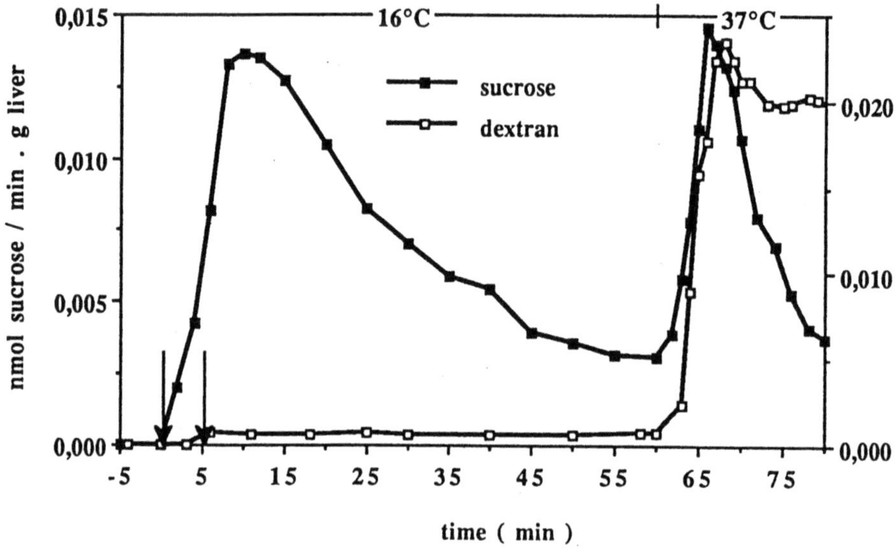

Figure 2: Influence of low perfusate temperature on biliary secretion of ^{14}C-sucrose and FITC-dextran after a 5-min pulse-load with the respective marker. The data presented are mean values from three independent experiments.

3) Kinetics of recycling and biliary secretion of fluid phase markers: In order to assess for intracellular compartments that could be responsible for biliary transport the kinetics of recycling of sucrose and FITC-dextran into the effluent perfusate was studied and compared to biliary secretion (Figure 3). Efflux of both markers could be reasonably fitted by assuming release from three independent compartments (designated 1, 2, 3). Withdrawal of the marker from the inflowing perfusate resulted in an initial rapid decrease of venous efflux, consistent with release from an extracellular (vascular) compartment at rates of 95% and 98% min^{-1} for sucrose and dextran, respectively. The second and third component of the curve represents release from two different (fast and slow) recycling compartments. Their content of marker and volumes (content/concentration of marker in the loading perfusate) are given in Table 1. The volumes and rates for marker release of the fast recycling compartments are similar for dextran and sucrose and their depletion is nearly complete within 55 minutes. Depletion of the slow recycling compartments is different for dextran and sucrose: 55 minutes after the loading pulse, this compartment contains still 41% of its initial content of dextran, but only 11% of sucrose, respectively. This observation suggests that release of dextran is inhibited to a larger extent than release of sucrose.

COMPARTMENT	VOLUME (µl)		DECREASE OF CONTENT(%/min)	
	SUCROSE	DEXTRAN	SUCROSE	DEXTRAN
2 (fast)	26,7 ± 8,1	20,4 ± 5,7	25,5 ± 2,6	22,5 ± 2,1
3 (slow)	20,8 ± 10	28,5 ± 16	4 ± 0,3	2,1 ± 0,9

COMPARTMENT	CONTENT (t=5,5 min)		CONTENT (t=60 min)	
	SUCROSE (µmol/g liver)	DEXTRAN (µg/g liver)	SUCROSE (nmol/g liver)	DEXTRAN (µg/g liver)
2 (fast)	0,05 ± 0,02	61,49 ± 17,35	0,0000113±0,000014	0,000375±0,00055
3 (slow)	0,04 ± 0,02	85,52 ± 48,61	4±2	35,28 ± 35,85

Table 1: Volumes and content of intracellular compartments involved in recycling of fluid phase markers during low temperature perfusion. Efflux of fluid phase markers into perfusate (see Figure 3) were analyzed as described in materials and methods.

As shown in Figure 3A, release of sucrose into bile exhibits the same rate constant as release into perfusate along the slow efflux component. These observations indicate that biliary secretion as well as recycling into the perfusate originate from the same intracellular compartment, which releases its content into perfusate and bile at a ratio of 100 : 1. Shifting the temperature to 37°C increases sucrose release into perfusate and bile. In contrast, rewarming decreases release of dextran into the perfusate, but induces biliary secretion (Figure 3B). We presume that by rewarming dextran is withdrawn from a pool that recycles at the sinusoidal membrane and shifts to a pool where it becomes available for biliary secretion.

In conclusion, sucrose transport differs from transport of dextran in two important aspects: 1) secretion into bile is not blocked by low temperature and 2) sinusoidal regurgitation of sucrose is enhanced by rewarming. Apparently there are at least two intracellular mechanisms for non-adsorptive transcytosis and recycling in polarized cells. The functional significance of these pathways remains to be elucidated.

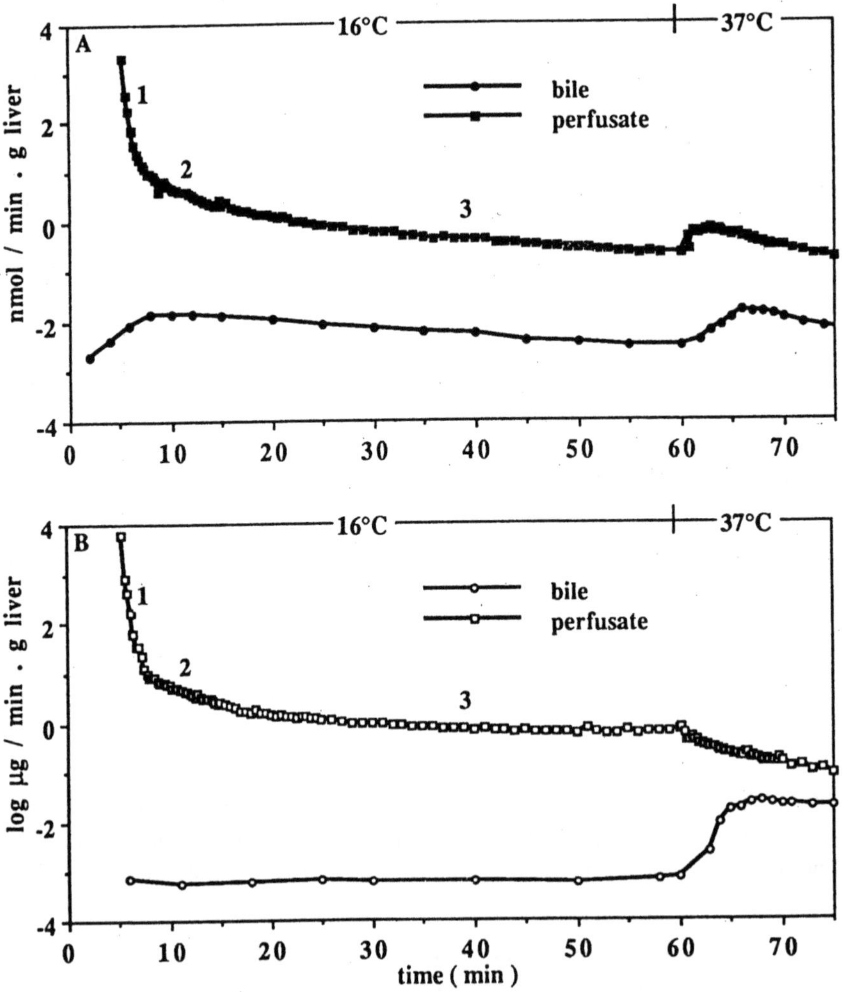

Figure 3: Kinetics of recycling (squares) and biliary secretion (circles) of sucrose (3A, closed symbols) and FITC-dextran (3B, open symbols) during perfusion at 16°C. The experimental design is described in Figure 2. Data are presented as semilogarithmic plots.

AKNOWLEDGEMENTS

This work was supported by Anton Dreher-Gedächtnisschenkung für Medizinische Forschung. The authors are greatly indebted to Dr V. Scheiber, Institut für Medizinische Statistik und Dokumentation der Universität Wien, for statistical calculations.

REFERENCES

Bradley, S. E., and Herz, R. (1978) Am. J. Physiol. **235(5)** : E570-E576

Brown, W. R., and Kloppel, T. M. (1989) Hepatology **9** : 763-784

Dunn, W. A., Hubbard, A. L., and Aronson, N. N. Jr. (1980) J. Biol. Chem. **255** : 5971-5978

Dunn, W. A., La Badie, J. H., and Aronson, N. N. Jr. (1979) J. Biol. Chem. **254** : 4191-4196

Fuchs, R., Klapper, H., Peterlik, M., and Mellman, I. (1988) J. Cell Biol. **107** : 118a

Graf, J. (1983) Am. J. Physiol. **244** : G233-246

Griffiths, G., Back, R., and Marsh,M. (1989) J. Cell Biol. **109** : 2703-2720

Hardison, W. G. M., Lowe, P. J., and Shanahan, M. (1989) Hepatology **9** : 866-871

Hayakawa, T., Cheng, O., Ma, A., and Boyer, J. L. (1990) Gastroenterology **99** : 216-228

Jaeschke, H., Krell, H., and Pfaff, E. (1987) Biochem. J. **241** : 635-640

Lake, J. R., Licko, V., Van Dyke, R. W., and Scharschmidt, B. (1985) J. Clin. Invest. **76** : 676-684

Mueller, S. C., and Hubbard, A. L. (1986) J. Cell Biol. **102** : 932-942

Reuben, A. (1984) Hepatology **4** : 46S-50S

Scharschmidt, B. F., Lake, J. R., Renner, E. L., Licko, V., and Van Dyke, R. W. (1986) Proc. Natl. Acad. Sci. U.S.A. **83** : 9488-9492

Schiff, J. M., Fisher, M. M., and Underdown, B. J. (1984) J. Cell Biol. **98** : 79-89

Schiff, J. M., Fisher, M. M., Jones, A. L., and Underdown, B. J. (1986) J. Cell Biol. **102** : 920-931

Fc RECEPTOR-MEDIATED ENDOCYTOSIS IN POLARIZED AND NON-POLARIZED CELLS

I. Mellman, H. Miettinen, P. Mâle and W. Hunziker
Department of Cell Biology
Yale University School of Medicine
333 Cedar Street, P.O. Box 3333
New Haven, Connecticut 06510 USA

Even in non-polarized cells, the organelles of the endocytic pathway generally assume a polarized distribution in the cytoplasm (Kornfeld and Mellman, 1989). Thus, endocytic coated vesicles and the early endosomes to which they deliver their contents are typically found in the cell periphery while late endosomes and lysosomes -- to which they deliver their contents -- are concentrated in the perinuclear region. The Golgi complex, from which newly synthesized lysosomal enzymes and membrane proteins are transported, is also found in perinuclear cytoplasm near the microtubule organizing center. Transport of internalized material from early to late endosomes would thus be expected to require the centripetal translocation of endosomes from the periphery, and that this occurs has been suggested by time lapse photomicroscopy (Albertini et al., 1984; Hirsch, 1968). Moreover, a number of studies have indicated that the translocation event is at least partly dependent on cytoplasmic microtubules (Gruenberg et al., 1989).

It seems likely that the characteristic spatial organization of endosomes and lysosomes is not accidental but is important for their biogenesis and function. Unfortunately, addressing this issue has proved problematic because organelle distribution can vary considerably from cell to cell, and because it is difficult to establish convenient quantitative or biochemical assays that relate function with location in the cytosol. For these reasons, we have recently turned towards the analysis of endocytosis in polarized Madin-Darby canine kidney (MDCK) cells. MDCK cells, like most epithelial cells, exhibit two separate endocytic pathways initiated at their distinct apical and basolateral plasma membranes (Bomsel et al., 1989; Parton et al., 1989). While this would appear to complicate any analysis of endocytosis relative to non-polarized cells, the fact that epithelial cells are capable of endocytic transport across the cell monolayer -- i.e., transcytosis -- creates an opportunity to more easily relate the contribution of spatial distribution to endocytic transport. This can be accomplished by simply monitoring the appearance of an endocytic tracer on the opposite pole from that to which it was originally applied.

NATO ASI Series, Vol. H 62
Endocytosis
Edited by P. J. Courtoy

To use MDCK cells for this purpose, however, it is necessary to provide for specific markers for receptor-mediated transcytosis. At present, transcytosis has been most extensively characterized in the case of immunoglobulin (Ig) transport across several different epithelia. In the case of basolateral to apical transcytosis, it is well known that hepatocytes and mammary epithelium express a receptor for polymeric Ig (pIgR) which mediates the transport of IgA or IgM from the blood to the bile or milk. From the work of Mostov and colleagues, it is clear that one can reconstitute basolateral to apical transcytosis in MDCK cells transfected with a cDNA encoding the rabbit pIgR (Mostov and Deitcher, 1986). Apical to basolateral transcytosis of Ig also occurs, but generally involves the transport of IgG via one or more classes of IgG Fc receptors (FcR). In the neonatal rat intestine, maternal IgG is bound by an FcR at the apical brush border of the ileal and jejeunal epithelium which then mediates its endocytosis and transport to lysosomes and to the basolateral surface (Abrahamson and Rodewald, 1981). Apical to basolateral transport of maternal IgG also occurs in the human placenta across the syncytiotrophoblast layer. While the receptor responsible for trans-placental transport has not been identified definitively, our recent investigations of FcR in the immune system has led to the finding that syncytiotrophoblasts express at high levels a receptor that is either very similar or identical to the major FcR expressed by human and mouse macrophages (Stuart et al., 1989). Indeed, as described below, when this receptor is expressed in MDCK cells, it mediates IgG transcytosis in the expected apical to basolateral direction (Hunziker and Mellman, 1989).

By stably transfecting MDCK cells with cDNAs encoding this FcR and/or the pIgR, we have produced a useful approach to studying the bi-directional endocytosis across polarized epithelial cells. In addition, it has recently become apparent that certain naturally occurring heterogeneity in the structure of FcR have yielded some unexpected insights into the mechanism of receptor-mediated endocytosis and the polarized expression of these molecules in MDCK cells. Thus, before describing our use of Ig receptors to study the requirements for endocytosis in polarized cells, we wish to summarize some of our recent results concerning some of the relevant features defined by the FcR itself.

The FcR cytoplasmic tail contains a unique, tyrosine-independent coated pit localization domain.

FcR are now known to comprise a large, multi-gene family whose products are expressed on a variety of cells in the immune system as well as on certain epithelia (Mellman, 1988). The most widely distributed and quantitatively most important class of receptor are designated FcRII, and are capable of mediating a variety of functions including endocytosis via coated pits and coated vesicles, phagocytosis of large particles, regulation of B-cell activation, and signalling macrophage secretion of a variety of important inflammatory mediators. cDNA cloning work from both human and mouse cells has demonstrated that even within the FcRII class, however, there is considerable structural heterogeneity (Hogarth et al., 1987; Lewis et al., 1986; Ravetch et al., 1986; Stuart et al., 1989). In the mouse, this leads to the production of distinct receptor isoforms which differ only in their cytoplasmic tails. As illustrated in Figure 1, two major isoforms are produced, designated FcRII-B1 and -B2 and expressed by lymphocytes and macrophages, respectively. The two receptors differ from each other due to alternative mRNA splicing which introduces an in-frame insertion of 47 amino acids at position 6 of the FcRII-B1 cytoplasmic tail. Their extracellular domains and membrane anchors are identical. Given that the expression of the two isoforms is cell type specific, we reasoned that it was likely for the two receptors to exhibit activities associated with the distinct functions of FcR on macrophages vs. B-cells.

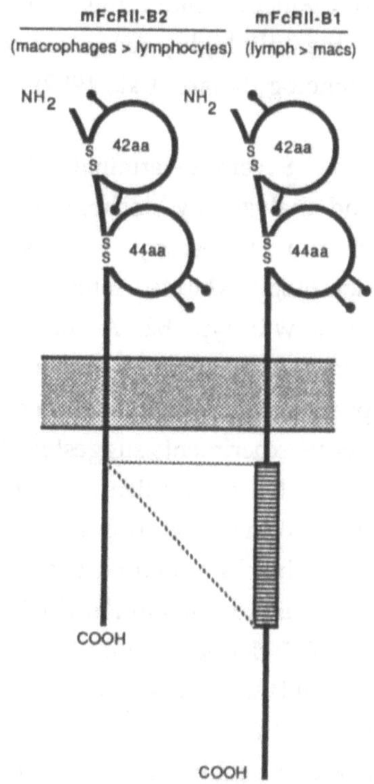

Following stable expression in FcR-negative cells (fibroblasts or MDCK cells), it became clear that only the macrophage isoform, FcRII-B2, could mediate the efficient internalization of IgG-containing antibody-antigen complexes. This difference was due to the fact that relative to FcRII-B2, the B1 isoform was <0.5% as efficient at localizing at clathrin coated pits on the surface of stably transfected CHO cells (Miettinen et al., 1989). This is in accord with our observation that unlike macrophages -- well known for their ability to internalize FcR-bound ligands -- FcR on mouse B-cell lines are almost completely unable to mediate endocytosis. Thus, the presence of the 47 amino acid insert in the FcRII-B1 cytoplasmic tail disrupts or otherwise inactivates a domain required for localization at coated pits.

To understand the mechanism of inhibition, we next sought to localize the putative coated pit localization domain in B2. The amino acid sequence of the B2 tail contains two tyrosine residues: at positions 26 and 43. The membrane proximal tyrosine (Y26) is contained within a sequence motif that can be predicted by at least one algorithim to correspond to the type of coated pit recognition signal found in other coated pit receptors (Ktistakis et al., 1990). The structure of this region is thought to involve a random coil or β-sheet structure ending in a turn at the tyrosine, presumably keeping it accessible for interaction with coated pit components. Interestingly, Y26 also resides within a stretch of 14 amino acids which is the only highly conserved region (95%) between mouse FcRII-B2 and its human homolog (Stuart et al., 1989).

Several C-terminal deletion mutants were prepared, expressed in CHO cells, and analyzed by biochemical assays for endocytosis and quantitative immuno-electron microscopy. As expected, mutant with a deletion just up to the region of homology exhibited internalization and coated pit localization phenotypes similar to the wild type B2. A mutant with a cytoplasmic tail deletion through the region of homology exhibited a greatly reduced ability to localize at coated pits, similar to the phenotype due to a deletion of the entire cytoplasmic tail (Miettinen et al., 1989). These experiments suggested that the region between amino acids 18 and 31 defined the coated pit localization domain in FcRII-B2. Although this region also contained the tyrosine at position 26, alteration of this tyrosine to a cysteine -- a mutation which blocks coated pit localization in LDL receptors -- surprisingly had no effect on the ability of the mutant FcRII-B2 to accumulate at coated pits. This finding suggests that FcR defines a second type of coated pit localization signal; one with superficial resemblance to the "classical" tyrosine-dependent signals, but which does not require the presence of the tyrosine. Conceivably, this variation reflects the fact that FcR only accumulate efficiently at coated pits *after* ligand binding; such receptors may thus have to regulate expression of the coated pit signal. In contrast, known tyrosine-dependent receptors (e.g., LDL and transferrin receptors) all accumulate at coated pits even in the absence of bound ligand.

The FcRII-B1 insert is a "retention signal" that prevents coated pit localization.

Since the FcRII-B2 coated pit localization domain maps between amino acids 18 and 31, it is not at all clear why the 47 amino acid insertion in FcRII-B1 -- which occurs at position 6 in the cytoplasmic tail -- should act to disrupt coated pit localization. One clue came from the quantitation of receptor distribution in transfected cells: the efficiency of FcRII-B1 localization at coated pits (<0.5% of B2) was far less than an FcR mutant entirely devoid of a cytoplasmic tail (~10% of B2) (Miettinen et al., 1989). Thus, sequences in the insertion may act to actively exclude FcRII-B1 from accumulating at coated pits. To test this possibility, polymerase chain reaction (PCR) was used to place the FcRII-B1 insertion at the carboxy terminus of

FcRII-B2. CHO cells expressing the B2/B1 mutant were produced and found to exhibit a phenotype similar to the FcRII-B1 transfectants (*i.e.*, a greatly reduced capacity for coated pit localization).

A likely mechanism by which the insertion acts as a "retention signal" to prevent the receptors from accumulating at coated pits was first suggested by immunofluorescence microscopy. Most of the receptor in cells transfected with FcRII-B1 (or the B2/B1 mutant) was found to occur in distinct linear arrays that aligned with cytoplasmic actin filaments (visualized using fluorescent phalloidin). The suggested association of the receptor with the cytoskeleton was then confirmed biochemically. Following ligand binding at 37°C (but not at 0°C), a significant fraction of the receptor became resistant to extraction by Triton X-100. While the precise mode of receptor interaction with one or more cytoskeletal elements remains to be defined, these results indicate that the B1 insertion at least in part acts to prevent coated pit localization by tethering the receptor at uncoated regions of the plasma membrane in a time, temperature, and ligand-dependent fashion. Thus, in B-cells which normally express FcRII-B1, the signalling functions of FcR (Hunziker et al., 1990a) may be enhanced by retaining the receptor on the cell surface and possibly by allowing its interaction with cytoskeleton.

Transcytosis of Fc receptors in MDCK cells.

As mentioned above, the human homolog of FcRII-B2 is expressed at high levels in syncytiotrophoblasts, the major IgG-transporting epithelium in human placenta (Stuart et al., 1989). Since this suggested a possible role for a macrophage-like FcR in IgG transcytosis, we recently transfected polarized MDCK cells with cDNAs encoding FcRII-B2, -B1 (Hunziker and Mellman, 1989) and the various receptor mutants. As would be expected for the directionality of the syncytiotrophoblast Fc receptor, FcRII-B2 was found to mediate the efficient apical to basolateral transcytosis of ligand across filter-grown MDCK; transcytosis via FcRII-B1 was far less efficient. Interestingly, the various receptors were also expressed with distinct polarities. At equilibrium, FcRII-B2 was found mostly on the basolateral surface while FcRII-B1, and the tail-minus receptor mutant, were mostly apical. The different polarities were established as a result of sorting in the Golgi since newly synthesized receptors reached the apical or basolateral surface in ratios that approximated their final distributions.

These results show that the presence or absence of a coated pit localization signal and/or cytoskeletal binding domain in the FcR cytoplasmic tail influences the polarized insertion of FcR in MDCK cells. They also suggest that the macrophage-type FcR expressed in human syncytiotrophoblasts may, in fact, be responsible for transplacental transport of IgG from the mother to the fetus. In addition, however, the expression of FcRII-B2 in MDCK cells provides an efficient and convenient marker for apical to basolateral transcytosis in these cells.

Spatial organization and transcytosis in MDCK cells.

As mentioned above, one of our major reasons for becoming interested in transcytosis in polarized cells was to develop new and more quantitative approaches to studying the role of spatial organization in controlling membrane traffic during endocytosis. Although it appears likely that transport of internalized markers from early to late endosomes and lysosomes in non-polarized cells correlates with the translocation of endocytic vesicles from the peripheral to the perinuclear cytoplasm, this is a difficult event to measure. Transcytosis, on the other hand, must involve a physical translocation of vesicles through the cytoplasm and can be assayed by the appearance of an endocytic tracer on the surface opposite from that to which it was added.

Extensive work from Keith Mostov's laboratory (Mostov and Deitcher, 1986) has established that it is possible to reconstitute the basolateral to apical transcytosis of IgA in MDCK cells transfected with the rabbit hepatic polymeric Ig receptor (pIg-R). We have, therefore, recently begun to compare the requirements for basolateral to apical transcytosis of IgA with the requirements for apical to basolateral transcytosis of IgG in filter-grown MDCK cells transfected with either the pIg-R or FcRII-B2. Our results show that both forms of transcytosis depend on at least two ATP-requiring steps, the first corresponding to internalization of the ligand at the basolateral or apical surfaces, the second corresponding to the final appearance IgA or IgG at its final destination. Transcytosis in both directions was also found to be temperature dependent. In cells incubated at 17°C, transcytosis -- but not the initial internalization -- was reversibly blocked (Hunziker et al., 1990b).

One factor that clearly distinguished apical to basolateral from basolateral to apical transcytosis, however, was their requirements for intact cytoplasmic microtubules (Hunziker et al., 1990b). Treatment of transfected cells with concentrations of nocodazole that disrupted microtubule networks completely blocked IgA transcytosis via the pIg-R but had no effect on transcytosis of IgG via FcRII-B2. Neither internalization nor recycling of either receptor or ligand was significantly affected. Importantly, the nocodazole block of basolateral to apical transyctosis could be shown to result entirely from the inability of transcytotic vesicles to translocate from the basolateral to the apical cytoplasm. This was demonstrated by confocal microscopy. Following nocodazole treatment, IgA-containing transport vesicles were found to accumulate in the basolateral cytoplasm, apparently unable to reach their target membrane. In untreated cells, small numbers of vesicles were scattered throughout both the basal and apical regions. Cells incubated at 17°C, however, showed a different pattern. Here, IgA-containing vesicles were enriched in the apical cytoplasm, suggesting that the microtubule-dependent basolateral to apical translocation had occurred but the final fusion with the apical plasma membrane was blocked. Indeed, warming these cells to 37°C reversed the inhibition and was accompanied by the immediate release of IgA into the apical medium. This reversal occurred even in the presence of nocodazole, further indicating that the vesicles that accumulated apically during the 17°C incubation had progressed beyond the microtubule-dependent step (Hunziker et al., 1990b).

While much work remains before one can completely appreciate the role for spatial organization in regulating the function and interactions of endocytic organelles, it is already clear from these results that the microtubule-dependent translocation of vesicles across MDCK cells is of critical importance in the basolateral to apical transcytosis of IgA. The fact that apical to basolateral transcytosis of IgG does not exhibit a similar requirement for microtubules can probably be understood from the overall organization of MDCK cells. Since the tight junctions in these cells are very close to the apical plasma membrane, endocytic vesicles in the apical cytoplasm do not have a great distance to traverse before encountering the basolateral membrane. In contrast, vesicles formed at the basal pole must traverse the entire height of the cells -- and negotiate around the nucleus *en route* -- prior to reaching the apical plasma membrane.

Abrahamson, D. R., and Rodewald, R. (1981). Evidence for the sorting of endocytic vesicle components during the receptor-mediated transport of IgG across the newborn rat intestine. J. Cell Biol. 91, 270-280.

Albertini, D. F., Herman, B., and Sherline, P. (1984). In vivo and in vitro studies on the role of HMW-MAPs in taxol-induced microtubule bundling. Eur. J. Cell Biol. 33, 134 - 143.

Bomsel, M., Prydz, K., Parton, R. G., Gruenberg, J., and Simons, K. (1989). Endocytosis in filter-grown Madin-Darby canine kidney cells. J. Cell Biol. 109, 3243-3258.

Gruenberg, J., Griffiths, G., and Howell, K. E. (1989). Characterization of the early endosome and putative carrier vesicles in vivo and with an assay of vesicle fusion in vitro. J. Cell Biol. 108, 1301-1316.

Hirsch, J. G., M.E. Fedorko, and Z.A. Cohn. (1968). Vesicle fusion and formation at the surface of pinocytic vesicles in macrophages. J. Cell Biol. 38, 629-632.

Hogarth, P. M., Hibbs, M. L., Bonadonna, L., Bernadette, M. S., Witort, E., Pietersz, G. A., and McKenzie, I. F. C. (1987). The mouse Fc receptor for IgG (Ly-17): molecular cloning and specificity. Immunogen. 26, 161-168.

Hunziker, W., Koch, T., Whitney, J. A., and Mellman, I. (1990a). Fc receptor phosphorylation during receptor-mediated control of B-cell activation. Nature 345, 628-632.

Hunziker, W., Mâle, P., and Mellman, I. (1990b). Differential microtubule requirements for transcytosis in MDCK cells. EMBO J. in press,

Hunziker, W., and Mellman, I. (1989). J. Cell Biol. 109:3291-3302.

Kornfeld, S., and Mellman, I. (1989). The biogenesis of lysosomes. Annu. Rev. Cell Biol. 5, 483-525.

Ktistakis, N. T., Thomas, D., and Roth, M. G. (1990). Characteristics of the tyrosine recognition signal for internalization of transmembrane surface glycoproteins. J. Cell Biol. in press.,

Lewis, V. A., Koch, T., Plutner, H., and Mellman, I. (1986). A complementary DNA clone for a macrophage-lymphocyte Fc receptor. Nature 324, 372-375.

Mellman, I. (1988). Relationships between structure and function in the Fc receptor family. Current Opinion in Immunology 1, 16-25.

Miettinen, H. M., Rose, J. K., and Mellman, I. (1989). Fc receptor isoforms exhibit distinct abilities for coated pit localization as a result of cytoplasmic domain heterogeneity. Cell 58, 317-327.

Mostov, K. E., and Deitcher, D. L. (1986). Polymeric immunoglobulin receptor expressed in MDCK cells transcytoses IgA. Cell 46, 613-621.

Parton, R. G., Prydz, K., Bomsel, M., Simons, K., and Griffiths, G. (1989). Meeting of the apical and basolateral endocytic pathways of the Madin-Darby canine kidney cell in late endosomes. J. Cell Biol. 109, 3259-3272.

Ravetch, J. V., Luster, A. D., Weinshank, R., Kochan, J., Pavlovec, A., Portnoy, D. A., Hulmes, J., Pan, Y.-C. E., and Unkeless, J. C. (1986). Structural heterogeneity and functional domains of murine immunoglobulin G Fc receptors. Science 234, 718-725.

Stuart, S., Simister, N. E., Clarkson, S. B., Kacinski, B. M., Shapiro, M., and Mellman, I. (1989). Human IgG Fc receptor (hFcRII; CD32) exists as multiple isoforms in macrophages, lymphocytes and IgG-transporting placental epithelium. EMBO J. 12:3657-3666.

ENDOCYTOSIS IN MDCK CELLS

Robert G. Parton, European Molecular Biology Laboratory,
Postfach 10.2209, D-6900 Heidelberg, F.R.G.

Introduction

In polarized cells the plasma membrane is divided into apical and basolateral domains. Membrane is internalized from both these domains and the internalized components can follow several different routes, for example, they can be transported to the opposite domain of the cell (transcytosis), routed to lysosomes or recycled back to the same domain. Despite these continuous processes the distinct protein and lipid composition of the two surface domains must be maintained. In this review I will discuss experiments which led us to formulate a model of the endocytic pathways in the Madin Darby Canine Kidney (MDCK) cell, a model epithelial cell system (reviewed by Simons and Fuller, 1985), and briefly consider transcytosis of endogenous glycoproteins in these cells.

Morphological characterization of endocytic pathways in the MDCK cell

In non-polarized cells the following pathway from the cell surface to lysosomes has been proposed (reviewed by Kornfeld and Mellman, 1989; Gruenberg and Howell, 1989). Ligands internalized via clathrin-coated pits enter early endosomes. The early endosomes are peripherally-located in the cell, have a pH of approximately 6.0 and show specific fusogenic activity in vitro. From this compartment internalized markers can be recycled back to the cell surface, or can be targeted to late endosomes (also called prelysosomes). In our working model transfer of material from the early endosome to the late endosome is mediated by a large endosome carrier vesicle (ECV) which is transported via a microtubule-dependent step to the perinuclear region of the cell where it fuses with late endosomes (Gruenberg et al., 1989; and see contribution by G. Griffiths in this issue). The late endosomes are enriched in the cation-independent mannose-6-phosphate receptor (CI-MPR) and contain lysosomal hydrolases (Griffiths et al., 1988). Finally markers enter CI-MPR negative lysosomes.

In a recent study (Parton et al., 1989; also see Bomsel et al., 1989) we investigated the organisation of the endocytic apparatus in the MDCK cell. Of particular interest was the site where the pathways from the apical and basolateral domains intersected. Using a variety of

NATO ASI Series, Vol. H 62
Endocytosis
Edited by P. J. Courtoy

fluid-phase and membrane markers we showed that after 5-10 minutes at 37°C no meeting of the apically and basolaterally-internalized markers occurred and the markers were restricted to peripherally-located early endosomes. Similar results were recently obtained using other endocytic markers in both MDCK cells (Van Deurs et al., 1990) and Caco-2 cells (Hughson and Hopkins, 1990). We showed that the early endosomes were acid-phosphatase negative and contained low levels of CI-MPR. After 15 minutes at 37°C the apical and basolateral markers started to appear within the same structures in the Golgi area of the cell. These structures were identified as late endosomes using antibodies raised against the CI-MPR and were shown to be acid phosphatase positive. Although meeting of the markers was initially evident only in those late endosomes found in the Golgi area of the cell, with increasing times of internalization the two markers were found in MPR-enriched structures basal to the nucleus. Finally, markers entered MPR-negative lysosomes which were electron-dense, essentially spherical structures as seen in other cell types (Griffiths et al., 1988). Using a non-degradable marker (gold particles coated with BSA), we quantitatively followed the transfer of a marker from the late endosomes to the lysosomes. The cells were incubated with gold in order to label the late endosomes and then were incubated in marker-free medium overnight. The decrease in the number of gold particles in the late endosomal compartment could quantitatively account for the increase in the number of gold particles in the lysosomes showing that late endosomes are an obligatory intermediate in the pathway to lysosomes.

Quantitative analysis of the endocytic pathways in MDCK cells

Using stereological methods in combination with parallel biochemical experiments we performed a quantitative analysis of the endocytic pathways in the MDCK cell (Parton et al., 1989; Bomsel et al., 1989). The basolateral early endosomes were shown to be approximately 4 times larger in surface area and volume than the apical early endosomes. The total endocytic rates from the two surfaces measured biochemically and the surface areas of the two domains also appear to be maintained in the same ratio (i.e. apical to basolateral approximately 1:4). Therefore the rate of internalization per surface area for the two domains is similar (an area equivalent to the surface area would be internalized in 2.5 hours). Biochemical studies (Bomsel et al., 1989) showed that the early endosomes were the sites where the pathways for recycling, transcytosis and delivery to late endosomes diverged. For every 100 units of fluid entering the basolateral early endosomes, 70 were shown to be retained within the cell and the remainder were released through transcytosis and recycling. In contrast, the apical early endosomes released 90 units of internalized fluid through transcytosis and recycling and only 10 were routed to late endosomes and retained in the cell. These results showed that there were functional differences between the apical and basolateral early endosomes and were the first indication that the two sets of early endosomes were functionally as well as topologically distinct. This has recently been confirmed by in vitro experiments (Bomsel et al., 1990). In these studies lateral fusion events between elements of the basolateral early endosomal compartment and within the apical early endosome compartment were reconstituted but no fusion occurred between apical and basolateral early endosomes. These results suggest that specific recognition components of the apical and basolateral early endosomes control their fusion.

The role of microtubules in polarized endocytic traffic

The microtubules in polarized cells are organized quite differently to those in non-polarized cells. Whereas in a non-polarized cell the microtubules radiate out from the centrally-located centrioles towards the cell periphery, in MDCK cells the bulk of the microtubules are orientated vertically through the cell (Bacallao et al., 1989). Other microtubules form a dense apical cap and a sparse basal network. The microtubules are not organized by the centrioles which lie under the apical surface. Most of the vertical microtubules have the same polarity with their plus (fast-growing) ends at the base of the cell. This organisation raises interesting questions about the microtubule dependent steps in membrane traffic. In non-polarized cells it has been postulated that a minus end-directed motor could be involved in the microtubule-dependent traffic between the early endosomes and the late endosomes (see review by Kelly, 1990). In the MDCK cell, however, the bulk of the late endosomes are not located at the minus ends of the polarized microtubules. Recently, we examined the role of microtubules in the transfer steps between the two sets of early endosomes and the late endosomes (Bomsel et al., 1990). Fluid-phase markers were internalized after cold-induced microtubule depolymerization in the presence of nocodazole. Under these conditions colocalization of apical and basolateral markers in CI-MPR-enriched perinuclear late endosomes was greatly reduced as compared to control cells. Apical and basolateral markers were visualized within apical and basolateral early endosomes, respectively, and within spherical vesicles with the morphology of ECVs. Chase experiments suggested that these structures were distinct from, and distal to, the early endosomes. On removing the nocodazole, colocalization of apical and basolateral markers was observed in late endosomes in the Golgi area of the cell. The results suggest, by analogy to the more extensive studies in non-polarized cells (Gruenberg et al., 1989), that transfer of markers from the two sets of early endosomes to the late endosomes is microtubule-dependent and mediated by apical and basolateral ECVs.

The late endosomes of the MDCK cell also appear to interact with microtubules. Under normal conditions the late endosomes exist as discrete structures throughout the cell. On microtubule depolymerization the individual late endosomal structures associate and fuse together. These results suggest that the distribution of late endosomes in the cell is maintained through interactions with microtubules. The late endosomes of MDCK cells may form one functional unit in which lateral transfer of material can occur and therefore may show similarities to the tubular lysosomes of macrophages described by Swanson et al. (1987) which form an extensive reticulum aligned on microtubules. In contrast to the late endosomes, the distribution of early endosomes is unaffected by microtubule-depolymerization.

Our current model for endocytosis in the MDCK cell is as follows (see figure 1). Markers internalized from the two surface domains enter apical and basolateral early endosomes which are functionally distinct organelles. From these compartments, routing to the same surface domain (recycling), to the other surface domain (transcytosis) or to late endosomes can occur. Carrier vesicles (apical, Ap, and basolateral, Bl, ECVs) bud off the apical and basolateral early endosomes and move in a microtubule-dependent fashion to the Golgi area of the cell where they fuse with late endosomes. This model is compatible with the vesicular shuttle model suggested by Helenius et al. (1983). In the alternative maturation model proposed by these authors, the carrier vesicles would need to fuse together and mature into late endosomes through acquisition of lysosomal enzymes from Golgi-derived vesicles. Although we cannot formally rule out this model, in vitro studies suggest that apically- and basolaterally-derived ECVs do not fuse together but fuse specifically with a common late endosomal compartment in a microtubule-dependent fashion (Bomsel et al., 1990). More insights into the mechanisms involved in vesicular traffic may come from studies of ras-like GTP-binding proteins (rab proteins). Based on studies of GTP-binding proteins in yeast (Salminen and Novick, 1987; Segev et al., 1988; Schmitt et al., 1988; Goud et al., 1988) and also on the effects of a non-hydrolysable analogue of GTP (GTPγS) on membrane traffic in vitro (e.g. Melançon et al., 1987; Baker et al., 1988) it has been suggested that these proteins may play a role in controlling specific steps of membrane traffic (Bourne, 1988). We have recently localized three such proteins in non-polarized mammalian cells and two of these were found to be associated with endocytic compartments; rab 5 with early endosomes and the cell surface, and rab 7 with CI-MPR enriched late endosomes (Chavrier et al., 1990). Functional assays are now required to investigate whether, as predicted, these proteins are involved in specific steps of membrane traffic (see contribution by J. Gruenberg in this issue).

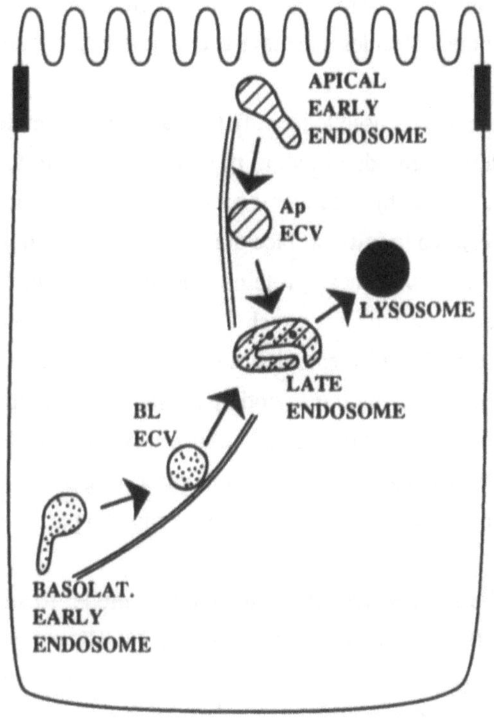

Fig. 1 Schematic model of the endocytic pathways in the MDCK cell

Transcytosis of endogenous MDCK glycoproteins

Transcytosis is a unique property of polarized cells which is important in the maintenance of the polarized phenotype and in the vectorial transport of ligands across the epithelium (for review see Mostov and Simister, 1985). MDCK cells are a useful system to study this process and they have been shown to transcytose fluid-phase markers (e.g. HRP; von Bonsdorff et al., 1985; Bomsel et al., 1989), implanted viral glycoproteins (Pesonen et al., 1983), ricin (Van Deurs et al., 1990) and transfected receptors (Mostov and Deitcher, 1986; Hunziker and Mellman, 1989). As yet few studies have addressed the transcytosis of endogenous MDCK proteins. In a recent study (Brändli et al., 1990) we looked at this process by introducing biotin into the membrane glycoproteins of either the apical or basolateral surface at 4°C. Internalization of the biotinylated proteins was allowed to occur by warming the cells before fixation. They were then processed for cryosectioning and the sections were labelled with streptavidin-gold. Using this procedure in combination with stereological techniques, we quantitated the amount of transcytosis in each direction and correlated these figures with the surface areas of the two domains. After 3 hours, approximately 10% of the biotinylated membrane components of each surface domain had been transferred to the opposite domain. This gives an indication of the magnitude of the membrane traffic across the cell in each direction. In fact, our values may well be an underestimate as some molecules may have been reinternalized after being transcytosed to the other surface.

A biochemical assay was used to identify the transcytosing molecules. Three classes of transcytosing glycoproteins were identified; a) proteins moving unidirectionally between the apical and basolateral surfaces, b) basolateral proteins moving to the apical surface, and c) glycoproteins moving between the two surface domains bidirectionally. The early endosomes presumably play a role in sorting these proteins, which are destined for transcytosis, from proteins such as the transferrin receptor which after basolateral internalization is efficiently recycled back to the same surface (Fuller and Simons, 1986). The sorting signals which may be involved in diverting proteins to the transcytotic pathway are presently unknown but for the polyIgA receptor transfected into MDCK cells phosphorylation of the cytoplasmic domain seems to be important (Casanova et al., 1990). One interesting possibility is that some of the endogenous transcytosing molecules mentioned above may be part of the "machinery" of the transcytotic pathway.

In summary, the MDCK system is an excellent model to study various aspects of epithelial membrane traffic. Earlier work showing that MDCK cells have polarized plasma membrane domains has now been extended by showing that the endocytic organelles of the two domains are also polarized and that microtubules play an important role in directed membrane traffic on the two endocytic pathways.

References

Bacallao, R., C. Antony, C. Dotti, E. Karsenti, E.H.K. Stelzer, and K. Simons. 1989. The subcellular organization of Madin-Darby canine kidney cells during the formation of a polarized epithelium. J. Cell Biol. 109: 2817-2832.

Baker, D., L. Hicke, M. Rexach, M. Schleyer, and R. Schekman. 1988. Reconstitution of *SEC* gene product-dependent intercompartmental protein transport. Cell 54: 335-344.

Bomsel, M., K. Prydz, R.G. Parton, J. Gruenberg, and K. Simons. 1989. Endocytosis in filter-grown Madin-Darby canine kidney cells. J. Cell Biol. 109: 3243-3258.

Bomsel, M., R.G. Parton, S.A. Kuznetsov, T.A. Schroer, and J. Gruenberg. 1990. Microtubule- and motor-dependent fusion in vitro between apical and basolateral endocytic vesicles from MDCK cells. Cell 62: 719-731.

Bourne, H.R. 1988. Do GTPases direct membrane traffic in secretion? Cell 53: 669-671.

Brändli, A. W., R.G. Parton and K. Simons. 1990. Transcytosis in MDCK cells: identification of glycoproteins transported bidirectionally between both plasma membrane domains. J. Cell Biol. In Press.

Casanova, J. E., P.P. Breitfeld, and K.E. Mostov. 1990. Phosphorylation is required for efficient transcytosis of the polymeric immunoglobulin receptor. Science 246: 742-745.

Chavrier, P., R. G. Parton, H.P. Hauri, K. Simons, and Marino Zerial.1990. Localization of low molecular weight GTP-binding proteins to exocytic and endocytic compartments. Cell 62: 317-329.

Fuller, S.D., and K. Simons. 1986. Transferrin receptor polarity and recycling accuracy in "tight" and "leaky" strains of Madin-Darby canine kidney cells. J. Cell Biol. 103: 1767-1779.
Goud, B., A. Salminen, N.C. Walworth, and P.J. Novick. 1988. A GTP-binding protein required for secretion rapidly associates with secretory vesicles and the plasma membrane in yeast. Cell 53: 753-768.
Griffiths, G., B. Hoflack, K. Simons, I. Mellman and S. Kornfeld. 1988. The mannose-6-phosphate receptor and the biogenesis of lysosomes. Cell 52: 329-341.
Griffiths, G., R. Matteoni, R. Back, and B. Hoflack. 1990. Characterization of the cation-independent mannose-6-phosphate receptor-enriched prelysosomal compartment. J. Cell Sci. 95: 441-461.
Gruenberg, J., G. Griffiths, K.E. Howell. 1989. Characterization of the early endosome and putative endocytic carrier vesicles in vivo and with an assay of vesicle function in vitro. J. Cell Biol. 108: 1301-1316.
Gruenberg, J. and K.E. Howell. 1989. Membrane traffic in endocytosis: insights from cell-free assays. Ann. Rev. Cell Biol. 5: 453-481.
Helenius, A., I. Mellman, D. Wall, and A. Hubbard. 1983. Endosomes. Trends Biochem. Sci. 8: 245-250.
Hughson, E. J. and C.R. Hopkins. 1990. Endocytic pathways in polarized Caco-2 cells: identification of an endosomal compartment accessible from both apical and basolateral surfaces. J.Cell Biol. 110: 337-348.
Hunziker W. and I. Mellman. 1989. Expression of macrophage- lymphocyte receptors in MDCK cells: polarity and transcytosis differ for isoforms with or without coated pit localization domains. J. Cell Biol. 109: 3291-3302.
Kelly, R.B. 1990. Microtubules, membrane traffic and cell organization. Cell 61: 5-7.
Kornfeld, S. and I. Mellman. 1989. The biogenesis of lysosomes. Ann. Rev. Cell Biol. 5: 483-526.
Melançon, P., B.S. Glick, V. Malhotra, P.J. Weidman, T. Serafini, M.L. Gleason, L. Orci, and J.E. Rothman. 1987. Involvement of GTP-binding "G" proteins in transport through the Golgi stack. Cell 51: 1053-1062.
Mostov K.E. and D.L Deitcher. 1986. Polymeric immunoglobulin receptor expressed in MDCK cells transcytoses IgA. Cell 46: 613-621.
Mostov, K.E. and N.E. Simister. 1985. Transcytosis. Cell 43: 389-390.
Parton, R.G., K. Prydz, M. Bomsel, K. Simons, and G. Griffiths. 1989. Meeting of the apical and basolateral endocytic pathways of the Madin-Darby canine kidney cell in late endosomes. J. Cell Biol. 109: 3259-3272.
Pesonen, M., W. Ansorge, and K. Simons. 1984. Transcytosis of the G protein of vesicular stomatitis virus after implantation into the apical plasma membrane of Madin-Darby Canine Kidney cells. I. Involvement of endosomes and lysosomes. J. Cell Biol. 99: 796-802.
Salminen, A., and P.J. Novick. 1987. A *ras*-like protein is required for a post-Golgi event in yeast secretion. Cell 49: 527-538.
Schmitt, H.D., P. Wagner, E. Pfaff, and D. Gallwitz. 1986. The *ras*-related *YPT1* gene product in yeast: a GTP-binding protein that might be involved in microtubule organization. Cell 47: 401-412.
Segev, N., J. Mulholland, and D. Bostein. 1988. The yeast GTP-binding YPT1 protein and a mammalian counterpart are associated with the secretion machinery. Cell 52, 915-924.
Simons K. and S.D. Fuller. 1985. Surface polarity in epithelia. Ann. Rev. Cell Biol. 1: 243-288.
Swanson, J.A., A. Bushnell, and S.C. Silverstein. 1987. Tubular lysosome morphology and distribution within macrophages depend on the integrity of cytoplasmic microtubules. Proc. Natl. Acad. Sci. USA 84: 1921-1925.
Van Deurs, B., S.H. Hansen, O.W. Petersen, E.L. Melby, and K. Sandvig. 1990. Endocytosis, intracellular transport and transcytosis of the toxic protein ricin by a polarized epithelium. Eur. J. Cell. Biol. 51: 96-109.

Von Bonsdorff, C.-H., S.D. Fuller, and K. Simons. 1985. Apical and basolateral endocytosis in Madin-Darby canine kidney (MDCK) cells grown on nitrocellulose filters. EMBO J. 4: 2781-2792.

PROTEIN HANDLING FROM APICAL AND BASOLATERAL SURFACES IN RAT AND RABBIT RENAL PROXIMAL TUBULES

Erik Ilsø Christensen and Søren Nielsen
Department of Cell Biology
Institute of Anatomy
University of Aarhus
DK-8000 Aarhus C
Denmark

INTRODUCTION

For several years it has been known that low molecular weight proteins after glomerular filtration are taken up by the renal proximal tubule by endocytosis and transferred into lysosomes (Straus 1964; Maunsbach 1966). However, the initial events i.e. the specificity of binding to the luminal plasma membrane are not very well understood.

It is also well known that several low molecular weight peptides are degraded by peptidases located in the brush border. However, the factors determining whether a peptide is degraded at the brush border or is taken up by endocytosis has not been clarified.

Another important issue which has been raised with respect to tubular protein handling is to what extent, if any, proteins can be transported more or less intact from the luminal aspect to the renal interstitium.

Finally, a growing interest has evolved in the tubular basolateral handling of protein since this aspect of the plasma membrane of tubule cells is the most likely localization for e.g. hormone receptors and since several hormones have been demonstrated to exert their hormonal actions on different parts of the nephron.

In the following we shall try to answer and clarify some of the controversial issues raised above using genuine proximal tubules in different experimental models.

BRUSH BORDER DEGRADATION

The brush border plasma membrane of the proximal tubule has been demonstrated to be rich in different peptidases (Kenny & Maroux 1982), and several peptides have been shown to be degraded by these peptidases (Carone & Peterson 1980; Carone et al 1987). We have shown that size and complexity of molecules decide whether a protein is taken up by endocytosis or is degraded at the brush border (Peterson et al 1982) and that e.g. neurotensin in addition is also degraded in the urine possibly by enzymes secreted by the tubule cells (Bjerke et al 1989). Furthermore, there seems to be distinct differences in the degradation pattern of neurotensin in the different segments of the proximal tubule (Bjerke et al 1990).

NATO ASI Series, Vol. H 62
Endocytosis
Edited by P. J. Courtoy

LUMINAL BINDING AND ENDOCYTOSIS

The efficiency with which proteins are taken up varies between proteins. Electrical charge seems to a certain degree to influence the uptake. Thus, we have shown that cationization of ferritin increased the uptake with a factor 8-9 (Christensen et al 1981) and anionizing proteins decreases the uptake (Christensen et al 1983). On the other hand, the isoelectric point is not the only determinant, - insulin and EGF has about the same isoelectric point and about the same m.w. and while 70% of insulin in micropunctured rats and in isolated perfused tubules are removed, only about 5% of EGF is absorbed under the same conditions (Bjerke et al 1988; Nielsen et al 1989).

Specific binding sites for insulin are present at the luminal cell membrane but recently we have demonstrated that these are not the genuine insulin receptor but low affinity high capacity receptors (Figs. 1 and 2) (Nielsen & Christensen 1989). Thus, the nature of these more or less specific binding sites remains to be clarified.

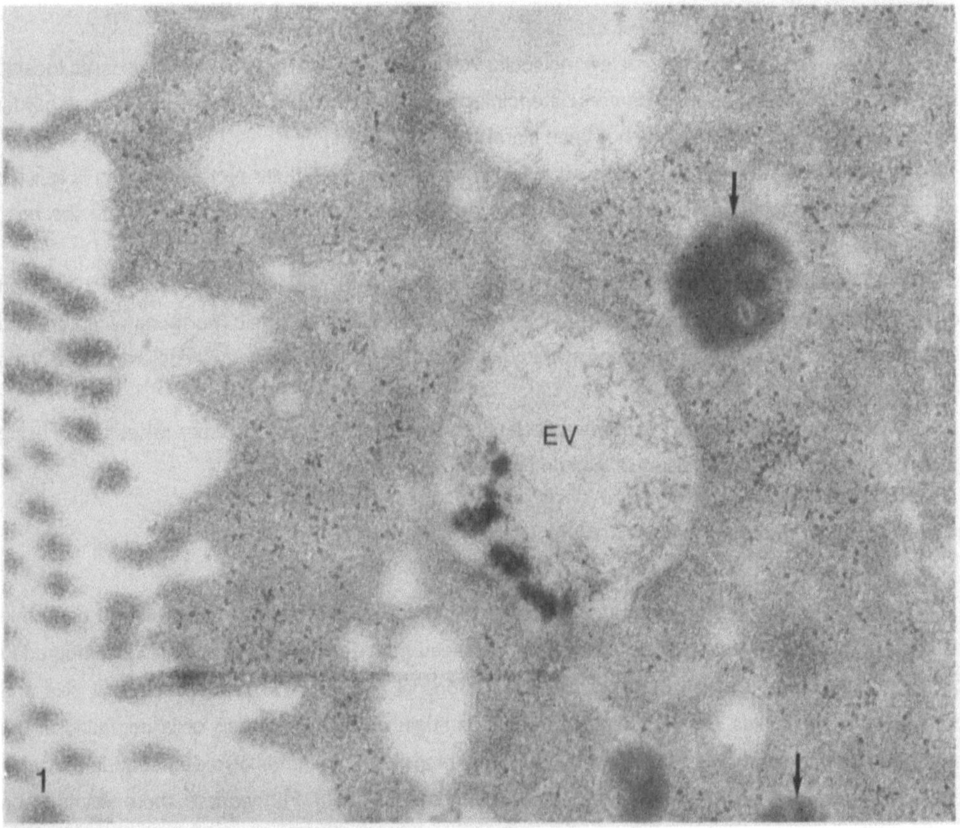

Fig. 1. Isolated rabbit proximal tubule, segment S2, perfused with native insulin, 58 ng/ml, for 30 min, fixed in glutaraldehyde and embedded at low temperature in Lowicryl K4M. The sections was incubated with insulin antiserum and pAg (6 nm particles). Labelling is seen over endocytic vacuoles (EV) and lysosomes (arrows). x 50,000. (From Nielsen & Christensen 1989).

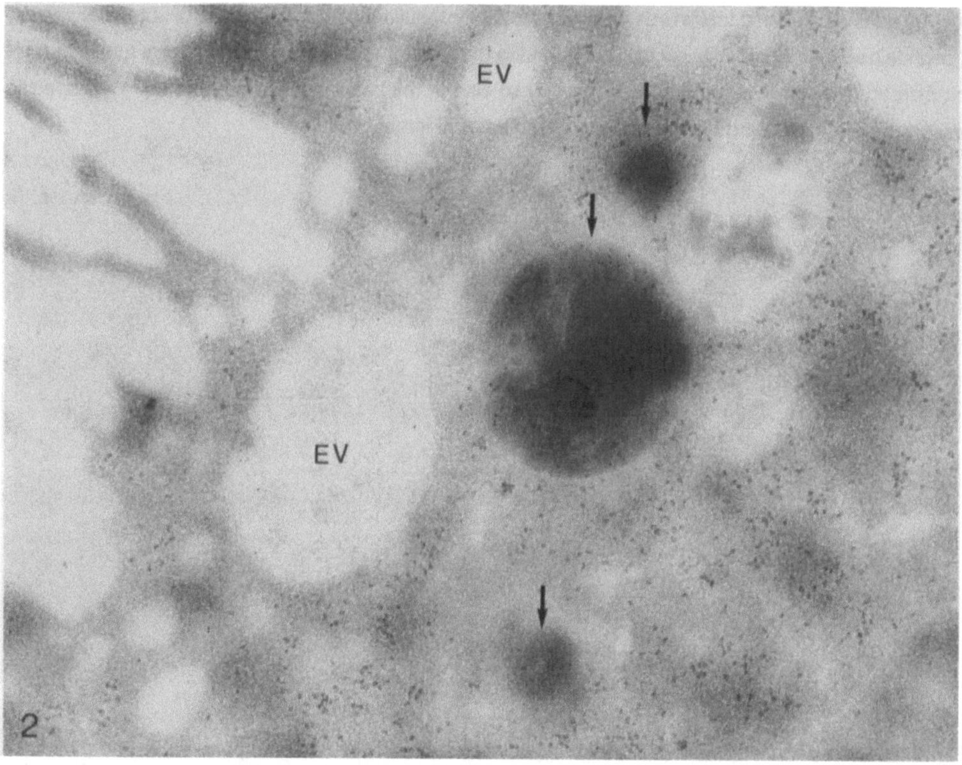

Fig. 2. Isolated proximal tubule as in Fig. 1 except it was perfused with (Leu^{B-25}) insulin (2% receptor binding ability and full immunological activity). Labelling is seen over endocytic vacuoles (EV) and lysosomes (arrows). The labelling density is similar to the one seen in Fig. 1. x 44,000. (From Nielsen & Christensen 1989).

BASOLATERAL ENDOCYTOSIS

At the basolateral cell membrane several hormone receptors have been demonstrated (for a review, see Rabkin & Mahoney 1988) and the existence of peritubular extraction of proteins have been subject to several investigations. Various methods including clinical and experimental have suggested peritubular extraction of various proteins (Chamberlain & Stimmler 1967; Feria-Velasco 1974; Kessel 1970), however, in these studies accidental luminal uptake may not with certainty be excluded. Using the technique with isolated and in vitro perfused proximal tubules it is possible to separate totally the luminal and the peritubular compartments and, thus, in a direct way characterize a possible peritubular extraction. HRP is taken up by endocytosis from the basolateral membranes (Fig. 3), as is ferritin (Nielsen & Christensen 1985; Nielsen et al 1985). ^{125}I-insulin was localized in proximal tubules after peritubular exposure by EM-autoradiography and was located in endocytic vacuoles and lysosomes. Quantitation of basolateral endocytosis of ^{125}I-insulin and ^{125}I-EGF demonstrated a small but significant uptake (Nielsen et al 1989; Nielsen et al 1987). The accumulated/bound amount of ^{125}I-insulin in proximal tubule could be reduced

50% by adding unlabeled insulin suggesting that specific insulin receptors mediate the uptake. However, a peptide hormone like PYY is exclusively bound to the S_1 segment of the rabbit proximal tubule but not internalized (Nielsen et al, unpublished results), so also at the basolateral membrane there are specific mechanisms which determine the endocytic uptake of different proteins.

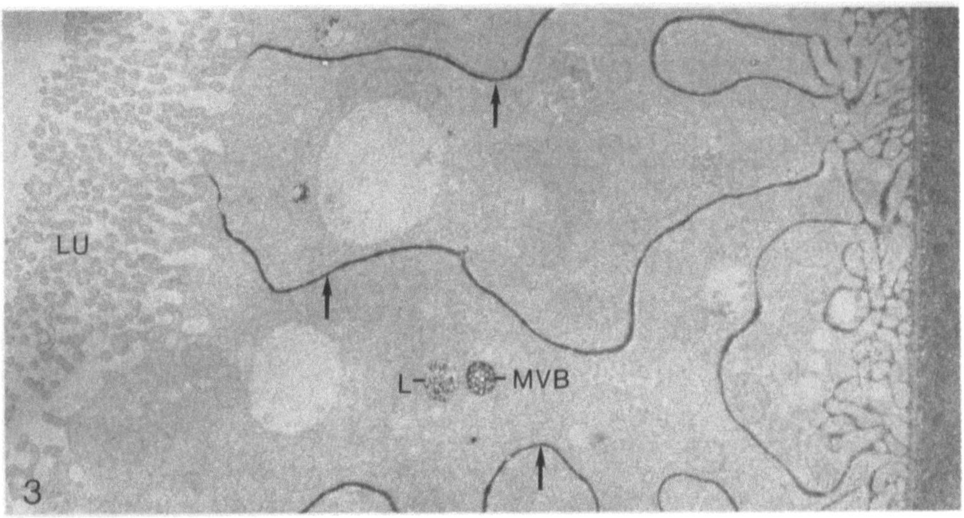

Fig. 3. Isolated perfused rabbit proximal tubules incubated with horseradish peroxidase in the bath for 30 min. The reaction product is seen in the intercellular spaces (arrows), in a lysosome (l) and a multivesicular body (MVB), no reaction is seen in the tubular lumen (LU). x 14,000. (From Nielsen & Christensen 1985).

TRANSTUBULAR TRANSPORT OF PROTEINS

From lumen to bath: Originally Maack and Kinter (Maack & Kinter 1969; Maack et al 1971) suggested that lysozyme may be subject to transtubular transport from lumen to bath in amounts constituting 50% of the absorbed amount. Other investigators later demonstrated that a transport of intact proteins is unlikely to exceed 22% (Just et al 1975), and 11% (Ottosen et al 1979). Due to the applied experimental conditions it was still unclear whether a transtubular transport existed or not and we therefore set out a series of experiments to study transcytosis. Isolated proximal tubules were perfused in vitro with cationized ferritin for 30 min and chased for 30-60 min (Nielsen et al 1985). Ferritin was found in apical and basal located endocytic vacuoles and lysosomes. Several small vacuoles were located close to the basolateral membranes and 0.5% of the absorbed amount of ferritin was transferred to the intercellular space (Figs. 4 and 5). A very large fraction of the transported ferritin was organized in a crystalline pattern (Figs. 4 and 5). This organization of ferritin was also occasionally found in cytoplasmic bodies (Fig. 5). However, a small fraction of transported ferritin was not organized in this crystalline pattern. Whether ferritin was transported in a non-crystalline form and later organized in the intercellular spaces is unclear. Thus, ferritin or at least the iron core was transcytosed in a vesicular process.

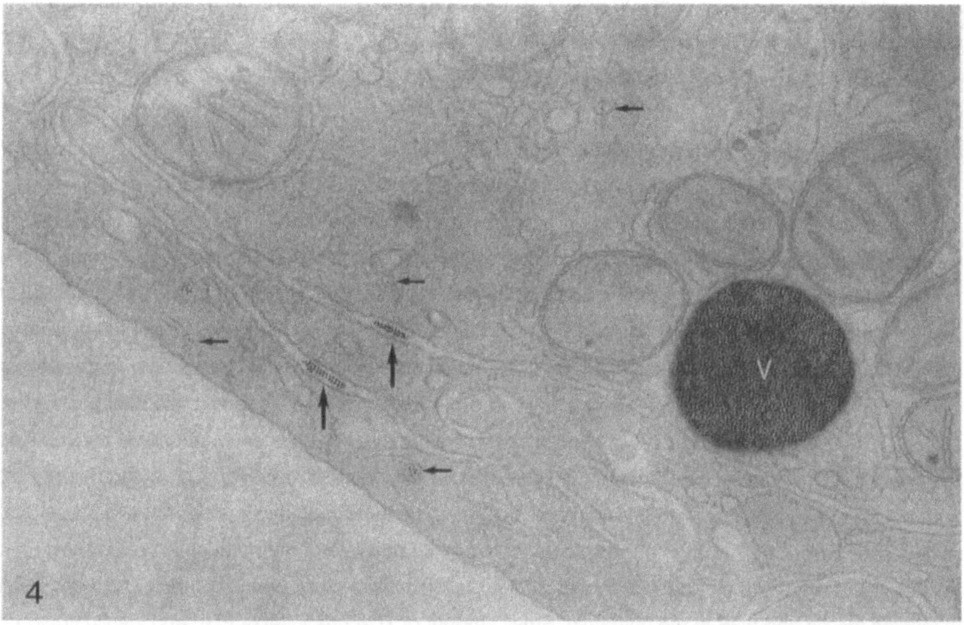

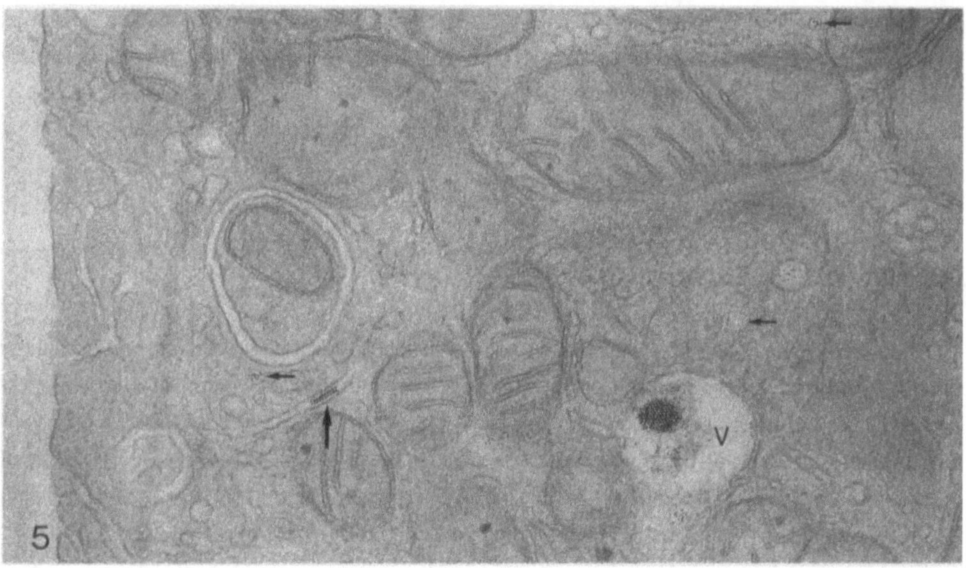

Figs. 4 and 5. Isolated rabbit proximal tubules perfused with cationized ferritin for 30 min and then chased for 30 min. Ferritin was seen in the intercellular space (large arrows) sometimes in a crystalline pattern. Ferritin is also seen in small vesicles (small arrows) and in larger vacuoles (V), occasionally also in a crystalline pattern (Fig. 5). Fig. 4, x 57,000 and Fig. 5 x 48,000.

To study the transtubular transport further, proximal tubules were perfused with ^{125}I-labeled lysozyme, insulin, or EGF. Lysozyme was not transported in amounts larger than 4.7% of the absorbed amount (Nielsen et al 1986), and insulin and EGF were transported in significant amounts (by means of TCA precipitable label measured in the bath fluid) which generally constituted 5% of the absorbed amount during 30 min (Nielsen & Nielsen 1988; Nielsen et al 1989).

Two mechanisms exist with respect to the nature of a transtubular transport: it may be 1) vesicular, 2) paracellular or 3) both. Ferritin is trancytosed by a vesicular transport (Nielsen et al 1985). Insulin transport is dependent on the absorbed amount rather than on the perfused amount (Nielsen 1990) during 30 min of perfusion suggesting a dominating vesicular transport. However, PYY a 36 residue linear peptide hormone is not endocytosed but degraded at the brush border level, and interrestingly a large fraction of perfused PYY is transported as TCA precipitable label from lumen to bath suggesting a paracellular transport (Nielsen S et al, unpublished observations). Thus, it seems that intact proteins or at least large degradation products may be transported either by a vesicular or a paracellular mechanism.

From bath to lumen: HRP and ^{125}I-insulin could be demonstrated by EM cytochemistry or autoradiography (Nielsen et al 1985; Nielsen et al 1987) to be translocated in vesicles from the basal part to the apical part of the cell after endocytic uptake from the basolateral membranes in proximal tubules. To investigate whether a transtubular transport of protein may take place proximal tubules were exposed to labeled insulin and EGF in the bath fluid. Small amount of TCA precipitable EGF and insulin could be detected in the perfusate after 30 min constituting 50 - 60% of the total uptake calculated as the sum of bound/accumulated and transported amount (Nielsen et al 1989; Nielsen et al 1987). Exposure for only 10 min reduced the transtubular transport of ^{125}I-insulin to only 30% of the uptake (Nielsen et al 1987). Thus, transtubular transport takes place both from the luminal to the basolateral compartment and in the opposite direction.

LUMINAL AND CONTRALUMINAL ABSORPTION OF PROTEINS

The relative contribution of luminal and peritubular extraction has been investigated quantitatively using ^{125}I-insulin in isolated perfused proximal tubules. The results demonstrate that at physiological concentrations peritubular extraction contribute 15.2% but at 20 times higher concentration only 1.8% due to limited capacity of peritubular extraction (Nielsen et al 1987). The relative contribution of basolateral endocytosis in MDCK was 3.7 times higher than the luminal (Bomsel et al 1989) suggesting that these cultured cells differ in this respect in comparison to renal proximal tubule cells.

The intracellular transport of proteins were studied after simultaneous exposure of HRP from the bath fluid and ferritin from lumen, and the proteins remained to a high degree in separate cytoplasmic bodies, however, occasional mixing of the two proteins was seen in multivesicular bodies after 60 min (Nielsen et al 1985). Thus, fusion of endocytic vacuoles derived from the two aspects of cells from isolated and perfused proximal tubules appears to be very limited. In contrast converging of the apical and the basolateral endocytic pathways in MDCK cells has been reported to constitute 50% of the luminal endocytic uptake (Bomsel et al 1989) and to take place in late endosomes (Parton et al 1989).

ACKNOWLEDGEMENTS

The authors thank Jytte Kragelund for typing the manuscript. This study was supported in part by grants from the Dansih Medical Research Council, the Danish Biomembrane Research Center, the Aarhus University Research Foundation, Ruth I.E. König-Petersen's Foundation, and Helen and Ejnar Bjørnow's Foundation.

REFERENCES

Bjerke T, Christensen EI, Boye N (1989) Tubular handling of neurotensin in the rat kidney as studied by micropuncture and HPLC. Am J Physiol 256 (Renal Fluid Electrolyte Physiol 25):F100-F106

Bjerke T, Nielsen S, Christensen EI, Sheikh MI (1990) Step-wise degradation of neurotensin in pars convoluta and pars recta proximal tubules. Abstracts, XIth Internat Congr of Nephrology, Tokyo, Japan, 467A

Bjerke T, Nielsen S, Hellfritzsch M, Christensen EI (1988) Renal tubular handling of epidermal growth factor in rat and rabbit. In Contributions to Nephrology: Kidney and Proteins in Health and Disease, Bianchi C, Bocci V, Carone FA, Rabkin R (eds) Karger Basel, 98-103

Bomsel M, Prydz K, Parton RG, Gruenberg J, Simons K (1989) Endocytosis in filter-grown Madin-Darby canine kidney cells. J Cell Biol 109:3243-3258

Carone FA, Christensen EI, Flouret G (1987) Degradation and transport of AVP by proximal tubule. Am J Physiol 253 (Renal Fluid Electrolyte Physiol 22):F1120-F1128

Carone FA, Peterson DR (1980) Hydrolysis and transport of small peptides by the proximal tubule. Am J Physiol 238 (Renal Fluid Electrolyte Physiol 7):F151-F158

Chamberlain MJ, Stimmler L (1967) The renal handling of insulin. J Clin Invest 46:911-919

Christensen EI, Carone FA, Rennke HG (1981) Effect of molecular charge on endocytic uptake of ferritin in renal proximal tubule cells. Lab Invest 44:351-358

Christensen EI, Rennke HG, Carone FA (1983) Renal tubular uptake of protein: effect of molecular charge. Am J Physiol 244 (Renal Fluid Electrolyte Physiol 13):F436-F441

Feria-Velasco A (1974) The ultrastructural bases of the initial stages of renal tubular excretion. Lab Invest 30:190-200

Just M, Röckel A, Stanjek A, Bode F (1975) Is there any transtubular reabsorption of filtered proteins in rat kidney? Naunyn-Schmiedeberg's Arch Pharmacol 289:229-236

Kenny AJ, Maroux S (1982) Topology of microvillar membrane hydrolases of kidney and intestine. Physiol Rev 62:91-128

Kessel RG (1970) The permeability of dragonfly Malphigian tubule cells to protein using horseradish peroxidase as a tracer. J Cell Biol 47:299-303

Maack T, Kinter WB (1969) Transport of protein by flounder kidney tubules during long-term incubation. Am J Physiol 216:1034-1043

Maack T, Mackensie DDS, Kinter WB (1971) Intracellular pathways of renal reabsorption of lysozyme. Am J Physiol 221:1609-1616

Maunsbach AB (1966) Absorption of ^{125}I-labeled homologous albumin by rat kidney proximal tubule cells. A study of microperfused single proximal tubules by electron microscopic autoradiography and histochemistry. J Ultrastruct Res 15:197-241

Nielsen JT, Christensen EI (1985) Basolateral endocytosis of protein in isolated perfused proximal tubules. Kidney Int 27:39-45

Nielsen JT, Nielsen S, Christensen EI (1985) Transtubular transport of proteins in rabbit proximal tubules. J Ultrastruct Res 92:133-145

Nielsen JT, Nielsen S, Christensen EI (1986) Handling of lysozyme in isolated perfused proximal tubules. Am J Physiol 251 (Renal Fluid Electrolyte Physiol 20):F822-F830

Nielsen S (1990) Time-course of proximal tubular processing of insulin. Abstracts, XIth Internat Congr on Nephrology, Tokyo, Japan, 468A

Nielsen S, Christensen EI (1989) Insulin absorption in renal proximal tubules: A quantitative immunocytochemical study. J Ultrastruct Molec Struct Res 102:205-220

Nielsen S, Nexø E, Christensen EI (1989) Absorption of epidermal growth factor and insulin in rabbit renal proximal tubules. Am J Physiol 256 (Endocrinol Metab 19):E55-E66

Nielsen S, Nielsen JT (1988) Influence of flow rate and perfused load on insulin absorption in isolated proximal tubules. Am J Physiol 254 (Renal Fluid Electrolyte Physiol 23):F802-F812

Nielsen S, Nielsen JT, Christensen EI (1987) Luminal and basolateral uptake of insulin in isolated, perfused, proximal tubules. Am J Physiol 253 (Renal Fluid Electrolyte Physiol 22):F857-F867

Ottosen PD, Bode F, Madsen KM, Maunsbach AB (1979) Renal handling of lysozyme in the rat. Kidney Int 15:246-254

Parton RG, Prydz K, Bomsel M, Simons K, Griffiths G (1989) Meeting of the apical and basolateral endocytic pathways of the Madin-Darby canine kidney cell in late endosomes. J Cell Biol 109:3259-3272

Peterson DR, Carone FA, Oparil S, Christensen EI (1982) Differences between renal tubular processing of glucagon and insulin. Am J Physiol 242 (Renal Fluid Electrolyte Physiol 11):F112-F118

Rabkin R, Mahoney CA (1988) Hormones and the kidney. In Diseases of the Kidney, Schrier RW, Gottschalk CW (eds). Little, Brown and Co Boston Toronto, 309-355

Straus W (1964) Cytochemical observations on the relationships between lysosomes and phagosomes in kidney and liver by combined staining for acid phosphatase in intravenously injected horseradish peroxidase. J Cell Biol 20:497-507

II. ENDOCYTOSIS IN HEALTH, DISEASE AND THERAPY

2.1. ENDOCYTOSIS IN THE IMMUNE RESPONSE

(Chairperson : Colin WATTS)

The role of endocytosis in antigen presentation is currently attracting considerable attention and was vividly discussed, both after each lecture and during a second roundtable, whose conclusions are included in the present overview. It is clear that antigen processing involves a partial intracellular hydrolysis of foreign antigens, producing fragments that are combined with MHC-II class molecules for presentation to T cells. However, the subcellular location of hydrolysis, dissociation of the invariant chain and fragment binding to MHC-II remained unclear.

Colin Watts presented a biochemical analysis of the fate of tetanus toxin in an antigen-specific B lymphoblastoid presenting cell. He reported that processing occurs late in the endocytic pathway, at least beyond an active recycling endosome, and involves cleavage of an immunoglobulin-antigen complex. Thus, protection of the buried domain of the particular protein bound would account for the specificity of antigen presentation by B cells. Toxin-derived peptides appear to bind to newly synthesized, rather than surface-derived MHC-II.

Other laboratories are attempting to localize by morphology the components required for presentation. Using confocal immunofluorescence on murine B lymphomas and transfected fibroblasts, and cryosection immunolabelling on human fibroblasts, respectively, **Jean Davoust** and **Hans Geuze** found that MHC-II are preferentially located in late endocytic compartments devoid of the cation-independent mannose 6-phosphate receptor, i.e. lysosomes. In addition, Geuze reported that lysosomal fractions of mouse lymphoblastoid cells can process antigens into peptides that can be effectively presented to T-cells in the MHC-II context.

In contrast, **Frances Brodsky**, using cryosection immunolabelling of Epstein Barr-virus transformed B lymphocytes (IM-9 cells), reported that α, β and invariant chains of MHC-II, internalized membrane Ig and the proteolytic enzymes cathepsin B and D are all found in early endocytic organelles that can be accessed either via endocytic coated vesicles or via a macropinocytic/phagocytic route. Aside from differences in cell type and

NATO ASI Series, Vol. H 62
Endocytosis
Edited by P. J. Courtoy

technique, the reasons for this serious discrepancy are unclear. The localization of MHC-II in lysosomes appears to support the biochemical evidence for a late processing. However, it cannot be excluded that those MHC-II molecules represent receptors that have failed to bind peptide earlier in the pathway and are destined for degradation. Alternatively, Brodsky suggested that MHC-II/invariant chain complexes may mature along the endocytic pathway and have maximal exposure to antigenic fragments at differnt stages.

Annegret Pelchen-Matthews, from the laboratory of **Mark Marsh,** studied the dynamics of the T-cell differentiation antigen CD4, which is the target for the human immunodeficiency virus. She reported that CD4 is internalized relatively efficiently in non-lymphoid cells but not in T cell lines, due to its interaction with a lymphoid specific tyrosine kinase, $P56^{lck}$. This observation suggests a novel function for membrane-asociated protein kinases, i.e. interaction with the cortical cytoskeleton, and has implications in lymphocyte activation.

Finally, **Jean-Pierre Kraehenbuhl** illustrated how a thorough understanding of a receptor and its ligand can be exploited to establish a protective immunity. B-cell mediated mucosal immunity requires two opposite transepithelial transport processes : the transfer of external antigens across the specialized M epithelial cells, to reach mucosal-associated lymphoid tissues, and the secretion of the corresponding IgA across mucosal and glandular epithelia, to reach the pathogens. He reported that IgA hybridomas produced from Peyer's patch lymphoblasts primed against *Vibrio cholerae* can be combined with recombinant secretory component for efficient passive oral prophylaxy. Alternatively, when implanted into recipient animals, the secreted antibodies are apparently exported and confer active protection against this oral pathogen.

ANTIGEN PROCESSING IN B LYMPHOCYTES

Colin Watts, Pamela A.Reid, Michele A.West and Howard W.Davidson
Department of Biochemistry, Medical Sciences Institute, University of Dundee, Dundee, DD1 4HN, U.K.

INTRODUCTION

There is considerable current interest in how the events occurring during antigen processing and presentation via class II MHC glycoproteins are organised with respect to the endocytic pathway. This follows the discovery that most antigens require an intracellular processing step in order to produce a suitable ligand for binding to MHC Class II glycoproteins and, as a complex, to the T cell receptor. The available evidence strongly suggests that following endocytosis, processing and unfolding take place in intracellular compartments and that suitable MHC/peptide complexes are then returned to the cell surface. Using T cells to assay for the appearance of processed determinants it has been established that these antigen processing events take approximately one hour, are sensitive to chloroquine and fixation of the antigen presenting cells (APC) and to inhibitors of endosomal/lysosomal proteases (Ziegler and Unanue 1982; Lanzavecchia 1985; Buus and Werdelin 1986).

The presumptive peptide products of antigen processing bound to MHC glycoproteins have been visualised at the structural level (Bjorkman et al. 1987) and at the biochemical level (Buus et al. 1988) and fragmented antigen or appropriate synthetic peptides can interact directly with MHC glycoproteins (Babbitt et al. 1985; Buus et al. 1986; Shimonkevitz et al. 1983) thereby by-passing the requirement for processing. Taken together these findings are suggestive of a sequence of events that we call 'antigen processing' but in fact, we have little information on what happens to protein antigens during physiological processing.

ANTIGEN UPTAKE AND PROCESSING

The availability of antigen specific B cell lines (Lanzavecchia, 1985) has made it

NATO ASI Series, Vol. H 62
Endocytosis
Edited by P. J. Courtoy

possible to follow the fate of a single population of receptor-bound antigen molecules. Using tetanus toxin/toxoid specific EBV lymphoblastoid cells we have found that the initial kinetics of endocytosis of pre-bound antigen were rapid, independent of receptor occupancy, did not require cross-linking of membrane IgG (mIgG)(Watts and Davidson 1988) and appeared to be mediated by cell surface coated pits (Watts et al 1989). The mIgG receptor for antigen on lymphoblastoid cells is endocytosed and recycled in the presence or absence of ligand; at any one time approximately 50% of the cycling receptors are inside the cell (Davidson et al 1990).

About 20-30 minutes after the initiation of antigen endocytosis we detected trichloroacetic acid soluble material in the medium as well as discrete antigen fragments resolvable on Tris-Tricine SDS gels (Davidson and Watts 1989). At the pH values found in the endosome/lysosome system (pH 4-6) the affinity of the anti-tetanus IgGs for antigen is still high which dictates that the substrate for processing in B lymphocytes is the ligand/receptor complex and not the dissociated ligand. A consequence of this is that those parts of the antigen in the vicinity of the epitope are 'shielded' from complete degradation giving rise to distinct stable fragments in clones recognising different antigenic epitopes. In addition when other epitopes on the antigen were complexed with anti-tetanus Fabs the pattern of fragmentation observed within a single clone, was altered (Davidson and Watts 1989).These results imply that high-affinity Igs could, by influencing the course of processing and by stabilising certain antigen fragments, also influence which parts of antigen can most readily associate with Class II MHC and become presented to T cells. This biochemical informatin on processing in B lymphocytes might explain the observed immunological 'connectivity' between T and B cell epitopes (Manca et al 1988).

Another consequence of the 'low pH' resistant binding of antigen to mIg is that the cells capacity for antigen uptake is limited compared with, for example, capacity for LDL uptake where receptors are reused many times. We have found that each mIg receptor is probably used only once and that receptor processing accompanies antigen processing. Even monovalent antigen appears to accelerate the turnover of the receptor. Experiments using antibodies to different domains on the mIg molecule established that the Ig-bound antigen fragments can, at early times (30mins) be detected bound to an intact, membrane bound Ig molecule while in the same clone no evidence was obtained for intact antigen bound to a cleaved receptor. In other words, antigen processing appears to precede receptor processing (Davidson et al 1990).

COMPARTMENTATION OF PROCESSING

Separation of recycling events from processing events is a hallmark of the endocytic pathway in many cell types. In this way processing is made selective being confined to material which fails to efficiently recycle. Three pieces of evidence suggest that the two alternative fates we have identified for monovalent antigen, i.e. recycling or processing, also take place in different compartments. First, equilibration with the recycling endosome compartment occurs within 5-10 minutes yet no antigen fragments or return of degraded material to the medium could be detected before 20-30 minutes (Watts and Davidson 1988; Davidson and Watts 1989). Second at 20°C, endocytosis of antigen still occured, albeit more slowly, yet no processing was observed. Thirdly, antigen fragments bound to intact mIg cannot be detected on the cell surface (unpublished observations) although, as indicated above, intact antigen can clearly be recycled. Although there are limits to the sensitivity of this experiment it does appear that for the most part there cannot be extensive recycling to the surface of intact mIg from the site where the Ag fragments are generated.

So far then endocytosis and processing of monovalent antigen in B lymphocytes is in many respects similar to the processing of other endocytosed ligands althogh the tight association between ligand and receptor and the clonotypically unique nature of the receptor has some potentially important consequences as noted above. Our studies suggest, as do several earlier studies on other endocytosed ligands, that recycling and processing of monovalent antigen take place in kinetically and functionally distinct sites. A recent study co-localised cross-linked surface Ig with several other markers relevant to antigen processing such as cathepsins and class II MHC after only 2 minutes at 37°C (Guagliardi et al 1990). This is not easy to reconcile with the data summarised above and suggests that the itineraries taken by monovalent and polyvalent antigens may differ.

TRANSFER OF PROCESSED ANTIGEN TO MHC CLASS II

One important facet to Class II restricted antigen preseentation not yet discussed is the biosynthesis, trafficking and peptide-loading of the MHC molecules themselves. The important questions here relate to the site of peptide-loading of MHC molecules, the physiological size, stability and heterogeneity of processed antigen and the source of the class II molecules involved. Progress here will require that the formation of peptide/MHC complexes be detectable at a biochemical level. Using the tetanus-specific lymphoblastoid cells we have recently been able to detect the transfer of radoiodinated peptides derived from processing of native antigen. These fragments, which can be displayed on Tris-Tricine SDS gels are specifically associated with Class II, appear with the expected kinetics (1hour), persist in association with Class II for at least 24 hours and their appearance on Class II is blocked by known inhibitors of antigen presentation such as chloroquine and leupeptin. The apparent molecular weights are distinct from the mIg-associated fragments suggesting that they may be produced in parallel to or following further processing of the Ig-bound fragments (Davidson and Watts, in preparation).

The ability to detect the loading of Class II under physiological conditions should facilitate the analysis of the site of peptide-loading and the source of the class II MHC molecules involved. There is currently much debate as to which population of class II MHC molecules is involved in capturing the processed peptides generated during processing of native antigen the candidates being those Class II molecules on the cell surface and those which are still undergoing processing along the biosynthetic pathway. Involvement of the surface population would presumably require co-internalization with antigen, transport to the processing compartment, loss of any existing peptides occupying the cleft formed by the α and β chains, binding of newly processed antigen peptides and then return to the cell surface. We have recently provided the first direct evidence that cell surface Class II (as well as Class I) MHC which becomes internalised can be returned efficiently to the cell surface (Reid and Watts, 1990) while other studies have demonstrated that under some conditions peptide/class II complexes can be suprisingly short-lived (Adorini et al 1989). Our results suggest that a cell surface Class II molecule could internalise and recycle as many as 60 times during its lifetime whereas it can of course be processed as a newly synthesised molecule only once. It remains to be shown however that this cycling population of surface class II molecules really is involved in binding processed antigen. Two key questions that remain to be answered are (a) does the itinerary of cycling Class II include those compartments where processing occurs and (b) are the peptide/MHC complexes generated by processing of native antigen, as opposed to synthetic peptides sufficiently labile to allow a single binding pocket to be used more than once.

ACKNOWLEDGEMENTS

We thank Antonio Lanzavecchia, Keith Guy and Mark Bretscher for reagents. Work in the authors' lab is supported by the Medical Research Council and the Wellcome Trust.

REFERENCES

Adorini L, Appella E, Doria G, Cardinaux F, Nagy ZA (1989) Competition for antigen presentation in living cells involves exchange of peptides bound by class II MHC molecules. Nature 343: 800-803

Bjorkman P.J., M.A. Saper, B Samraoui, W.S. Bennett, J.L. Strominger and D.C. Wile (1987) Structure of the human class I histocompatibility antigen, HLA-A2. Nature 329: 506.

Buus, S., and O. Werdelin. 1986. A group-specific inhibitor of both proteolytic degradation and presentation of the antigen dinitrophenyl-poly-L-lysine by guinea pig accessory cells to T cells. J. Immunol. 136: 452.

Buus, S., A. Sette, S.M. Colon, D.M. Jenis, and H.M. Grey. 1986. Isolation and characterization of Antigen-Ia complexes involved in T cell recognition. Cell 47: 1071.

Buus, S., A. Sette, S.M. Colon and H.M. Grey. 1988. Autologous peptides constitutively occupy the antigen binding site on Ia. Science 242: 1045.

Davidson H.W. and C. Watts. (1989) Epitope-directed processing of specific antigen by B-lymphocytes. J. Cell Biol. 109: 85-92

Davidson HW, West MA, Watts C (1990) Endocytosis, intracellular trafficking and processing of membrane IgG and monovalent antigen/membrane IgG complexes in B lymphocytes. J.Immunol. 144: 4101-4110

Guagliardi LE, Koppelman B, Blum JS, Marks MS, Cresswell P, Brodsky F (1990) Co-localization of molecules involved in antigen procesing and presentation in an early endocytic compartment. Nature 343:133-139.

Lanzavecchia, A. 1985. Antigen-specific interaction between T and B cells. Nature 314: 537.

Manca, F., D. Feneglio, A. Kunkl, C. Cambiaggi, M. Sasso, and F. Celada. (1988) Differential activation of T cell clones stimulated by macrophages exposed to antigen complexed with monoclonal antibodies. A possible influence of paratope specificity on the mode of antigen processing. J. Immunol. 140: 2893.

Ozaki, S., and J.A. Berzofsky (1987) Antibody conjugates mimic specific B cell presentation of antigen: relationship between T and B cell specificity. J. Immunol. 138: 4133.

Reid PA, Watts C (1990) Cell surface MHC glycoproteins cycle through primaquine sensitive intracellular compartments. Nature 346: 655-657

Shimonkevitz, R., J. Kappler, P. Marrack, and H.M. Grey. 1983. Antigen recognition by H-2 restricted T cells. II. Cell-free antigen-processing. J. Exp. Med. 158: 303.
Watts, C., and H.W. Davidson, 1988. Endocytosis and recycling of specific antigen by human B cell lines. EMBO J. 7: 1937.
Watts C, West MA, Reid PA, Davidson HW (1989) Processing of immuoglobulin associated antigen in B lymphocytes. Cold Spring Harb. Symp. Quant. Biol. 54, 345-352
Ziegler HK, Unanue ER (1982) Decreasein macrophage antigen catabolism caused by ammonia and chloroquine with inhibiition of antigen presentation to T cells. Proc. Natl. Acad. Sci. USA 79:175-178

ENDOCYTOSIS AND BIOSYNTHETIC TRANSPORT OF MURINE MHC CLASS II MOLECULES IN ANTIGEN PRESENTING CELLS

J. DAVOUST, P. COSSON, J.M. ESCOLA, J. HENRY, M. HUMBERT, G. RAPOSO and J. SALAMERO.

Centre d'Immunologie de Marseille-Luminy INSERM-CNRS, Case 906, 13288 Marseille Cédex 09.

In B lymphocytes, the processing of exogenous proteins and the subsequent binding of antigenic peptides to MHC class II molecules occurs most likely within endocytic compartments. Using various biochemical and morphological techniques, we show that the endosomal compartments are loaded with a pool of MHC class II molecules which are permanentely endocytosed and recycled to the cell surface in murine B lymphoma cells. L fibroblasts transfected with $I\text{-}A_{\alpha\beta}{}^{k}$ or $I\text{-}E_{\alpha}{}^{k}{}_{\beta}{}^{k/d}$ fail to internalize their surface class II molecules even when expressed in conjunction with the class II associated invariant chain (Ii). We suggest that in murine B lymphoma cells, antigenic peptides can gain access to a pool of recycling class II molecules whereas in L cells they meet newly synthesized class II molecules targeted to the endosomal compartments. In B lymphocytes and macrophages, the invariant chain (Ii) is synthesized in large excess with respect to MHC class II α and β chains. Ii binds to MHC class II molecules in endoplasmic reticulum and is partially degraded presumably at the level of the trans-Golgi network or of prelysosomal compartments. To estimate the role fo Ii molecules in the function and the transport of class II molecules, we have surtransfected murine L fibroblasts expressing IAk molecules using as controls L fibroblasts expressing only IA^k or Ii. As pointed out before for the presentation of cytochrome B5 in the context of IE^d (Stockinger et al., 1989 Cell 56, 683-689) we found that Ii expression enhance by a hundred fold the capacity of IA^k positive L cell to present hen egg lysozyme. From radiolabeling followed by immunoprecipitation of Ii or IAk and

NATO ASI Series, Vol. H 62
Endocytosis
Edited by P. J. Courtoy

1D or 2D gel analysis, we found that Ii expression does not affect the rate of biosynthetic transport of class II molecules to the cell surface as well as their accessibility to in situ iodination in endosomal compartments. However Ii sialilation occured efficiently only in MHC class II positive cells and a fraction of mature Ii-IA^k complexes are detected in endosomal compartments by in situ iodination. Laser scanning confocal microscopy revealed that MHC class II molecules were present in intracellular vesicles that contain fluid phase markers internalized for 20 min in Ii positive and negative cells. This late endosomal compartment contains most of the terminally glycosilated MHC class II molecules. It is distinct from the Golgi apparatus, from the transferrin positive early endosomes and from the Mannose 6 phosphate receptor rich polysosomal compartment. We conclude that the invariant chain does not influences the transport and the steady state endosomal distribution of MHC class II molecules. Since Ii expression can enhance the antigen presentation of L cells, Ii is likely to act as a modulator of the binding capacity of the newly synthetized MHC class II molecules.

References :

Salamero, J., Humbert, M., Cosson, P., and J. Davoust.
Mouse B lymphocyte specific endocytosis and recycling of MHC class II molecules. The EMBO J. 9, 3489-3496.

Humbert, M., Raposo, G., Cosson, P., Gerlier, D., Davoust, J., and J. Salamero.
Delivery of MHC class II molecules into the endosomal pathway is independant of the invariant chain. 1991 (submitted for publication).

ENDOCYTOSIS: MOLECULAR MECHANISM AND ROLE IN THE IMMUNE RESPONSE

F.M. Brodsky, L.E. Guagliardi, C.J. Galloway, B. Koppelman, S.L. Acton, D.H. Wong and I. Näthke
Departments of Pharmacy and Pharmaceutical Chemistry
School of Pharmacy, University of California
513 Parnassus Avenue
San Francisco, California 94143-0446
U.S.A.

Clathrin-mediated endocytosis is responsible for selective and facilitated internalization of cell surface receptors (Brodsky, 1988). This process contributes to the uptake of nutrients and hormones and targets internalized proteins and their receptors to degradative and/or recycling intracellular compartments. Antigenic fragments of internalized proteins are generated as a byproduct of this pathway and bind to histocompatibility molecules which present them to T cells, thereby stimulating the immune response. The research summarized below investigates the molecular control of clathrin-mediated endocytosis and its participation in the intracellular events that lead to antigen processing and presentation.

Biochemical properties of clathrin light chains suggest a regulatory role

The clathrin light chains LC_a and LC_b are encoded by two distinct genes that undergo differential RNA splicing in neurons generating higher molecular weight forms. There are two higher molecular weight forms of LC_a with 30- and 18-amino acid inserts respectively and one of LC_b with a different but homologous 18-amino acid insert (Jackson et al., 1987; Kirchhausen et al., 1987). These inserted sequences are found at corresponding positions in both light chains and are exposed to the cytoplasm in assembled coated vesicles (Brodsky et al., 1987).

The primary structure of the clathrin light chains suggests they are organized into an array of linear domains, some of which are shared by LC_a and LC_b and others which are unique to each light chain. Recent work in the

NATO ASI Series, Vol. H 62
Endocytosis
Edited by P. J. Courtoy

laboratory has identified the light chain domain responsible for calcium binding and has analyzed the tissue distribution of different forms of the light chains. These studies have suggested that the light chains play a regulatory role in clathrin function.

Mapping the calcium binding site (Figure 1) has indicated two ways in which calcium could influence clathrin function. The calcium binding site was found to resemble the EF-hand loop characteristic of a family of calcium-binding proteins. Secondary structure predictions for the clathrin light chains (Chou and Fasman, 1978; Garnier et al., 1978) indicate that the 85-96 loop could be part of a helix-loop-helix motif. The C-terminal helix flanking the calcium-binding loop coincides with the heavy chain binding region (Scarmato and Kirchhausen, 1990). This arrangement is similar to the second EF-hand loop in calmodulin where the calcium chelating loop is flanked by a long alpha helix that interacts with the target enzymes of calmodulin (Babu et al., 1988). In the clathrin triskelion, the helical region of heavy-light chain binding has been shown to participate in triskelion-triskelion interactions during assembly of the clathrin coat (Blank and Brodsky, 1987). Calcium influences triskelion assembly (Keen et al., 1979) and may do so through binding adjacent to this helical region of triskelion interaction. At the N-terminal side of the calcium binding site is a region of the clathrin light chains that is involved in binding hsc70 and stimulating ATP hydrolysis (DeLuca-Flaherty et al., 1990). Accessibility of this region is increased by calcium, demonstrating that calcium has local structural effects on the light chains.

Calcium binding by free light chains and whole triskelions has been demonstrated at 1μM calcium and 3mM magnesium (Mooibroek et al., 1987). This is the upper range of physiological calcium concentrations (0.1-1μM) (Alberts et al., 1989) and exceeds intracellular magnesium concentrations (0.5mM) (Alberts et al., 1989). It has also been shown that the calcium concentration close to the plasma membrane or membranes of other cellular compartments can be higher than the average cytoplasmic calcium concentration (Foskett et al., 1989; Rasmussen et al., 1984). Since clathrin operates in close proximity to the plasma membrane it could experience fluxes of calcium concentration that would be sufficient to regulate clathrin assembly or disassembly through binding of calcium to light chains. Clathrin light chains bind calcium with a K_D of 25μM, which means 50% of clathrin light chains have calcium bound at 25μM. Thus at physiological concentrations of calcium at least 1-5% of light chains have calcium bound, depending on local calcium concentrations. The average clathrin coated vesicle is composed of 60-140

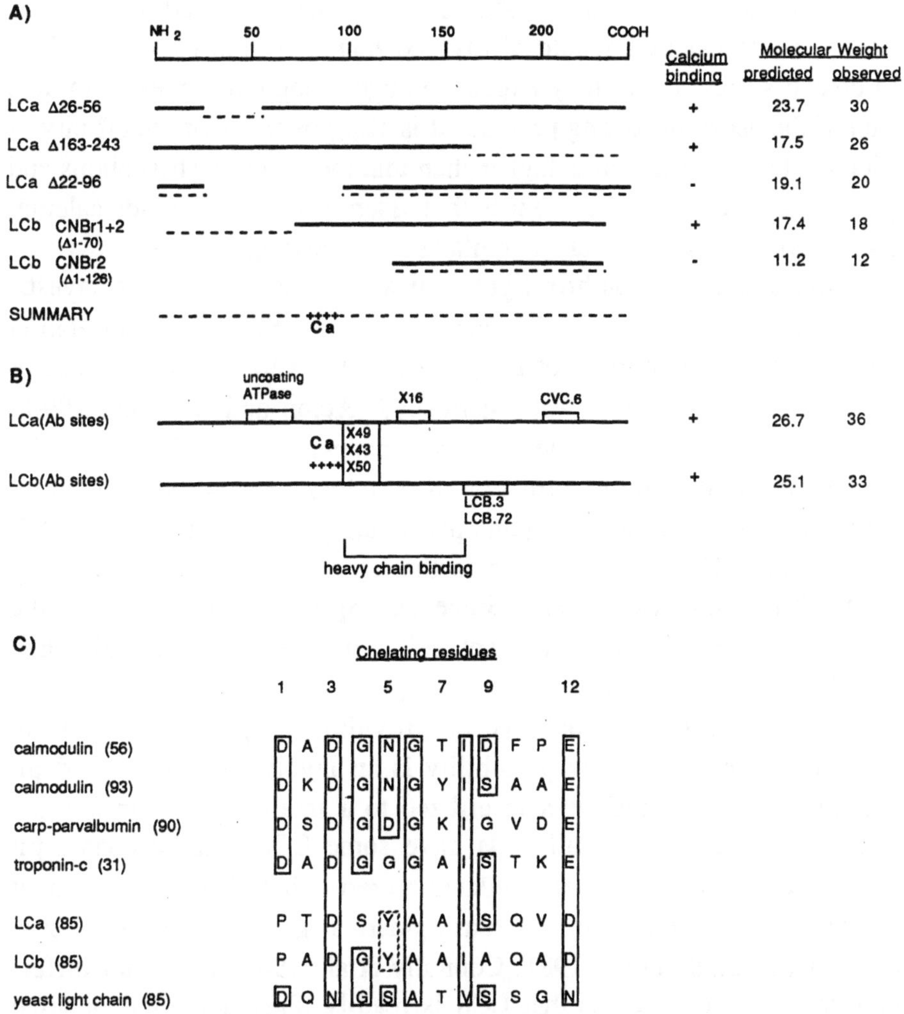

Figure 1: The calcium-binding site of clathrin light chains (Näthke et al., 1990).

A) Calcium-binding properties and molecular weights of clathrin light chains and light chain deletion mutants. The minus signs under each sequence indicate residues not involved in calcium binding. These are summarized at the bottom, with the calcium-binding sequences shown with plus signs. B) The proximity of the calcium-binding site to the heavy chain binding region is shown. The target sequences for several anti-light chain monoclonal antibodies are mapped. Binding of the X49, X43 and X50 antibodies causes a reduction in calcium binding. C) Alignment of calcium-binding loop sequences from known EF-hand structures with the predicted calcium-binding loop sequences in clathrin light chains. Identical or homologous residues are framed with solid lines. Additional residues with side chains that are predicted to chelate calcium are framed with dashed lines.

clathrin triskelions (Heuser and Kirchhausen, 1985) with 180-420 light chains. Statistically, some calcium would always be bound by the triskelions in a clathrin coated structure and may expose enough recognition sites to initiate the action of the hsc70 uncoating protein. It is also possible that the affinity of assembled clathrin for calcium is higher than that measured for triskelions and free light chains, increasing the chances that assembled clathrin binds calcium at physiological concentrations and stimulates the uncoating ATPase.

Expression patterns of clathrin light chains, LC_a and LC_b, were investigated to determine their differential influence on clathrin function. Correlation between an increased proportion of LC_b in a tissue or cell and the presence of a regulated secretory pathway was established (Acton and Brodsky, 1990). There are two stages in the regulated secretory pathway that might specifically require LC_b. The first is in the formation of secretory granules where clathrin has been implicated in concentrating and packaging proteins (Orci et al., 1985; Tooze and Tooze, 1986). The second is the coordinated rapid retrieval of granule membrane after exocytosis. Since LC_b specific antibodies stain the Golgi and periphery of many different cell types (S. Acton unpublished results), LC_b is probably not restricted to specialized functions. Rather its presence could confer the capability of providing a specialized clathrin function when needed. A likely possibility is an involvement in coated pit upregulation, a process which is also utilized to a small capacity in non-secretory cells, explaining why all cells have some LC_b. Rapid coated pit upregulation occurs not only after regulated secretion but also after treatment of some cells with growth factors such as epidermal growth factor and nerve growth factor (Connolly et al., 1981; Connolly et al., 1984). A major difference between LC_b and LC_a is that LC_b is readily phosphorylated *in vitro* (Usami et al., 1985) and *in vivo* (Bar-Zvi et al., 1988). Phosphorylation of LC_b could provide a signal influencing regulation and recruitment of triskelions for coated pit formation.

Studies of the distribution of the high molecular weight forms of the clathrin light chains showed they are expressed in a neurons (Wong et al., 1990). The concentration of neuron-specific clathrin light chains at the synaptic terminal suggests that they play a role in clathrin-mediated membrane retrieval. It is possible that the inserted sequence could be part of a mechanism for coordinating clathrin-mediated endocytosis following synaptic vesicle fusion. The short hydrophobic stretch introduced by the insertion sequence has been proposed to be a binding site for a cytoplasmic protein that might control mobilization of clathrin for this specialized function (Jackson et al.,

1987). Axonal transport may play a role in the recruitment of clathrin for membrane recycling. In this scheme, a neuron-specific protein might bind to the inserted sequences to promote clathrin light chain association with cytoskeletal elements and prevent triskelia from assembling into cages during axonal transport. After synaptic vesicle fusion, the proposed axonal binding protein would be released from the inserted sequence allowing triskelia to polymerize and form coated vesicles for membrane retrieval. Electron microscopy and immunocytochemistry studies suggest that clathrin is transported in soluble form (Cheng et al., 1980; Morré, 1982) and other studies have shown it is in the SCb component of axonal transport (Gower and Tytell, 1987), which translocates soluble proteins rather than organelles. The 100kD components of clathrin coated vesicles, known as adaptins (Glickman et al., 1989), also have "brain-specific" insertion sequences (Ponnambalam et al., 1990; Robinson, 1989) which might play a role in mediating cytoskeletal binding and/or preventing polymerization during axonal transport for these proteins as well.

Endocytosis in lymphocytes: antigen presentation and receptor regulation

Histocompatibility molecules are peptide-binding proteins and present antigenic peptides to T cells. In spite of similarity in structure and peptide specificity, Class I and Class II histocompatibility molecules bind peptides at different intracellular locations (Brodsky and Guagliardi, 1991). Class I molecules present peptides derived from endogenous proteins but Class II molecules can present peptides from endocytosed antigen as well as endogenous antigen. To determine whether intracellular trafficking and the associated invariant chain could be responsible for the versatility of peptide binding displayed by Class II histocompatibility molecules, their intracellular location was mapped by immunoelectron microscopy. The pathway for antigen uptake was labeled by internalization of colloidal gold bound to surface immunoglobulin on human B cells (Figure 2). Endosomes formed after uptake of cross-linked gold (Guagliardi et al., 1990) or monovalent gold were analyzed. Although the former was taken up by a non-coated pathway and the latter was taken up by coated pits, the endosome composition with respect to Class II molecules and proteases was similar for either internalization mechanism.

Class II molecules associated with invariant chain, *en route* to the cell surface were observed to intersect both antigen uptake pathways in an early (2-5 minute) endocytic compartment. These immature Class II molecules were also

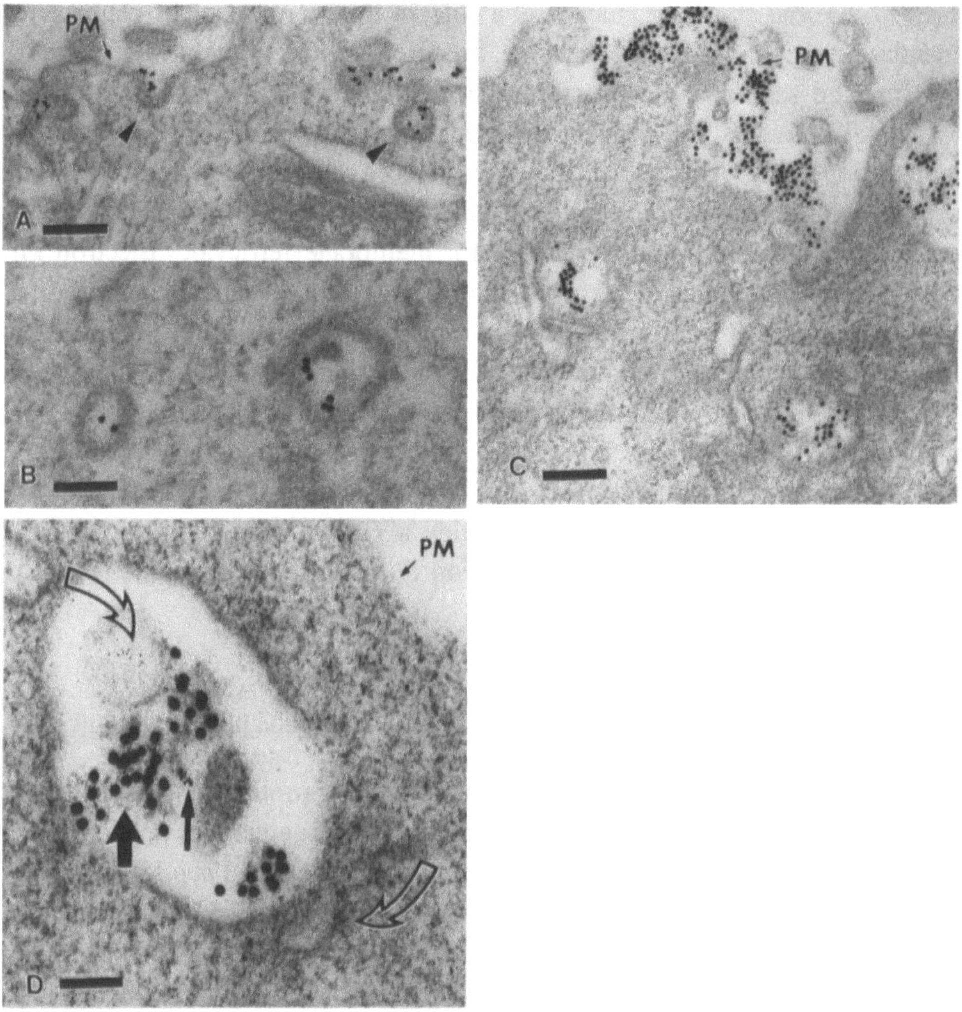

Figure 2: Endocytosis of surface immunoglobulin in human B-lymphoblastoid cells.

Surface immunoglobulin on IM-9 cells was labeled with goat anti-human Ig, followed by 15nm gold conjugated to rabbit anti-goat Ig (cross-linked) (Guagliardi et al., 1990) or with biotinylated anti-human Ig followed by 15nm gold conjugated to streptavidin (minimally cross-linked). (A) Uptake of minimally cross-linked immunoglobulin in clathrin coated pits (arrowheads). Bar = 0.3077μm. (B) Endosomes formed after uptake of minimally cross-linked immunoglobulin. Bar = 0.2667 μm. (C) Uptake of cross-linked surface immunoglobulin by non-coated membrane. Bar = 0.2778 μm. (D) Endosome formed after 20 minutes uptake of cross-linked immunoglobulin (15 nm gold, large black arrow), labeled with antibody to invariant chain (MATILDA, 2 nm gold, clear arrow) and antibody to mature Class II molecules (L243, 5 nm gold, small black arrow). Bar = 0.1667μm. This demonstrates maturation of the Class II molecule in a later endocytic compartment. Previous studies have provided evidence for maturation in an early endocytic compartment as well (Guagliardi et al., 1990).

observed in late (20-30 minute) endosomes formed after uptake of crosslinked Ig. Proteolytic enzymes Cathepsin B and Cathepsin D were localized to both early and late endocytic compartments formed after Ig uptake, along with mature Class II molecules. Since very little mature Class II is internalized by these cells, the endosomal mature Class II is probably generated by dissociation of invariant chain, which requires activity of proteases Cathepsin B and D. Dissociation of invariant chain is most likely accompanied by peptide binding, as associated invariant chain obscures the peptide binding site (Roche and Cresswell, 1990). In addition, dissociation from invariant chain allows the peptide-Class II complex to traffic to the cell surface. This suggests a mechanism by which Class II molecules can bind peptides generated in several stages of the endocytic pathway. The immature Class II molecules may be delivered to early endocytic compartments from the Golgi and then mature within the endocytic pathway, passing into late endosomes if invariant chain removal is incomplete in early endosomes. Thus these Class II molecules would become competent to bind antigen at different stages of antigen degradation. This would account for the differences in antigen processing sites observed with different antigens, and allow Class II molecules to acquire peptides from different antigens in different compartments. Class I molecules were not observed in endocytic compartments, suggesting that a trafficking pathway specific to invariant chain-associated Class II molecules contributes to their antigen presenting function.

Future studies on antigen presenting cells will require kinetic analysis of trafficking of the molecules involved and identification of the organelles in which antigen processing and Class II maturation takes place. It will also be of interest to identify the transport vesicles responsible for targeting the immature Class II molecules and proteases to endosomes. One approach to these questions is to isolate organelles using antibodies to proteins on their cytoplasmic face and to analyze them for the presence of biosynthetically pulse-labeled histocompatibility molecules and proteases. As a model system, a method was developed for purification of clathrin-coated vesicles from lymphocytes and for detecting receptors inside them. Lymphocyte surface molecules were labeled by iodination. Then, cells were homogenized and clathrin-coated vesicles were isolated with anti-clathrin monoclonal antibody bound to iron-containing microspheres, using a magnet. The isolated vesicles were solubilized with detergent and analyzed for the presence of receptors by immunoprecipitation with anti-receptor antibodies. Clathrin-association

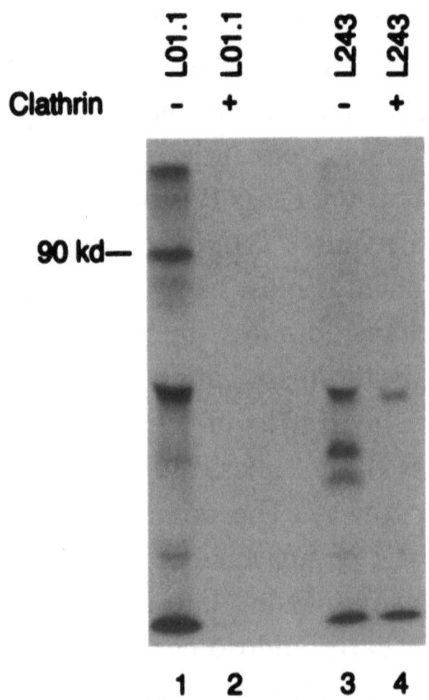

Figure 3: B lymphocyte receptors in immunoisolated clathrin coated vesicles. ^{125}I-labeled cell homogenate was exposed to anti-clathrin antibody attached to iron-containing microspheres in the presence (+) and absence (-) of purified clathrin. Microspheres were separated from the homogenate using a magnet and associated vesicles solubilized with detergent. Then receptors were immunoprecipitated from the isolated material using antibodies to transferrin receptor (lanes 1 and 2) and Class II MHC molecules (lanes 3 and 4). Note the absence of receptors when immunoisolation was inhibited with purified clathrin (+ lanes).

patterns of receptors on lymphocytes provided insight into their immunological and biological functions. Surface immunoglobulin had an extremely high level of clathrin association, explaining its efficient endocytosis of antigen. Lack of clathrin association of the CD8 molecule on T cells explained its limited internalization and its poor efficacy as a target for immunotoxins. CD8 could be excluded from clathrin-coated pits as a result of its cytoplasmic tail binding to tyrosine kinase p56^{lck} in T cells (Turner et al., 1990). Clathrin-mediated endocytosis was found to contribute to regulation of other lymphocyte receptors, including CD3, CD5, CD21, CD45 and Class I and Class II histocompatibility molecules (Figure 3). To prove that the receptors isolated were derived from clathrin-coated membranes and not from organelles sticking non-specifically to the immunoaffinity beads, the isolation procedure was inhibited by including an excess of purified clathrin during exposure to the cell homogenate. This completely inhibited detection of receptors in material eluted from the beads, demonstrating specificity of the receptor isolation from the organelle recognized. This technique should be applicable to determining whether Class II molecules, exported from the Golgi, are targeted to endosomes in clathrin-coated vesicles. With appropriate antibodies, this procedure could also be used to isolate endosomes for analysis of events relevant to antigen processing and presentation.

Acknowledgments

This work was supported by National Institutes of Health grants AR20684, GM26691, GM38093, National Science Foundation grant DCB8711317, the Pew Charitable Trusts and Fellowship grants from the Leukemia Society of America (CG), the Cancer Research Institute (BK), and the Pharmaceutical Manufacturers Association Fund (IN).

References

Acton S, Brodsky FM (1990) Predominance of clathrin light chain LC_b correlates with the presence of a regulated secretory pathway. J. Cell Biol. 111:1419-1426

Alberts B, Bray D, Lewis J, Raff M, Roberts K, Watson JD (eds) (1989) Molecular Biology of the Cell, 2nd ed., Garland Publishing, Inc., New York and London

Babu YS, Brugg CE, Cook WJ (1988) in Three dimensional structure of calmodulin (Cohen, P. and Klee, C. B., eds) pp. 83-89, Elsevier, Amsterdam

Bar-Zvi D, Mosley ST, Branton D (1988) *In vivo* phosphorylation of clathrin-coated vesicle proteins from rat reticulocytes. J. Biol. Chem 263:4408-4415

Blank GS, Brodsky FM (1987) Clathrin assembly involves a light chain-binding region. J. Cell Biol. 105:2011-2019

Brodsky FM (1988) Living with clathrin: Its role in intracellular membrane traffic. Science 242:1396-1402

Brodsky FM, Galloway CJ, Blank GS, Jackson AP, Seow H-F, Drickamer K, Parham P (1987) Localization of clathrin light-chain sequences mediating heavy-chain binding and coated vesicle diversity. Nature 326:203-205

Brodsky FM, Guagliardi L (1991) The cell biology of antigen processing and presentation. Annu. Rev. Immunol. 9: in press

Cheng TP, Byrd FI, Whitaker JN, G. J (1980) Immunocytochemical localization of coated vesicle protein in rodent nervous system. J. Cell Biol. 86:624-633

Chou PY, Fasman GD (1978) Prediction of the secondary structure of proteins from their amino acid sequence. Adv. Enzymol. 47:45-148

Connolly JL, Green SA, Greene LA (1981) Pit formation and rapid changes in surface morphology of sympathetic neurons in response to nerve growth factor. J. Cell Biol. 90:176-180

Connolly JL, Green SA, Greene LA (1984) Comparison of rapid changes in surface morphology and coated pit formation of PC12 cells in response to nerve growth factor, epidermal growth factor, and dibutyryl cyclic AMP. J. Cell Biol. 98:457-465

DeLuca-Flaherty C, McKay DB, Parham P, Hill BL (1990) Uncoating protein (hsc70) binds a conformationally labile domain of clathrin light chain LC_a to stimulate ATP hydrolysis. Cell 62:875-887

Foskett JK, Gunter-Smith PJ, Melvin JE, Turner RJ (1989) Physiological localization of an agonist sensitive pool of Ca^{2+} in parotid acinar cells. Proc. Natl. Acad. Sci. USA 86:167-171

Garnier J, Osguthorpe JD, Robson B (1978) Analysis of the accuracy and implications of simple methods for predicting the secondary structure of globular proteins. J. Mol. Biol. 120:97-120

Glickman JN, Conibear E, Pearse BMF (1989) Specificity of binding of clathrin adaptors to signals on the mannose-6-phosphate/insulin-like growth factor II receptor. EMBO J. 8:1041-1047

Gower DJ, Tytell M (1987) Axonal transport of clathrin-associated proteins. Brain Research 407:1-8

Guagliardi L, Koppelman B, Blum JS, Marks MS, Cresswell P, Brodsky FM (1990) Co-localization of molecules involved in antigen processing in an early endocytic compartment. Nature 343:133-139

Heuser J, Kirchhausen T (1985) Deep-etch view of clathrin assembly. J. Ultrastructural Research 92:1-27

Jackson AP, Seow H-F, Holmes N, Drickamer K, Parham P (1987) Clathrin light chains contain brain-specific insertion sequences and a region of homology with intermediate filaments. Nature 326:154-159

Keen JH, Willingham MC, Pastan IH (1979) Clathrin-coated vesicles: Isolation, dissociation, and factor-dependent reassociation of clathrin baskets. Cell 16:303-312

Kirchhausen T, Scarmato P, Harrison SC, Monroe JJ, Chow EP, Mattaliano RJ, Ramachandran KL, Smart JE, Ahn AH, Brosius J (1987) Clathrin light chains LCa and LCb are similar, polymorphic, and share repeated heptad motifs. Science 236:320-324

Mooibroek MJ, Michiel DF, Wang JH (1987) Clathrin light chains are calcium-binding proteins. J. Biol. Chem. 262:25-28

Morré DJ (1982) in Intracellular transport: vehicles, guide elements, and mechanisms (Weiss, D., ed) pp. 2-14, Springer-Verlag, Berlin

Näthke I, Hill BL, Parham P, Brodsky FM (1990) The calcium binding site of clathrin light chains. J. Biol. Chem. 265:18621-18627

Orci L, Ravazzola M, Amherdt M, Louvard D, Perrelet A (1985) Clathrin immunoreactive sites in the Golgi apparatus are concentrated at the trans pole in polypeptide hormone-secreting cells. Proc. Natl. Acad. Sci. USA 82:5385-5389

Ponnambalam S, Robinson MS, Jackson AP, Peiperl L, Parham P (1990) Conservation and diversity in families of coated vesicle adaptins. J. Biol. Chem. 265:4814-4820

Rasmussen H, Kojima I, Kojima K, Zawalich W, Apfeldorf W (1984) Calcium as intracellular messenger: sensitivity, modulation, C-kinase pathway, and sustained cellular response. Adv. Cycl. Nucleotide and Prot. Res. 18:159-193

Robinson MS (1989) Cloning of cDNAs encoding two related 100-kD coated vesicle proteins (α-Adaptins). J. Cell Biol. 108:833-842

Roche PA, Cresswell P (1990) Invariant chain association with HLA-DR molecules inhibits immunogenic peptide binding. Nature 345:615-618

Scarmato P, Kirchhausen T (1990) Analysis of clathrin light chain-heavy chain interactions using truncated mutants of rat liver light chain LCB3. J. Biol. Chem. 265:3661-3668

Tooze J and Tooze SA (1986) Clathrin-coated vesicular transport of secretory proteins during the formation of ACTH-containing secretory granules in AtT20 cells. J. Cell Biol. 103:839-850

Turner JM, Brodsky MH, Irving BA, Levin SD, Perlmutter RM, Littman DR (1990) Interaction of the unique N-terminal region of tyrosine kinase $p56^{lck}$ with the cytoplasmic domains of CD4 and CD8 is mediated by cysteine motifs. Cell 60:755-765

Usami M, Takahashi A, Kadota T, Kadota K (1985) Phosphorylation of a clathrin light chain of coated vesicles in the presence of histones. J. Biochem. 92:1819-1822

Wong DH, Ignatius MJ, Parosky G, Parham P, Trojanowski JQ, Brodsky. FM (1990) Neuron-specific expression of high molecular weight clathrin light chain. J. Neurosci. 10:3025-3031

ENDOCYTOSIS OF CD4 IN LYMPHOID AND NON-LYMPHOID CELLS

Annegret Pelchen-Matthews, Jane E. Armes and Mark Marsh
The Institute of Cancer Research, Chester Beatty Laboratories
Fulham Road, London SW3 6JB, U.K.

The lymphoid differentiation antigen CD4 is a glycoprotein which binds to non-polymorphic regions on MHC class II molecules and is expressed primarily on helper/inducer T-lymphocytes and on some cells of the macrophage/monocyte lineage. CD4 is involved in T-cell ontogeny in the thymus and in the activation of peripheral blood lymphocytes (Robey and Axel, 1990). The specific association of CD4 with the cytoplasmic protein tyrosine kinase $p56^{lck}$ (Turner *et al.*, 1990) suggests that CD4 may be directly involved in signal transduction. In addition CD4 is used as a receptor by the human immunodeficiency viruses HIV-1 and 2. In order to gain a better understanding of the mechanisms underlying the normal function of CD4 in T-cells and in the pathway of HIV entry, we have studied the endocytic trafficking of CD4 in non-lymphoid cells transfected with the CD4 gene and in a number of leukaemia/lymphoma-derived T-cell lines.

Constitutive Endocytosis of CD4

In order to follow the internalization of CD4 from the cell surface, we have developed two endocytosis assays (Pelchen-Matthews *et al.*, 1989, 1990). In both assays, CD4 is labelled in the cold, when endocytosis is inhibited, either with a specific radioiodinated anti-CD4 monoclonal antibody or its Fab fragments, or by covalent modification with biotin. After allowing endocytosis at 37°C, cells are cooled and label at the cell surface is removed by acid-stripping or digestion with trypsin, and the levels of intracellular label are analysed either by γ-counting or, in the biotinylation assay, by immunoprecipitation with a specific anti-CD4 antiserum. Both assays gave comparable results, indicating that the antibody ligands did not induce CD4 endocytosis.

These assays revealed that CD4, when expressed in transfected

NATO ASI Series, Vol. H 62
Endocytosis
Edited by P. J. Courtoy

HeLa-CD4 or NIH-3T3-CD4 cells, was taken up at rates of 2-3% per min (Fig. 1, Table I). After 30-60 min, the level of internalized CD4 reached a plateau with about 40% of the CD4 labelled initially being re-located inside the cells. This plateau has been shown to result from re-cycling of the internalized CD4 (Pelchen-Matthews *et al.*, 1989). While the rates

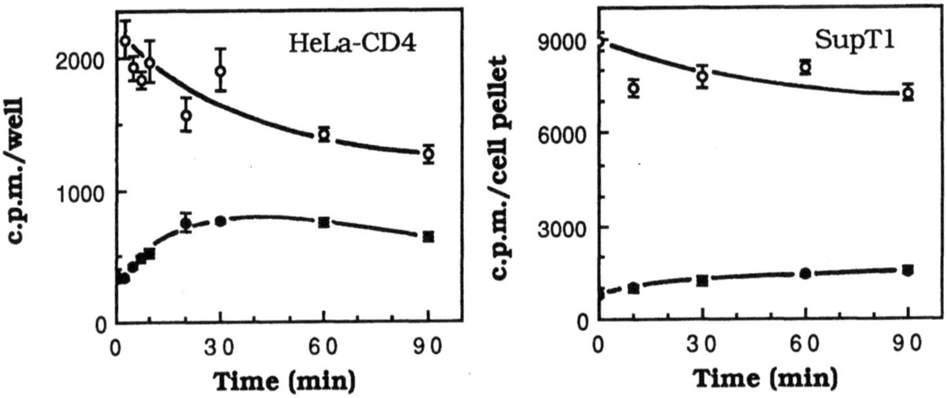

Figure 1: Endocytosis of CD4 in HeLa-CD4 and SupT1 cells. CD4 expressed on HeLa-CD4 and SupT1 cells was labelled by incubation at 0-4°C with 0.3 nM ^{125}I-labelled anti-CD4 monoclonal antibody, before cells were washed and incubated at 37°C for various times. Cells were then harvested directly for determination of the total cell-associated ligand (○). Alternatively, cell surface label was removed by washing at 0-4°C in medium at pH 2, leaving only the acid-resistant, internalized counts (●).

Table I: Endocytosis of CD4 in lymphocytic and CD4-transfected non-lymphocytic cells

Cell line	No. of experiments	Rate of endocytosis (% per min)	% internal at equilibrium*
HeLa-CD4	6	1.9 ± 0.9	38 ± 3
NIH-3T3-CD4	2	3.6 ± 0.1	47 ± 6
HL-60	1	3.4	41
SupT1	2	0.3 ± 0.2	6.8 ± 2.8
CEM	2	0.2 ± 0.03	9.6 ± 6.9
Jurkat	2	0.4 ± 0.05	4.4 ± 0.2
MOLT-4	1	0.2	4.2
RPMI-CD4	3	0.5 ± 0.45	13.4 ± 0.6

All endocytosis experiments used ^{125}I-labelled anti-CD4 monoclonal antibodies. * To determine % internalized ligand at equilibrium, the level of acid-resistant activity at in cells held at 0-4°C has been subtracted.

and levels of CD4 endocytosis in the transfected cells were not as great as those of transport molecules such as the transferrin or LDL receptors, they were nonetheless significant and reproducibly observed in the absence of any stimulation of the cells, and thus represent constitutive trafficking of CD4. Similar results were obtained with the promyelocytic cell line HL-60 (Table I). Electron microscopic immunogold localization studies have demonstrated that CD4 endocytosis in HeLa-CD4 cells takes place via coated pits and coated vesicles (Pelchen-Matthews *et al.*, 1990). Approximately 4.5% of the CD4 was located in coated pits, which accounted for only 1.2% of the total plasma membrane area in these cells.

In contrast to the results with the non-lymphocytic cell lines, the CD4 expressed naturally on lymphocytic cell lines was only internalized very slowly, and the internal pool of CD4 never reached more than 14% (Fig. 1, Table I). The difference between the lymphoid and non-lymphoid cells was not due to the nature of the CD4 construct used to transfect HeLa cells, since CD4 expressed from the same construct in the CD4-negative T-cell line RPMI was also poorly endocytosed (Table I). Overall, the rates of endocytosis of CD4 were about 10-fold lower on lymphocytic cell lines compared with non-lymphocytic cells.

The Endocytic Capacity of HeLa-CD4 and Lymphoid Cells

In order to examine whether the differences in CD4 endocytosis between HeLa-CD4 and lymphoid cells may be due to inherent differences in the endocytic capacity, we measured the rates of fluid phase endocytosis of horseradish peroxidase in HeLa-CD4, CEM and SupT1 cells (Fig. 2). This revealed that the lymphocytic cell lines internalized about half as much fluid per μg cell protein as the HeLa-CD4 cells. Assuming that the internal diameter of a coated vesicle is 100 nm and hence the internal volume 5.24×10^{-10} nl (Griffiths *et al.*, 1989), HeLa-CD4 cells internalized a volume equivalent to about 930 coated vesicles/min, while the lymphocytic cell lines only took up around 70 vesicles/cell/min. Estimates of the volumes of the cells by morphometry or analysis using a Coulter counter channelizer indicated that the HeLa-CD4 cells were much larger than the lymphocytic cells (the volumes of SupT1 and HeLa-CD4 cells were 500 and 3000 μm^3, respectively). Thus

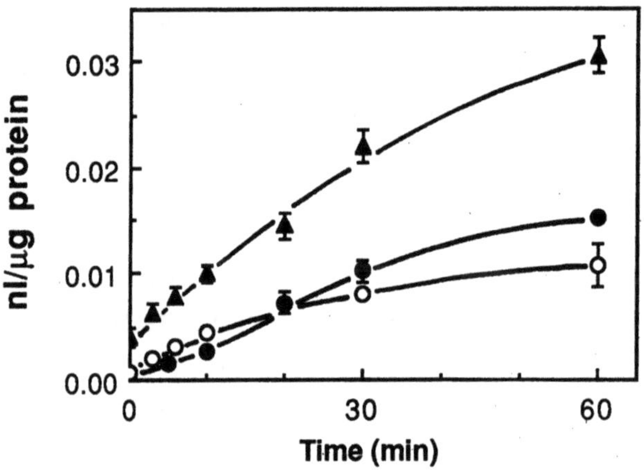

Figure 2: Fluid phase endocytosis of media containing horseradish peroxidase by HeLa-CD4 (▲), SupT1 (●) and CEM cells (○). Error bars indicate standard deviations for triplicate cell samples. The initial rates of fluid uptake measured over the first 10 min were 0.036, 0.035 and 0.46 fl/min/cell for the CEM, SupT1 and HeLa-CD4 cells, respectively.

the differences in fluid phase endocytosis per unit cell volume were at most two-fold greater in the HeLa-CD4 cells, and not sufficient to explain the much higher CD4 endocytosis.

To further define endocytic capacity in HeLa-CD4 and lymphoid cells, we also measured the proportion of the plasma membrane containing clathrin coats. This revealed that on SupT1 cells 5.2 μm^2 of the mean surface area of 326 μm^2 were coated (1.6%), while on HeLa-CD4 cells, 23.84 of 2020 μm^2 (1.2%) were coated. Since a 100 nm coated vesicle (above) would have a surface area of 0.0314 μm^2, these coated areas are sufficient to cover 165 and 760 coated vesicles in SupT1 and HeLa-CD4 cells, respectively. Based on the predicted $t_{1/2}$ of a coated pit at the cell surface of 1-2 min, these numbers can account for all of the fluid phase uptake measured. Further morphometric measurements showed that in SupT1 cells both the volume of the endosome compartment as a proportion of the volume of the cytoplasm (0.6%) and the surface areas of the endosome compartment compared to the plasma membrane (16%) were similar to values obtained for a variety of other cell lines (e.g. BHK cells, Griffiths *et al.*, 1989).

Together, these results indicate that the endocytic capacities of the lymphoid and non-lymphoid cells are not sufficiently different to account for the variations in CD4 uptake. Furthermore, endocytosis experiments with ^{125}I-transferrin demonstrated that the coated vesicle pathway in the lymphoid cells is functional. Thus the reduced internalization of CD4 on lymphoid cells appears to be due to exclusion of CD4 from coated pits.

The Role of the Cytoplasmic Domain of CD4 in Endocytosis

CD4 contains a 38 amino acid cytoplasmic domain. To determine the role of this domain in endocytosis, we studied both HeLa and lymphocytic cell lines transfected with cytoplasmic-deleted CD4 molecules (Table II). With both cell types, endocytosis of the truncated CD4 could readily be detected. Internalization occurred at rates of around 1% per minute, and at steady state about 20% of the antibody-labelled CD4 had been internalized. Given the density of coated pits (about 1.2-1.6%) and the size of the endosome compartment as a proportion of the plasma membrane (16-20%), these endocytosis kinetics suggest that the tailless CD4 molecules are internalized non-concentratively as part of the bulk membrane turnover. The similarity in the behaviour of tailless CD4 on lymphoid and non-lymphoid cells confirms that there are no major differences in the endocytic capacities of these cells.

The observation that the cytoplasmic deleted CD4 molecules are internalized in both transfected HeLa and lymphoid cells suggests that

Table II: Endocytosis of mutant CD4 molecules lacking the cytoplasmic domain

Cell line	No. of experiments	Rate of endocytosis (% per min)	% internal at equilibrium *
HeLa/CD4-cyt402	3	0.76 ± 0.20	17 ± 1.9
A2.01/CD4-cyt399	1	1.30	29
A2.01/CD4-cyt401	1	0.91	21.7

All endocytosis experiments used ^{125}I-labelled anti-CD4 monoclonal antibodies. The HeLa and A2.01 cell lines have been transfected with truncated CD4 molecules terminating at the indicated amino acid residues and thus lacking 31, 34 and 32 amino acids of the cytoplasmic domain, respectively. * To determine % internalized ligand at equilibrium, the level of acid-resistant activity at in cells held at 0-4°C has been subtracted.

sequences in the cytoplasmic domain are involved in mediating the different patterns of CD4 endocytosis. In HeLa-CD4 cells, the cytoplasmic domain allows a limited enrichment of CD4 in coated pits. In the lymphocytic cell lines, in contrast, the cytoplasmic domain is responsible for the exclusion of CD4 from coated pits. As CD4 is known to associate with a T-cell specific protein tyrosine kinase, p56*lck*, it is possible that this molecule may regulate CD4 endocytosis in T-cells. This hypothesis has been confirmed by our recent observation that transfection of the *lck* gene into NIH-3T3-CD4 cells specifically inhibited endocytosis of CD4 (manuscript in preparation) without affecting either fluid phase endocytosis or the uptake of ^{125}I-transferrin in the doubly transfected NIH-3T3-CD4/p56*lck* cells. The mechanism by which p56*lck* would interfere with CD4 endocytosis is not clear at present. Since p56*lck* is a relatively large molecule, it may sterically prevent access of the cytoplasmic domain of CD4 to the coated vesicle adaptor proteins. However, growth factor receptors such as the EGF, PDGF and insulin receptors, which contain tyrosine kinase domains as part of their cytoplasmic sequences, are efficiently endocytosed through coated pits after binding the relevant growth factors. Thus a large kinase cytoplasmic domain is not sufficient *per se* to prevent endocytosis. Instead, p56*lck* could be involved in anchoring the CD4 molecule in non-coated regions of the plasma membrane. Indeed, solubilization studies with a number of detergents have suggested that p56*lck* may be associated with the cytoskeleton (Louie *et al.*, 1988).

Since the interaction of p56*lck* with the cytoplasmic domain of CD4 is non-covalent, it may be reversible under certain circumstances, thus modulating the endocytic trafficking of CD4. Indeed, when T cells are treated with phorbol esters, p56*lck* and CD4 dissociate, allowing CD4 to be internalized and down-regulated from the cell surface (Hurley *et al.*, 1989). Thus, in lymphocytic cells, p56*lck* holds CD4 at the cell surface, which is its functional site both for interaction with MHC class II molecules and for signal transduction. During T cell stimulation, dissociation of p56*lck* would permit CD4 endocytosis. This may help in the dissociation of T cells from the MHC-II expressing antigen presenting cells, and make T cells refractory to further stimulation, while the p56*lck* molecules are released, perhaps to allow them to interact with specific substrates. The analysis of the mechanisms and regulation of

CD4 endocytosis will therefore help in understanding the role of CD4 and p56^{lck} in T cell activation.

Acknowledgements

We thank Dr Gareth Griffiths for the morphometry. This work was supported by grants to the Institute of Cancer Research from the Cancer Research Campaign and the Medical Research Council.

References

Griffiths G, Back R Marsh M (1989) A quantitative analysis of the endocytic pathway in baby hamster kidney cells. J. Cell Biol. 109: 2703-2720.
Hurley TR, Luo K, Sefron BM (1989) Activators of protein kinase C induce dissociation of CD4, but no CD8, from p56^{lck}. Science 245:407-409
Louie RR, King CS, MacAuley A, Marth JD, Perlmutter RM, Eckhart W, Cooper JA (1988) p56^{lck} protein-tyrosine kinase is cytoskeletal and does not bind to polyomavirus middle T antigen. J. Virol. 61:4673-4679
Pelchen-Matthews A, Armes JE, Marsh M (1989) Internalization and recycling of CD4 transfected into HeLa and NIH3T3 cells. EMBO J. 8:3641-3649.
Pelchen-Matthews A, Armes JE, Griffiths G, Marsh M (1990) Differential endocytosis of CD4 in lymphocytic and non-lymphocytic cells. Submitted for publication.
Robey E, and Axel R (1990) CD4: Collaborator in immune recognition and HIV infection. Cell. 60:697-700.
Turner JM, Brodsky MH, Irving BA, Levin SD, Perlmutter RM, Littman DR (1990) Interaction of the unique N-terminal region of tyrosine kinase p56^{lck} with cytoplasmic domains of CD4 and CD8 is mediated by cysteine motifs. Cell. 60:755-765.

ROLE OF TRANSEPITHELIAL TRANSPORT IN TRIGGERING A MUCOSAL IMMUNE RESPONSE AND IN DELIVERY OF MUCOSAL ANTIBODIES INTO SECRETIONS

J.P. Kraehenbuhl+, Pierre Michetti+•, Robert Hirt+, Christine Perregaux+
John Mekalanos°·and Marian Neutra•.

+Swiss Institute for Experimental Cancer Research and Institute of Biochemistry, University of Lausanne, CH-1066 Epalinges, Switzerland, and •GI Cell Biology Laboratory, Children's Hospital and °Department of Microbiology and Molecular Genetics, Harvard Medical School, MA, USA.

Many pathogens, including viruses, bacteria and parasites as well as toxins and allergens, gain entry into the organism by crossing the epithelia of the digestive, respiratory or genital tracts. Mucosal surfaces are protected against environmental pathogens by non immune and immune mechanisms (Mayrhofer, 1984). A distinct "mucosal" immune system plays a major role in protection, and an important component of this protection is production of secretory antibodies of the IgA isotype (sIgA) (Russel and Mestecky, 1988).

Uptake of luminal antigens is required to trigger the B-cell-mediated mucosal immune response, and export of IgA antibody molecules is necessary to deliver these effector molecules to their site of action on the mucosal surface (Solari and Kraehenbuhl, 1987). Thus two transepithelial transport processes, operating in opposite directions, are required for immune protection of mucosal surfaces.

Proteins and microorganisms that can adhere to epithelial mucosal surfaces tend to be effective immunogens (Mayrhofer, 1984). Binding and uptake of macromolecules and microorganisms occurs at specific "sampling sites" on mucosal surfaces (Owen et al., 1986; Wolf et al., 1983; Neutra et al., 1987). Such sampling sites consists of epithelial cells - the M cells, specialized in transcytosis of antigens. M-cells are found in the follicle epithelium associated with the mucosal lymphoid follicles that are present under mucosal surfaces in the intestinal tract (rectum, colon, appendix, Peyer's patches), the oral cavity (tonsils), the respiratory tract (bronchi) and perhaps in the genital tract (endocervix). Some pathogens, including poliovirus (Sicinski et al., 1990), reovirus (Wolf et al., 1983), *V. cholerae* (Owen et al., 1986), and *Shigella* have exploited the transcytotic machinery of M-cells to invade and infect the host.

Once the antigens have crossed the epithelial barrier through the M-cells at these sites, they elicit an immune response in the mucosal lymphoid follicles. The primed B-cells, which express IgA receptors at their surfaces, eventually leave the follicles, are drained by the lymphatic vessels, enter the circulation and home to local and distant mucosal or glandular sites (McDermott and Bienenstock, 1979), where they differenciate into IgA producing plasma cells. The polymeric IgA antibodies are transported across glandular and mucosal epithelia via the polymeric Ig receptor. The receptor is cleaved

NATO ASI Series, Vol. H 62
Endocytosis
Edited by P. J. Courtoy

either during transport or at the luminal surface, releasing into secretions the ectoplasmic receptor fragment known as secretory component bound to IgA. The complex cellular routing of the receptor has been analyzed in transfected MDCK cells (Mostov and Deitcher, 1986) and in mammary cells (Schaerer et al., 1990). The sorting and addressing information appears restricted to the cytoplasmic tail of the receptor and phosphorylation modulates membrane trafficking of the receptor (Casanova et al., 1990).

To clarify the role of sIgA in mucosal protection we have developed a novel approach which allows for efficient production of polymeric monoclonal IgA antibodies from hybridoma cells (Weltzin et al., 1989) and their stabilization by recombinant secretory component (SC) (Michetti et al., 1989 and ms in preparation). Following mucosal immunization of Balb-c mice with various enteropathogens or protein antigens, lymphoblasts are recovered from Peyer's patches and are fused with myeloma cells. The resulting hybridomas are screened for microbial specificity and IgA isotype. Stabilization with SC is achieved *in vitro*, by combining IgA monoclonal antibodies with SC produced in transfected myeloma cells. Alternatively, we have produced sIgA in which the SC is covalently linked to IgA by a novel system in which IgA-producing hybridoma cells are co-cultured with MDCK cells that express the poly-Ig receptor. MDCK cells were transfected with polymeric Ig receptor cDNA inserted into an inducible expression vector and grown on Transwell filters. When stimulated with dexamethasone, these cells synthesize high levels of the poly Ig-receptor and secrete its ectodomain (SC) vectorially into the apical medium. When cultivated on top of a collagen layer containing hybridoma cells and then hormonally-stimulated, sIgA is recovered from the apical medium. sIgA produced by such a system is resistant to proteolytic degradation, and survives a 4 hour incubation with trypsin at pH 8.0, or pepsin at pH 2.0.

To test whether monoclonal sIgA alone can protect the host against viral or bacterial infection, we have used invasive viral pathogens, and enteric bacterial pathogens that colonize the mucosal surface. We have generated monoclonal IgA antibodies after intestinal immunization, and tested them in various protection assays. Our investigations have been concentrated on three pathogens: reovirus type I, mouse mammary tumor virus (MMTV) and *Vibrio cholerae*. In this report we shall restrict ourselves to the two latter organisms.

MMTV, a retrovirus that causes mammary tumors, is shed into the milk of infected mothers, crosses the intestinal barrier via M cells of suckling neonates, infects mucosal T cells (Tsubara et al., 1988), and spreads to the mammary glands of female offspring (Hainaut et al., 1985). *Vibrio cholerae* causes diarrhea by colonizing the mucosal surface and producing an enterotoxin which irreversably activates adenylate cyclase in the epithelial cells and results in opening of apical chlorid channels. (Miller et al., 1989).

One of the anti-MMTV IgA monoclonals recognized a continuous epitope associated with the N-terminus of the major envelope glycoprotein, gp 52. Several anti-cholera IgAs reacted with a strain-specific carbohydrate epitope on the lipopolysacchoride (LPS) exposed on the bacterial surface.

To test the protective effects of anti-cholera IgA antibodies, an *in vivo* hybridoma tumor system was developed that results in continuous delivery of monoclonal sIgA antibodies into intestinal secretions of mice via the normal transcytotic pathmay (Winner et al., 1989). This "backpack" system has been used to demonstrate that secretion of sIgA antibodies directed against a single LPS carbohydrate epitope was able to protect suckling mice against a lethal oral challenge with virulent *V. cholerae*. Oral administration of these monoclonal IgAs also protected neonates against a lethal oral challenge with virulent organisms. To test protection against MMTV, we had first to dermine the dose of virus required for infection and the duration of vulnerability in susceptible, healthy Balb/c female neonates. We found that a single dose (20μl) of milk from infected mothers (C3H) was consistently able to induce mammary tumors in the female offspring (Balc/c) at 6-8 months of age. Protection experiments in which female neonates were given anti-gp52 IgA antibodies orally 2 h prior to MMTV challenge, are underway.

Conclusions

1. Hybridomas producing monoclonal dimeric IgA directed against viral and bacterial surface molecules can be generated from Peyer's patch lymphoblasts.

2. Monoclonal IgA antibodies can be combined with recombinant secretory component, to produce engineered sIgA for passive protection of mucosal surfaces.

3. Monoclonal dimeric IgA produced by hybridoma "backpack" tumors *in vivo* is efficiently delivered into intestinal secretions as sIgA. This method allows identification of protective epitopes in the mucosal system.

4. Monoclonal IgA directed against a single microbial surface epitope can protect against enteric viral or bacterial challenge.

Acknowledgements.

The authors thank Ms.Marie-Thérèse Bayard for expert preparation of the manuscript. This work was supported by Swiss National Science Fondation grant 31 26404 89 (JPK), by Swiss League against Cancer grant FOR 37389.2 (JPK), and by NIH Research grant HD 17557.

References

Casanova, J.E., Breitfeld, P.P., Ross, S.A. and Mostov, K.E. (1990) Phosphorylation of the polymeric immunoglobulin receptor required for its efficient transcytosis. Science 248:742-745.

Hainaut, P., Vaira, D., Francois, C., Calberg Bacq, C.M. and Osterrieth, P.M. (1985) Natural infection of Swiss mice with mouse mammary tumor virus (MMTV): viral expression in milk and transmission of infection. Arch.Virol. 83:195-206 .

Mayrhofer, G. (1984) Physiology of the intestinal immune system.In "Local Immune Responses of the Gut" Eds Newby, T.N., and C.R. Stokes, CCR Press, Boca Raton, Florida pp 1-98.

McDermott, M.R. and Bienenstock, J. (1979) Evidence for a common mucosal immunologic system I. Migration of B immunoblasts into intestinal, respiratory, and genital tissues. J.Immunol. **122**:1892-1898.

Michetti, P., Perregaux, C., Weltzin, R.A., Fasel, N., Neutra, M.R. and Kraehenbuhl, J.P. (1989) The Use of Secretory IgA Monoclonal Antibodies to Test Protection Against Retrovirally-Induced Mammary Tumors. J.Cell Biol. **109**:295a.

Miller, J.F., Mekalanos, J.J. and Falkow, S. (1989) Coordinate regulation and sensory transduction in the control of bacterial virulence. Science **243**:916-922.

Mostov, K.E. and Deitcher, D.L. (1986) Polymeric immunoglobulin receptor expressed in MDCK cells transcytoses IgA. Cell **46**:613-621.

Neutra, M.R., Phillips, T.L., Mayer, E.L. and Fishkind, D.J. (1987) Transport of Membrane-bound Macromolecules by M cells in Follicle-associated Epithelium of Rabbit Peyer's Patch. Cell Tissue Res. **247**:537-546.

Owen, R.L., Pierce, N.F., Apple, R.T. and Cray, W.C.Jr. (1986) M cell transport of Vibrio cholerae from the intestinal lumen into Peyer's patches: a mechanism for antigen sampling and for microbial transepithelial migration. J.Infect.Dis. **153**:1108-1118.

Russell, M.W. and Mestecky, J. (1988) Induction of the mucosal immune response. Rev.Infect.Dis. **10**:440-446.

Schaerer, E., Verrey, F., Racine, L., Tallichet, C., Rheinhardt, M. and Kraehenbuhl, J.P. (1990) Polarized transport of the polymeric immunoglobulin receptor in transfected rabbit mammary epithelial cells. J.Cell Biol. **110**:987-998.

Sicinski, P., Rowinski, J., Warchol, J.B., et al. (1990) Poliovirus type 1 enters the human host through intestinal M cells. Gastroenterology. **98**:56-58.

Solari, R. and Kraehenbuhl, J.P. (1987) Receptor-mediated transepithelial transport of polymeric immunoglobulins. In: The Mammary Gland. Development, Regulation and Function, edited by Neville, M.C. and Daniel, C.W. New-York and London: Plenum Press p. 269-298.

Tsubura, A., Inaba, M., Imai, S., et al. (1988) Intervention of T-cells in transportation of mouse mammary tumor virus (milk factor) to mammary gland cells in vivo. Cancer Res. **48**:6555-6559.

Weltzin, R.A., Lucia Jandris, P., Michetti, P., Fields, B.N., Kraehenbuhl, J.P. and Neutra, M.R. (1989) Binding and transepithelial transport of immunoglobulins by intestinal M cells: demonstration using monoclonal IgA antibodies against enteric viral proteins. J.Cell Biol. **108**:1673-1685.

Winner, L.S.III, Weltzin, R.A., Mekalanos, J.J., Kraehenbuhl, J.P. and Neutra, M.R. (1989) A Novel in Vivo System to Assess Secretory IgA mediated Protection against Enteric Pathogens : Transepithelial Transport of Monoclonal anti-Vibro Cholera IgA from Hybridoma Tumors. J.Cell Biol. **109**:295a.

Wolf, J.L., Kauffman, R.S., Finberg, R., Dambrauskas, R. and Fields, B.N. (1983) Determinants of reovirus interaction with the intestinal M cells and absorptive cells of murine intestine. Gastroenterology **85**:291-300.

2.2. PHAGOCYTOSIS AND INTRACELLULAR PATHOGENS

(Chairperson : Antoinette RYTER)

When microbes penetrate our organism, they are generally rapidly destroyed by phagocytes. However, several pathogens have developed strategies to escape phagocytic killing. This is overviewed in the introduction by **Antoinette Ryter**. Some patogens avoid ingestion by macrophages or kill them by excreting toxins (see session on toxins). Others actually multiply inside the cells, because they are able to avoid or resist lysosomal toxicity. Three stratagems have been described into more detail in the subsequent chapters.

The first stratagem is the inhibition of fusion with lysosomes, as examplified by **Chantal de Chastellier**. She reported that, compared with the phagocytosis of non-pathogenic bacteria or the non-degradable latex beads, macrophage infection by *Mycobacterium avium* dramatically slows down the exchange of content and membrane along the endocytic apparatus. In addition, if fusion with lysosomes still occurs, *M. avium* is protected by its thick hydrophobic capsule.

The second stratagem is a direct resistance to lysosomal hydrolysis, as shown by **Jean-Claude Antoine** for *Leishmania amazonensis*. This parasite does not inhibit lysosomal fusion but can multiply in the phagolysosome without any visible protective layer. It does actually induce the formation of a giant acidic vacuole, considered as a modified lysosome. Interestingly, the limiting membrane of this parasitophorous vacuole is extremely rich in MHC-II molecules (Ia), mostly in the region of *Leishmania* binding sites.

The third stratagem is the destruction of the phagocytic membrane before fusion with lysosomes, as illustrated by **Philippe Sansonetti** for *Shigella flexneri*. After apical phagocytosis by enterocytes, the phagosome membrane is lyzed, the bacteria grows in the cytosol, and eventually spreads to adjacent cells. The latter movement results from the formation of a long F-actin tail, which seems to propel the bacterium in a long cell protrusion. The genetic analysis of the virulence plasmid is making great progress.

NATO ASI Series, Vol. H 62
Endocytosis
Edited by P. J. Courtoy

OVERVIEW ON THE RELATIONS OF PATHOGENIC MICROORGANISMS WITH THEIR HOST CELL

Antoinette Ryter
Département de Biologie Moléculaire,
Institut Pasteur
25, rue du Dr. Roux
75724 Paris Cedex 15, France

When microbes penetrate our organism, they are generally rapidly eliminated by polymorphonuclear leukocytes or macrophages. However, pathogenic bacteria and protozoars have developed different strategies to escape the microbicidal properties of these phagocytes. Some avoid ingestion by macrophages or kill them by excreting toxins, whereas others are ingested by the macrophage, in which they are able to survive and multiply despite the microbicidal properties of this phagocyte.

Several stratagems used by intracellular pathogens have been discovered during the last 15 years. Intracellular pathogens confront first with the production of toxic oxydative products triggered by their adhesion onto the macrophage surface. Certain pathogens seem to prevent the triggering of the oxidative burst as observed with Toxoplasma (Wilson et al. 1980). The second microbicidal step corresponds to the fusion of lysosomes with the phagocytic vacuole leading to the release of acid hydrolases and cationic proteins into this vacuole. Three different stratagems allowing to escape from lysosomal toxicity have been found. The first one consists in inhibiting phagosome-lysosome fusions. Such an inhibition was observed with *Mycobacterium tuberculosis, M. avium, M. leprae, Nocardia, Legionella pneumonphila* and *Toxoplasma* (Armstrong et al.

NATO ASI Series, Vol. H 62
Endocytosis
Edited by P. J. Courtoy

1875, Fréhel et al. 1986a, Fréhel and Rastogi 1987, HorWitz 1983b, Jones and Hirsch 1972). The reasons for such an inhibition have not yet been elucidated. In the case of *Mycobacteria* which are surrounded by a layer of complex peptidoglycolipids, it is possible that the insertion of some of these compounds into the phagosomal membrane could modifiy its fusion property. It is also likely that fusion inhibition is induced by products excreted by the pathogen (see Ryter and de Chastellier, 1983) because in many cases, the inhibition was not observed when bacteria had been killed prior to ingestion .

The second stratagem to avoid the lysosomal toxicity resides in the resistance of the pathogen to lysosomal hydrolases. This resistance could be related to the presence of the thick peptidoglycolipidc capsule of *M. avium* and *M. leprae* (Fréhel et al., 1986 b). This resistance is especially efficient in the case of *M. leprae*, since even heat killed bacteria are not degraded after several months of incubation. However, the mode of resistance of intracellular pathogens, which do not possess any visible protective layer is not yet understood.

One of the interesting problems related to the survival properties of intraphagosomal pathogens concerns the communications of pathogens with the other compartments of the endocytic pathway (endosomes or lysosomes) and also the extracellular medium. Two different behaviours will be described in this session. Chantal de Chastellier will show that pathogenic mycobacterial species enclosed in their phagocytic vacuoles do not communicate very actively with the other cellular compartments and the extracellular medium. A quite different situation will be described by Jean-Claude Antoine for *Leishmania*, the phagocytic vacuole of which seems to correspond after some hours to a giant and unique lysosome implicated in the endocytic activity of the macrophage.

The third stratagem which avoids lysosomal toxicity consists in the rapid destruction of the phagocytic membrane. This property, observed for *Rickettsia tzutsugamushi* (Rikihisa and Ito, 1979), *Shigella* (Sansonetti et al. 1986) and *Listeria monocytogenes*, Mounier et al .,1990, Tilney and Portnoy ,1989)was shown to be mediated by phospholipase-A or by others kinds of hemolysins. The destruction of the phagocytic membrane allows bacteria to multiply freely in the cytoplasm of the host cell. The rate at which the phagosomal membrane destruction will occur is a key factor to survival, especially in macrophages, which are very rich in lysosomes. *Shigella* , *L. monocytogenes Rickettsia tsutsugamushi* possess the ability to induce their own ingestion by non-phagocytic cells, such as enterocytes or fibroblasts. Their phagocytosis requires the active participation of the bacterium, since killed bacteria are not phagocytosed. The genetic analysis of *Shigella flexneri* suggests that the bacterial hemolysin, responsible for the destruction of the phagocytic vacuole is also implicated in the induction of this process. (P. Sansonetti, personnal communication)

It was recently shown that after destruction of the phagocytic vacuole, *Shigella* and *Listeria monocytogenes* move inside enterocytes or macrophages (Sansonetti et al 1986, Mounier et al. 1990, Tilney and Portnoi 1989). Philippe Sansonetti will show in this session how these bacteria mobilize the host cell actin to move and propagate from cell to cell and so to reach the deep cellular layers of the intestin and probably the blood stream.

The adhesion of *Salmonella typhimurium* onto intestinal cells induces the disorganization of microvilli and the subsequent formation of the phagocytic cup, but contrary to *Shigella*, *Salmonella* penetration does not seem to result from the action of an hemolysin. This bacterium remains enclosed in the phagocytic vacuole and is transported by transcytosis from the apical face to the baso-lateral face of the intestinal cell (Popiel and Turnbull , 1985). The fusion of phagosomes with the baso-lateral membrane leads to the release of bacteria in the extra-cellular medium . Their subsequent phagocytosis by macrophages, in which they can survive, probably allows them to reach the liver and spleen.

Finally, phagocytosis of *Legionella pneumophila* occurs by a peculiar mechanism corresponding to the formation of a unique pseudopod which coils around the bacterium (Horwitz,1984). Bacteria are finally found in a vacuole, the membrane of which is covered with ribosomes (Horwitz, 1983), as if they had penetrated the rough endoplasmic reticulum. A similar location was also found with *Rickettsia rickettsii* (Silverman and Wisseman 1979)

In conclusion, the different behaviours of intracellular pathogens and their interaction with the host cell machinery represent an interesting field of cell biology and the association of bacterial genetics, molecular biology and electron microscopy represents a powerful tool for the study of these complex survival processes .

References

Armstrong J.A. & d'Arcy Hart P.D. (1975) Phagosome-lysosome interactions in cultured macrophages infected with virulent tubercle bacilli. Reversal of the usual non-fusion pattern and observations on bacterial survival. J. Exp. Med. 142, 1-16

Fréhel C, de Chastellier C., Lang T. & Rastogi N. (1986) Evidence for inhibition of fusion of lysosomal and prelysosomal compartments with phagosomes in macrophages infected with pathogenic *Mycobacterium avium.* Infect. Immun. 52, 252-262.

Fréhel C., & Rastogi N. (1987) *Mycobacterium leprae* surface components intervene in the early phagosome-lysosome fusion inhibition event. Infect. Immun. 55, 2916-2921.

Fréhel C., Ryter A., Rastogi N., & David H. (1986 b) The electron transparent zone in phagocytized *Mycobacterium avium* and other *Mycobacteria* : Formation, persistence and role in bacterial survival Ann. Inst. Pasteur. 137 B, 238-257.

Horwitz M. (1983a) Formation of a novel phagosome by the Legioniaires' disease bacterium (*Legionella pneumophila*) in humanmonocytes. J. Exp. Med. 158, 1319_1331

Horwitz M. (1983 b) The Legiuonaires' bacterium *(Legionella pneumonphila*) inhibits phagosome-lysosome fusion in human monocytes. J. Exp. Med. 158, 2108-2126.

Horwitz M (1984) Phagocytosis of Legionaires' disease bacterium (*Legionella pneumophila*) occurs by a novel mechanism: engulfment within a pseudopod coil. Cell. 36, 27-33.

Jones T.C. & Hirsch J.G. (1872) The interaction between Toxoplasma gondii and mammalian cells. II The absence of lysosomal fusion with phagocytic vacuoles containing living parasites. J. Exp. Med. 136, 1173-1180

Mounier J., Ryter A., Coquis-Rodon M. & Sansonetti P.J. (1990) Intracellular and cell-to-cell spread of *Listeria monocytogenes* involves interaction with F-actin in the enterocytelike cell line Caco-2. Infect. Immun. 58, 1048-1058.

Popiel I. & Turnbull P. (1985) Passage of *Salmonella enteriditis* and *Salmonella thompson* through chick ileovecal mucosa. Infect. Immun. 47, 786-792.

Rikihisa Y., & Ito S. (1979) Intracellular localization of Rickettsia tsutsugamushi in polymorphonuclear leukocytes. J. Exp. Med. 150, 703-708.

Ryter A., & de Chastellier C. (1983) Phagocyte-pathogenic microbe interactions. Intern. Rev. Cytol. **85**, 287-327.

Sansonetti P.J., Ryter A., Clerc P.; Maurelli A.T. & Mounier J. (1986) Multiplication of *Shigella flexneri* within HeLa cells: lysis of the phagocytic vacuole and plasmid mediated contact hemolysis. Infect. Immun. 51, 461-469.

Silverman D.J.,& Wisseman C.L. (1979) *In vitro* studies of rickettsia-host cell interactions: ultrastructural changes induced by *Rickettsia rickettsii* infection of chicken embryo fibroblasts. Infect. Immun. 26, 714-727.

Tilney L.G. & Portnoy D.A. (1989) Actin filaments and the growth, movement, and spread of the intracellular bacterial parasite, *Listeria moncytogenes* . J. Cell Biol. 109, 1597-1608.

Wilson C.B., van Tsai V. & Remington J.S. (1980). Failure to trigger the oxidative metabolic burst by normal macrophages. Possible mechanism for survival of intracellular pathogens. J. Exp. Med. **151**, 328-334.

INTRACELLULAR GROWTH OF *MYCOBACTERIUM AVIUM* IN MACROPHAGES : CONSEQUENCES ON MEMBRANE TRAFFIC AND EXCHANGE OF CONTENTS BETWEEN ENDOSOMES, LYSOSOMES AND PHAGOSOMES

Chantal de Chastellier (1), Claude Fréhel (1) and Thierry Lang (2)
(1) Faculté de Médecine Necker-Enfants Malades - Laboratoire de Bactériologie
(2) Institut Pasteur
75730 Paris Cédex 15
France

INTRODUCTION

Macrophages represent major effector cells in the defense against a wide variety of microorganisms. However, several pathogens, such as mycobacteria (d'Arcy Hart 1982), *Legionella pneumophila* (Horwitz 1983), *Toxoplasma gondii* (Jones and Hirsch 1972), resist to the microbicidal activity of macrophages by blocking phagosome-lysosome fusions. Among the non tuberculous mycobacteria, *M. avium* emerges as a major pathogen in AIDS patients (Young *et al.* 1986). This organism survives within in vitro grown bone marrow-derived macrophages, remaining confined to the phagosome compartment in which it multiplies (Ryter *et al.* 1984). It resists to the hydrolytic activity of macrophages by inhibiting phagosome-lysosome (P-L) fusions (Fréhel *et al.* 1986a) and thanks to its capsule that seems to reduce the diffusion of lysosomal enzymes (Fréhel *et al.* 1986b).

The present work was undertaken to explore other modifications of the endocytotic process related to *M. avium* survival. Exchanges of membrane and contents between the different endomembrane compartments (endosomes, lysosomes) and phagosomes were therefore determined in *M. avium* -infected macrophages. The same experiments were carried out after phagocytosis of indigestive latex beads to discriminate modifications due to the non digestibility of phagocytic particles from those related to properties specific to pathogenic *M. avium* . Results are compared to those obtained in the same cells after phagocytosis of a rapidly degraded bacterium , *B. subtilis* (Lang *et al.* 1988).

NATO ASI Series, Vol. H 62
Endocytosis
Edited by P. J. Courtoy

RESULTS AND DISCUSSION

A) Transfer of contents from endosomes and lysosomes to *M. avium*-containing phagosomes is reduced.

Ten-day-old bone marrow-derived mouse macrophages, cultured as described before (Fréhel *et al.* 1986a), were infected for 4 hours with *M. avium* TMC 724 (multiplicity of infection of 1-5 bacteria per macrophage). This strain, highly virulent for mice, was obtained from the Trudeau Mycobacterial Culture Collection, kindly provided by Dr Franck Collins. Cells were thoroughly washed and refed with fresh medium. Immediately after infection (0 min) and 1 or 15 days later, cells were stained for acid phosphatase (Fréhel *et al.* 1986a) to assess transfer of contents from lysosomes. In parallel experiments, infected cells were incubated for 60 min in presence of horseradish peroxidase (HRP). This pinocytotic marker, uptaken via mannose receptor-mediated endocytosis (Lang and de Chastellier 1985), served to assess transfer of contents from all endomembrane compartments to phagosomes. The same studies were carried out with macrophages that had phagocytized indigestive latex beads (25 particles per cell) for 45 min.

We have checked that *M. avium* TMC 724 survives and multiplies inside macrophages grown in culture, as observed before for another strain of *M. avium* (Fréhel *et al.* 1986a).

Table I : Percentages of phagosomes that showed fusion with AcPase-positive (A) or HRP-positive (B) vesicles at different times after phagocytosis of bacteria or latex beads.

Time / Particle	A : AcPase-positive phagosomes			B : HRP-positive phagosomes			
	0 min	day 1	day 14	0 min	day 1	day 14	day 28
M. avium	25	55	40	ND[a]	83	69	63
Latex beads	71	82	65	78	82	ND[a]	ND[a]
B. subtilis [b]	84	-	-	98	-	-	-

[a] ND, not determined ;[b] Lang *et al.* 1988

Bacteria remained confined to the phagosome compartment and more than 99 % of the bacilli encountered under the electron microscope were morphologically intact throughout infection.

The percentage of AcPase-positive *M. avium* -containing phagosomes (Table IA) was particularly low immediately after infection. It increased to 55 % on day 1 and then decreased to 40 % in the following days. Bacteria enclosed in AcPase-positive phagosomes showed no signs of degradation (Fig. 1a). The amount of AcPase-positive *M. avium* -containing phagosomes was 30 to 60 % lower than their latex counterparts, depending upon the time of reincubation. It is important to note that *B. subtilis* -containing macrophages displayed the highest percentage of AcPase-positive phagosomes (84 %) (Fréhel *et al.* 1986a).

Intramacrophagic growth of *M. avium* did not impair subsequent uptake of HRP. At least 95 % of the macrophages displayed the same HRP-positive compartments as uninfected cells (de Chastellier *et al.* 1987). At the early time points (0 min, day 1) cells displayed the same percentage of HRP-positive phagosomes whether they contained latex particles or *M. avium* . This percentage was, however, 15 % lower than that encountered in cells that had phagocytized rapidly degraded *B. subtilis* (Lang *et al.* 1988). During the course of infection, the percentage of HRP-positive *M. avium* phagosomes decreased as bacteria multiplied. Fig. 1b shows the appearance of cells that had pinocytosed HRP for 1 hour on day 28 after infection with *M. avium* .

These results provide evidence that the important reduction in the transfer of contents from acid phosphatase positive vesicles to *M. avium* phagosomes is mostly due to the expression of properties specific to this organism. However, the state of indigestibility of phagocytic particles also takes part in the reduction of this transfer. The transfer of contents from HRP-positive vesicles to *M. avium* phagosomes is also reduced. Experiments in course will allow to determine the respective roles of pathogenic properties and the state of non degradability of this organism.

B) *M. avium* modifies endocytic membrane traffic with respect to phagosomes.

Endocytic membrane traffic was studied in macrophages containing phagosomes either with *M. avium* strain ATCC15769 (kindly provided by Dr Hugo David, Institut Pasteur) that had multiplied within host cells for 15 days, or with latex beads.

Cell surface glycoproteins were labelled with (^{3}H)-galactose which provided a covalent membrane marker (Thilo 1983). The redistribution of marker between the plasma membrane and the different endomembrane compartments, including phagosomes, was analyzed by autoradiography during subsequent uptake of HRP. The composition of endocytic

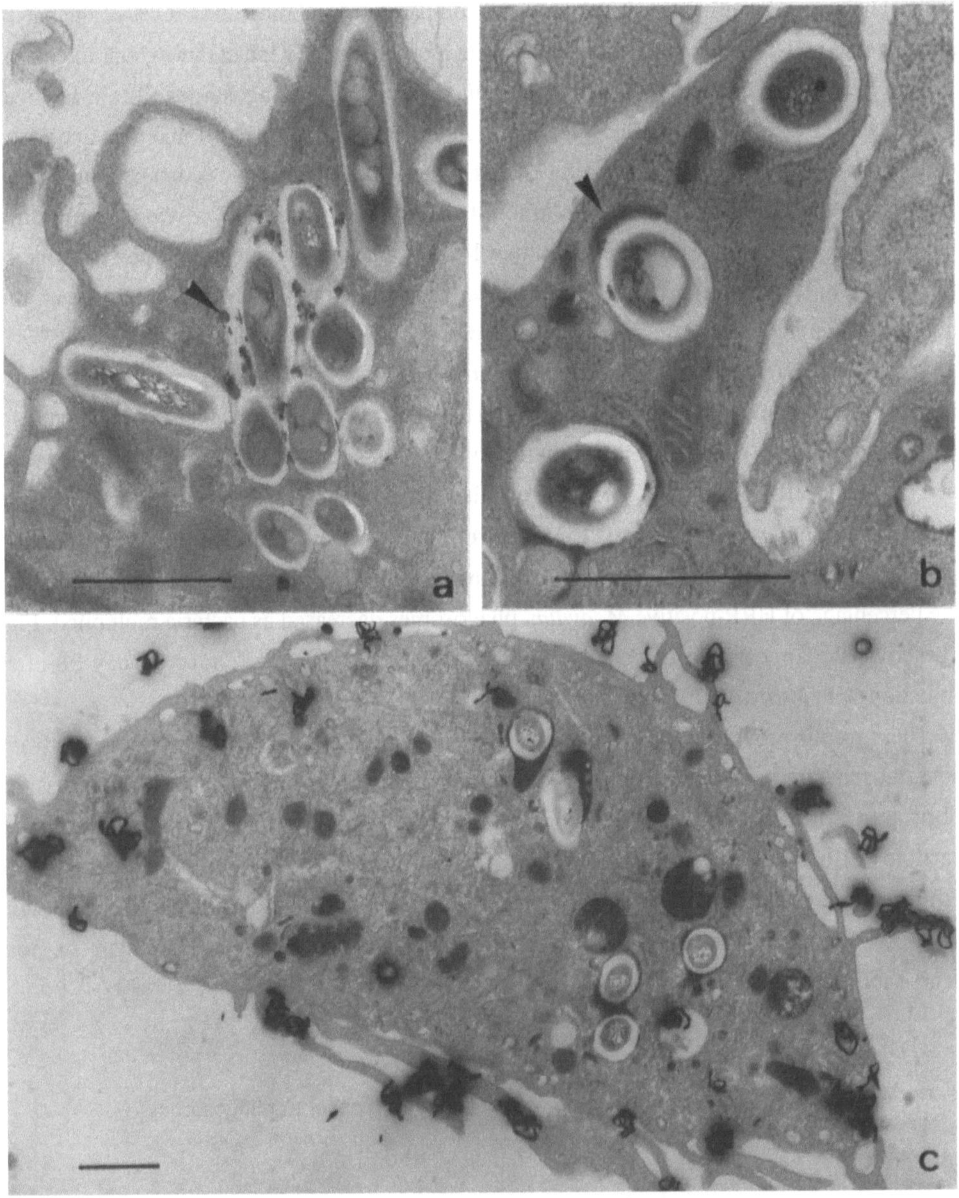

Fig. 1. Thin sections of macrophages infected with *M. avium* . Fourteen days after infection, cells were either stained for AcPase (a), exposed for 1hr to HRP (b), or labelled with [^{3}H] galactose and observed after exposure to HRP for 2hr (c). Bar : 1 μm

membranes in terms of (^{3}H)-gal-labelled plasma membrane constituents was determined by measuring the membrane grain density (ratio of autoradiographic grain number to membrane area measured morphometrically) of each compartment.

After PM constituents have been labelled at the cell surface, subsequent membrane internalization and recycling leads to a redistribution of label between the cell surface and intracellular membranes (Fig. 1c). An indication of the relative composition (in terms of labelled cell-surface-derived membrane constituents) of the endocytic membranes is given by the steady state value observed for the grain density as explained before (de Chastellier *et al.* 1987).

We had shown previously that after phagocytosis of *B. subtilis* and during a further 60 min of subsequent pinocytosis the cell surface-derived membrane of phagosomes was mostly replaced by membrane of lysosomal origin (Lang *et al.* 1988). We had proposed that this depended on the process of degradation of phagosome contents.

The present work is in accord with this proposal since, at steady state, latex-containing phagosomes displayed a grain density very similar to that of endosomes, distinct from the grain density of lysosomes (Table II).

Concerning the *M. avium* -phagosomes, exchange of label (membrane) even with endosomes is quite reduced since the grain density represented less than 10 % that of the cell surface even after 2 hours of pinocytosis (Table II).

Table II : Relative steady state composition, in terms of labelled cell surface-derived membrane constituents, of the different endomembrane compartments in macrophages after phagocytosis of different particles.

Membrane / Phagocytic particle	Plasma membrane	Endosomes	Lysosomes	Phagosomes
None	100	50	30	-
M. avium	100	50	20	8
Latex beads	100	45	30	40

Our actual hypothesis is that the phagosome membrane is modified in such a way that it can no longer exchange membrane with the plasma membrane via the endosome compartment, nor can it exchange membrane with the lysosome compartment. Whether this modification occurs during or immediately after entry of this bacterium in macrophages or is gradually acquired in the course of bacterial multiplication is under examination.

In conclusion, *M. avium* modifies the endocytic process of macrophages by reducing the transfer of contents between endomembrane compartments and phagosomes and by strongly inhibiting the exchanges of membrane with the phagosome compartments. These modifications, which are the expression of the pathogenicity of this organism, could prevent the transfer of certain drugs that could no longer exert their bactericidal effects. This could in part explain the high resistance of *M. avium* to a variety of antimicrobial drugs.

LITERATURE CITED

de Chastellier C, Lang T, Ryter A, Thilo L (1987) Exchange kinetics and composition of endocytic membranes in terms of plasma membrane constituents : a morphometric study in macrophages. Eur J Cell Biol 44 : 112-123

Fréhel C, de Chastellier C, Lang T, Rastogi N (1986a) Evidence for inhibition of fusion of lysosomal and prelysosomal compartments with phagosomes in macrophages infected with pathogenic *Mycobacterium avium*. Infect Immun 52 : 252-262

Fréhel C, Ryter A, Rastogi N, David H (1986b) The electron-transparent zone in phagocytized *Mycobacterium avium* and other mycobacteria : formation, persistence and role in bacterial survival. Ann Inst Pasteur/Microbiol 137b : 239-257

Hart d'Arcy P (1982) Lysosome fusion responses of macrophages to infection : behaviour and significance. *In* Karnovsky ML, Bolis L (eds) Phagocytosis : past and future. Academic Press New York

Horwitz MA (1983) The Legionnaire's disease bacterium (Legionella pneumophila) inhibits phagosome-lysosome fusion in human monocytes. J Exp Med 158 : 2108-2126

Jones TC and Hirsch JG (1972) The interaction between Toxoplasma gondii and mammalian cells. II The absence of lysosomal fusion with phagocytic vacuoles containing living parasites. J Exp Med 136 : 1173-1194

Lang T and de Chastellier C (1985) Fluid phase and mannose receptor-mediated uptake of horseradish peroxidase in mouse bone marrow-derived macrophages. Biochemical and ultrastructural study. Biol Cell 53 : 149-154

Lang T, de Chastellier C, Ryter A, Thilo L (1988) Endocytic membrane traffic with respect to phagosomes in macrophages infected with non-pathogenic bacteria : phagosomal membrane acquires the same composition as lysosomal membrane. Eur J Cell Biol 46 : 39-50

Ryter A, Fréhel C, Rastogi N, David HL (1984) Macrophage interaction with mycobacteria including *M. leprae*. Acta Leprologica 2 : 211-226

Thilo L (1983) Labeling of plasma membrane glycoconjugates by terminal glycosylation (galactosyl transferase and glycosidase). Methods Enzymol 98 : 415-421

Young LS, Inderlied CB, Berlin OG, Gottlieb MS (1986) Mycobacterium infections in AIDS patients with an emphasis on the *Mycobacterium avium* complex. Rev Infect Dis 8 : 1024-1029

MORPHOLOGY, COMPOSITION AND FUNCTIONS OF THE ENDOCYTIC COMPARTMENT HOUSING *Leishmania amazonensis* IN INFECTED MACROPHAGES

Jean-Claude Antoine and Eric Prina
Unité d'Immunophysiologie cellulaire de l'Institut Pasteur et du C.N.R.S., UA 1113, Institut Pasteur, 25 rue du Dr. Roux, 75724 Paris Cedex 15, France

INTRODUCTION.

Leishmania are protozoan of the trypanosomatidae family that infect certain mammals including man. These parasites have a digenetic life cycle. The flagellated motile promastigote form, which develops in the digestive tract of diptera vectors, is transmitted to mammalian hosts by the bite of the insects. Taken up by host macrophages through a receptor-mediated phagocytic process, it transforms into the nonmotile amastigote form which multiplies within membrane-bound endocytic compartments termed parasitophorous vacuoles (p.v.) (Peters and Killick-Kendrick 1987). Studies of these organelles are of paramount importance to understand how these parasites inhibit or withstand microbicidal mechanisms of macrophages and how they can escape immune responses of the host animals. Such studies are also essential for the development of new drugs against the parasite.

We have applied various microscopical and biochemical methods to assess certain properties of p.v. The results obtained are summarized in this report. *In vitro* models of infection which consisted of rat or mouse bone marrow-derived macrophages parasitized with amastigotes of *Leishmania amazonensis* (strain MPRO/BR/72/M1841) were used in these experiments. They were described in detail elsewhere (Antoine et al 1987, 1990, manuscript submitted ; Prina et al 1990).

NATO ASI Series, Vol. H 62
Endocytosis
Edited by P. J. Courtoy

RESULTS AND DISCUSSION

Course of amastigote infection and morphology of p.v. Rat or mouse macrophages were infected at a multiplicity of 4 amastigotes/host cell and then incubated at 34°C (the optimal temperature for growth of this *Leishmania* strain). Cultures of uninfected macrophages were run in parallel. Microscopic examination showed that amastigotes are rapidly phagocytized by most of the macrophages. Two h after adding parasites about 70% of macrophages are infected and the degree of infection reaches 80-90% at 24 h. Furthermore, amastigotes multiply quite well within these host cells. Thus, these *in vitro* systems are well suited for comparative studies of uninfected and infected macrophages.

Early after infection (2 to 6 h), p.v. are of relatively small size and their membranes closely surround the parasites, but they considerably enlarge later on. Twenty-four h after adding parasites, infected macrophages contain on average 1 to 2 p.v. which occupy 70 to 80% of the total cell volume. Amastigotes are not free inside p.v., but are tightly linked to p.v. membrane through their posterior pole (opposite the flagellar pocket). The mechanism of adhesion is unknown (Benchimol and de Souza 1981).

Fusion competence and luminal content of p.v. Alexander and Vickerman (1975) were the first to demonstrate the fusion of vacuoles containing *Leishmania* amastigotes with lysosomes prelabelled with endocytic tracers. This process is so efficient and apparently not followed by intense fission events in macrophages infected with *Leishmania amazonensis*, that most if not all secondary lysosomes progressively disappear (Barbieri et al 1985). The huge p.v. which house this *Leishmania* species can thus be regarded as the modified lysosomal compartment of the host cell. Even after depletion of lysosomes (24 to 48 h after infection), macrophages still accumulate fluid phase endocytic tracers or ligands into p.v. The p.v. have the characteristics of late endocytic compartments and are most likely able to directly fuse with endosomes or endosome-derived vesicles

(Rabinovitch 1985 ; Antoine et al 1990) (Fig. 1).

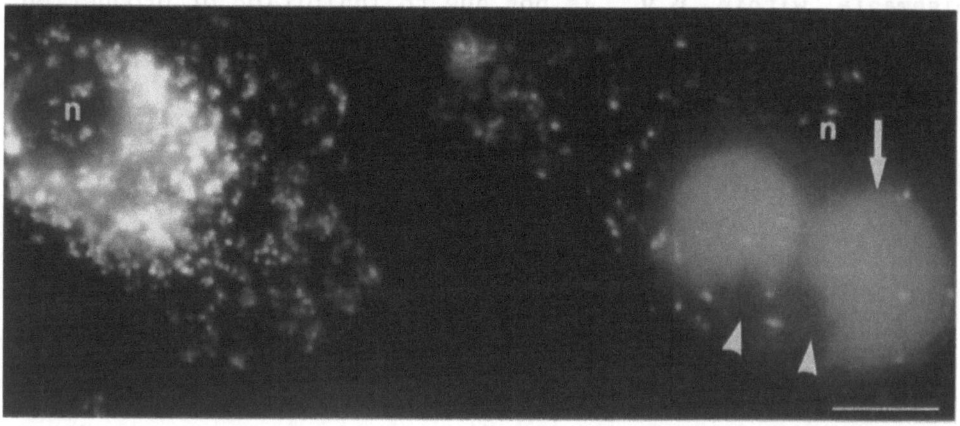

Fig. 1. Detection of endocytic compartments in uninfected (left side) and infected (right side) rat bone marrow-derived macrophages. Living cells were photographed 51 h after infection and 3 h after adding 5 mg/ml fluoresceinated dextran (F-Dex). Culture and incubation with F-Dex were performed at 34°C. In uninfected cell, the marker is present mainly in perinuclear granules (probably lysosomes for the most part). In infected cells, it is detected mainly in p.v. (arrow). Amastigotes are unstained (arrowheads). The stained peripheral granules detected in these conditions disappear after a 90 min chase in F-Dex-free medium. n, nucleus. Bar, 10 µm.

Fusion of lysosomes is accompanied by the translocation of lysosomal enzymes into the lumen of the p.v. Host cell-derived acid phosphatase, trimetaphosphatase, arylsulfatase and cathepsins (cat) B, L, H and D have been localized in these compartments by cytochemistry or immunocytochemistry (Antoine et al 1987 ; Prina et al 1990). Furthermore, the activities of eight lysosomal enzymes (acid phosphatase, trimetaphosphatase, arylsulfatase, dipeptidyl peptidases I and II, cat B, H and D), mainly of host cell origin, are either unaffected or rise progressively after infection. For instance, at 72 h post-infection, cat B and cat D activities are 5-to 6-fold higher in infected than in uninfected rat macrophages. The rise in these enzyme activities appears to be linked with a greater amount of enzymes in infected macrophages (Fig. 2). It is not as yet known whether this increase is due to an enhanced enzyme synthesis in infected macrophages and/or to a reduced enzyme degradation in p.v.

Overall, these results imply that survival of intracellular *Leishmania* within p.v. is not due to inhibition of production and/or activity of lysosomal enzymes.

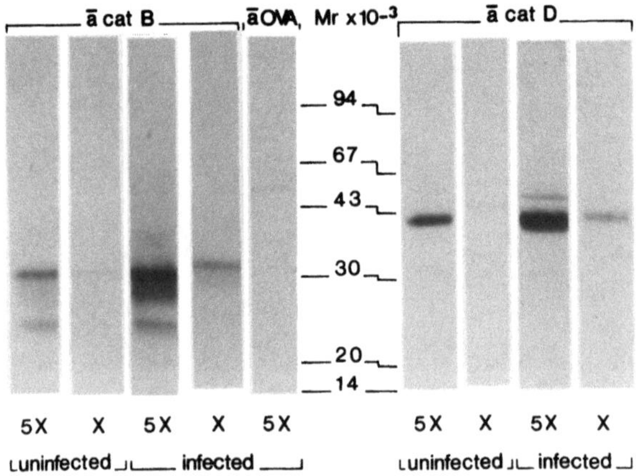

Fig. 2. Cat B and D are more abundant in infected than in uninfected rat macrophages as shown by Western blot analysis. Cell lysates, prepared 53 h after infection or the beginning of the culture at 34°C (uninfected cells) were subjected to SDS-PAGE and transferred to nitrocellulose. Blots were successively incubated with anti-cat B or anti-cat D rabbit IgG and with anti-rabbit Ig antibodies linked to horseradish peroxidase. As a negative control, a blot incubated with anti-ovalbumin (OVA) antibodies instead of anti-cat IgG is shown. Lanes 5 X : 2.2×10^5 lysed cells/lane, lanes X : 4.4×10^4 lysed cells/lane.

As pH is a key parameter which controls the activity of lysosomal enzymes and of other microbicidal mechanisms, we have also addressed the issue of p.v. pH (Antoine et al 1990). A qualitative approach using 3-(2,4-dinitroanilino)-3'-amino-N-methyldipropylamine (DAMP) as a pH probe showed that p.v. are the main acidic compartments of infected rat macrophages. P.v. pH was also estimated quantitatively in living cells loaded with the pH-sensitive endocytic tracer F-Dex. We found statistically different mean pH values of 5.32 for secondary lysosomes of uninfected macrophages and 4.94 for p.v. of macrophages infected for 24 to 48 h. *Leishmania* amastigotes thus appear to be acidophilic organisms resistant to the hydrolytic conditions of p.v. Data from other laboratories seem to indicate that this environment is needed for long-term survival and growth of these parasites (cited in Antoine et al 1990).

Fate of different organelle markers in infected macrophages. In order to define more precisely the origin of the p.v. membrane, we investigated by immunocytochemistry the distribution of several membrane markers in infected mouse macrophages (Antoine et al manuscript submitted). Most of the experiments were performed on cells infected for 24 to 48 h. Neither F4/80, a plasma membrane antigen, nor the cation-independent mannose-6 phosphate receptor (CI-MPR), a marker of endosomes and trans-Golgi network, are detected in p.v. The p.v. , however, do contain lysosomal glycoproteins lgp110 and lgp120. Thus, p.v. exhibit features typical of lysosomes. This suggests that endosomal membranes either do not participate in the genesis of mature p.v. or else that the transfer of some of their components is a selective process.

Localization of MHC class II molecules (Ia) in p.v. Infected macrophages could potentially act as antigen-presenting cells for CD4-positive T lymphocytes which recognize *Leishmania* antigens coupled to Ia molecules. P.v. which are strongly acidic and protease-enriched might be sites for processing of parasite antigens. Furthermore, the detection of high amounts of Ia molecules whithin p.v. was recently documented in infected mouse macrophages stimulated with interferon-γ (IFN-γ) (Antoine et al submitted) which might indicate that p.v. are also involved in the coupling of parasite antigens and Ia. As shown in Fig. 3, Ia molecules which are localized on the plasma membrane and in perinuclear vesicles of uninfected macrophages (Fig. 3A) undergo a dramatic redistribution after infection. Most of internal molecules become associated with p.v. membranes (Fig. 3B) and are fairly often polarized towards amastigote-binding sites (data not shown). These latter observations suggest the existence of interactions between amastigote components and Ia. The Ii invariant chain, which is transiently associated with Ia during their intracellular transport, apparently does not reach p.v. This indicates that p.v. Ia are probably mature molecules. The possible implications of these findings are the followings : 1) Ia located in p.v. could originate from secondary lysosomes involved in the biogenesis of

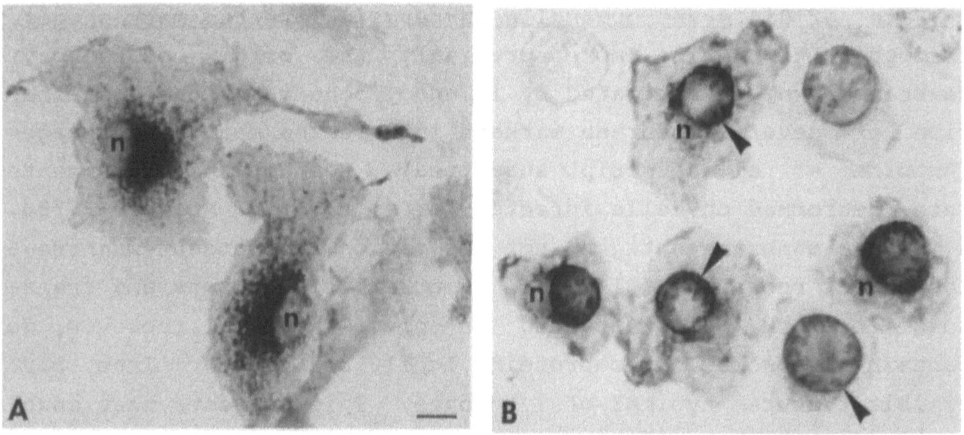

Fig. 3. Immunocytochemical localization of Ia molecules in uninfected (A) and infected (B) macrophages from BALB/c mice. Eight h after infection or the beginning of the culture at 34°C (uninfected macrophages), 25 U/ml IFN-γ were added. Forty h later, cells were fixed, permeabilized with saponin and incubated with anti-Ia M5/114 monoclonal antibody and with horseradish peroxidase-linked secondary antibodies. Staining appears as black deposits. Arrowheads in B point to Ia staining associated with the periphery of p.v. n, nucleus. Bar, 10 μm.

this compartment or circulate in several endocytic organelles including lysosomes ; 2) p.v. could play a role in antigen presentation ; 3) alternatively, the presence of high amounts of Ia in p.v. could be induced by *Leishmania* itself in order to evade an immune response. Studies of the ability of infected macrophages to present exogenous soluble antigens or parasitic antigens are under investigation and we hope they will help to resolve these issues.

ACKNOWLEDGEMENTS.

This work was supported by the Institut Pasteur, the CNRS and the UNDP/World Bank/WHO Special Program for Research and Training in Tropical Diseases. Eric Prina is a recipient of a Fondation Marcel Mérieux student fellowship. The studies reviewed here were performed in collaboration with C. Jouanne, A. Ryter and T. Lang (Institut Pasteur, Paris, France), B. Wiederanders and H. Kirschke (Martin Luther University, Halle-Wittenberg, GDR), P. Bongrand (Hôpital de Sainte-Marguerite, Marseille, France), C. de Chastellier and C. Frehel (Faculté de médecine Necker-Enfants Malades, Paris, France). The authors thank Dr. G.R. Adolf (Ernst-Boehringer Institut für Arzneimittelforschung, Vienna, Austria) for IFN-γ, Drs. S. Gordon (University of Oxford, U.K.), B. Hoflack (EMBL, Heidelberg, FRG), N. Koch (Institut für Immunologie und Genetik, Heidelberg, FRG), I. Mellman (Yale University, New Haven, USA), M. Pierres (Centre d'Immunologie de Marseille-Luminy, France) for providing hybridomas and immune sera, Dr. A. Szabo (Hôpital Cochin, Paris, France) for the critical reading of the manuscript, Danielle Antoine for the iconographic work and Chantal Maczuka for manuscript preparation.

LITERATURE REFERENCES

Alexander J, Vickerman K (1975) Fusion of host cell secondary lysosomes with the parasitophorous vacuoles of *Leishmania mexicana*-infected macrophages. J Protozool 22:502-508.

Antoine J C, Jouanne, C, Lang T, Prina E, de Chastellier C, Frehel C. Localization of MHC class II molecules in phagolysosomes of murine macrophages infected with *Leishmania amazonensis*. manuscript submitted.

Antoine J C, Jouanne C, Ryter A, Zilberfarb V (1987) *Leishmania mexicana* : a cytochemical and quantitative study of lysosomal enzymes in infected rat bone marrow-derived macrophages. Exp Parasitol 64:485-498.

Antoine, JC, Prina E, Jouanne C, Bongrand P (1990) Parasitophorous vacuoles of *Leishmania amazonensis*-infected macrophages maintain an acidic pH. Infect Immun 58:779-787.

Barbieri CL, Brown K, Rabinovitch M (1985) Depletion of secondary lysosomes in mouse macrophages infected with *Leishmania mexicana amazonensis*. Z Parasitenkd 71:159-168.

Benchimol M, De Souza W (1981) *Leishmania mexicana amazonensis* : attachment to the membrane of the phagocytic vacuoles of macrophages *in vivo*. Z Parasitenkd 66:25-29.

Peters W, Killick-Kendrick R (eds) (1987) The leishmaniases in biology and medicine I. Academic Press London UK.

Prina E, Antoine JC, Wiederanders W, Kirschke H (1990) Localization and activity of various lysosomal proteases in *Leishmania amazonensis*-infected macrophages. Infect Immun 58: 1730-1737.

Rabinovitch M (1985) The endocytic system of *Leishmania*-infected macrophages pp611-619. in "Mononuclear phagocytes. Characteristics, physiology and function" van Furth R (ed) Martinus Nijhoff Dordrecht The Netherlands.

ENTRY INTO EUCARYOTIC CELLS AND INTRACELLULAR MOVEMENT OF *SHIGELLA FLEXNERI*

P. J. Sansonetti, N. High, T. Vasselon, P. Clerc, M. L. Bernardini, A. Allaoui, H. d'Hauteville and J. Mounier
Unité de Pathogénie Microbienne Moléculaire and U 199 Inserm,
28, Rue du Dr. Roux. 75724 Paris Cédex 15,
France

INTRODUCTION

Bacteria belonging to the genus *Shigella* cause bacillary dysentery, an invasive disease of the human colon which reflects the capacity of these bacteria to invade intestinal epithelial cells (LaBrec et al 1964; Takeuchi et al 1968). "Invasion" encompasses several stages: entry of the bacterium into epithelial cells, intracellular growth, intracellular movement and cell to cell spread, and eventually death of the host cell (Maurelli et al 1988). This process can be analysed in cell assay systems (LaBrec et al 1964; Hale et al 1979$_{(a)}$; Hale et al 1979$_{(b)}$; Oaks et al 1985), knowing that any phenotype recognized *in vitro* has to be assessed in more definitive virulence assays such as the Sereny test in which invasive bacteria elicit a keratoconjunctivitis in Guinea pigs (Sereny 1957), in the rabbit ligated ileal loop model (Formal et al 1961), or in macaque monkeys, following intragastric inoculation.

ENTRY OF *SHIGELLA FLEXNERI* INTO EPITHELIAL CELLS

The entry process of *S. flexneri* within mammalian cells grown in monolayers requires energy production by bacteria and the host cell (Hale et al 1979$_{(a)}$; Hale et al 1979$_{(b)}$) and is inhibited by cytochalasins. Accumulation of polymerized actin (F-actin) and myosin at the site of bacterial entry into HeLa cells has been directly demonstrated using specific probes such

NATO ASI Series, Vol. H 62
Endocytosis
Edited by P. J. Courtoy

as NBD-phallacidin and an anti-myosin monoclonal antibody respectively (Clerc et al 1987). S. flexneri therefore induces cells which are not professionnal phagocytes to direct an endocytic process similar to classical phagocytosis (Stendahl et al 1980; Sheterline et al 1984). Host- cell receptors involved in this process as well as the transmembrane signal inducing reorganization of the cytoskeleton are as yet unknown. Variations in cytosolic calcium concentrations ($Ca^{++}i$) do not seem to be involved (Clerc et al 1989). Binding to host-cell integrins has been recently demonstrated in related models such as the entry of Yersinia pseudotuberculosis within epithelial cells (Isberg 1988) and the phagocytosis of Bordetella pertussis by macrophages (Hynes 1987). Studies are currently underway in our laboratory to explore these possibilities in S. flexneri.

On the other hand, progresses have been made in the identification of the bacterial genes and their products which trigger the entry process. These genes are carried on a 220 kb plasmid in S. flexneri (Sansonetti et al 1982). Five contiguous loci on a 25 kb sequence (Baudry et al 1987; Watanabe et al 1986; Sasakawa et al 1988) are required. One of these, "locus 2", has been particularly studied. It contains six genes organized as an operon in the following order of transcription: ipp17, ipp21, ipaB, ipaC, ipaD, ipaA. The four ipa genes encode polypeptides of 62, 48, 37, 78 kD respectively. These Ipa (invasion plasmid antigens) are particularly immunogenic since they are the major protein antigens recognized by sera from macaque monkeys or human beings convalescing of shigellosis. We have recently studied the functions of each of the four Ipas by gene destruction in pWR100, the virulence plasmid of S. flexneri serotype 5 (M90T) (High and Sansonetti, submitted). ipaB and ipaC mutants are no longer invasive but demonstrate a strong capacity to bind to the surface of HeLa cells without triggering phagocytosis. They are at the moment considered as the "invasins" of S. flexneri. These mutations have allowed to sort out an adhesion and an entry step. The ipaD mutant is neither adhesive, nor invasive. It may therefore be the adhesin or a protein involved in the organization of the adhesin/entry complex. Finally, ipaA mutants are slightly less invasive than the wild type strain M90T thus IpaA does not seem to have a critical function in entry. Except "locus 1" which appears to contain a gene called *virB* which is a positive regulator of entry functions (Sasakawa et al 1989), functions of loci 3, 4 and 5 are not yet fully characterized except that they may contain genes necessary for proper translocation and organization of Ipas during the entry process (Hromockyj et al 1989).

INTRACELLULAR MULTIPLICATION OF *S. FLEXNERI*

S. flexneri has a tremendous capacity to multiply intracellularly within minutes following entry (Sansonetti et al 1986). Other enteroinvasive pathogens such as *Salmonella* and *Yersinia* grow slowly in the intracellular compartment during the same period (Sansonetti et al 1986; Small et al 1987; Finlay et al 1988). Neither production of Shiga or Shiga-like toxins (Clerc et al 1987; Fontaine et al 1988), nor production of aerobactin (Nassif et al 1987; Lawlor et al 1987) account for rapid intracellular growth. A key factor is the capacity of *S. flexneri* to lyse its phagocytic vacuole early after entry (Sansonetti et al 1986). This does not occur with *Salmonellae* and *Yersiniae*.

INTRACELLULAR MOVEMENT AND CELL TO CELL SPREAD OF *S. FLEXNERI*

Once free within the cytoplasm, *S. flexneri* expresses a capacity to spread within the intracellular compartment. Movements are rapid and random when observed in phase contrast microcinematography (Ogawa et al 1968; Sansonetti, unpublished results). These movements also allow cell to cell spread of the bacterium without passage in the extracellular medium. A mutant has been obtained which was described as unable to pass from one cell to another while invading the Guinea-pig's cornea in the Sereny test (Makino et al 1986). The gene has been called *virG*. It is located on the virulence plasmid and encodes a 120 kD outer membrane protein (Lett et al 1989). The molecular basis of this movement have recently been analysed in more details (Pal

et al 1989; Bernardini et al 1989). Staining of infected cells with
NBD-phallacidin has shown that intracellular bacteria were often covered with F-actin and that some of these bacteria were followed by a thick trail of F-actin (Bernardini et al 1989). Some of these trails, with a bacterium at their extremity, allowed formation of protrusions ensuring passage of the bacteria from one cell to another. Listeria monocytogenes displays a similar behavior (Tilney et al 1989; Mounier et al 1990). icsA, a mutant which does not express this phenotype has been obtained (Bernardini et al 1989). The gene identified appears to be similar to virG (Makino et al 1986). This mutant forms microcolonies of bacteria close to the nucleus, it does not move intracellularly and is unable to spread from one cell to another although it still lyses its phagocytic vacuole.

Based on these results, it becomes obvious that once released in the cytoplasm, bacteria such as S. flexneri and L. monocytogenes have the capacity to interact with host cell cytoskeleton components in order to spread in the intracellular compartment. This Ics phenotype appears essential for proper development of the infection process. Intragastric infection of macaque monkeys with the icsA mutant of S. flexneri causes only slight symptoms or no symptoms at all as compared with animals infected with the wild type strain. Endoscopic examination of the rectum and sigmoïd colon shows only a small number of slightly ulcerated lesions of the mucosa whereas animals infected with the wild type strain presented large abscesses, hemorrhagic ulcerations and purulent membranes. Therefore, once S. flexneri has penetrated into enterocytes, it spreads from one cell to the others in the epithelium without "seeing" the extracellular medium again. It probably remains protected against host specific and non specific immune defenses. Rapid intracellular growth and cell to cell spread, both allowed by lysis of the phagocytic vacuole, appear to be major components of efficient invasion of the intestinal epithelium.

CONCLUSION

S. flexneri appears to be a good model to study invasion of eucaryotic cells and of mucosal tissues. It turns out to be also an interesting probe to study some fundamental properties of the cell response to invading agents. Design of future vaccines will undoubtly bear on these notions.

REFERENCES

Baudry B, Maurelli AT, Clerc P, Sadoff JC, Sansonetti PJ (1987) Localization of plasmid loci necessary for the entry of *Shigella flexneri* into HeLa cells, and characterization of one locus encoding four immunogenic polypeptides. J Gen Microbiol 133: 3409-3413

Bernardini ML, Mounier J, d'Hauteville H, Coquis-Rondon M, Sansonetti PJ (1989) Identification of *icsA*, a plasmid locus of *Shigella flexneri* that governs intra- and intercellular spread through interaction with F-actin. Proc Natl Acad Sci USA 86: 3867-3871

Clerc P, Sansonetti PJ (1987) Entry of *Shigella flexneri* into HeLa cells: evidence for directed phagocytosis involving actin polymerization and myosin accumulation. Infect Immun 55: 2681-2688

Clerc P, Ryter A, Mounier J, Sansonetti PJ (1987) Plasmid mediated early killing of eucaryotic cells by *Shigella flexneri* as studied by infection of J774 macrophages. Infect Immun 55: 521-527

Clerc P, Berthon B, Claret M, Sansonetti PJ (1989) Internalization of *Shigella flexneri* into HeLa cells occurs without an increase in Ca++ concentration. Infect Immun 57: 2919-2922

Finlay BB, Falkow S (1988) A comparison of microbial invasion strategies of *Salmonella, Shigella* and *Yersinia* species. UCLA Symp. Mol Cell Biol 64: 227-243

Fontaine A, Arondel J, Sansonetti PJ (1988) Role of *Shiga* toxin in the pathogenesis of shigellosis as studied using a Tox^- mutant of *Shigella dysenteriae* 1. Infect Immun 56: 3099-3109

Formal SB, Kundel D, Schneider H, Kunev N, Sprinz H (1961) Studies with *Vibrio cholerae* in the ligated loop of the rabbit intestine. Br J Exp Pathol 42: 504-510

Hale TL, Bonventre P.F, (1979a) Shigella infection of Henle intestinal epithelial cells: role of the bacteria. Infect Immun 24: 879-886

Hale TL, Morris RE, Bonventre PF (1979b) Shigella infection of Henle intestinal epithelial cells: role of the host cell. Infect. Immun. 24: 887-894

Hynes RD (1987) Integrins, a family of cell surface receptors. Cell 48: 549-554

Hromockyj AE, Maurelli AT (1989) Identification of *Shigella* invasin genes by isolation of temperature-regulated *inv::lacZ* operon fusions Infect Immun 57: 2963-2970

Isberg RR, Leong JM (1988) Cultured mammalian cells attach to the invasin protein of *Yersinia pseudotuberculosis*. Proc Natl Acad Sci USA 85: 6682-6686

LaBrec EH, Schneider H, Magnani TJ, Formal SB (1964) Epithelial cell penetration as an essential step in the pathogenesis of bacillary dysentery. J Bacteriol 88:1503-1518

Lawlor KM, Daskaleros PA, Robinson RE, Payne S M (1987) Virulence of iron transport mutants of *Shigella flexneri* and utilzation of host iron compounds. Infect Immun 55: 594-599

Lett M, Sasakawa C, Okada N, Sakai T, Makino S, Yamada M, Komatsu K, Yoshikawa M (1989) *VirG*, a plasmid-coded virulence gene of *Shigella flexneri*:identification of the VirG protein and determination of the complete coding sequence. J Bacteriol 171: 353-359

Makino S, Sasakawa C, Kawata K, Kurata T, Yoshikawa M (1986) A genetic determinant required for continuous reinfection of adjacent cells on large plasmid in *Shigella flexneri* 2a. Cell 46: 551-555

Maurelli AT, Sansonetti PJ (1988) Genetic determinants of *Shigella* pathogenicity. Ann Rev Microbiol 42: 127-15

Mounier J, Ryter A, Coquis-Rondon M, Sansonetti PJ (1990) Intracellular and cell-to-cell of *Listeria monocytogenes* involves interaction with F-actin in the enterocytelike cell line Caco-2. Infect Immun 58: 1048-1058

Nassif X, Mazert MC, Mounier J, Sansonetti PJ (1987) Evaluation with an *iuc*::Tn*10* mutant of the role of aerobactin production in the virulence of *Shigella flexneri*. Infect Immun 48:124-129

Oaks EV, Wingfield ME, Formal SB (1985) Plaque formation by virulent *Shigella flexneri*. Infect Immun 53: 57-63

Ogawa H, Nakamura A, Nakaya R (1968) Cinematographic study of tissue cell cultures infected with *Shigella flexneri*. Japan J Med Sci Biol 21: 259-273

Pal T, Newland JW, Tall BD, Formal SB, Hale T L (19) Intracellular spread of *Shigella flexneri* associated with the *kcp*A locus and a 140 kilodalton protein. Infect Immun 57: 477-486

Relman D, Tuomanen E, Falkow S, Golenbock DT, Saukkonenk K, Wright SD (1990) Recognition of a bacterial adhesin by an integrin: macrophage CR_3 (aMb2, CD11b (CD18) binds filamentous hemagglutinin of *Bordetella pertussis*. Cell 61: 1375-1382

Sansonetti PJ, Kopecko DJ, Formal SB (1982) Involvement of a plasmid in the invasive ability of *Shigella flexneri*. Infect Immun 35: 852-860

Sansonetti PJ, Ryter A, Clerc P, Maurelli AT, Mounier J (1986) Multiplication of *Shigella flexneri* within HeLa cells: lysis of the phagocytic vacuole and plasmid mediated contact hemolysis. Infect Immun 55: 521-527

Sasakawa C, Kamata K, Sakai T, Makino S, Yamada M, Okada N, Yoshikawa M (1988) Virulence-associated genetic region comprising 31 kilobases of the 230 kilobase plasmid in *Shigella flexneri* 2a. J Bacteriol 170: 2480-2484

Sasakawa C, Adler B, Tobe T, Okada N, Nagai S, Komatsu K, Yoshikawa M (1989) Functionnal organization and nucleotide sequence of virulence region-2 on the large virulence plasmid in *Shigella flexneri* 2a. Mol Microbiol 3: 1191-1201

Sereny B (1957) Experimental keratoconjunctivitis shigellosa. Acta Microbiol Acad Sci Hung 4: 367-376

Small PLC, Isberg RR, Falkow S (1987) Comparison of the ability of enteroinvasive *Escherichia coli*, *Salmonella typhimurium*, *Yersinia pseudotuberculosis* and *Yersinia enterocolitica* to enter and replicate within HEp2 cells. Infect Immun 55: 1674-1679

Stendahl OI, Hartwig JH, Brotschi EA, Stossel TP (1980) Distribution of acting-binding protein and myosin in macrophages during spreading and phagocytosis.J Cell Biol 84: 215-224

Takeuchi A, Formal SB, Sprinz H (1968) Experimental acute colitis in the rhesus monkey following peroral infection with *Shigella flexneri*, Am J Pathol 52: 503-519

Tilney LG, Portnoy DA (1989) Actin filaments and the growth, movement and spread of the intracellular bacterial parasite, *Listeria monocytogenes*. J Cell Biol 109: 1597-1608

Watanabe H, Nakamura A (1986) Identification of *Shigella sonnei* form I plasmid genes necessary for cell invasion and their conservation among *Shigella* species and enteroinvasive *Escherichia coli*. Infect Immun 53: 352-358

2.3. ENTRY OF VIRUSES AND TOXINS

(Chairperson : Sjur OLSNES)

Opportunistic ligands such as viruses and toxins subvert endocytosis to cross a membrane and penetrate into the cytosol. **Mark Marsh** reviewed the mode of entry of animal viruses. The so-called pH-dependent viruses rely on the acidic pH of endosomes to undergo some conformational change, so as to fuse with the endosomal membrane and transfer nucleic acids into the cytosol. He reported that viruses such as the immunodeficiency virus, whose entry is not strongly inhibited by agents neutralizing the pH of cytoplasmic organelles, can fuse with the cell surface without necessarily leading to productive viral infection. This leads to the suggestion that endocytosis confers another advantage, possibly a by-pass of the dense cortical cytoskeleton.

Kirsten Sandvig compared the mode of entry of ricin and shigella toxin. Both are binary toxins, consisting of a binding subunit B and an enzymatic subunit A that penetrates alone the cytosol and inactivates protein synthesis on ribosomes by ADP-ribosylation. Experiments using K^+-depletion show that coated pits are not required for ricin entry but are essential for shigella toxin entry. This is intriguing since shigella binds to glycolipids of the outer membrane leaflet, which are unable to interact directly with the adaptin complexes. Several lines of evidence lead to the conclusion that both toxins are transported to the Golgi region, where translocation appears to occur.

Patrice Boquet described the penetration of another binary toxin from *Clostridium spiroforme*, which acts by ADP ribosylation on G-actin. He showed that entry requires coated pits like shigella toxin, but that toxicity is not arrested at 15°C and is not sensitive to weak bases. This indicates a pH-independent penetration via early endosomes, possibly into the peripheral cytosol, i.e. at the vicinity of the target. Those studies thus provide both interesting insights into the endocytic pathway and essential information for the construction of efficient immunotoxins.

NATO ASI Series, Vol. H 62
Endocytosis
Edited by P. J. Courtoy

NATO ASI Series, Vol. H [illegible]
[illegible]
Edited by [illegible]
© Springer-Verlag Berlin Heidelberg 1992

On the Role of Endocytosis in the Entry of Animal Viruses

Mark Marsh and Annegret Pelchen-Matthews.
Institute of Cancer Research, Chester Beatty Laboratories
Fulham Road
London SW3 6JB
United Kingdom

In the last ten years an understanding of the molecular mechanisms involved in the entry of animal viruses has begun to emerge. It is now clear that many enveloped (viruses containing a lipid/protein membrane) and non-enveloped (viruses containing a protein shell) viruses need to be exposed to acidic pH for effective entry into a cell (Table 1). These so-called pH-dependent viruses rely on endocytosis for their delivery to acidic endocytic organelles and their penetration can be blocked by reagents, such as weak bases and carboxylic ionophores, that neutralise the pH in cytoplasmic organelles (see Marsh and Helenius 1989). However, a number of other viruses, both enveloped and non-enveloped, are not affected by weak bases or ionophores and it appears that pH is not an important factor in their entry (Choppin and Scheid 1980; McClure et al. 1988; 1990). Consequently these 'pH-independent' viruses can potentially penetrate the host cell at the plasma membrane and/or from acidic endocytic organelles.

The plasma membrane is a barrier which viruses must penetrate in order to infect a cell. For enveloped viruses, at least, the principle is understood - penetration occurs by membrane fusion. As with other biological membrane fusions, viral fusion reactions are tightly controlled. Fusion is mediated by glycoproteins contained in the viral membrane, and the fusion activities of these proteins must be activated. In the case of the pH-dependent viruses it is known to be an acid-induced conformational change that triggers fusion. However, for the pH-independent enveloped viruses, such as Sendai virus or the human immunodeficiency viruses, the requirement for and mechanism of activation remain unclear. Syncytia formation is frequently observed following infection by many pH-independent enveloped viruses indicating that fusion events involving the

NATO ASI Series, Vol. H 62
Endocytosis
Edited by P. J. Courtoy

plasma membrane can occur. Nevertheless, it remains unknown whether or not fusion at the cell surface gives rise to productive infection. The experiments described here with Semliki forest virus suggest that endocytosis may be necessary for the productive entry of many pH-dependent and pH-independent viruses.

TABLE 1 pH-DEPENDENCE OF VIRAL ENTRY*

Acid-dependent viruses	Possibly acid-dependent	Acid-independent viruses
Alpha	Corona	Paramyxo
Orthomyxo	Baculo	Retro+
Flavi	Pox	Lenti
Rhabdo	Hepadna	Herpes
Bunya		Papova
Irido		Rota
Picorna		
Adeno		
Reo		

*Based on Marsh and Helenius 1989
+With the exception of mouse mammary tumour virus.

Much of the current knowledge of virus entry and fusion derives from research on three families of enveloped viruses:- the influenza (flu) viruses, the alphaviruses (e.g. Semliki Forest virus [SFV]) and the rhabdoviruses (e.g. vesicular stomatitis virus [VSV]). These are all pH-dependent enveloped viruses and for each the fusion reactions are activated only when a virion enters a mildly acidic environment. Fusion does not occur at neutral pH and does not occur until the pH has dropped below the fusion threshold for a particular strain of virus. As fusion is acid dependent it is impossible, under normal circumstances, for these viruses to penetrate the cell through the plasma membrane. Studies with SFV indicated that virions can bind to cell surface components that are capable of bringing about the endocytosis of intact virus particles. For SFV endocytosis occurs through coated pits and, in baby hamster kidney (BHK) cells, results in the uptake of up to 3000 virions/cell/min (Marsh and Helenius 1980). Coated vesicle-mediated uptake from the cell surface results in the delivery of virions to endosomes. As indicated elsewhere in this volume endosomes have an acidic lumen and it is within these organelles that viruses may undergo fusion or penetration.

Thus for pH-dependent viruses endocytosis is essential for successful entry into the cell. Does endocytosis play a role in the entry of the pH-independent viruses, or can these viruses productively infect a cell by fusion at the cell surface? As the fusion activity of SFV can be triggered at low pH it is possible in tissue culture to ask whether virions fused at the plasma membrane are capable of infecting a cell. By binding viruses to cells at 0°C and briefly dropping the pH of the medium to <6.0, fusion can be induced at the plasma membrane. The ability of the cell surface-fused viruses to establish infection can then be determined. As previously described, cell surface fusion can lead to the productive infection of BHK cells (Helenius et al. 1982). However, when the same experiment is carried out with chinese hamster ovary (CHO) cells, no infection is observed. Figure 1 shows that when SFV is allowed to enter CHO cells through the endocytic pathway there is a high level of infection, as judged by the incorporation of ^{3}H-uridine into newly synthesised viral RNA. When the cells with bound virus are treated with low pH medium there is a decrease in ^{3}H-uridine incorporation. During this low pH incubation some viruses will enter and infect the cell through endocytosis. If the endocytic route is blocked by including 10mM monensin in the medium, infection is almost completely inhibited and the incorporation of ^{3}H-uridine is similar to that of cells incubated in the absence of virus. When the monensin-treated cells are exposed to acidic medium there is no significant increase in the level of infection, indicating that viruses fused at the cell surface during the low pH incubation are unable to infect the cell. Control experiments indicate that approximately 50% of the bound virions fuse at the plasma membrane during the low pH incubation.

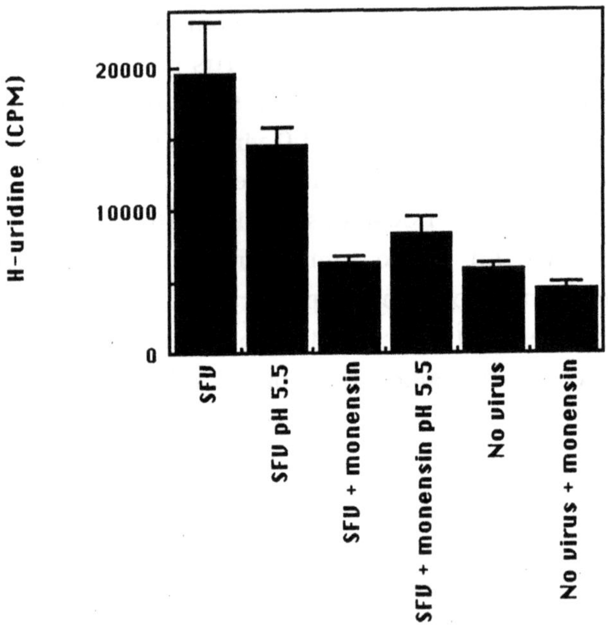

Figure 1: CHO cells grown on glass coverslips were incubated with 10pfu/cell SFV for 1 hr on ice. The cells were then warmed to 37°C in medium adjusted to pH 7.0 or 5.5 with or without 10mM monensin, as indicated. After 1min the cells were returned to neutral pH medium with or without monensin and incubated for 2hrs at 37°C prior to labelling with ^{3}H-uridine as described (Helenius et al. 1982).

Thus fusion at the cell surface can lead to infection in some, but not all, tissue culture cell types. Fusion of SFV at the surface of CHO cells fails to give rise to productive infection. Similarly, we have observed that fusion of VSV with CHO cells fails to initiate infection, even though the normal endocytic route is permissive, and others have failed to observe infection following the fusion of VSV, flu or west nile virus at the surface of MDCK and P388 D1 cells. Why fusion at the cell surface does not result in infection is unclear. With SFV at least it is known that replication of the viral RNA and synthesis of viral structural proteins occurs in complexes associated with cytoplasmic aspect of membranes of endosomes, lysosomes and the endoplasmic reticulum (Froshauer and Helenius 1988). Presumably fusion at the cell surface may result in the viral genome being in the wrong cellular location and endocytosis is therefore necessary to ensure delivery of the viral capsid to the correct site for replication.

If endocytosis is essential for the delivery of viral genomes to a correct cytoplasmic location, it would be required not only for the entry of pH-dependent viruses but also for pH-independent viruses, i.e. endocytosis maybe essential for the entry of many enveloped and non-enveloped viruses regardless of their pH-dependence. Moreover, the viruses may have evolved different mechanisms of biasing the likelihood that they will enter the endocytic pathway prior to fusion. One such mechanism would be to have a strict dependence on the acid environment of endocytic organelles. Alternative mechanisms may involve dependencies on other properties of endosomes such as proteolytic activity (Andersen 1983; Mizzen et al. 1987), or kinetically slow fusion reactions which result in the virions being internalised into endocytic organelles before they undergo fusion. Significantly, the pH-dependent SFV is believed to bind to MHC class I antigens on certain cell types (Helenius et al. 1978). Under normal conditions MHC class I antigens exhibit only limited constitutive internalisation (Reid and Watts 1990), which suggests that virus binding might stimulate the endocytosis of these molecules. In contrast HIV-1, which is pH-independent for fusion (McClure et al. 1988), binds to CD4 which can be efficiently internalised in many cell types (see Pelchen-Matthews et al. this volume). Further work on both the mechanisms involved in viral fusion and the cell biology of virus entry will establish the full requirements for endocytosis in productive virus infection.

Acknowledgements

The authors are supported by grants to the Institute of Cancer Research from the Cancer Research Campaign and the Medical Research Council, and by a grant from the Medical Research Council's AIDS Directed Programme.

References

Andersen KB (1983) Leupeptin inhibits retrovirus infection in mouse fibroblasts. J. Virol. 48: 765-769

Choppin PW and Scheid A (1980) The role of viral glycoproteins in adsorption, penetration and pathogenicity of virus. Rev. Infect. Dis. 2: 40-61.

Helenius A, Marsh M and White J (1982) Inhibition of Semliki Forest virus penetration by lysosomotropic weak bases. J. gen. Virol. 58: 47-61.

Helenius A, Morien B, Fries E, Simons K, Robinson P, Schirrmacher V, Terhorst C and Strominger JL (1978) Human (HLA-A and HLA-B) and murine (H-2K and H-2D) histocompatibility antigens are cell surface receptors for Semliki Forest virus. Proc. Natl. Acad. Sci. USA 75: 3846-3850

Froshauer S, Kartenbeck J and Helenius A (1988) Alphavirus RNA replication occurs on the cytoplasmic surface of endosomes and lysosomes. J. Cell Biol. 107: 2075-2086.

Marsh M and Helenius A (1989) Virus entry into animal cells. Advances in Virus Res. 36: 107-151.

Marsh M and Helenius A (1980) Adsorptive endocytosis of Semliki Forest virus. J. Mol. Biol. 142: 439-454.

McClure MO, Marsh M and Weiss RA (1988) Human immunodeficiency virus infection of CD4-bearing cells occurs by a pH-independent mechanism. EMBO J. 7: 513-518.

McClure MO, Sommerfelt MA, Marsh M and Weiss RA (1990) The pH independence of mammalian retrovirus infection. J. gen. Virol. 71: 767-773.

Mizzen L, Daya M and Andersen R (1987) The role of protease dependent cell membrane fusion in persistent and lytic infections of murine hepatitis virus. Adv. Exp. Med. Biol. 48: 175-186.

Reid PA and Watts C (1990) Cycling of cell-surface MHC glycoproteins through primaquine-sensitive intracellular compartments. Nature 346: 655-667.

ENDOCYTIC UPTAKE OF RICIN AND SHIGA TOXIN

K. Sandvig[1], K. Prydz[1], and B. van Deurs[2]
[1]Institute for Cancer Research
The Norwegian Radium Hospital
Montebello
0310 Oslo 3
Norway

INTRODUCTION

Protein toxins which efficiently kill eukaryotic cells are found in plants and produced by bacteria. Examples of such toxins are the plant toxins ricin, abrin, modeccin, viscumin and volkensin, and the bacterial toxins diphtheria toxin and Shiga toxin (for review, see Olsnes and Sandvig 1988). Schematic structures of toxins are shown in Fig. 1. All these toxins kill cells in the following manner: They bind to cell surface receptors by their B-chains, they are internalized by endocytosis, and then an enzymatically active part of the molecule, the A-chain, enters the cytosol where it inhibits protein synthesis, either by inactivation of the 60 S subunit of the ribosome or by inactivation of elongation factor 2. In spite of their structural similarities, these protein toxins enter the cytosol from different intracellular compartments, and they have different requirements for entry.

Studies of the entry mechanisms of protein toxins are of interest for several reasons. First of all, studies of the entry of the A-chains through the membrane may give information of relevance to transport of other proteins. Due to the high toxicity, transport of a few molecules across the cell membrane is sufficient to monitor an inhibition of protein

---

[2]Structural Cell Biology Unit, Department of Anatomy, The Panum Institute, University of Copenhagen, DK-2200 Copenhagen N, Denmark.

NATO ASI Series, Vol. H 62
Endocytosis
Edited by P. J. Courtoy

synthesis. Secondly, the protein toxins have during the last few years been used to make so-called "immunotoxins", i.e. molecules consisting of the enzymatically active part of the toxin (the A-chain) or a larger part of the toxin molecule coupled to an antibody directed against certain types of cancer cells (Olsnes et al 1989). In order to construct efficient immunotoxins it is useful to know as much as possible about the normal mechanism of toxin transport. The third point is that the protein toxins have proven useful as tools in the study of endocytosis and intracellular transport (van Deurs et al 1989; Sandvig et al 1989). The toxin ricin binds to both glycoproteins and glycolipids with terminal galactose and can thus serve as a marker for surface molecules in general. It is bound even when exposed to low intracellular pH and can therefore be used to follow the transport of endocytosed molecules. In contrast to ricin, the bacterial toxin Shiga toxin binds to a much more limited number of molecules at the cell surface, it binds to glycolipids with the sequence galα1-4gal (Lindberg et al 1987). This toxin can therefore be used to study routes taken by glycolipids.

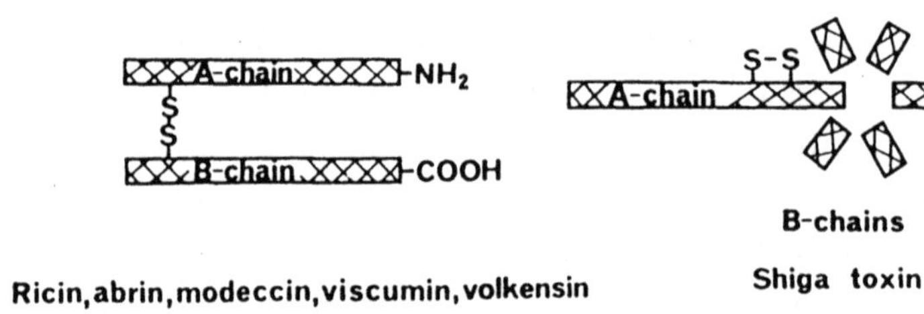

Fig. 1. Schematic structure of protein toxins.

We will in the present paper concentrate on studies of the uptake, intracellular transport and the intoxication of cells with ricin and Shiga toxin. We will present evidence that ricin is endocytosed by two different mechanisms, and we will show quantitative data on the amount of Shiga toxin transported to the Golgi apparatus in polarized cells. Furthermore, evidence that the transport of these toxins to the Golgi apparatus is required for intoxication of cells is shown.

ENDOCYTIC UPTAKE OF RICIN AND SHIGA TOXIN

Endocytosis of cell-surface bound ricin seems to occur both from clathrin-coated pits and from uncoated areas at the cell surface. Evidence for clathrin-independent uptake comes from different types of experiments. In Hep2 cells ricin is still endocytosed when clathrin-coated pits are removed from the cell surface by potassium-depletion (Moya et al 1985; Madshus et al 1987). Also, acidification of the cytosol inhibits formation of coated vesicles from coated pits, but does not block uptake of ricin (Sandvig et al 1987). Recently, we found that it is possible to modulate the endocytic uptake of ricin without alteration of transferrin endocytosis (Sandvig and van Deurs 1990). Incubation of Vero cells with cytochalasin D reduced ricin endocytosis without affecting the uptake of transferrin (Fig. 2A) or the number of coated pits at the cell surface (data not shown), suggesting that cytochalasin D inhibits uptake from non-clathrin coated areas at the cell membrane. On the other hand, in A431 cells ricin endocytosis was stimulated by the tumor promoter TPA (12-O-tetradecanoylphorbol-13-acetate) whereas transferrin endocytosis was unchanged (Fig. 2B). The cellular structures responsible for non-clathrin-coated endocytosis have not yet been identified.

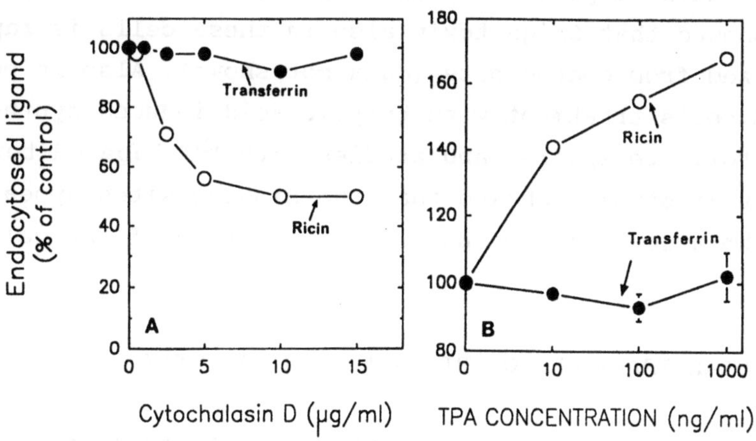

Fig. 2. Modulation of ricin endocytosis in Vero cells (A) and in A431 cells (B). Endocytosis was measured as earlier described (Sandvig and van Deurs, 1990).

The numerous small invaginations observed at the cell membrane are candidates for such a mechanism, but also other types of invaginations and membrane ruffles may be involved in such a process (see however, van Deurs et al, this volume).
Shiga toxin binds evenly to the cell surface at 0°C, but is rapidly endocytosed from clathrin-coated pits upon incubation at 37°C (Sandvig et al 1989). This finding was somewhat surprising since the toxin binds to glycolipids. It is not known how the toxin-receptor complex is concentrated in coated pits, but interactions with proteins may be involved. Recently it was also shown that glucosylceramide is concentrated in coated pits, but also in that case the mechanism is unknown (Kok et al 1989). That Shiga toxin is internalized from coated pits is supported by the finding that acidification of the cytosol which blocks this pathway protects cells against the toxin.

In order to use Shiga toxin to study lipid-transport in polarized MDCK (Madin-Darby canine kidney) cells we developed methods to induce synthesis of Shiga toxin receptors in these cells, which originally do not bind the toxin. After incubation of MDCK cell for 48 h in the presence of low concentrations of butyric acid, the cells bind large amounts of Shiga toxin. These binding sites are functional in the sense that they mediate the transport of Shiga toxin to the cytosol where the toxin inhibits protein synthesis. Electron microscopical studies showed that Shiga toxin also in these cells is rapidly internalized from coated pits (data not shown). Also in polarized MDCK cells treatment with butyric acid induced synthesis of Shiga toxin receptors, and studies with ^{125}I-labeled toxin as well as EM studies showed that the binding sites appeared both at the apical and the basolateral surface of the cells.

INTRACELLULAR TRANSPORT OF RICIN AND SHIGA TOXIN

Endocytosed ricin follows a number of different intracellular pathways. It is recycled to the cell surface, and it is transported to the lysosomes and to the Golgi apparatus from where it is believed to enter the cytosol (see below). Studies on BHK (baby hamster kidney) cells revealed that about 5 % of the endocytosed ricin is transported to the trans-Golgi network (van Deurs et al 1988). In polarized cells ricin is also transcytosed, the transcytotic process beeing most efficient from the apical pole of the cells (van Deurs et al 1990). Interestingly, ricin intoxicates polarized MDCK cells equally well from the apical and the basolateral side, and EM studies showed that ricin coupled to horseradish peroxidase is transported to the Golgi apparatus from both sides of the cells. Furthermore, studies with ricin added basolaterally and cationized ferritin added basolaterally revealed that the two endocytic pathways actually meet. This is in agreement with studies performed by Parton et al (1989).

Like ricin, Shiga toxin is after internalization found in endosomes, lysosomes and in the Golgi apparatus both in polarized and non-polarized cells. By using ^{125}I-labeled Shiga toxin and subcellular fractionation methods we have quantitated the amount of Shiga toxin which is transported to the Golgi apparatus from the apical and basolateral side of MDCK cells after induction of Shiga toxin receptors with butyric acid. As shown in Fig. 3, about the same fraction of Shiga toxin is trans-

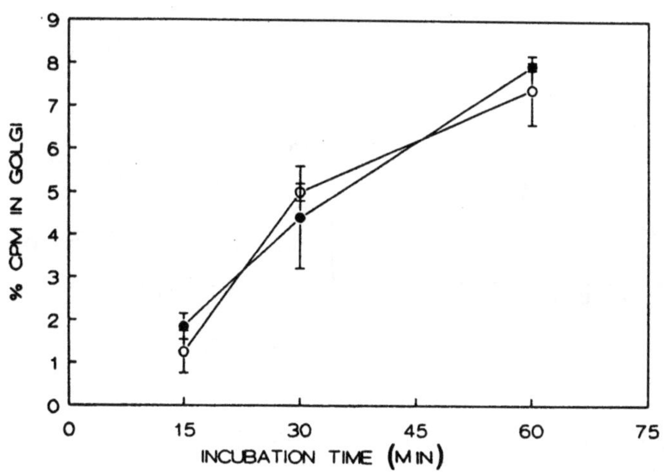

Fig. 3. Transport rate of ^{125}I-Shiga toxin to the Golgi apparatus from the apical (o) and the basolateral side (●) of MDCK cells. The data are expressed as per cent of total cell-associated radioactivity.

ported to the Golgi apparatus from the two sides of the polarized cells. After 1 h at 37°C ca. 10 % of the cell-associated toxin is found in the Golgi apparatus. This is to our knowledge the first quantitative measurement of the transport of a glycolipid-binding ligand to the Golgi apparatus.

EVIDENCE FOR INVOLVEMENT OF THE GOLGI APPARATUS IN THE TRANSPORT OF RICIN AND SHIGA TOXIN TO THE CYTOSOL

Several results suggest that transport of ricin and Shiga toxin to the Golgi apparatus may be required for translocation of the toxins to the cytosol (for review, see van Deurs et al 1989). Youle and Colombatti (1987) found that hybridoma cells producing antiricin were protected against ricin, and their results suggested that the neutralization occurred in the Golgi apparatus. Furthermore, low concentrations of monensin

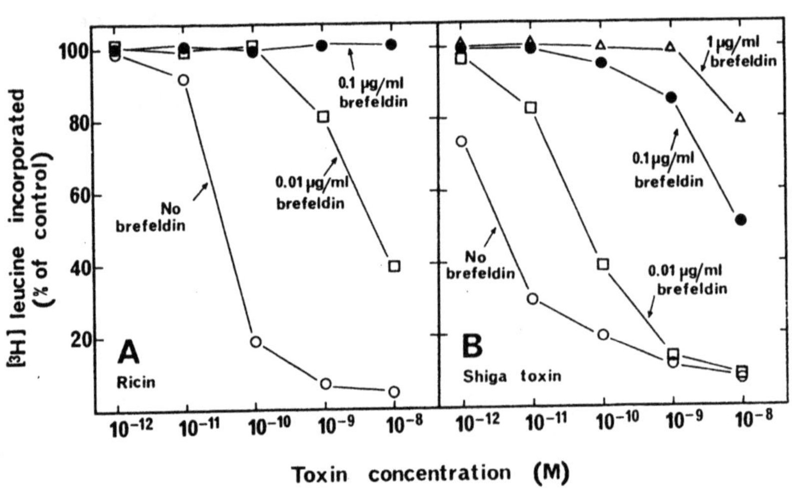

Fig. 4. Ability of brefeldin A to protect Vero cells against ricin and Shiga toxin. Vero cells were incubated with and without brefeldin A for 30 min at 37 °C. Then increasing concentrations of toxin were added, and the protein synthesis was measured 3 h later.

which cause swelling of the Golgi apparatus sensitize cells to ricin (Sandvig and Olsnes 1982). Also, cycloheximide which blocks synthesis of new proteins, and thereby affect transport of molecules through the Golgi apparatus, sensitize cells to both ricin and Shiga toxin (Sandvig et al 1986). Low temperature (18-20 °C) both inhibits transport of the toxins to the Golgi apparatus and protect the cells against intoxication (Sandvig et al 1986, 1989; van Deurs et al 1987). Furthermore, it has recently been shown that the drug brefeldin A affects Golgi function and structure (Lippincott-Schwartz et al 1990), and it turns out that this drug efficiently protects Vero cells against ricin and Shiga toxin (Fig. 4), supporting the view that tranport to the Golgi apparatus may be required to intoxicate cells with ricin and Shiga toxin.

Work from our laboratories referred to in this paper has been supported by grants from The Norwegian and Danish Cancer Societies, The Danish Medical Research Counsil, The NOVO Foundation, and by a NATO Collaborative Research Grant (CRG 900517).

REFERENCES

Kok JW, Eskelinen S, Hoekstra K, Hoekstra D (1989) Salvage of glucosylceramide by recycling after internalization along the pathway of receptor-mediated endocytosis. Proc Natl Acad Sci USA 86:9896-9900

Lindberg AA, Brown JE, Strömberg N, Westling-Ryd M, Schultz JE, Karlsson KA (1987) Identification of the carbohydrate receptor for Shiga toxin produced by Shigella dysenteriae type 1. J Biol Chem 262:1779-1785

Lippincott-Schwartz J, Donaldson JG, Schweizer A, Berger EG, Hauri HP, Yuan LC, Klausner RD (1990) Microtubule-dependent retrograde transport into the ER in the presence of Brefeldin A suggests an ER recycling pathway. Cell 60:821-836

Madshus IH, Sandvig K, Olsnes S, van Deurs B (1987) Effect of reduced endocytosis induced by hypotonic shock and potassium depletion on the infection of Hep 2 cells by picornaviruses. J Cell Physiol 131:6-13

Moya M, Dautry-Varsat A, Goud B, Louvard D, Boquet P (1985) Inhibition of coated pit formation in Hep 2 cells blocks the cytotoxicity of diphtheria toxin but not that of ricin toxin. J Cell Biol 101:548-559

Olsnes S, Sandvig K, Petersen OW, van Deurs B (1989) Immunotoxins - entry into cells and mechanism of action. Immunol Today 10:291-295

Parton RG, Prydz K, Bomsel M, Simons K, Griffiths G (1989)

Meeting of the apical and basolateral endocytic pathways of the Madin-Darby canine kidney cell in late endosomes. J Cell Biol 109:3259-3272
Sandvig K, Olsnes S (1982) Entry of the toxic proteins abrin, modeccin, ricin, and diphtheria toxin into cells. II. Effect of pH, metabolic inhibitors, and ionophores and evidence for toxin penetration from endocytic vesicles. J Biol Chem 257:7504-7513
Sandvig K, Olsnes S, Brown JE, Petersen OW, van Deurs B (1989) Endocytosis from coated pits of Shiga toxin: A glycolipid-binding protein from Shigella dysenteriae 1. J Cell Biol 108:1331-1343
Sandvig K, Olsnes S, Petersen OW, van Deurs B (1987) Acidification of the cytosol inhibits endocytosis from coated pits. J Cell Biol 105:679-689
Sandvig K, Tønnessen TI, Olsnes S (1986) Ability of inhibitors of glycosylation and protein synthesis to sensitize cells to abrin, ricin shigella toxin and pseudomonas toxin. Cancer Res 46:6418-6422
Sandvig K, van Deurs B (1990) Selective modulation of the endocytic uptake of ricin and fluid phase markers without alteration in transferrin endocytosis. J Biol Chem 265: 6382-6388
van Deurs B, Hansen SH, Petersen OW, Melby EL, Sandvig K (1990) Endocytosis, intracellular transport and transcytosis of the toxic protein ricin by a polarized epithelium. Eur J Cell Biol 51:96-109
van Deurs B, Petersen OW, Olsnes S, Sandvig K (1987) Delivery of internalized ricin from endosomes to cisternal Golgi elements is a discontinuous, temperature sensitive process. Exp Cell Res 171:137-152
van Deurs B, Petersen OW, Olsnes S, Sandvig K (1989) The ways of endocytosis. Int Rev Cyt 117:131-177
van Deurs B., Sandvig K, Petersen, OW, Olsnes S, Simons K, Griffiths G. (1988) Estimation of the amount of internalized ricin that reaches the trans-Golgi network. J Cell Biol 106:253-267
Youle RJ, Colombatti M (1987) Hybridoma cells containing intracellular anti-ricin antibodies show ricin meets secretory antibody before entering the cytosol. J Biol Chem 262:4676-4682

ENDOCYTOSIS OF CLOSTRIDIUM SPIROFORME BINARY TOXIN BY HEP2 CELLS

Patrice Boquet and Michel R.Popoff
Unité des Antigènes Bactériens, UA CNRS 557
Unité des Anaérobies
Institut Pasteur, 28 rue du Docteur Roux, 75724 PARIS CEDEX 15, France

SUMMARY

C. spiroforme toxin acted on cells by disrupting microfilaments due to ADP-ribosylation of G actin. Toxicity is not blocked by 10 or 20 mM ammonium chloride and only moderately inhibited by 30 mM NH_4Cl. Inhibition of coated-pit formation in Hep2 cells by potassium depletion strongly protected against the effect of C. spiroforme toxin. Finally, toxicity was not blocked by incubating Hep2 cells and spiroforme toxin at 15°C. These results suggest that this new binary toxin enters the cell via the coated pits coated vesicle pathway and might reach the cytoplasm concomitantly with or before transfer to early endosomes.

INTRODUCTION

Several species of Clostridium produce binary toxins (composed of two independent proteins) which ultimately disrupt the microfilaments of animal cells. The molecular basis of this effect is the ADP-ribosylation of G actin which impedes assembly of this molecule into F actin (Aktories et al., 1986, Ohishi and Tsuyama, 1986, Reuner et al., 1986). These binary toxins are divided into two groups based on their immunological reactivities (Popoff and Boquet, 1988). The first group contains the C2 toxin of C. botulinum types C and D, while the second

NATO ASI Series, Vol. H 62
Endocytosis
Edited by P. J. Courtoy

consists of *C. perfringens* type E iota toxin (Simpson *et al.*, 1987) and *C. spiroforme* toxin (Popoff and Boquet, 1988, Simpson *et al.*, 1989, Stiles and Wilkins, 1986). In addition we have recently found an actin ADP-ribosylating protein in a strain of *C. difficile* (Popoff *et al.*, 1988). This last molecule immunologically cross-reacted with the ADP-ribosylating moieties of the iota-spiroforme group of binary toxins (Popoff *et al.*, 1988).

C. spiroforme binary toxin is composed of two independent polypeptide chains called Sa (light chain) and Sb (heavy chain) (Popoff and Boquet, 1988); these are not associated by either covalent or non-covalent bonds. In this they resemble the *C. botulinum* C2 toxin (Ohishi *et al.*, 1980). The light chain of *C. spiroforme* toxin has been shown to be an ADP-ribosyl transferase which covalently modifies G actin (Popoff and Boquet, 1988, Simpson *et al.*, 1989). The heavy chain, on the other hand, is required for the penetration of Sa into the cytosol (Popoff and Boquet, 1988, Simpson *et al.*, 1989, Stiles and Wilkins, 1986). Like C2 toxin component II, Sb must undergo limited proteolysis to be functionally active.

In this paper, we have studied the interaction of this toxin with cultured Hep2 cells and investigated its uptake by an endocytotic process.

RESULTS

We have examined the effect on toxicity of ammonium chloride, a drug well known to raise endosomal, trans Golgi and lysosomal pHs, on *Clostridium spiroforme* cell toxicity. Hep2 cells were first treated for 30 min with medium containing either 10, 20, or 30 mM NH_4Cl. *C. spiroforme* toxin (Sa plus Sb in equimolar amounts) was then added to the cell culture at various concentrations in the presence of NH_4Cl (10, 20, or 30 mM) and incubated at 37°C. After 6 h, the cells were harvested and tested for *in vitro* actin ADP-ribosylation by *C. spiroforme* toxin Sa chain as described (Popoff *et al.*, 1988). Ammonium

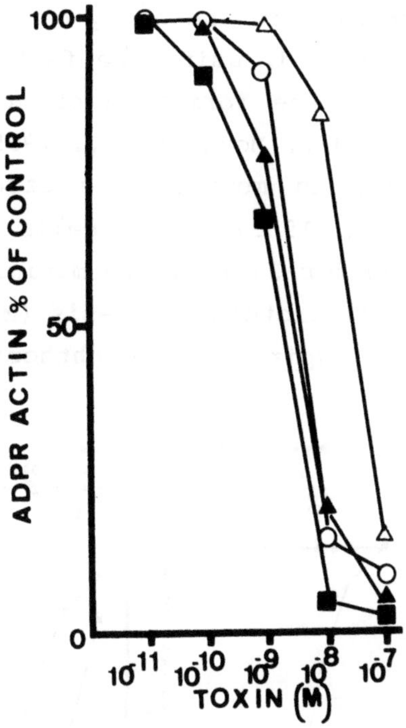

Fig. 1: Effect of NH_4Cl on intoxication of Hep2 cells by C. spiroforme toxin.

Hep2 cells grown in 24 well microplates were first incubated with 10, 20 or 30 mM NH_4Cl for 30 min. Then, spiroforme toxin dilutions were applied to the monolayer. After 6 h at 37°C the cells were processed for in vitro ADP-ribosylation of actin by Sa spiroforme toxin chain as described (Popoff et al., 1988). Controls without toxin but with ammonium chloride at the different concentrations were ran at the same time. ■---■ ; Hep2 cells with spiroforme toxin without NH_4Cl, ▲---▲ with 10 mM NH_4Cl, o---o with 20 mM NH_4Cl, Δ---Δ toxin with 30 mM NH_4Cl.

chloride afforded no protection against C. spiroforme toxin at 10 or 20 mM (Fig. 1). However, at 30 mM a slight protection (especially at 10^{-8} M toxin concentration) was observed (Fig. 1). We have to stress that at 30 mM NH_4Cl a reduction of 70% in actin ADP-ribosylation was observed in control preparations (not treated with toxin). We do not know the mechanism of action of NH_4Cl on the size of the cellular G actin pool. However this large reduction of actin ADP-ribosylation in vitro by 30 mM NH_4Cl makes interpretation of the protection against C. spiroforme toxin difficult.

The role of coated pits in the entry of C. spiroforme toxin into Hep2 cells was examined, using the fact that depletion of potassium totally blocks the formation of coated pits in Hep2 cells (Madshus et al., 1987, Moya et al., 1985). Diphtheria and ricin toxins were used to monitor the effects of potassium depletion (Moya et al., 1985). Hep2 cells were depleted of potassium and incubated for 30 min. (to minimize the effects of lack of potassium on cell protein synthesis (Moya et al., 1985)) with various doses of C. spiroforme, diphtheria or ricin toxin.

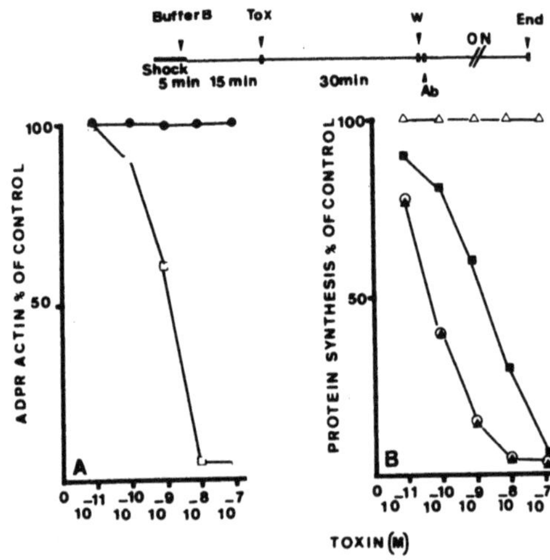

Fig. 2: Effects of Clostridium spiroforme toxin on Hep2 cell actin ADP-ribosylation after intracellular K^+ depletion; comparison with diphtheria and ricin toxins effects on protein synthesis.

Experiments were conducted as represented in the schematic drawing shown above the figure. After isotonic K^+-free medium (buffer B), various concentrations of C. spiroforme toxin (panel A) were added to the cells. After 30 min at 37°C cells were carefully washed 3 times with buffer B (W) and then incubated overnight in DMEM medium containing 50 µg/ml of rabbit anti-C. spiroforme toxin component Sb serum (panel A) with 20 flocculation units of horse anti-diphtheria toxin serum per ml or 10 µl of rabbit anti-ricin serum per ml (panel B) (Ab). ADP-ribosylation of actin in vitro (panel A) was performed as described (Popoff et al., 1988). Protein synthesis (panel B) was assayed by incorporation of [^{14}C] leucine. Panel A; ●---● cells without K^+ □---□ cell with K^+, panel B; △---△ diphtheria toxin on cells without K^+, ▲---▲ ricin toxin on cells without K^+, ■---■ diphtheria toxin on cells with K^+, o---o ricin toxin on cells with K^+.

After this step, the cells were carefully washed with medium without potassium and incubated with DMEM containing antibodies against C. spiroforme Sb component, diphtheria toxin, or ricin toxin to block further entry (control experiments confirmed that the antibodies were effective). In this way, only molecules taken up during the potassium depletion step (in the absence of coated pits) could intoxicate the cells.

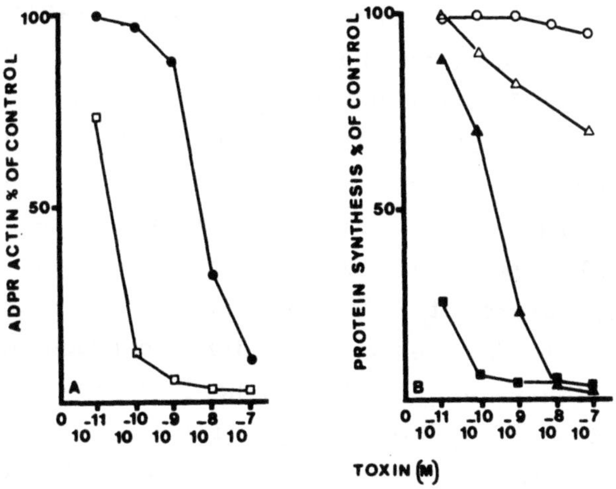

Fig. 3: Effects of low temperature (15°C) on actin ADP-ribosylation of Hep2 cells by C. spiroforme toxin; comparison with diphtheria and ricin toxins effects on protein synthesis.

Hep2 cell monolayers after 48 h of growth were washed 3 times with ice cold DMEM (to block endocytosis). The toxin dilutions were first incubated with cells at 4°C in DMEM medium adjusted to pH 7.4 with 10 mM Hepes buffer. The microplates were then wrapped into sealed plastic bags and incubated for 10 h into a water bath (either at 37°C or 15°C). At the end of incubation cell monolayers were washed 3 times with ice cold in DMEM and incubated at 4°C for 20 min DMEM containing the specific antibodies against each toxin (as described in fig. 6). For ADP-ribosylation in vitro of actin (panel A) the cells were processed as described (Popoff et al., 1988). For estimation of protein synthesis (panel B) cell monolayers were incubated further for 10 min at 37°C with 0.5 µCi/ml of [^{14}C]leucine and processed as previously described (17). Panel A; □---□ C. spiroforme toxin on cells at 37°C, ●---● C. spiroforme toxin on cells at 15°C, Panel B; ○---○ ricin toxin on cells at 15°C, △---△ diphtheria toxin on cells at 15°C, ▲---▲ ricin toxin on cells at 37°C, ■---■ diphtheria toxin on cells at 37°C.

The effect of C. spiroforme on actin ADP-ribosylation was blocked (as was the cytotoxicity of diphtheria toxin (Fig. 2B)) when coated-pit formation was inhibited (Fig. 2A). As we have already reported (16), ricin toxicity was not affected by inhibition of receptor mediated endocytosis, involving clathrin (Fig. 2B).

Finally, we have examined the possibility that once taken up into vesicles, C. spiroforme binary toxin could travel to the trans-Golgi network (Griffiths and Simons, 1986), the site at which ricin is thought to cross the membrane to the cytosol (van Deurs et al., 1987, van Deurs et al., 1988). A temperature below 20°C (van Deurs et al., 1987, van Deurs et al., 1988) has been shown to block the traffic from endosomes to the trans-Golgi network, probably by disrupting the microtubule meshwork which tracks vesicles between endosomes to the trans-Golgi network area. A temperature of 15°C does not block receptor mediated endocytosis (i.e. the vesicular traffic between plasma membrane to endosomes (Marsh et al., 1983). Hep2 cells were incubated for 10 h at 15°C with C. spiroforme, diphtheria, or ricin toxins. The same experiments were done in parallel with cells incubated at 37°C. After incubation (at 15 or 37°C), the in vitro ADP-ribosylation of actin was tested and protein synthesis (for diphtheria and ricin toxins) was estimated. Even when incubated with cells at 15°C C. spiroforme binary toxin was able to reduce the G actin available for ADP-ribosylation in vitro, although less efficiently (Fig. 3A).

The same results were obtained with diphtheria toxin (Fig. 3B). At 15°C ricin toxin was without effect on Hep2 cell protein synthesis (Fig. 3B) as previously shown by van Deurs et al. (1987).

DISCUSSION

C. spiroforme binary toxin was active on cultured cells (Hep2 cells). Interesting was the finding that ammonium chloride, a chemical known to increase the pH of endosomes and lysosomes (Marsh et al., 1987, Maxfield, 1982) does not block the cytotoxic effect of C. spiroforme toxin (except for a moderate inhibition at 30 mM). Ammonium chloride (a lysomotropic amine) is well known to block the toxic effect of several bacterial and plant toxins (diphtheria toxin, modeccin, Pseudomonas aeruginosa exotoxin A). It is thought that toxins which are blocked by weak bases, such as ammonium chloride, must reach an acidic cell compartment to be translocated (or to translocate a fragment) into the cytosol (Olsnes and Sandvig, 1988). These acidic compartments are either endosomes, trans Golgi, or of course, lysosomes (Anderson et al., 1985, Mellman et al., 1986, Tycko and Maxfield, 1982). A new acidic cell compartment required for sorting lysosomal proteins using the mannose-6-phosphate receptors has been described (Griffiths et al., 1988), but so far no toxin has been shown to be routed to this organelle. Ricin is known not to be affected by lysomotropic amines; instead, these drugs increase its toxicity for cells (Mekada et al., 1981) and recently this toxin has been shown to reach the trans Golgi compartment (van Deurs et al., 1987, van Deurs et al., 1988). The results we obtained with ammonium chloride suggest that for C. spiroforme toxin, an acidic compartment is not required for translocation of the molecule through the membrane. A comparable non-acidic mechanism for the entry of Bordetella pertussis adenylate cyclase, has been reported recently by Gordon et al. However, Bacillus anthracis adenylate cyclase cell intoxication was found to be acid dependent (Gordon et al., 1988).

At 15°C, Hep2 cells had their G actin ADP-ribosylated by C. spiroforme binary toxin. This result suggests that the toxin may not have to move from the cell periphery to the organelles close to the nucleus (lysosomes, mannose-6-phosphate receptor uncoupling compartment or trans-Golgi) since a temperature below 20°C is supposed to affect the traffic of vesicles between late endosomes and the trans-Golgi network and has been shown to block ricin toxin effects (van Deurs et al., 1987).

Inhibition of coated pit formation by potassium depletion seemed to block the cytotoxicity induced by C. spiroforme toxin as it does for diphtheria toxin (Moya et al., 1985). Such treatment has no effect on ricin toxin (Moya et al., 1985, Madshus et al., 1987). It has been inferred from previous data that diphtheria toxin must use coated pits (Morris et al., 1985) to make a productive entry into the cells whereas ricin toxin was supposed to use an alternative pathway of endocytosis (smooth pits, smooth vesicles) (Moya et al., 1985, Madshus et al., 1987, Sandvig et al., 1987). On this basis, C. spiroforme toxin could behave like diphtheria toxin (i.e. entry via coated pits-coated vesicles). However, we are not entirely certain that potassium depletion is without effect on toxin binding to the cell membrane. Indeed, incubation at 4°C of Hep2 cells with C. spiroforme toxin, in a medium with or without potassium, followed by repeated washing and incubation at 37°C in the presence of antibodies against C. spiroforme Sb toxin chain, did not induce a profond cytopathic effect (data not shown). Therefore it is difficult to determine the influence of potassium depletion on C. spiroforme toxin binding.

On the basis of the data presented in this paper, we can tentatively suggest the location for cell entry of C. spiroforme binary toxin. Since the toxin requires neither an acidic pH nor vesicle traffic from the periphery to the cell center and yet does seem to require the coated pit coated vesicle pathway of endocytosis, it might well enter the cytosol from uncoated vesicles (coated vesicles which have lost their clathrin coat) or during fusion of uncoated vesicles with early endosomes (Schmid et al., 1988). This location of toxin entry would differ from that of diphtheria toxin in an acidic environment (Draper and Simon, 1980, Kim and Groman, 1965, Marnell et al., 1984, Sandvig and Olsnes, 1980) (early endosome (Olsnes and Sandvig, 1988)) or that of ricin toxin (trans Golgi (van Deurs et al., 1988) and could therefore be comparable to that of shigella toxin which is not inhibited by lysomotropic amines (Sandvig and Brown, 1987) and requires coated pits (Sandvig et al., 1989). This site of cell penetration would be, for a toxin which ultimately acts on globular actin, quite useful since it is known that formation of F actin from G-actin is achieved mainly in the periphery of the cell (sub cortical actin web) by gelsolin or villin (Stossell et al., 1985) and in the vicinity of the cytoplasmic domain of Integrins, the specialized transmembrane proteins involved in the binding of extra cellular matrix (Buck and Horwitz, 1987). Therefore, in this cell domain there probably exists a higher concentration of G-actin molecules. That way, the ADP-ribosylation of G actin by Sa could be quite efficient.

REFERENCES

Aktories K, Bärmann I, Ohishi I, Tsuyama, S Jakob, KH, Habermann E (1986) Botulinum toxin C2 ADP-ribosylates actin. Nature (London) 322:390-392

Anderson RGW, Pathak RK (1985) Vesicles and cisternae in trans Golgi apparatus of human fibroblasts are acidic compartments. Cell 40:635-643

Buck CA, Horwitz AF (1987) Cell surface receptors for extracellular matrix molecules. Ann Rev Cell Biol 3:179-205

Draper RK, Simon AI (1980) The entry of diphtheria toxin into mammalian cell cytoplasm: evidence for lysosomal involvement. J Cell Biol 87:849-854

Gordon VM, Leppla SH, Hewlett EL (1988) Inhibitors of receptor mediated endocytosis block the entry of Bacillus anthracis adenylate cyclase but not that of Bordetella pertussis adenylate cyclase toxin. Infect Immun 56:1066-1069

Griffiths G, Hoflack B, Simons K, Mellmann I, Kornfeld S (1988) The mannose 6-phosphate receptor and the biogenesis of lysosomes. Cell 52:329-341

Griffiths G, Simons K (1986) The trans Golgi network: sorting at the exit site of the Golgi complex. Science 234:438-443

Kim K, Groman NB (1965) In vitro inhibition of diphtheria toxin action by ammonium salts and amines. J Bacteriol 90:1552-1562

Madshus F, Sandvig K, Olsnes S, van Deurs BO (1987) Effect of reduced endocytosis induced by hypotonic shock and potassium depletion on the infection of Hep2 cells by Picornaviruses. J Cell Physiol 131:14-22

Marnell MH, Shia SP, Stookey M, Draper RK (1984) Evidence for penetration of diphtheria toxin to the cytosol through a prelysosomal membrane. Infect Immun 44:145-150

Marsh M, Bolzau E, Helenius A (1983) Penetration of Semliki Forest Virus from acidic prelysosomal vacuoles. Cell 32:931-940

Maxfield FR (1982) Weak bases and ionophores rapidly and reversibly raise the pH of endocytic vesicles in cultured mouse cells. J Cell Biol 95:676-681

Mekada E, Uchida T, Okada Y (1981) Methylamine stimulates the action of ricin toxin but inhibits that of diphtheria toxin. J Biol Chem 256:1225-1228

Mellman I, Fuchs R, Helenius, A (1986) Acidification of the endocytic and exocytic pathways. Ann Rev Biochem 55:663-700

Morris RE, Gerstein AS, Bonventre PF, Saelinger CB (1985) Receptor entry of diphtheria toxin into monkey kidney (Vero) cells: electron microscopic evaluation. Infect Immun 50:721-727

Moya M, Dautry-Varsat A, Goud B, Louvard D, Boquet P (1985) Inhibition of coated pit formation in Hep2 cells blocks the cytotoxicity of diphtheria toxin but not that of ricin toxin. J. Cell. Biol. 101:548-559

Ohishi I, Iwasaki M, Sakaguchi G (1980) Purification and characterization of two components of botulinum C2 toxin. Infect Immun 30:668-673

Ohishi I, Tsuyama S (1986) ADPribosylation of non-muscle actin with component I of C2 toxin. Biochem Biophys Res Comm 136:802-806

Olsnes S, Sandvig K (1988) How protein toxins enter and kill cells, pp. 39-73. In AE Frankel (ed.), Immunotoxins, Kluwer Academic Publishers.

Popoff MR, Boquet P (1988) Clostridium spiroforme toxin is a binary toxin which ADP-ribosylates cellular actin. Biochem Biophys Res Comm 152:1361-1368

Popoff MR, Rubin EJ, Gill DM, Boquet, P (1988) Actin-specific ADP-ribosyltransferase produced by a Clostridium difficile strain. Infect Immun 56:2299-2306

Reuner KH, Presek P, Boschek C, Aktories K (1987) Botulinum C2 toxin ADP-ribosylates actin and disorganizes the microfilament network in intact cells. Eur J Cell Biol 43:134-140

Sandvig K, Brown JE (1987) Ionic requirement for entry of Shiga toxin from Shigella dysenteriae 1 into cells. Infect Immun 55:298-303

Sandvig K, Olsnes S (1980) Diphtheria toxin entry into cell is facilitated by low pH. J Cell Biol 87:828-832

Sandvig K, Olsnes S, Brown JE, van Deurs B (1989) Endocytosis from coated pit of Shiga toxin: a glycolipid binding protein from Shigella dysenteriae 1. J Cell Biol 108:1331-1343

Sandvig K, Olsnes S, Petersen OW, van Deurs B (1987) Acidification of the cytosol inhibits endocytosis from coated pits. J. Cell. Biol. 105:679-689.

Schmid SL, Fuchs R, Male P, Mellman I (1988) Two distinct subpopulation of endosomes involved in membrane recycling and transport to lysosomes. Cell 52:73-83

Simpson LL, Stiles BG, Zepeda H, Wilkins T (1989) Production by Clostridium spiroforme of an iota-like toxin that possesses mono(ADP-ribosyl) transferase activity: identification of a novel class of ADP-ribosyl transferases. Infect Immun 57:255-261

Simpson LL, Stiles BG, Zepeda H., Wilkins TD (1987) Molecular basis for the pathological actions of Clostridium perfringens iota toxin. Infect Immun 55:118-122

Stiles BG, Wilkins TD (1986) Clostridium perfringens iota toxin: synergism between two proteins. Toxicon 24:767-773

Stossel TP, Chaponnier C, Ezzel RM, Hartwig JH, Janmey PA, Kwiatkowski DJ, Lind SE, Smith DB, Soutwick FS, Lin HL, Zaner, KS 1985. Non-muscle actin-binding proteins. Ann Rev Cell Biol 1:353-402

Tycko B, Maxfield FR (1982) Rapid acidification of endocytic vesicles containing .2 macroglobulin. Cell 28:643-651

Van Deurs B, Petersen, OW, Olsnes S, Sandvig K (1987) Delivery of internalized ricin from endosomes to cisternal Golgi element is a discontinuous temperature-sensitive process. Exp Cell Res 171:137-152

van Deurs B, Sandvig K, Petersen OW, Olsnes S, Simons K, Griffiths G (1988) Estimation of the amount of internalized ricin that reaches the trans-Golgi network. J Cell Biol 106:253-267

2.4. LYSOSOMAL DISEASES

(Chairperson : Konrad SANDHOFF)

Chronic lysosomal storages due to inborn errors of metabolism are well-known, but whether acute lysosomal overload or a rupture of the lysosomal membrane may actually be the cause of acute cell suffering is still debated. **Konrad Sandhoff** discussed the effects of inherited defects of lysosomal catabolism, using glycolipid storage diseases as an example. Glycolipid lysosomal overload results from a defective lysosomal hydrolase, arylsulfatase, or from a defective glycolipid-binding protein acting as "substrate extractor" and thereby activating the hydrolase. Both defects result in the progressive accumulation of undegraded substrates and of co-precipitating material. His clinical studies stress the importance of minor differences in residual activity for the clinical onset of Tay-Sachs disease.

Robert Wattiaux examined whether labilization of the lysosomal membrane may jeopardize cell survival. He first reported that transitory liver ischaemia causes a release of lysosomal enzymes, which can be prevented by mild hypothermia or by scavengers of oxygen radicals. Both treatments improve cell survival. He also showed that low-molecular substrates can reach lysosomes as weak bases and be hydrolyzed into membrane-impermeant derivatives, whose accumulation causes the osmotic rupture of lysosomes, and cell death.

Paul Tulkens finally showed that medical treatments can cause acute lysosomal storage diseases. Therapeutic doses of gentamicin not infrequently induce a lysosomal phospholipidosis associated with acute tubular cell necrosis : the antibiotic passes the glomerular barrier, is reabsorbed by proximal tubular cells and inhibits phospholipases. Phospholipidosis can be prevented by stable acidic poly-L-peptides. Stable poly-D-peptides also protect against phospholipidosis but cause a spectacular thesaurismosis of a different nature.

NATO ASI Series, Vol. H 62
Endocytosis
Edited by P. J. Courtoy

CORRELATION BETWEEN RESIDUAL ACTIVITY OF A LYSOSOMAL HYDROLASE AND THE SEVERITY OF THE RESULTING STORAGE DISEASE

K. Sandhoff, E. Conzelmann, S. Michel, P. Leinekugel
Institut für Organische Chemie und Biochemie der Universität Bonn
Gerhard Domagk-Str. 1
5300 Bonn 1, FRG.

Glycosphingolipids (GSL) are components of the outer leaflet of animal plasma membranes. Sialoglycosphingolipids (gangliosides) are especially enriched in neuronal cell surfaces. Biosynthesis of GSL is catalyzed by enzymes bound to the membranes of the endoplasmic reticulum and the Golgi stacks whereas the final degradation of these amphiphilic GSL takes place in secondary lysosomes. Electron spin resonance studies with nitroxide labelled gangliosides (Schwarzmann et al. (1983); Schwarzmann et al. (1984)) and metabolic studies (Sonderfeld et al. (1985) suggest that gangliosides added to the medium of cultured human fibroblasts insert into the plasma membrane from where they can be transported into the lysosomal compartment for sequential degradation by exohydrolases. In mutant cells, with a block in lysosomal GSL degradation, undegradable intermediates accumulate, for example ganglioside G_{M2} in Tay Sachs disease. Blocs in lysosomal GSL catabolism are caused by mutations either in GSL hydrolases or in GSL-binding proteins which stimulate the degradation of some membrane-bound GSL by lysosomal hydrolases (Sandhoff et al. (1989); Conzelmann and Sandhoff (1987); Fürst et al. (1988); Schröder et al. (1989); Nakano et al. (1989)).
Inherited deficiencies of one and the same GSL hydrolase may cause quite different clinical phenotypes, e.g. the severe infantile, the juvenile and adult forms as well as the late adult variants with onset of symptoms in adulthood, slow progression of the disease and almost normal life span.

NATO ASI Series, Vol. H 62
Endocytosis
Edited by P. J. Courtoy

Clinical healthy probands with severely reduced hydrolase activity (e.g. 10 - 20 % of normal arylsulfatase activity) were also identified.
Identification of the different allelic mutations at the genomic level may explain the existence of different forms but is not sufficient to understand the correlation between primary defect and clinical picture. It seems reasonable to assume that one of the major factors that determine the severity and course of the disease is the residual activity of the affected enzyme. In the case of lysosomal storage diseases, however, the residual activities found in different variants are generally very small and only minute differences (if any) are found between them.
We proposed a simple kinetic model that describes the correlation between the residual activity of an enzyme and the turnover rate of its substrate in a limited compartment such as the lysosome (Conzelmann and Sandhoff (1983/84).
For the sake of simplicity, we assume that influx rate of the substrate and enzyme activity remain constant over time, there is only one metabolic pathway and side reactions or excretion of the storage compound can be neglected and there is no feedback inhibition of substrate influx. The substrate concentration in the compartment under study will then usually attain an equilibrium at a level [S]eq, at which the degradation rate is equal to the influx rate. (Normal enzyme activities must be sufficiently high to keep [S]eq of all substrates reasonably low to avoid congestion of the lysosome.) A reduction of the activity of one enzyme will, within wide limits, be compensated by an increase of [S]eq so that, due to the higher degree of saturation of the remaining enzyme, the turnover rate will still be equal to the influx rate (Figure 1).

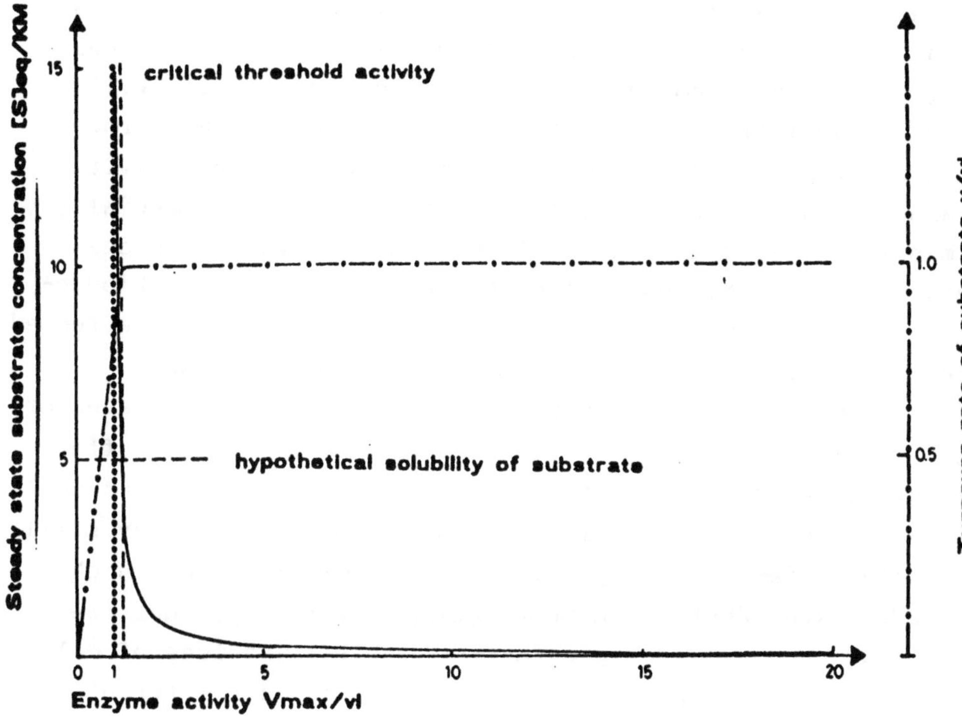

Figure 1: Theoretical correlation between the residual activity of an enzyme and the steady state concentration and turnover rate of its substrate in a defined compartment (Conzelmann and Sandhoff (1983/84)). The substrate concentration is expressed as multiples of K_M, turnover rate and enzyme activity (V_{max}) as multiples of the influx rate V_i.

____________	steady state substrate concentration [S]eq
.-.-.-.-.-.	turnover or degradation rate
....	critical threshold activity
- - - - - -	critical threshold activity, taking limited solubility of substrate into account

Only when the residual activity falls below a critical threshold (i.e., when [S]eq mathematically becomes infinite), will the turnover rate be reduced below influx rate. The difference between influx rate and degradation rate is assumed to constitute the rate of accumulation; a small variation of the residual activity in this critical region between zero and critical threshold activity will have large influence on the velocity of the storage process and hence on the development of the disease. The activity range around the critical threshold activity is especially important. Cells having about the same levels of residual hydrolase activity but different rates of substrate biosynthesis (corresponding to the influx rate v_i) will be affected quite differently. E.g. at a given level of residual hydrolase activity only neurons with specially high rates of ganglioside biosynthesis will accumulate. They will suffer first, start to dysfunction and eventually die, whereas cells with a rather low rate of ganglioside biosynthesis may still be able to degrade all gangliosides formed at rather high steady state substrate concentrations. Assuming, that the clinical phenotypes originate from the dysfunctions of cells or specific subsets of cells, it is clear that different levels of residual enzyme activities will give rise to different clinical syndromes.

With these basic considerations, at least the temporal variability of the onset of the lysosomal storage disease could be understood as the consequence of minor variations in the residual activity of the affected enzyme. They could also explain the occurrence of pseudodeficient probands (as persons with a residual activity still above the critical threshold). It seemed therefore desirable to test the underlying assumptions as well as the predictions of the model experimentally.

We performed model studies to verify the postulated correlation between residual activity and turnover of the substrate in cultured cells obtained from patients with G_{M2}-gangliosidosis, metachromatic leukodystrophy and, obligate heterozygotes and normal controls (Leinekugel et al. (1990) submitted). The radiolabeled substrates ganglioside G_{M2} and sulfatide were added to cultures of skin fibroblasts with different activities of β-hexosaminidase A (Figure 2) or arylsulfatase A, respectively, and

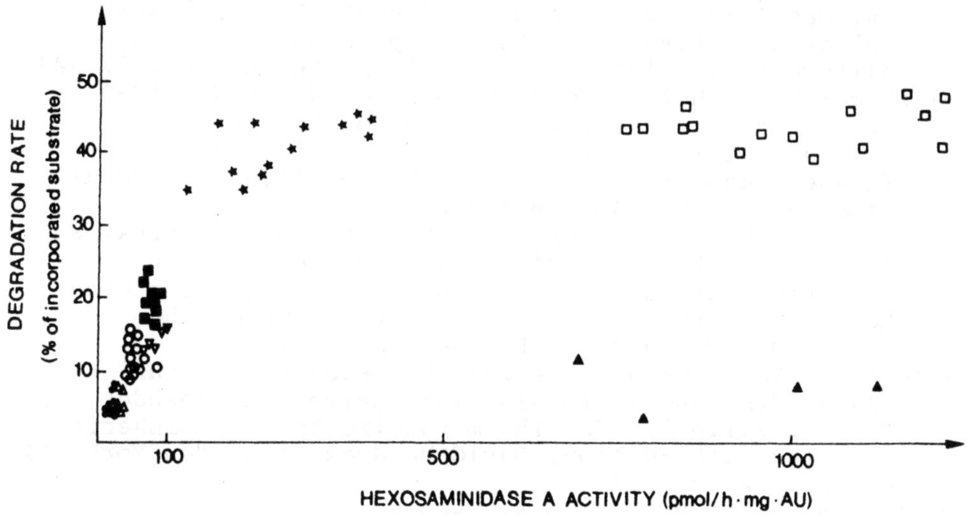

Figure 2: Hexosaminidase A activity and degradation rate of ganglioside G_{M2} in cultured fibroblasts (Leinekugel et al. (1990) submitted). Ganglioside G_{M2} was added to the culture medium of skin fibroblasts from patients with various clinical forms of G_{M2} gangliosidosis, from heterozygotes and from normal control subjects. Cellular uptake of ganglioside G_{M2} and its degradation rate were assayed after 72 h. Residual hexosaminidase A activity was determined with ganglioside G_{M2} as substrate, in the presence of the G_{M2} activator protein, in aliquots of the cell homogenate after harvesting.

● : infantile B-variant	△: infantile O-variant
O : juvenile B-variant	▽: juvenile O-variant
■ : adult B-variant	✱: heterozygote
▲: activator deficient	□: controls

their degradation rate measured. Comparison of the results (residual activity of the enzyme as well as degradation rate of the respective substrate) with the clinical status of the donor of each cell line basically confirmed our notions but also revealed the limitations of this experimental approach.

References

Conzelmann E, Sandhoff K (1983/84) Partial enzyme deficiencies: Residual activities and the development of neurological disorders. Dev Neurosci 6:58-71

Conzelmann E, Sandhoff K (1987) Glycolipid and glycoprotein degradation. In: Meister A, ed. Advances in Enzymology, John Wiley & Sons, Inc 60:89-216

Fürst W, Machleidt, W, Sandhoff K (1988) The precursor of sulfatide activator protein is processed to three different proteins. Biol Chem Hoppe-Seyler 369:317-328

Leinekugel P, Michel S, Conzelmann E, Sandhoff K. (1990 submitted) Quantitative correlation between the residual activity of β-hexosaminidase A and arylsulfatase A and the severity of the resulting lysosomal storage disease. Am J Hum Genet

Nakano T, Sandhoff K, Stümper J, Christomanou H, Suzuki K (1989) Structure of full-length cDNA coding for sulfatide activator, a co-β-glucosidase and two other homologous proteins: Two alternate forms of the sulfatide activator. J Biochem 105:152-154

Sandhoff K, Conzelmann E, Neufeld EF, Kaback M, Suzuki K (1989) G_{M2}-Gangliosidosis. In: Scriver Ch, Beaudet AL, Sly WS, Valle D, eds. The metabolic basis of inherited diseases, 6th Edition, David Mc Graw Hill, New York, II Chapter 72:1807-1839

Schröder M, Klima H, Nakano H, Kwon H, Quintern LE, Gärtner S, Suzuki K, Sandhoff K (1989) Isolation of a cDNA encoding the human G_{M2} activator protein. FEBS Lett 251:197-200

Schwarzmann G, Hoffmann-Bleihauer P, Schubert J, Sandhoff K, Marsh D (1983) Incorporation of ganglioside analogues into fibroblast cell membranes, A spin label study. Biochem 22:5041-5048

Schwarzmann G, Sonderfeld S, Conzelmann E, Marsh D, Sandhoff K (1984) Insertion into cultured cells and metabolism and intracellular transport of exogenous gangliosides. Colloque INSERM/CNRS - Aspects cellulaires et pathologiques de metabolism des glycoconjugues - Cellular and pathological aspects of glycoconjugate metabolism, INSERM 126:195-210

Sonderfeld S, Conzelmann E, Schwarzmann G, Burg J, Hinrichs U, Sandhoff K (1985) Incorporation and metabolism of ganglioside G_{M2} in skin fibroblasts from normal and G_{M2} gangliosidosis subjects. Eur J Biochem 149:247-255

LYSOSOMES AS SUICIDE BAGS

R. Wattiaux, S. Wattiaux-De Coninck, M. Jadot, I. Hamer, V. Bielande and V. Beauloye
Laboratoire de Chimie Physiologique, Facultés Universitaires N D de la Paix, 61, rue de Bruxelles, 5000 Namur,Belgium.

INTRODUCTION.

Early after the discovery of lysosomes, their involvement in pathological phenomena was proposed. It was supposed that if the organelle membrane could be disrupted inside the cell, hydrolases would have access to cytosol with as a result a dramatic degradation of vital components leading to cell death. That led to the concept of lysosomes as "suicide bags".

Many works were performed to test that hypothesis.with two types of experimental approache. In one type of experiment, cells are subject to chemical, physical agents able to cause their death and the intracellular release of lysosomal enzymes is investigated during the progression of cell damage. In a second type of experiments, it is searched if substances that bring about a disruption of lysosomal membrane "in vitro" can induce cell death and if a relationship exists between the two phenomena. Obviously, the main problem is to distinguish if the alteration of lysosomes that apparently takes place in an event leading to cell death is the causal agent of the death, thus occurs before the "point of no return" or results from the cell death and therefore can not be a causal agent of the death.

In this paper, we would like to illustrate the two experimental approaches we have just mentioned by showing 1) how lysosomes behave during ischemia, 2) how the cytotoxicity of a substance glycyl-D-phenylalanine 2 naphthylamide could be related to its effects on lysosomes.

LYSOSOMES AND ISCHEMIA

Many years ago, de Duve and Beaufay (1959) found that acid hydrolase unsedimentable activity was increased in homogenates of liver lobes subject to ischemia. Their results indicated that lysosomes were altered by ischemia but did not show if that alteration occured before or after cell death. Moreover, they did not allow to distinguish if the release of lysosomal enzymes took place inside the cells or resulted from the fact that lysosomes of ischemic cells are more susceptible to homogeneisation procedure and thus more easily disrupted "in vitro".More recently we have tried to answer these questions by measuring the latency of lysosomal enzymes after ischemia and after reperfusion to determinate to what extent the apparent lysosomal alteration caused by ischemia was

NATO ASI Series, Vol. H 62
Endocytosis
Edited by P. J. Courtoy

reversible. In agreement with the Beaufay and de Duve results,we found that a certain proportion of lysosomal enzymes was released in homogenate of liver lobe made ischemic for one hour. However, the latency change is reversible after readmission of the blood flow (Wattiaux and Wattiaux-De Coninck 1981). Accordingly, it is highly probable that the release of hydrolases observable in this conditions does not take place inside the cell but results from fragilisation of lysosomes causing them to be less resistant to homogeneisation procedure. Indeed it would be difficult to imagine by what mechanism hydrolases free in the cytosol could be reassociated with lysosomes no more than one hour after blood reflow. Therefore one hour ischemia does not probably cause a release of lysosomal enzymes inside the cell, but causes a release in the homogenate.

2h ischemia causes a drastic increase of unsedimentable acid hydrolases. However,again an important recovery is observed after reperfusion, indicating that probably the lysosomal enzyme release does not take place inside the cell. But an unexpected phenomeniom can now be seen; a secondary rise of hydrolase free activity during reperfusion.It is interesting to consider here the effect of chlorpromazine administration to the animals. The drug has only a slight effect on the hydrolase latency change observed during ischemia. On the contrary, it prevents quasi totally the latency loss caused by reperfusion. That suggests that probably the two kinds of lysosome alteration we have shown are not directly linked, the first (largely reversible) is caused by ischemia, the second (irreversible) by reperfusion. The first is only slightly affected by chlorpromazine, the second can be prevented by the drug. In fact, we have found that the effect of chlorpromazine results from hypothermia it induces.(Wattiaux and Wattiaux-De Coninck 1984) It can be observed if hypothermia is generated by other drugs than chlorpromazine. It is worthwhile to mention that hypothermia has to be maintained during the ischemic period; hypothermia is without effect if induced during reperfusion. That suggests that lysosome disruption occuring during reperfusion depends on something that takes place during the ischemic period. Moreover, the lowering of the animal temperature that prevents hydrolase release is only of 3 to 4 degree Celsius. In fact a striking parallelism exists between the effect of temperature on the lysosomal enzyme release during blood flow readmission and its effect on cell injury, caused by reperfusion known as the O_2 paradox (Hearse et al 1978). That strongly suggests that the lysosome alteration that we observe during the reperfusion originates from the reperfusion.

Now what could be the processus that takes place during ischemia, leading to an alteration of lysosomes during reperfusion. During the recent years, the involvement of oxygen free radicals in cell injury caused by reperfusion following ischemia has been emphasized. The role of xanthine oxidase could be primordial in the production of these radicals (Roy and Mc Cord 1983). Briefly, xanthine dehydrogenase would be converted into xanthine oxidase in ischemic cells. As a result, when oxygen becomes available during reperfusion, a lot of free radicals would be generated, causing severe damage to the cells. Lysosomal membrane can be disrupted by free radicals as illustrated in Fig.1.

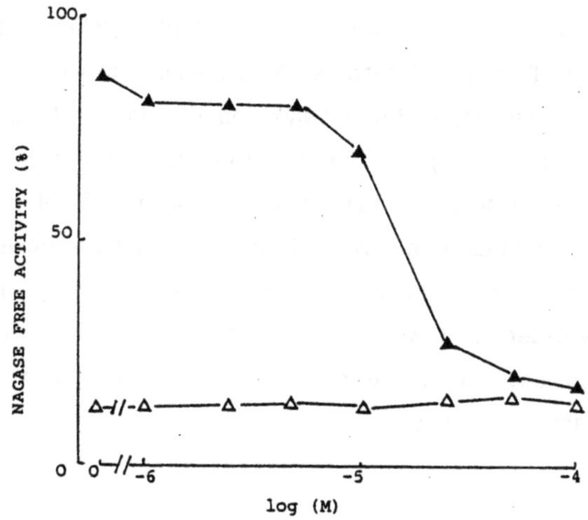

Fig.1. Effect of free radicals on lysosomes. A mitochondrial fraction from rat liver was incubated for 60 min. at 37°C in a medium containing 0.01M Tris buffer pH 8, 0.25M sucrose O.5 mM xanthine, 2 mM ADP, 0,2 mM FeCl3, d catechin in concentration indicated in abscissa, without (Δ) or with (▲) 0,018 U (Sigma) xanthine oxidase. After that free and total activities of N acetylglucosaminidase were measured. Free activity is given as percentage of total activity.

A release of N acetylglucosaminidase is observed when lysosomes are incubated in presence of xanthine/xanthine oxidase and FeADP. Such release can be prevented by adding to the medium,catechin a strong free radical scavenger. Now, we have found that, in our experimental conditions, a conversion of xanthine dehydrogenase to xanthine oxidase occurs in ischemic liver lobe, that can be prevented to some extent by hypothermia. Therefore, a plausible explanation of the lysosomal alteration that is observed during reperfusion,is that it originates from the attack of lysosomal membrane by oxygen free radicals produced by xanthine oxidase generated by ischemia.

These results suggest: 1) That at least in the case of liver ischemia, there is no intracellular release of lysosomal enzymes for up to 1 to 2 hour ischemia, and accordingly, that lysosomes probably do not behave as "suicide bags" during that pathological process. Nevertheless they are surely subject to some alterations in these conditions, as ascertained by the fact that they are less stable in the homogenate; 2) That a strong alteration of lysosomal membrane probably occurs during reperfusion, on condition that the ischemic period be long enough. Such lysosome deterioration could originate from an attack of lysosomal membrane by oxygen free radicals. Obviously that does not prove that the irreversible alteration of lysosomes that occurs during reperfusion is the causal agent of the death of cells caused by reperfusion.

CYTOTOXICITY OF GLYCYL D PHENYLALANINE 2 NAPHTHYLAMIDE (GdPN)

Several years ago we have shown that glycyl-L-phenylalanine 2 naphthylamide (GPN) was able to specifically disrupt lysosomes by a mechanism involving the penetration of the compound into the organelles and its intralysosomal hydrolysis by cathepsin C (Jadot et al.1984). Obviously if such a compound could disrupt lysosomes in the cell, it could be interesting for investigating the possible relationship of lysosomal membrane disruption and cell death. In fact, we found that GPN was cytotoxic.for cells in culture. However, when added to cell culture it is quickly hydrolysed probably by cathepsin C present in the culture medium. In some control experiments we used the D isomer of the dipeptidyl-naphthylamide and we found that it was also endowed with cytotoxicity. The cytotoxicity was reduced in presence of nigericin (Jadot et al.1990).

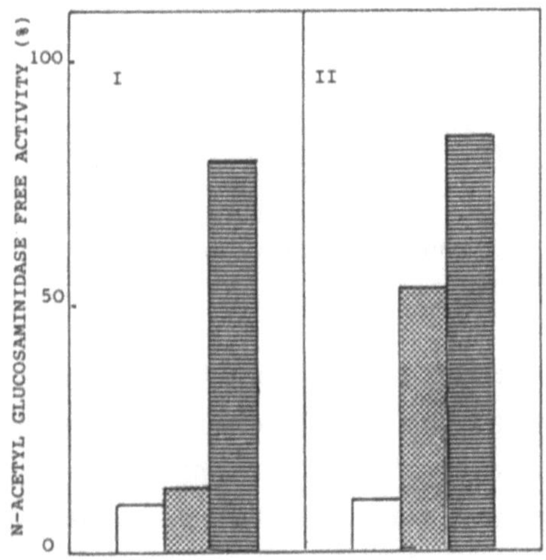

Fig.2. Effect of glycyl L phenylalanine 2-naphthylamide and glycyl D phenylalanine 2 naphtylamide on the free activity of N acetylglucosaminidase. A mitochondrial fraction from rat liver was incubated for 10 min in presence 0.25mM of glycyl D phenylalanine 2-naphthylamide (▒) or 0.25 mM of glycyl L Phenylalanine 2-naphthylamide (▤), 0.25 M sucrose, 5 mM acetate buffer pH 6 (I) or 5 mM Tris-HCl buffer pH 7.5 (II). After that free and total activities of N acetylglucosaminidase were measured. (□) free activity without incubation .

Like the L isomer, GdPN is able to unmask lysosomal enzymes "in vitro" ,as shown by Fig.2 However the effect of the D derivative is markedly less pronounced at pH 6 but is well apparent at pH 7.5. GdPN is a weak base. To see if it has to enter lysosomes to disrupt their membrane, incubation with the naphthylamide was achieved in the presence of molecules known to increase the intralysosomal pH. We found that NH4Cl, chloroquine and nigericin are able to protect lysosomes against the loss of latency of N acetyl glucosaminidase induced by GdPN. (Jadot et al. 1990) In addition GdPN has not effect on the latency of enzymes located in other subcellular structures than lysosomes. It is probable that GdPN disrupts the lysosomal membrane by a direct attack of the membrane and not as a result of an osmotic unbalance, causing the organelles to swell.

These results suggest that the cytotoxicity of GdPN is caused by its lytic effect on lysosomes. Both phenomena occur in the same concentration range, moreover, amongst the subcellular structures investigated only lysosomes are disrupted by that compound; finally nigericin opposes both lysosome lysis and cell killing by the naphthylamide.

ACKNOWLEDGEMENTS

This work was supported by the Fonds de la Recherche Scientifique, the Fonds de la Recherche Scientifique médicale,the Institut pour la Recherche Scientifique dans l'Industrie et l'Agriculture (IRSIA), the ASBL Air Escargot and the Ligue Cardiologique Belge.

REFERENCES

-de Duve,C. and Beaufay,H. (1959) Tissue fractionation studies.10-Influence of ischaemia on the state of some bound enzymes in rat liver. Biochem.J. 73: 610-616.

-Hearse,D.T., Humphrey,S.M. and Bullock,G.R. (1978) The oxygen paradox and the calcium paradox: two facets of the same problem. J.Mol.Cell.Cardiol. 10: 641-650.

-Jadot,M., Colmant,C., Wattiaux-De Coninck,S. and Wattiaux,R. (1984) Intralysosomal hydrolysis of glycyl-L-phenylalanine 2-naphthylamide. Biochem.J. 219: 965-970.

-Jadot,M., Bielande,V., Beauloye,V., Wattiaux-De Coninck,S. and Wattiaux,R. (1990) Biochim.Biophys.Acta 1027: 205-209.

-Roy,R.J. and McCord,J.M. (1983) Superoxide and ischaemia: conversion of xanthine dehydrogenase to xanthine oxidase in "oxyradicals and their scavenger system" (R.A.Greenwald and G.Cohen, eds, Elsevier Science Publ. Cy.New-York) 145-153.

-Wattiaux,R. and Wattiaux-De Coninck,S. (1981) Effect of a transitory ischaemia on the structure-linked latency of rat liver acid phosphatase and ß-galactosidase. Biochem.J. 196: 861-866.

-Wattiaux,R. and Wattiaux-De Coninck,S. (1984) Effect of ischaemia on lysosomes. Intern.Rev.Exptl.Pathol. 26: 85-106.

LYSOSOMAL STORAGE DISORDERS AND TOXICITY INDUCED BY AMINOGLYCOSIDE ANTIBIOTICS AND POLYANIONIC PEPTIDES

B.K. Kishore, P. Maldague and P.M. Tulkens
Lab. Physiol. Chem., and Unit of Exptl. Pathol. & Cytology,
Catholic University of Louvain and
International Institute of Cellular and Molecular Pathology,
avenue Hippocrate 75, Bte 75.39
B-1200 Bruxelles, BELGIUM

INTRODUCTION

Kidney plays a major role in the excretion of various end products of metabolism as well as drugs and toxins, and in the maintenance of water and electrolyte homeostasis of body fluids. Yet, the filtered proteins, peptides and other constituents may be reabsorbed by the epithelial cells of the kidney, especially in the proximal tubules, which show an intense endocytic activity. This exposes the proximal tubular cells to the insults of various exogenous and endogenous toxic substances, and often leads to acute lysosomal storage conditions. We describe hereunder the acute lysosomal storage disorders induced by aminoglycoside (AG) antibiotics and polyanionic peptides, in relation to lysosomal dysfunction, cell necrosis and cell proliferation.

LYSOSOMAL THESAURISMOSIS INDUCED BY AMINOGLYCOSIDES

AG antibiotics, such as gentamicin, are highly active against Gram (-) bacteria and hold a prominent place in the control of severe, life-threatening infections caused by these organisms. However, the use of AGs is associated with adverse effects such as nephro- and oto-toxicities and, under certain conditions, neuromuscular blockade. The incidence of nephrotoxicity in patients varies between 10 to 20% or even higher, depending upon the patient- or treatment-related risk factors (for reviews see, Tulkens 1986 & 1989). This, actually, repre-

NATO ASI Series, Vol. H 62
Endocytosis
Edited by P. J. Courtoy

sents a major limitation in the use of AGs in the clinics.

AGs are low molecular weight drugs (ca. 450 Da), made of an aminocyclitol (2-deoxystreptamine) linked by glycosidic bonds to aminated sugars. This linkage is, however, resistant to lysosomal glycosidases. AGs are also insensitive to microsomal detoxifying system. They are excreted exclusively through the kidneys by glomerular filtration in an intact form. A small, but sizeable proportion (3-5%) of the administered drug is taken up into the proximal tubular cells by adsorptive endocytosis, after binding to a putative receptor on the brush border membrane. The endocytosed drug rapidly reaches the secondary lysosomes, where it is sequestered for long duration and accumulates to a high conc. (100- to 200-fold higher than that of serum). The acidic environment (pH about 5.0 to 5.4) existing in the lysosomes causes an almost complete protonation of these polyaminated molecules. In vitro, the protonated drug tightly binds to anionic phospholipids, such as phosphatidylinositol and phosphatidylserine. Using liposomes, it was shown that this binding hampers the degradation of both acidic (phosphatidylinositol) and neutral (phosphatidylcholine, PC) phospholipids by lysosomal extracts. Inhibition of phospholipase A1 towards PC is thought to result from the fact that this enzyme requires negative charges on the bilayer to fully express its activity (Mingeot-Leclercq et al, 1988; Hostetler and Jellison, 1990). Consistent with these in vitro observations, AGs cause in vivo, an accumulation of all major phosphoglycerides with, however, a preferential increase of phosphatidylinositol in relative value. Morphologically, AGs cause the accumulation of electron-dense, osmiophilic, multilamellar, myelin-like structures, so called 'myeloid bodies' (Figure 1B; for a comprehensive description of this lysosomal phospholipidosis, see Laurent et al, 1990 and papers cited therein).

Although, various other effects of AGs on the functional integrity of plasma membranes, mitochondrial respiration, microsomal protein synthesis, prostaglandin synthesis, phosphoinositide-cascade etc... have been reported, lysosomal alterations appear to be the earliest demonstrable event in intact cortical cells both in experimental animals and humans treated with AGs. Below a critical threshold, the lysosomal altertions are apparently well tolerated by the cells. However, at levels above this threshold, phospholipidosis is conductive to the (i) lack or slowing down of lysosome-endosome fusion, (ii) release of lysosomal content, including phospholipids, into the urinary fluid and (iii) tubular necrosis leading to kidney dysfunction (Tulkens, 1989). It is assumed, but not proven that tubular necrosis occurs through the rupture and/or gross dysfunction of lysosomes. Whatever might be its mechanism, tubular necrosis is quickly associated with proliferation of tubular cells, many of which appear immature (peroxysome-poor). Most conspicuously, animals treated with aminoglycosides also show a marked interstitial proliferation of fibroblasts, even though no primary lesion is seen in that part of the kidney (Laurent et al, 1988).

PROTECTION AGAINST AG-INDUCED TOXICITY BY POLY-L-ASPARTIC ACID

Synthetic homopolymers of acidic aminoacids can modulate the AG nephrotoxicity. Thus co-administration of poly-L-aspartic acid (poly-L-Asp) with an AG completely protects rats against not only the functional and histopathological signs of nephrotoxicity (Williams et al, 1986; Gilbert et al, 1989), but also largely prevents the development of phospholipidosis (Beauchamp et al, 1990; Figure 1C). In vitro studies have demonstrated that poly-L-Asp binds AGs and also displaces them from anionic phospholipids under acidic conditions such as those prevailing in lysosomes (Kishore et al, 1990a). It also relieves the inhibition of phospholipases induced by AGs. Fractionation studies suggest that in vivo poly-L-Asp accumulates in lysosomes as the AGs do, and thereby is capable of binding the drug and preventing the development of phospholipidosis. An indirect proof that this mechanism is operating in vivo is given by the observation that poly-L-glutamic acid (poly-L-Glu), which is more than 10-fold susceptible to hydrolysis by lysosomal extracts in vitro than poly-L-Asp, offers no protection against AG toxicity in vivo (Figure 1D), despite its similar behaviour in binding and displacing the drug from anionic phospholipids in vitro (Kishore et al, 1990b). In a model of cultured fibroblasts, we have been able to make poly-L-Glu protective against AG-induced phospholipidosis if cells are incubated in the presence of leupeptin, a protease inhibitor (Kishore et al, 1990c).

LYSOSOMAL THESAURISMOSIS INDUCED BY PROTEASE-RESISTANT ACIDIC POLYPEPTIDES

Poly-D-glutamic acid (poly-D-Glu), which is highly resistant to lysosomal digestion, confers protection against the development of AG-induced phospholipidosis. But, it induces yet another, hitherto undescribed lysosomal thesaurismosis. The lysosomal storage condition induced by poly-D-Glu, either alone or in combination with AG, consists of marked enlargement and distortion of these organelles, which stain deeply with Giemsa in paraffin sections. In the electron microscope, these organelles display an osmiophilic, non-lamellar, granular content, the electron density of which becomes very high in animals co-administered with an AG (Figure 1E & 1F). The interstitium in the poly-D-Glu treated animals is often studded with a number of young, immature fibroblast-like cells. Histoautoradiographic studies reveal a conspicuous increase of the labelling index in kidney of animals receiving poly-D-Glu, with as many as 80% of the S-phase cells localised in the interstitium. Surprisingly, and in contrast to animals receiving the AG alone, the labelling index of proximal tubular cells remains low, suggesting that tubular necrosis is minimal.

SUMMARY AND CONCLUSIONS

The actual cause of cell necrosis in acquired lysosomal thesaurismoses in general, and in AG nephrotoxicity in particular, is ill-defined. However, studies reported here using either AG or poly-D-Glu indicate that the qualitative and quantitative nature of the lysosomal thesaurismosis probably plays a critical role in this respect. Thus, a generalised accumulation of phospholipids in the lysosomes as induced by AGs quickly causes tubular necrosis and repair. A similar effect was described in male F-344 rats exposed to certain petroleum hydrocarbons (2,2,4-trimethylpentane, TMP), and in which a conspicuous accumulation of alpha-2u-globulin, a low molecular weight protein, occurs in the lysosomes of renal tubular cells (Short et al, 1987). This accumulation results from the complexation of this protein with a metabolite of TMP making the protein resistant to lysosomal hydrolysis. In contrast, peptides resistant to lysosomal hydrolases per se, such as poly-D-Glu cause marked thesaurismosis without, apparently, marked tubular necrosis, but associated with a spectacular peritubular proliferation. These observations and their potential underlying mechanisms are illustrated schematically in Figure 2. They undoubtedly constitute interesting topics for future research.

REFERENCES

- Beauchamp D, Laurent G, Maldague P, Abid S, Kishore BK and Tulkens PM (1990) Protection against gentamicin-induced early renal alterations (phospholipidosis, increased DNA synthesis) by co-administration of poly-L-aspartic acid. J. Pharmacol. Exptl. Ther. vol. 255, no. 2 (in press)

- Gilbert DN, Wood CA, Kohlepp SJ, Kohnen PW, Houghton DC, Finkbeiner HC, Lindsley J and Bennett WM (1989) Polyaspartic aicd prevents experimental aminoglycoside nephrotoxicity. J. Infect. Dis. 159:945-953

- Hostetler KY and Jellison EJ (1990) Inhibition of kidney lysosomal phospholipase A1 by aminoglycosides is a novel variant of substrate depletion inhibition. J. Pharmacol. Exptl. Ther. 254:188-191

- Kishore BK, Kállay Z, Lambricht P, Laurent G and Tulkens PM (1990a) Mechanism of protection afforded by polyaspartic acid against gentamicin-induced phospholipidosis. I. Polyaspartic acid binds gentamicin and displaces it from negatively-charged phospholipid layers in vitro. J. Pharmacol. Exptl. Ther. vol. 255, no. 2 (in press)

- Kishore BK, Lambricht P, Laurent G, Maldague P, Wagner R and Tulkens PM (1990b) Mechanism of protection afforded by polyaspartic acid against gentamicin-induced phospholipidosis. II. Comparative in vitro and in vivo studies with poly-L-aspartic, poly-L-glutamic and poly-D-glutamic acids. J. Pharmacol. Exptl. Ther. vol. 255, no. 2 (in press)

- Kishore BK, Piret J, Lambricht P and Tulkens PM (1990c) Leupeptin, an inhibitor of cathepsin B, prevents gentamicin-induced lysosomal phospholipidosis (LPL) in cultured fibroblasts and aids poly-L-glutamic acid to become protectant against LPL. In: abstracts of the 30th Inter. Sci. Conf. Antimicrob. Agents Chemother. (ICAAC), Atlanta, GA, abstract no. 752

- Laurent G, Toubeau G, Heuson-Stiennon JA, Tulkens PM and Maldague P (1988) Kidney tissue repair after nephrotoxic injury: Biochemical and morphological characterisation. CRC Crit. Rev. Toxicol. **19**:147-183

- Laurent G, Kishore BK and Tulkens PM (1990) Aminoglycoside-induced renal phospholipidosis and nephrotoxicity. Biochem. Pharmacol. (in press)

- Mingeot-Leclercq MP, Laurent G and Tulkens PM (1988) Biochemical mechanism of aminoglycoside-induced inhibition of phosphatidylcholine hydrolysis by lysosomal phospholipases. Biochem. Pharmacol. **37**:591-599

- Short BG, Burnett VL, Cox MG, Bus JS and Swenberg JA (1987) Site-specific renal cytotoxicity and cell proliferation in male rats exposed to petroleum hydrocarbons. Lab. Invest. **57**:564-577

- Tulkens PM (1986) Experimental studies on nephrotoxicity of aminoglycosides at low doses. Mechanisms and perspectives. Am. J. Med. **80** (Suppl. 6B): 105-114

- Tulkens PM (1989) Nephrotoxicity of aminoglycoside antibiotics. Toxicol. Letters **46**:107-123

- Williams PD, Hottendorf GH and Bennett DB (1986) Inhibition of renal membrane binding and nephrotoxicity of aminoglycosides. J. Pharmacol. Exptl. Ther. **237**:919-925

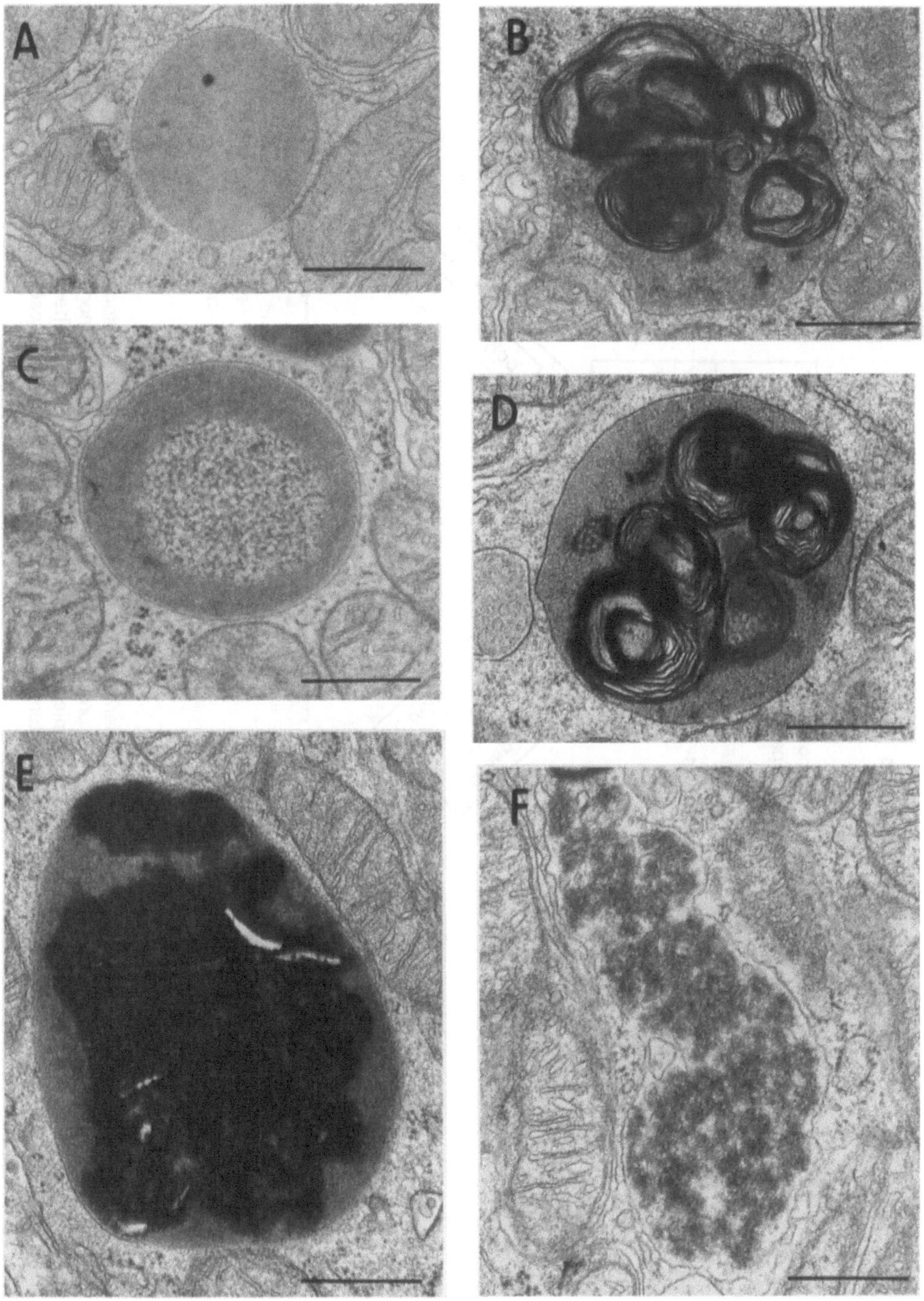

Figure 1: Electronmicroscopic appearance of lysosomes in renal proximal tubular cells of rats that were administered with saline (A), gentamicin (GM) alone (B), GM + poly-L-Asp (C), GM + poly-L-Glu (D), GM + poly-D-Glu (E) or poly-D-Glu alone (F). Bars are 0.5 μm.

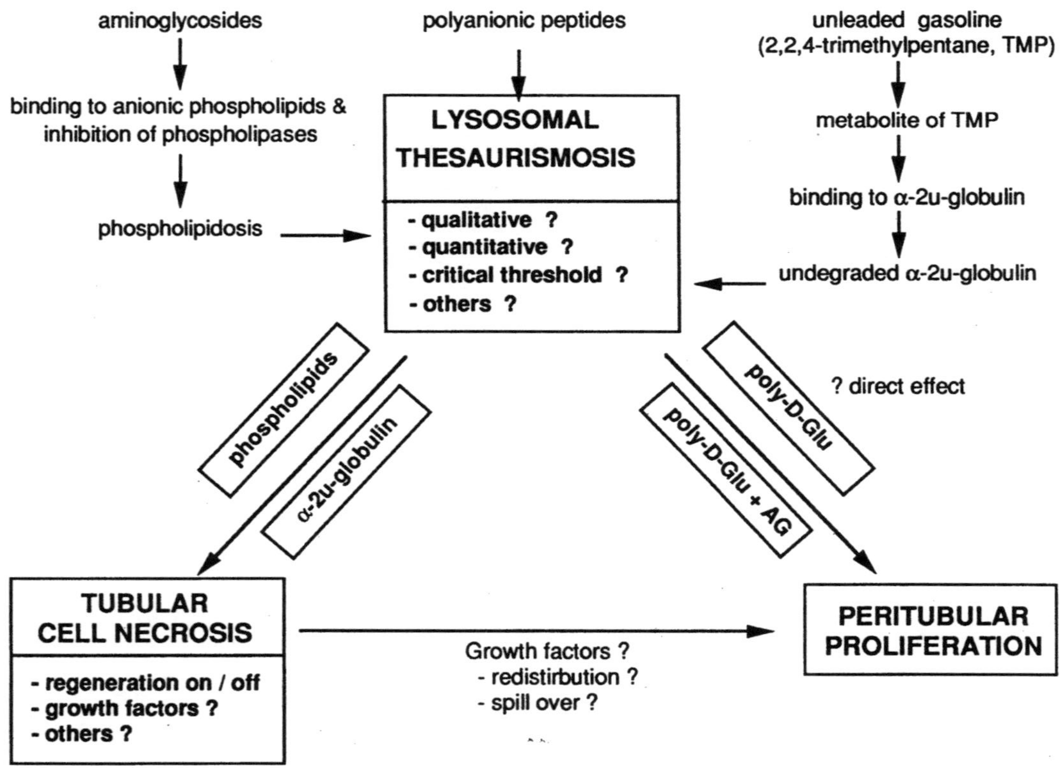

Figure 2 : Schematic representation of the relationship between lysosomal thesaurismosis, tubular cell necrosis and peritubular cell proliferation induced or modulated by aminoglycosides, polyanionic peptides and certain components of unleaded gasoline (TMP) in kidney cortex. The scheme also shows the various hypothetical or ill-defined factors controlling or responsible for each of these events.

2.5. ENDOCYTOSIS IN NON-MAMMALIAN CELLS

(Chairperson : David ROBINSON)

In a plenary lecture, the Nobel laureate **Christian de Duve** speculated that the acquisition of endocytosis, a virtually universal activity of eukaryotic cells, could have been essential for two critical steps in the emergence of our form of life. According to his views, phagocytosis and digestion of the preys first led to independence from the cocoon of favorable but restricted nutritive niches, by providing the capacity of "hunting and developing in the unknown world". In a second stage, phagocytes somehow realized the advantage of collaboration. Some preys were not destroyed but became the endosymbiotic precursors of mitochondria, chloroplasts and perhaps peroxisomes. It is to be hoped that evolutionary dating of well-conserved proteins will allow to test these fascinationg hypotheses.

In contrast to the large body of knowledge on endocytosis in mammalian cells, little work is currently being performed on other eukaryotic cells. **David Robinson** described the peculiarities of endocytosis in plants. In contrast to animal cells, plant cells develop a hydrostatic pressure (turgor) which has to be overcome for endocytosis to take place. Nevertheless, fluid-phase endocytosis can be demonstrated in turgescent plant tissues with lucifer yellow, and shows the same requirements as in mammalian cells. Receptor-mediated endocytosis also occurs for auxin, fungal toxins and defense elicitors. The endocytic apparatus of plant cells comprises endocytic coated pits, a "partially coated reticulum", multivesicular bodies and the central vacuole. Plants should not be overlooked as a convenient source of coated vesicles.

Howard Riezman has chosen to concentrate on yeast, with the long-term goal of genetic studies of the regulation of endocytosis. He reported that α-factor and its receptor are internalized by receptor-mediated endocytosis by *Saccharomyces cerevisiae* a cells. Internalization can be uncoupled from signal transduction. Two mutants, afi1 and afi2 are defective for alpha-factor internalization. The delivery to the vacuole involves an endosomal intermediate which has been recently purified.

Receptor-mediated endocytosis can be exploited by parasites in the bloodstream to efficiently capture essential nutrients from their hosts. **Isabelle Coppens** reported that *Trypanosoma brucei brucei*, the agent of sleeping sickness in animals called nagana, does not synthesize its own cholesterol, but avidly takes up LDL particles of its hosts. The intriguing

NATO ASI Series, Vol. H 62
Endocytosis
Edited by P. J. Courtoy

characteristics in this cell is a very high number of receptors on a cell mass basis, their concentration in the tiny flagellar pocket where endocytosis only takes place, and a presumably very fast recycling. More fascinating, there are great similarities with the LDL receptor of the hosts, despite the fact that evolutionary divergence occurred about one billion year ago. The LDL receptor of the parasite also differs from the major cell surface protein in being a stable surface antigen and could thus be used for a vaccine against sleeping sickness, if parasite-specific epitopes are identified.

ENDOCYTOSIS IN THE EMERGENCE OF EUKARYOTIC LIFE

Christian de Duve
International Institute of Cellular and Molecular Pathology
Avenue Hippocrate 75, Boîte 75.50, 1200 Brussels, Belgium
and The Rockefeller University, 1230 York Avenue, New York, N.Y. 10021, U.S.A.

The findings of modern biology, in particular the growing number of results provided by the comparative sequencing of proteins and nucleic acids, support the conclusion that all known living organisms are descendants from a single ancestral form of life. This common ancestor was probably a fairly typical prokaryote, perhaps not very different from some present-day bacteria. According to geochemical and other evidence, this organism arose some time between 4.0 and 3.5 billion years ago. How it did has been discussed elsewhere (de Duve, 1990, 1991) and will not be considered in this paper, which is restricted to some aspects, directly related to the topic of this symposium, of the further evolution of this common ancestral form. A few key steps in this evolution are shown schematically in Figure 1.

According to the latest phylogenetic reconstructions (Woese, 1987; Sogin, 1989), the first documented bifurcation in the tree of life occurred some 3.5 billion years ago, perhaps earlier, and gave rise to two divergent bacterial lines. One of these led to the Eubacteria, which comprise most of the more common bacterial forms found in the world today. A particularly important development in the evolution of this line was the appearance of the first oxygen-producing phototrophs, the ancestors of present-day cyanobacteria. This event led to a progressive rise in the oxygen content of the atmosphere, starting about 2.0 billion years ago, and forced all living organisms that did not take refuge in an oxygen-free niche to adapt to aerobic conditions. Some did so in a primitive fashion by simply acquiring the ability to detoxify oxygen. Others took full advantage of the situation by utilizing oxygen as the final electron acceptor of elaborate, phosphorylating, respiratory chains.

NATO ASI Series, Vol. H 62
Endocytosis
Edited by P. J. Courtoy

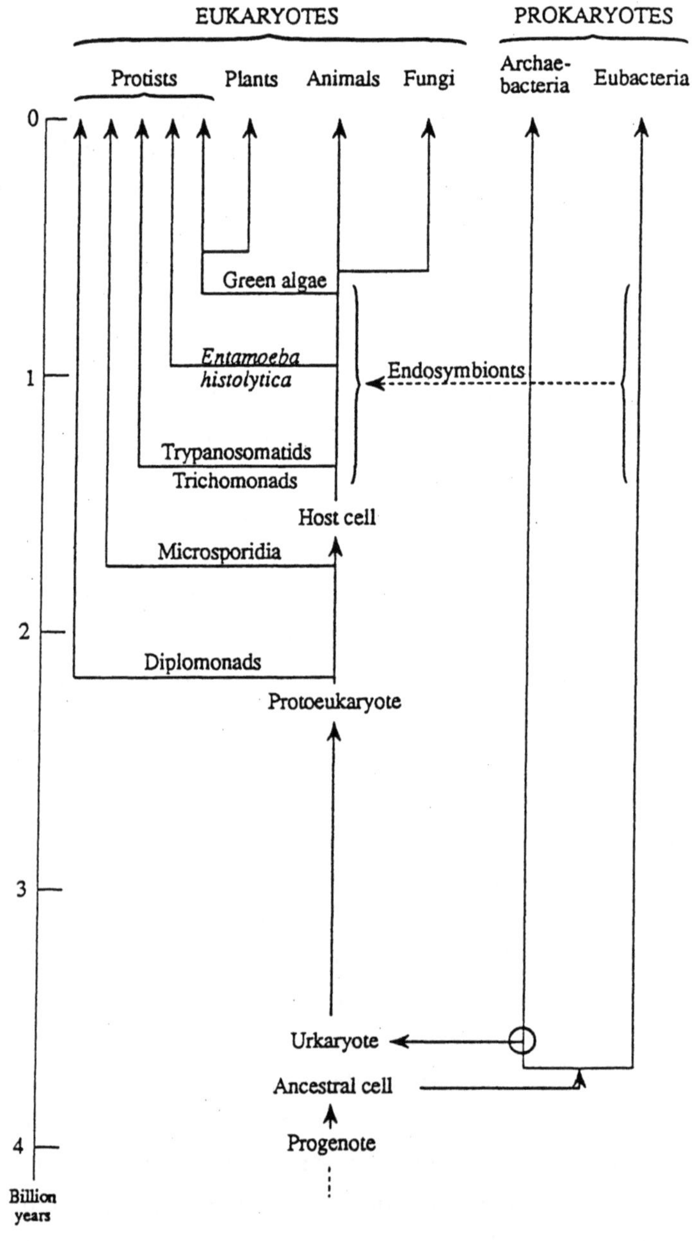

Figure 1. Simplified phylogenetic tree of living organisms according to Woese (1987) and Sogin (1989). Reprinted from de Duve (1991) with permission of the publisher.

The second bacterial line originating from the universal ancestor soon split again into two lines, one of which led to the other main prokaryotic branch : the Archaebacteria, among which are found a number of rather exotic species, including sulfur-utilizing thermophiles, halophiles, and methanogens. The other line arising from this fork developed into the whole family of Eukaryotes, which comprise all the protists, plants, fungi, and animals, including humans. The rest of this paper will be devoted to the remarkable prokaryote-eukaryote transition and to the important roles that endocytosis may have played in its unfolding.

When one compares the structural and functional organization of a typical animal or plant cell with that of a bacterium, one is hard put to imagine how the one can possibly have originated from the other. In recent years, a partial clue to this problem has been provided by the growing evidence that supports an endosymbiont origin of mitochondria and plastids, perhaps also of microbodies (The evidence for the latter is less compelling as they do not contain any genetic system). A possible scenario, known as serial endosymbiosis, pictures the first endosymbionts as primitive aerobic bacteria that established themselves within some host cells and eventually developed into peroxisomes and other related microbodies. Later, highly perfected aerobic bacteria were similarly adopted and gave rise to the mitochondria. Finally, one or more cell lines possessing both microbodies and mitochondria acquired a third kind of endosymbionts -- phototrophic, oxygen-releasing cyanobacteria -- which became the chloroplasts. Green algae and plants descend from cells that have adopted all three organelles. Fungi, animals, and many protists come from cells that have adopted only the former two. These events probably took place between about 1.5 and 1.0 billion years ago, i.e., after oxygen had started appearing in the atmosphere, a point that may be of some significance, as I shall point out later.

Thus, one major characteristic of eukaryotic cells, namely the presence of membrane-bounded organelles of energy metabolism in their cytoplasm, is explained by a relatively simple and direct derivation from prokaryotic precursors. However, there can be no endosymbionts without a host cell. For obvious reasons, this cell could not have been a simple prokaryote. For one thing, it must have been much larger than its prokaryotic guests; for the other, it must have had the means to let the guests in. Most authors believe that the bacterial ancestors of cytoplasmic organelles were originally taken up by phagocytosis. After that, they escaped intracellular destruction, as do many pathogenic bacteria, but without endangering the viability of their host cells. Instead, they progressively established a mutually beneficial relationship with the latter. This is not an unusual phenomenon. It still occurs on a large scale today.

An interesting question, in the framework of this hypothesis, concerns the fate of the original phagocytic vacuole. According to one school, this structure has been preserved in the outer membranes of mitochondria and of chloroplasts. Another alternative, defended by Cavalier-Smith (1987), is that the vacuolar membrane was lost and that the outer membranes of the organelles are derived from the outer membranes of their gram-negative ancestors. In either case, the inner membranes of the organelles are seen as originating from the plasma membranes of the ancestral prokaryotes. As to the microbodies, their single membrane could be derived from the vacuolar membrane, if their plasma membrane was lost; or it could, as is believed by Cavalier-Smith, originate from the plasma membrane of a gram-positive ancestor. There is at present no definitive solution to the problem except, perhaps, in the case of chloroplasts. The outer membrane of these organelles has a number of special lipids in common with the outer membrane of cyanobacteria, strongly indicating an evolutionary relationship between the two (Joyard et al., 1990).

The endosymbiont origin of cytoplasmic organelles brings to light one key evolutionary role of endocytosis. Without this process, some of the hallmarks of all but the most primitive eukaryotic cells would not have been acquired; plants and animals presumably would never have arisen. Note, however, that we are dealing with special endocytic events that, unlike their more common forms, led to the demise of neither the prey nor the captors. They resulted, instead, in an increasingly intimate integration of the captives with the captors. I shall not consider these fascinating phenomena further, as they fall outside the topic of this symposium. Let me now turn to a second question, which as I hope to show, illustrates another important evolutionary role of endocytosis. This question is the actual origin of the primitive phagocyte that first engulfed and adopted endosymbionts.

By definition, that cell must have been large and phagocytically active. Most likely, it possessed the kind of differentiated cytomembrane system generally associated with endocytosis, including an appropriately equipped plasma membrane, perhaps fitted with receptors; a lysosome system, for the performance of intracellular digestion; and a secretory machinery, possibly subdivided into a series of connected compartments, for supplying the lysosomes with digestive enzymes. It seems reasonable to assume that the genome of this large cell was centrally located and enclosed within an envelope derived from the cytomembrane system. It is also likely that the cell contained such cytoskeletal and motor elements as were needed to provide it with structural support, to guide and drive its intracellular traffic, and to assist in mitosis. In other words, it seems probable that the original host cell already possessed the main characteristics of eukaryotic cells, except for the absence of membrane-bounded organelles of energy metabolism. This hypothetical picture is supported by what is known of the most ancient protists, the diplomonads and the microsporidia, which are believed to have branched from the main eukaryotic line before endosymbionts were adopted. The question is how such an elaborately constructed cell could possibly have originated from a prokaryote.

In all likelihood, a key, early mechanism in this transformation was a progressive expansion of the plasma membrane of the ancestral prokaryote, leading to the development of an increasingly elaborate system of intracellular invaginations and, eventually, pockets derived from such invaginations. There are so many structural, functional, and biochemical similarities between parts of the eukaryotic cytomembrane system and the prokaryotic plasma membrane as to make this hypothesis very probable. One can readily visualize that a progressive differentiation of such an intracellular system could have led to the segregation of different plasma membrane functions in different parts of the system. For example, only the deepest recesses of the system would have kept ribosome-binding sites and associated protein-translocating machineries. Enzymes of lipid metabolism and glycosylating systems would have become concentrated in other parts. The site of anchoring of the bacterial chromosome to the plasma membrane and the connected replicating systems would have been driven inward, eventually to become enclosed by a membrane-derived envelope. And so on. Most likely, the cells undergoing this kind of transformation would, at some stage, have lost their outer walls, thereby gaining greater flexibility. Such wall-less forms (L-forms) exist in nature today.

One consequence of the postulated membrane-expansion phenomenon is that it would have increased the surface area available for exchanges between the cell and its environment, thereby allowing a commensurate increase in cell volume. It is, indeed, conceivable that the volume of bacterial cells surrounded by a smooth membrane is limited by the requirements of these exchanges. Supporting such a hypothesis is the example of Thiovulum majus Hinze, studied by Fauré-Frémiet and Rouiller (1958). This is a large, sulfur-utilizing bacterium, some 20 micrometers in diameter. Its plasma membrane shows a considerable degree of expansion, forming evaginations flattened against the cell wall and, especially, deeply indented invaginations, apparently studded with ribosomes and resembling rough-surfaced endoplasmic reticulum. This organism, however, does not show any sign of membrane differentiation. It is simply a very large prokaryote, surrounded by a correspondingly expanded plasma membrane and characteristically supported by a rigid outer wall.

If, as suggested above, only a wall-less form undergoing a concerted increase in cell volume and in membrane surface area could have differentiated further, it would most likely have required cytoskeletal elements to act as internal props. Furthermore, for functional differentiation of the membrane system to take place as I visualize, means of vesicular transport would have been necessary. Therefore, transition from a wall-less prokaryotic ancestor to a large, compartmentalized phagocyte by the mechanism envisaged could have taken place only through the simultaneous, co-evolutionary development of cytomembranes and of cytoskeletal and motor systems. This could have been a very long process, as it required the evolutionary appearance and development of highly specialized new proteins that are absent in bacteria, for example actin and tubulin, as well as myosinlike and dyneinlike ATPases. According to phylogenetic reconstructions, the whole transition could have taken as long as one billion years. The process, nevertheless, remains plausible, because it could have followed a smooth, progressive, stepwise course, each new development of the membrane system awaiting some improvement of the cytoskeletal and motor systems and vice versa. One key question remains, however, : What were the selective driving forces of the transformation ? What advantages did the cells gain, at every step of this lengthy process, that compensated for their bulk, sluggishness, and fragility ? The answer to this question, at least as I see it, brings us back to the topic of this symposium, namely endocytosis and its ancillary function, intracellular digestion.

Bacteria do not have the ability to engulf food and to digest it intracellularly. They can only digest food particles extracellularly by means of secreted exoenzymes. Gram-positive bacteria secrete their exoenzymes directly into the outside medium, so that they can take advantage of the activity of these enzymes only if they occupy a stagnant and confined environment. They are, so to speak, condemned to live within their food supply. Gram-negative bacteria are naturally surrounded by a stagnant and confined environment within which they can carry out digestion, namely the periplasmic space, which exists between the plasma membrane and the outer membrane. This advantage is, however, limited by the poor accessibility of the periplasmic space to exogenous food particles, even macromolecules.

Consider now what happened to the bacterial digestive system once the cell wall was lost and the membrane-expansion process that I postulate was initiated. Each invagination of the plasma membrane would become a segregated site, within which trapped food particles and macromolecules would become exposed to locally secreted digestive enzymes. The narrower the connection between the invagination and the surrounding medium, the tighter the segregation. Let the walls of this gullet fuse and the invagination detach from the plasma membrane to form an intracellular vacuole, and an authentic primitive form of endocytosis and of intracellular digestion would have been inaugurated. The site of digestion would have been an intracellular pocket combining the properties of endosomes, rough-surfaced endoplasmic reticulum, and lysosomes.

Cells capable of carrying out such processes, even in a very rudimentary and haphazard fashion, would have enjoyed a considerable advantage. No longer chained to their food supply, they would have acquired the freedom to move out and explore the outside world, surviving on the bacteria and other nutritive materials that they could engulf and digest intracellularly. This development could well have heralded the beginning of cellular emancipation. Once it was initiated, any evolutionary innovation that would have improved the efficiency of the uptake mechanism, or of the digestive process, or of both would have been the source of an additional selective advantage. However long the time it required, however many successive mutations were needed, the development could only move from improvement to improvement, driven by the nutritive advantages it conferred. Thus, if these guesses are correct, endocytosis would also have provided the main driving force in the early part of the prokaryote-eukaryote transition, up to the appearance of the large, primitive phagocyte that, in turn, used endocytosis to acquire endosymbiont-derived organelles.

It is conceivable that the cells that underwent the postulated changes lost, or failed to develop, adequate membrane-bounded electron-transport chains and associated phosphorylation systems, being busy, so to speak, differentiating their membranes in other directions. As a result, they may have been poorly prepared for the progressive rise in atmospheric oxygen that started some 2.0 billion years ago. The early endosymbiotic adoption of primitive aerobes capable of detoxifying oxygen, the possible ancestors to peroxisomes and other related microbodies, could have been the means that saved the phagocytes from the oxygen threat. The rescued cells may later have turned the menace into a major advantage by enslaving the most advanced aerobic organisms present, thereby acquiring mitochondria. Finally, cells thus equippped could have afforded to host the actual generators of the poison and to exploit for their own benefit the ability of their captives to utilize sunlight.

Let me conclude with a general remark, relevant to the development of science in our days. Thirty years ago, the kind of story that I have briefly recounted would have sounded like science-fiction, possibly entertaining but lacking any solid basis. It is still highly speculative today but no longer pure invention. One of the lessons of modern biology is that living cells do not only yield the secrets of their structural and functional organization to our increasingly incisive molecular dissections. They also reveal their past history. If my attempted reading of the document is correct, endocytosis was invented more than 3.0 billion years ago and played a decisive role in the evolution of life.

References

Cavalier-Smith, T (1987) The Simultaneous Symbiotic Origin of Mitochondria, Chloroplasts, and Microbodies. Ann. N.Y. Acad. Sci., 503, 55-71

de Duve, C (1990) Construire une Cellule. Bruxelles: De Boeck-Wesmael; Paris: InterEditions (1990)

de Duve, C (1991) Blueprint for a Cell. Burlington NC: Neil Patterson Publishers

Fauré-Frémiet, E and Rouiller, C (1958) Etude au Microscope Electronique d'une Bactérie Sulfureuse, Thiovulum majus Hinze. Exp. Cell Res., 14, 29-46

Joyard, J, Block, MA, Dorne, A-J and Douce, R (1990) Comparison of Plastid Envelope Membranes and Outer Membranes from Cyanobacteria: Relevance to Chloroplast Evolution. in: Endocytobiology IV (Nardon, P. et al., editors). Paris: Institut National de la Recherche Agronomique, 527-536

Sogin, ML (1989) Evolution of Eukaryotic Microorganisms and their Small Subunit Ribosomal RNAs. Amer. Zool., 29, 487-499

Woese, CR (1987) Bacterial Evolution. Microbiol. Rev., 51, 221-271 .

ENDOCYTOSIS IN PLANTS: PROBLEMS AND PERSPECTIVES

D.G. Robinson, R. Hedrich, B. Herkt, W. Diekmann and M. Robert-Nicoud

Abteilung Cytologie, Pflanzenphysiologisches Institut, Universität Göttingen. W - 3400 Göttingen. FRG.

Studies on endocytosis in plants are still in an elementary, descriptive stage, because both theoretical and technical difficulties, which are absent in the case of animal cells, have had and still need to be overcome.

The turgor problem

Unlike animals, plant cells develop a hydrostatic pressure termed turgor. Because of this it has been widely held that vesiculation at the plasma membrane (PM) against this pressure is energetically not feasible. Thermodynamical arguments indeed suggest that coated vesicle-mediated endocytosis may not be possible at turgor pressures greater than 1 bar (Gradmann and Robinson 1989). Textbooks normally give values for turgor ranging from 0.5 bar (root hairs) to over 30 bar for guard cells in leaves (Raschke 1979). "Typical" turgescent plant tissues have turgor pressures between 5-10 bar implying that endocytosis would not be a "typical" feature of plants. Many of these reservations against endocytosis are, in our opinion, unfounded, since they fail to take into accout the huge fluctuations and variability in turgor pressure which can be encountered in plants. Thus there can be rapid changes (10 bar in 1 h) in leaf turgor as well as considerable diurnal oscillations (Cutler et al 1980); leaf turgor can drop to 0.7 bar under saline conditions (Neumann et al 1988); and zero (wilting!) or even negative turgor can often be measured in leaves in field conditions (e.g. Turner 1974).

NATO ASI Series, Vol. H 62
Endocytosis
Edited by P. J. Courtoy

The cell wall problem

As discussed by Robinson and Hillmer (1990a) plant cell walls can easily allow the passage of molecules of up to 60 kDa (Stokes radius approx. 4 nm). The problem with the cell wall is not so much its presence (given time even electron dense tracers like cationic ferritin reach the PM and are endocytosed, L.C. Fowke personal communication), but what happens when it is removed. It is known that phosphorylation pattern of PM proteins are changed within minutes after the addition of protoplasting enzymes to plant cells (Blowers et al 1988), and it is conceivable that the quantity and quality of cell surface receptors may also be affected by such treatments.

Another problem when using protoplasts is that the enzymic breakdown of the cell wall causes the release of oligosaccharide fragments which are potential ligands for cell surface receptors (see below). As a result protoplasts may not respond to the addition of new ligand because many receptors may be blocked and the protoplasts may already be in a hypersensitized condition. One of the few, well-documented, examples where this is not the case is that of protoplasts prepared from suspension-cultured parsley cells, where the addition of elicitor (a 42 kDa glycoprotein, Parker et al 1990) still gives rise to a clear biochemical response (Dangl et al 1987).

Fluid phase endocytosis (FPE)

When presented to plant cells the membrane impermeant fluorochrome Lucifer Yellow-CH (LY) is taken up and accumulates in the vacuole. This phenomenon had been demonstrated with suspension cultured cells (Hillmer et al 1989), with intact roots (Oparka et al 1988), and with excised tissue (Oparka and Prior 1988; Hillmer et al 1990). Two lines of evidence indicate that LY enters the cell and reaches the vacuole via FPE. Firstly LY is not taken up by isolated vacuoles (Hillmer et al 1989); secondly, when introduced into the cytosol of protoplasts via patch pipettes (Hillmer et al 1990; Fig. 1d-g) LY does not gain access to the vacuole. As in animal cells, LY uptake in plants is non-saturable, energy and temperature dependent.

Unequivocal proof of the operation of FPE in plant cells can probably only be given by microinjecting antibodies against LY into the cytosol of a cell or protoplast prior to exposing to exogenous LY. Should a vacuolar accumulation of LY still occur this would indicate that the

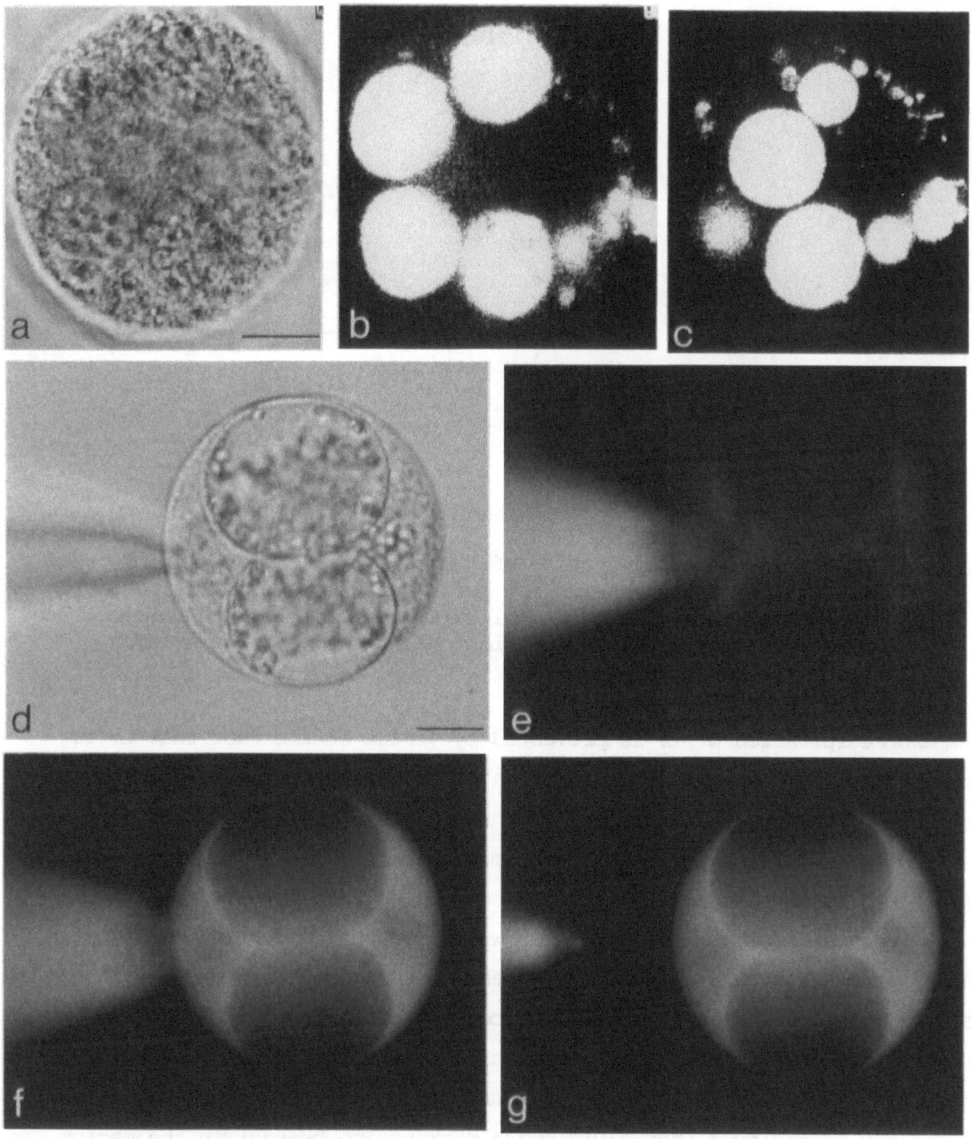

Fig. 1 Lucifer yellow-CH uptake into barley aleurone protoplasts (Developmental stage nomenclature after Bush et al 1986. During development protein bodies develop into vacuoles which fuse with one another).
a-c: Endocytic uptake of exogenous LY (2 $mg.ml^{-1}$) into vacuoles (Stage II/III Protoplasts). a. normal optics, b.,c. laser scan fluorescence with confocal optics of the same protoplast as in a. but at two different levels (Protoplasts immobilized in agarose).
d-g: Cytosolic LY loading with patch clamp pipettes (Stage III/IV Protoplasts). e. Patch pipette attached to protoplast, f. "whole-cell" configuration, begin of loading (few seconds), g. end of LY loading (5-6 min), h. Protoplast 2h 44min after release from patch pipette. LY has not gained access to the vacuole.
Magnification bar = 10 um.

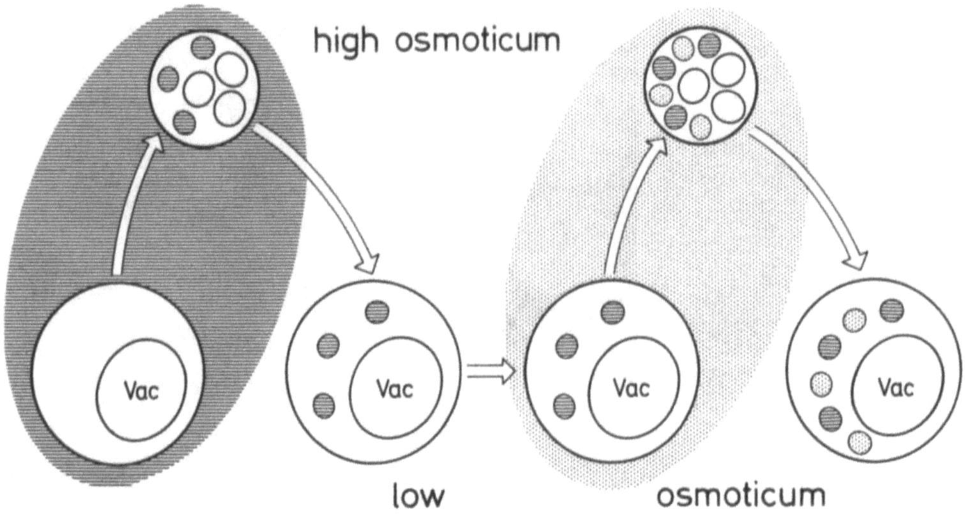

Fig. 2. Diagrammatic representation of "osmocytosis". Two cycles of osmotic contraction and expansion are portrayed with cycle 1 being performed in the presence of Lucifer yellow CH and cycle 2 with Cascade blue. The final protoplast at the right has two types of fluorescing endocytic vesicles.

immunoprecipitation of LY is prevented by virtue of the fact that LY is transported to the vacuole in vesicles. The nature of these vesicles is at the moment unknown. If they are clathrin-coated, their transport will, in analogy to similar experiments on animal cells (Doxsey et al 1987), be inhibited by microinjecting clathrin antibodies into the cytosol.

An unusual form of FPE in plants, which does not occur in animal cells, deserves mention here. Protoplasts behave as osmometers: they swell and shrink in accordance with the solute concentration in the surrounding medium. Contraction of protoplasts is accompanied by the formation of large, smooth-surfaced endocytic vesicles (Gordon-Kamm and Steponkus 1984). This event may be followed when fluid phase markers are present in the medium, and can in fact be repeated sequentially when a second marker is present (Fig. 2). As a result, it is possible to have two sets of endocytic vesicles existing side by side in the same cell (Oparka et al 1990). Two aspects of this phenomenon are of interest: these endocytic vesicles do not fuse with one another, nor do they fuse with the central vacuole. Moreover the membrane of these vesicles is not reinserted into the PM within several hours following cell expansion. Because of these differences to normal FPE we would prefer to term this phenomenon "osmocytosis".

Receptor-mediated endocytosis (RME)

There are a number of substances present in the apoplast (extracellular or cell wall space) of the higher plant which can give rise to a biochemical or morphogenetic response of one kind or another. Representatives of such "signalling" molecules which appear to interact with the PM, are to be found among the plant growth substances, fungal toxins and plant defense elicitors. Thus both auxin (hormone) and fusicoccin (toxin) are known to cause changes in the transmembrane potential of protoplasts (Barbier-Brygoo et al 1989), and to bind to the PM in vitro (Dohrmann et al 1978; De Michaelis et al 1989). In addition defense elicitors which are either oligosaccharides or glycoproteins derived from fungal or plant cell walls (e.g. Scheel and Parker 1990), have also been shown to bind with high affinity to the PM (Corsio et al 1988). These elicitors clearly have quite distinct targets in the PM since several species-specific elicitors have been isolated from one and the same phytopathogen (Parker et al 1988).

Work towards the characterization of the corresponding receptors is in progress. The fusicoccin receptor appears to be a trimeric integral membrane protein with subunits of around 30 kDa (De Boer et al 1989; Meyer et al 1989). Identification of the auxin receptor is complicated by the fact that there are three different binding sites for auxin in the cell (Dohrmann et al 1978). In addition to the PM ("site III"), auxin also binds to the tonoplast ("site II") and endoplasmic reticulum, ER, ("site I"). The largest amounts of auxin binding proteins (abp) in the cell are located at site I and have been shown to be a 22 kDa, high mannose glycoprotein having an ER-retention (KDEL) sequence (Hesse et al 1989). On the other hand site III appears to have 2 abp: one a polypeptide of 40 kDa and the other at 42 kDa (Hicks et al 1989).

Results of experiments designed to demonstrate ligand or ligand-receptor internalization in plants have so far only been published in the case of defense elicitors. Horn et al (1989) have coupled fluorescein isothiocyanate (FITC) to two different elicitor preparations and were able to demonstrate the presence of the FITC-elicitor conjugates in the vacuoles of suspension-cultured soybean cells after several hours. Uptake appears to be saturable after about 3 h and is elicitor-specific since neither FITC-BSA nor FITC-insulin (both capable of traversing the cell wall) were internalized. Interestingly vacuolar accumulation of the two elicitor conjugates proceeded at different rates suggesting either different endocytic routes or different rates of processing along the same route. Unclear remains the nature of the fluorescent signal in the vacuole in these experiments: it is assumed that the FITC-conjugate is intact there but evidence for this is lacking in the paper of Horn et al (1989).

Ultrastructural studies: the endocytic pathway in plants

Studies on protoplasts with unspecific, electron-dense, markers have established that at least 3 major compartments are present in the plant endocytic pathway (reviewed by Robinson and Hillmer 1990b). After internalization via coated pits and coated vesicles these tracers are first detected in a structure termed the "partially coated reticulum". This is a structure which is equivalent to the trans Golgi network of animal cells (Hillmer et al 1988), and bears large numbers of budding coated vesicles. The second compartment is the multivesicular body, which closely resembles the "early endosome" of animal cells. The final "resting place" of endocytosed substances in plants is the central vacuole. Fluid phase markers, in contrast to animal cells, do not seem to be released once they have been deposited in the vacuole. Assuming RME does occur in plant cells, we do not know, as yet, in which of these compartments ligands become separated from their receptors. Whether receptors are recycled back to the PM, and from which compartment, also remains a matter for speculation.

Acknowledgements

Heike Freundt is thanked for her help in preparing this manuscript and Bernd Rauffeisen for the art-work. Dr. T. Jovin is thanked for the use of his CLSM facilities. Useful discussions with Dr. K. Raschke are gratefully acknowledged. Work out of our own laboratory was supported by the Deutsche Forschungsgemeinschaft.

References

Barbier-Brygoo H, Ephritikikhine G, Klämbt D, Ghislain M, Guern J (1989) Functional evidence for an auxin receptor at the plasmalemma of tobacco mesophyll protoplasts. Proc Nat Acad Sci (USA) 86: 891-895

Blowers DP, Boss WF, Trewavas AJ (1988) Rapid changes in plasma membrane protein phosphorylation during initiation of cell wall digestion. Plant Physiol 86: 505-509

Boer AH de, Watson BA, Cleland RE (1989) Purification and identification of the fusicoccin binding protein from oat root plasma membrane. Plant Physiol 89: 250-259

Bush DS, Cornejo M-J, Huang CN, Jones RL (1986) Ca^{2+}-stimulated secretion of -amylase during development in barley aleurone protoplasts. Plant Physiol 82: 566-574

Corsio EG, Popperyl H, Schmidt WE, Ebel J (1988) High affinity binding of ß-glucan fragments to soybean Glycine max L. microsomal fractions and protoplasts. Eur J Biochem 175: 309-315

Cutler JM, Steponkus PL, Wach MJ, Shahan KW (1980) Dynamic aspects and enhancement of leaf elongation in rice. Plant Physiol 66: 147-152

Dangl JL, Hauffe KD, Lipphardt S, Hahlbrock K, Scheel D (1987) Parsley protoplasts retain differential responsiveness to U.V. light and fungal elicitor. EMBO J 6: 2551-2556

Dohrmann UC, Hertel R, Kowallik H (1978) Properties of auxin binding sites in different subcellular fractions from maize coleoptiles. Planta 140: 97-106

Doxsey SJ, Brodsky FM, Blank GS, Helenius A (1987) Inhibition of endocytosis by anti-clathrin antibodies. Cell 50: 453-463

Gordon-Kamm WJ, Steponkus PL (1984) The behaviour of the plasma membrane following osmotic contraction of isolated protoplasts: Implications in freezing injury. Protoplasma 123: 83-94

Gradmann D, Robinson DG (1989) Does turgor prevent endocytosis in plant cells? Plant, Cell and Envir 12: 151-154

Hesse T, Feldwisch J, Balshüsemann D, Bauw G, Puype M, Vanderkerckhove J, Löbler M, Klämbt D, Schell J, Palme K (1989) Molecular cloning and structural analysis of a gene from Zea mays (L.) coding for a putative receptor for the plant hormone auxin. EMBO J 8: 2453-2461

Hicks GR, Rayle DL, Jones AM, Lomax T (1989) Specific photoaffinity labelling of two plasma membrane polypeptides with an azido auxin. Proc Nat Acad Sci USA 86: 4948-4952

Hillmer S, Hedrich R, Robert-Nicoud M, Robinson DG (1990) Uptake of Lucifer yellow CH in leaves of Commelina communis is mediated by endocytosis. Protoplasma (in press)

Hillmer S, Quader H, Robert-Nicoud M, Robinson DG (1989) Lucifer yellow uptake in cells and protoplasts of Daucus carota visualized by laser scanning microscopy. J Exp Bot 40: 417-423

Horn MA, Heinstein PF, Low PS (1989) Receptor-mediated endocytosis in plant cells. Plant Cell 1: 1003-1009

Meyer C, Feyerabend M, Weiler E (1989) Fusicoccin-binding proteins in Arabidopsis thaliana L. Heynh. Plant Physiol 89: 692-699

Michaelis ML de, Pugliarello MC, Rasi-Caldogno R (1989) Fusicoccin binding to its plasma membrane receptor and the activation of the plasma membrane ATPase. I. Characteristics and intracellular localization of the fusicoccin receptor in microsomes from radish seedlings. Plant Physiol 90: 133-139

Neumann PM, Volkenburg R van, Cleland RE (1988) Salinity stress inhibits bean leaf expansion by reducing turgor, not wall extensibility. Plant Physiol 88: 23-237

Oparka KJ, Prior DAM (1988) Movement of Lucifer yellow CH in potato storage tissues: A comparison of symplastic and apoplastic transport. Planta 176: 533-540

Oparka KJ, Prior DAM, Harris N (1990) Osmotic induction of fluid-phase endocytosis in onion epidermal cells. Planta 180: 555-561

Oparka KJ, Robinson D, Prior DAM, Derrick P, Wright K (1988) Uptake of Lucifer yellow CH into intact barley roots: Evidence for fluid phase

endocytosis. Planta 176: 541-547
Parker JE, Hahlbrock K, Scheel D (1988) Different cell wall compartments from Phytophthora megasperma f.sp. glycinea elicit phytoalexin production in soybean and parsley. Planta 176: 75-82
Parker JE, Schulte W, Hahlbrock K, Scheel D (1990) An extracellular glycoprotein from Phytophthora megasperma f.sp. glycinea elicits phytoalexin synthesis in cultures parsley cells and protoplasts. Mol Plant-Microbe Interac (in press)
Raschke K (1979) Movements of stomata. In: Physiology of Movements (Haupt W, Feinleib E, eds.). Encycl Plant Physiology 7: 383-441. Springer Verlag Berlin
Robinson DG, Hillmer S (1990a) Coated pits. In: The Plant Plasma Membrane (Larsson C, Moller IM, eds.) pp. 233-255. Springer Verlag Berlin
Robinson DG, Hillmer S (1990b) Endocytosis in plants. Physiologia Plantarum 79: 96-104
Scheel D, Parker JE (1990) Elicitor recognition and signal transduction in plant defense gene activation. Zeitschr für Naturforsch 45c: 569-575
Turner NC (1974) Stomatal behaviour and water status of maize, sorghum, and tobacco under field conditions. II. At low soil water potential. Plant Physiol 53: 360-365

RECEPTOR-MEDIATED ENDOCYTOSIS OF α-FACTOR BY S. CEREVISIAE a CELLS

Howard Riezman, Susan Raths, Monique Rissi, Birgit Singer and Bettina Zanolari
Biocenter of the University Basel, Department of Biochemistry, Klingelbergstrasse 70, CH - 4056 Basel, Switzerland

INTRODUCTION

S. cerevisiae exists in two haploid cell types, a and α, which mate with each other to form the diploid a/α cell. Essential to the mating process is the reciprocal action of two peptide mating factors called a- and α-factor. The factors induce changes in the pattern of mRNA and protein synthesis, an arrest in the G1 phase of the cell cycle and a morphological change called the "shmoo". This oriented projection is the site where the two haploid cells fuse (see Cross *et al* 1988 for a review). The a- and α-factor receptors have been identified as the products of the STE3 and STE2 genes respectively (Jenness *et al* 1983; Nakayama *et al* 1985; Hagen *et al* 1986). These two receptors are polytopic membrane proteins that share a common basic structure with the β-adrenergic receptor (Dixon *et al* 1986). Like this receptor, the STE2 and STE3 gene products are coupled to a G protein (Dietzel and Kurjan 1987; Miyajima *et al* 1987; Whiteway *et al* 1989). The enzyme controlled by the G-protein has not been identified although several other genes acting downstream in the signal transduction pathway are known (Nakayama *et al* 1988; Blinder *et al* 1989).

Upon binding to its receptor α-factor is rapidly internalized by a time-, temperature- and energy-dependent process (Jenness and Spatrick 1986; Chvatchko *et al* 1986). A concommitent decrease in active cell-surfce receptors is seen in the presence, but not absence of α-factor (Jenness and Spatrick 1986) suggesting that the receptor is down-regulated. Once internalized the α-factor is degraded (Chvatchko *et al*

NATO ASI Series, Vol. H 62
Endocytosis
Edited by P. J. Courtoy

1986). The cytoplasmic tail of the receptor is necessary for its down-regulation, but not for signal transduction (Reneke *et al* 1988). It is believed that the integrity and possibly phosphorylation of the cytoplasmic tail (Konopka *et al* 1988; Reneke *et al* 1988) is necessary for adaptation to the pheromone. It is clear that S. cerevisiae is a promising system to study G-protein-coupled signal transduction and receptor trafficking as well as the relationship between the two processes.

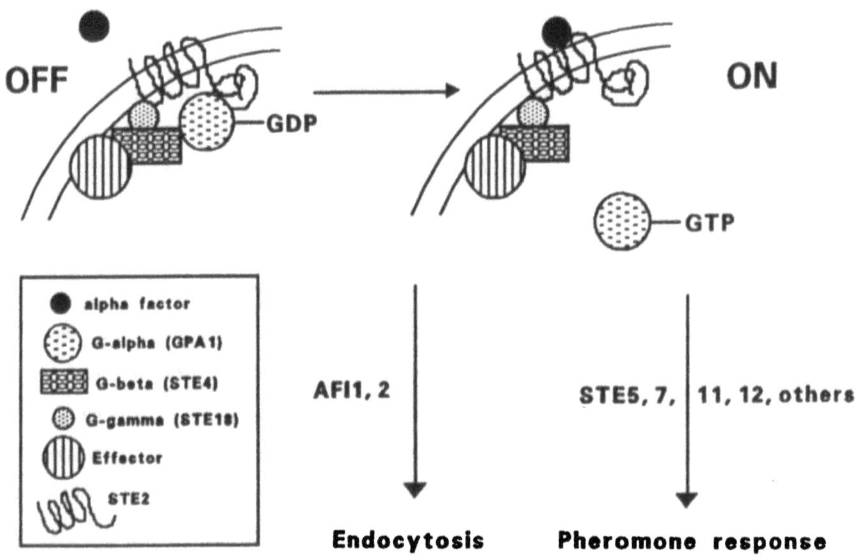

Figure 1. α-factor endocytosis and signal transduction. A schematic view of the initial steps of α-factor endocytosis and signal transduction is shown. Both processes are "triggered" by α-factor and probably occur independently (see below). Some of the genes involved in the two pathways are indicated.

SIGNAL TRANSDUCTION IS NOT NECESSARY FOR PHEROMONE INTERNALIZATION

Since pheromone addition triggers the disappearance of active cell-surface receptors which are otherwise fairly stable we asked whether signal transduction is necessary for pheromone internalization and clearance of cell-surface receptors. In order to do this we introduced a null mutation into the STE4 gene in our strain. The STE4 gene encodes a subunit of the signal transducing G-protein (see fig.1) and this STE4 null allele is completely defective in signal transduction (Whiteway *et al* 1989). We then measured the uptake of ^{35}-S-α-factor. Internalization is indistinguishable from that found in wild-type cells. In order to know whether any of the G-protein subunits are necessary for α-factor internalization we measured α-factor uptake in diploid cells. The pheromone receptors and G-protein subunits are only expressed in haploid yeast cells (Nakayama *et al* 1985; Hagen *et al* 1986). In order to express the α-factor receptor under these conditions we placed the STE2 gene behind the GAL1 promotor and grew the diploid cells on galactose. α-factor uptake by these diploid cells is similar to wild-type cells. We also measured the disappearance of cell-surface receptors in the presence and absence of α-factor in diploid cells. We observed a similar α-factor induced down-regulation of cell surface binding activity as in haploid cells. These results show that neither signal transduction nor the tripartite G-protein are necessary for α-factor uptake or the α-factor triggered disappearance of cell surface receptors.

THE AFI1 AND AFI2 GENES ARE NECESSARY FOR ALPHA FACTOR INTERNALIZATION

In order to identify genes that are necessary for receptor-mediated endocytosis in yeast we screened a bank of mutants that are temperature-sensitive for growth for their ability to internalize α-factor at non-permissive (37°C) temperature. So far we have found two mutants called afi1 and afi2 that are defective for α-factor internalization. Afi1 is defective for α-factor internalization at permissive (24°C) and non-permissive temperatures whereas afi2 is only defective above 30°C. Neither mutant strain is defective for vacuole biogenesis or secretion showing that the products of these two genes are not necessary for these two other pathways. Surpizingly, both mutants, afi1 at 24 and 37°C, afi2 at 37°C, accumulate internal membrane structures whose appearance depends upon the secretory pathway. These results raise the possibility that there is a coupling between the endocytic and secretory pathways. A block in the endocytic pathway could result in the overexpression of the secretory pathway.

Since the afi2 mutant is defective for α-factor uptake only above 30°C we were able to ask if the mutation affects later stages of the endocytic pathway such as delivery from endocytic compartments to the vacuole. A portion of the α-factor was allowed to enter afi2 cells at 15°C then the mutant was shifted to 34°C. Further α-factor uptake stopped immediately whereas all of the internalized α-factor continued along the pathway and was degraded. This suggests that only the internalization step is defective in this mutant. The afi2 mutant phenotype is reversible, but depends on new protein synthesis suggesting that α-factor accumulates along the endocytic pathway at a pre-internalization step and that the Afi2 protein function is lost irreversibly at non-permissive temperature.

α-FACTOR IS TRANSPORTED FROM THE PLASMA MEMBRANE TO THE VACUOLE VIA VESICULAR INTERMEDIATES

The internalization of α-factor with the simultaneous disappearance of active surface receptors is consistent with the model that the process represents a pathway of receptor-mediated endocytosis in yeast. In order to explore this further we designed experiments that would detect vesicular intermediates in the pathway. We showed that the degradation of α-factor by a cells depends upon the action of the PEP4 gene product (Singer and Riezman 1990). This protein, proteinase A, is a vacuolar enzyme that is also responsible for activating the other vacuolar hydrolases (Woolford *et al* 1986; Ammerer *et al* 1986). Our results strongly suggest that α-factor is delivered to the vacuole where it is degraded. In order to trap α-factor in an intermediate compartment on its way to the vacuole we investigated α-factor uptake and degradation at various incubation temperatures. At temperatures of 10-15°C the degradation of α-factor is slowed down with respect to its uptake. This α-factor could be trapped by adding energy poisons and direct evidence was obtained demonstrating that the α-factor was in an intermediate compartment on its way to the vacuole. First, quantitative differential centrifugation studies showed that the α-factor internalized at 15°C did not cofractionate with the plasma membrane nor with the vacuole. Second, the α-factor was in a structure(s) that protected it from protease digestion in the absence, but not presence of detergent. Third, this structure(s) floated in a Nycodenz gradient to give two peaks (Singer and Riezman 1990). These results show clearly that α-factor is transported to the vacuole via vesicular intermediates, a hallmark of endocytosis.

Further evidence for a vesicular mechanism comes from studies of α-factor uptake and degradation in a sec18 mutant. SEC18 encodes the yeast homolog of the NEM-sensitive fusion protein (NSF) (Eakle *et al* 1988; Wilson *et al* 1989) that functions in vesicle fusion in the secretory (Block *et al* 1988) and endocytic (Diaz *et al* 1989) pathways in animal cells. When sec18 cells are incubated with α-factor at 0°C then washed and shifted to non-permissive temperature, the α-factor is internalized normally, but it is not degraded. This phenotype is corrected by the wild-type SEC18 gene on a plasmid. These results show that the SEC18 gene product is essential for α-factor delivery to the degradative compartment and argue very strongly that this transport is vesicle-mediated.

REFERENCES

Ammerer G, Hunter CP, Rothman JH, Saari GC, Valls LA and Stevens TH (1986) *PEP4* gene of *Saccharomyces cerevisiae* encodes proteinase A, a vacuolar enzyme required for processing of vacuolar precursors. Mol. Cell. Biol. 6:2490-2499

Blinder D, Bouvier S, Jenness DD (1989) Constitutive mutants in the yeast pheromone response: ordered function of the gene products. Cell 56:479-486

Block MR, Glick BS, Wilcox CA, Wieland FT, Rothman JE (1988) Purification of an N-ethylmaleimide-sensitive protein catalyzing vesicular transport. Proc. Natl. Acad. Sci. USA 85:7852-7856

Chvatchko Y, Howald I, Riezman H (1986) Two yeast mutants defective in endocytosis are defective in pheromone response. Cell 46:355-364

Cross F, Hartwell LH, Jackosn C, Konopka JB (1988) Conjugation in *Saccharomyces cerevisiae*. A. Rev. Cell Biol. 4:429-457

Diaz R, Mayorga LS, Weidman PJ, Rothman JE, Stahl PD (1989) Vesicle fusion following receptor-mediated endocytosis requires a protein active in Golgi transport. Nature 339:398-400

Dietzel C, Kurjan J (1987) The yeast *SCG1* gene: a G_α-like protein implicated in the a- and α-factor response pathways. Cell 50:1001-1010

Dixon RA, Kobilka BK, Stadler DJ, Benovic JL, Dohlman HG, Frielle T, Bolanowski MA, Bennettt CD, Rands E, Diehl RE, Mumford RA, Slater EE, Sigal IS, Caron MG, Lefkowitz RJ, Strader CD (1986) Cloning of the gene and cDNA for mammalian beta-adrenergic receptor and homolgy with rhodopsin. Nature 321:75-79

Eakle KA, Bernstein M, Emr SD (1988) Characterization of a component of the yeast secretion machinery: identification of the *SEC18* gene product. Mol. Cell. Biol. 8:4098-4109

Hagen DC, McCaffrey G, Sprague GF (1986) Evidence the yeat *STE3* gene encodes a receptor for the peptide pheromone a factor. Gene sequence and implications for the structure of the presumed receptor. Proc. Nat. Acad. Sci. USA 83:1418-1422

Jenness DD, Burkholder AC, Hartwell LH (1983) Binding of α-factor pheromone to yeast a cells: chemical and genetic evidence for an α-factor receptor. Cell 35:521-529

Jenness DD, Spatrick P (1986) Down regulation of the α-factor pheromone receptor in *S. cerevisiae*. Cell 46:345-353

Konopka JB, Jenness DD, Hartwell LH (1988) The C-terminus of the *S. cerevisiae* α-pheromone receptor mediates an adaptive response to pheromone. Cell 54:609-620

Miyajima I, Nakafuka M, Nakayama N, Brenner C, Miyajima A, Kaibuchi K, Arai K, Kaziro Y, Matsumoto K (1987) *GPA1*, a haploid specific essential gene, encodes a yeast homolog of mammalian G protein which may be involved in mating factor signal transduction. Cell 50:1011-1019

Nakayama N, Miyajima A, Arai K (1985) Nucleotide sequences of STE2 and STE3, cell type-specific sterile genes from *Saccharomyces cerevisiae*. EMBO J. 4:2643-2648

Nakayama N, Kaziro Y, Arai K, Matsumoto K (1988) Role of *STE* genes in the mating factor signalling pathway mediated by *GPA1* in *Saccharomyces cerevisiae*. Mol.Cell.Biol. 8:3777-3783

Reneke JE, Blumer KJ, Courchesne WE, Thorner J (1988) The carboxy-terminal segment of the yeast α-factor is a regulatory domain. Cell 55:221-234

Singer B, Riezman H (1990) Detection of an intermediate compartment involved in transport of α-factor to the vacuole in yeast. J. Cell Biol. 110:1911-1922

Whiteway M, Hougan L, Dignard D, Thomas DY, Beli L, Saari GC, Grant FJ, O'Hara P, MacKay VL (1989) The *STE4* and *STE18* genes of yeast encode potential β and γ subunits of the mating factor receptor-coupled G protein. Cell 56:467-477

Wilson DW, Wilcox CA, Flynn GC, Chen E, Kuang WJ, Henzel WJ, Block MR, Ullrich A, Rothman JE (1989) A fusion protein required for vesicle-mediated transport in both mammalian cells and yeast. Nature 339:355-359

Woolford CA, Daniels LB, Park FJ, Jones EW, Van Arsdell JN, Innis MA (1986) The *PEP4* gene encodes an aspartyl protease implicated in the posttranslational regulation of *Saccharomyces cerevisiae* vacuolar hydrolases. Mol.Cell.Biol. 6:2500-2510

RECEPTOR-MEDIATED ENDOCYTOSIS IN TRYPANOSOMA BRUCEI

Isabelle Coppens[1,2], Philippe Bastin[1,2], Pierre Baudhuin[1], Frederik R. Opperdoes[2] and Pierre J. Courtoy[1]
University of Louvain Medical School & International Institute of Cellular and Molecular Pathology, (1) Cell Biology Unit & (2) Research unit for Tropical Diseases,
Avenue Hippocrate 75, 1200 Brussels,
Belgium

INTRODUCTION

African Trypanosomatidae, the causative agents of sleeping sickness in humans and cattle, are extracellular hemoflagellates found in the tissue fluids of the mammalian hosts, where they divide every 7 h. Such a rapid growth requires an efficient supply of nutrients. Most of them, such as amino acids and carbohydrates, are free in the body fluids and directly permeate the trypanosome membrane by simple or facilitated diffusion, or by active transport. Other nutrients are obtained from macromolecules made available by hydrolysis at the cell surface, as exemplified by nucleosides split from RNA by the plasma membrane-associated 3'-nucleotidase, or sugars released from oligosaccharides and glycoproteins by plasma membrane glycosidases.

However, other cell nutrients such as cholesterol or metals circulate tightly bound to carrier proteins and can only be used by the trypanosome after endocytosis of their respective transport protein and intracellular release. Many types of eukaryotic cells exploit receptor-mediated endocytosis for the uptake of carrier proteins. In this paper, we summarize morphological and biochemical evidences that the bloodstream form of *Trypanosoma brucei brucei*, although having diverged from mammals by evolution for one billion years, utilizes the host carrier proteins through receptor-mediated endocytosis. This process is essential for the optimal growth of the parasite. We further suggest that the receptors involved could escape antigenic variation and thus, be good candidates for vaccines against trypanosomiasis.

THE ENDOCYTOTIC APPARATUS OF *TRYPANOSOMA BRUCEI*

The bloodstream form of *T. b. brucei* is a spindle-shaped organism, 30 µm long and 2 µm wide. The cell surface is composed of the plasma membrane and a continuous layer formed by subpellicular microtubules, following a longitudinal helical pathway in the protozoan cortex. Where the flagellum emerges from the cell, the plasma membrane invaginates, forming a reservoir called flagellar pocket. This pocket is devoid of the subpellicular microtubules and is the only plasma membrane domain showing some invaginations coated by a dense material, resembling clathrin-coated pits of mammalian cells (Fig. 1).

NATO ASI Series, Vol. H 62
Endocytosis
Edited by P. J. Courtoy

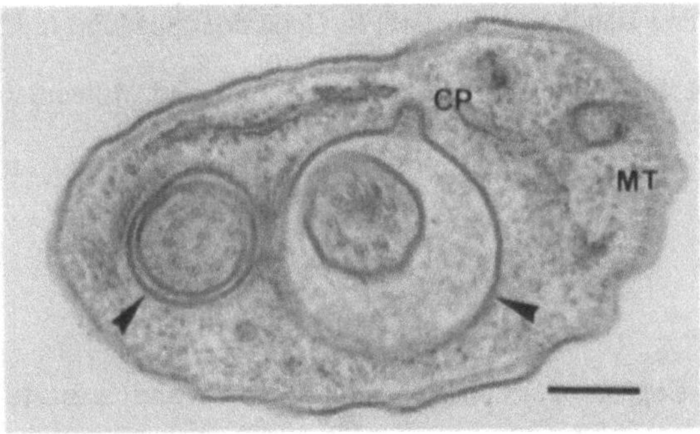

Fig.1 Coated pit-like structure on the flagellar pocket membrane of *Trypanosoma brucei brucei*, bloodstream form.
The pericellular membrane is underlined by pellicular microtubules (MT), except at the flagellar pocket membrane (arrows), where a coated pit can be seen (CP). Bar is 0.3 µm.

Previous electron-microscopic studies have shown that Trypanosomatidae can internalize exogenous macromolecules, such as ferritin (Steinert and Novikoff 1960), peroxidase (Herbert 1965), polyvinylpyrrolidone (Fairlamb and Bowman 1980) and Triton WR-1339 (Opperdoes and Van Roy 1982). The macromolecular material first accumulates in the lumen of the flagellar pocket. The pocket membrane invaginates, generates endocytotic vesicles, and the foreign material ends up in large digestive vacuoles resembling lysosomes (Langreth and Balber 1975). The presence of a variety of acidic hydrolases, anchored in the flagellar pocket membrane (phosphatase, pyrophosphatase, adenylate cyclase; Walter and Opperdoes 1982) or concentrated in lysosomal-like structures (proteinase, mannosidase, phospholipase A1, RNAse; Steiger et al 1980) argues for a large capacity of digesting biological macromolecules.

RECEPTOR-MEDIATED ENDOCYTOSIS

Plasma proteins accumulate at various rates in *T. brucei* . The clearance of the major plasma protein albumin is very low (Table I), corresponding to about 1 % of cell volume per hour, and the uptake shows no saturation. This suggests a mechanism of fluid-phase endocytosis. In contrast, two carrier proteins, present at lower concentrations in the serum, Low-Density Lipoprotein (LDL) particles and transferrin, are taken up at rates respectively 3 and 2 orders of magnitude higher than calculated for fluid endocytosis.

TABLE I. Clearance of serum proteins by *Trypanosoma brucei*

Protein	Clearance (µl/mg cell protein/h)		
Total plasma	0.07	±	0.01
Albumin	0.1	±	0.05
Transferrin	20.5	±	1.7
LDL	215.3	±	23.1

mean ± S.D., n =10. Adapted from Coppens et al (1987).

This observation suggests that *T. brucei* has developed selective mechanisms of endocytosis of these transport proteins. Incubation of parasites with proteins adsorbed onto gold particles confirms the difference in rates of uptake at the ultrastructural level. Numerous LDL- or transferrin-gold particles are accumulated in endocytotic structures (Fig. 2), while albumin-gold is exceptionally seen. LDL or transferrin-gold particles first enter the flagellar pocket and are visible in coated pit-like structures. No evidence has been found for any binding of the proteins elsewhere at the pericellular membrane, confirming the monopoly of the pocket as the port of entry.

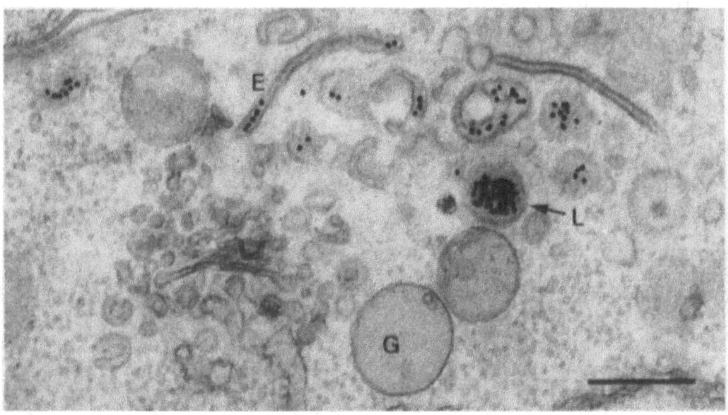

Fig. 2. Accumulation of transferrin-gold complexes in *T. brucei*. Trypanosomes were incubated for 10 min at 30°C in presence of transferrin-gold (15 nm particles in diameter,15 µg protein/ml). Gold particles are visible in endocytic structures tentatively identified as endosomes (E) and lysosomes (L). G, Glycosomes. Bar is 0.3 µm.

The avid uptake of LDL particles may be explained by the requirement of large amounts of cholesterol for rapid growth. The main sterol of the trypanosome found in the bloodstream form is cholesterol, assumed to be obtained intact from the host., in the absence of evidence for de novo cholesterol synthesis (Dixon et al 1971). In the case of mammalian cells, receptor-mediated endocytosis of LDL particles is essential in making cholesterol available (Brown and Goldstein, 1974). Similarly, the growth rate of trypanosomes is reduced by 50 % upon cultivation in cholesterol-free medium and is restored by the addition of LDL particles to the medium. As far as transferrin is concerned, there is no evidence that *T. brucei* has an intracellular pool of ferritin, although it should require iron for the synthesis of several iron-containing proteins. This also suggests that plasma transferrin could be the source of iron for the bloodstream trypanosomes.

Additional lines of evidence suggest that *T. brucei* has specific receptors for LDL particles and transferrin. Binding shows saturation at 4°C, is calcium-dependent (Fig. 3) and, in the case of LDL, is blocked by suramin. Endocytosis of both proteins is specifically competed for by the homologous protein and is sensitive to trypsin.

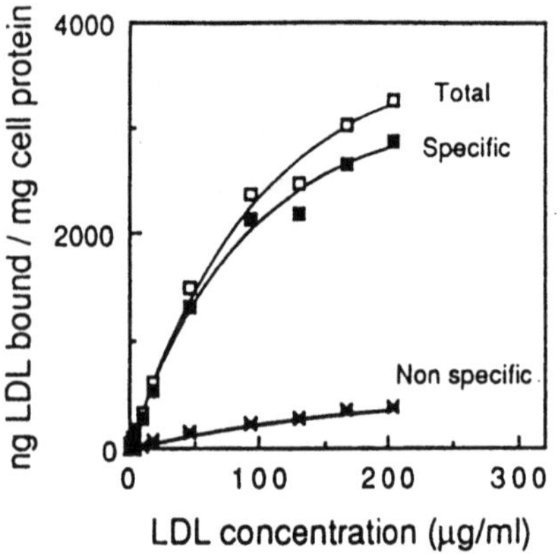

Fig. 3. Binding of LDL particles to intact trypanosomes. Trypanosomes were incubated for 6 h at 4°C in presence of increasing concentrations of radioactive LDL. Total binding (□) was corrected for nonspecific binding (∗), measured in presence of 4 mM EDTA, to yield specific binding (■).
Reproduced from Coppens et al (1988).

LDL RECEPTORS

When preincubated in the absence of lipoproteins, each trypanosome exposes about 50,000 to 90,000 LDL receptors (15,000 to 150,000 copies on human fibroblasts: Goldstein and Brown 1977), with a constant of dissociation close to 1 µM (K_d of 10 nM for fibroblasts: Goldstein and Brown 1977). The LDL receptor of *T. brucei* purified in the presence of PMSF alone corresponds to a single subunit with an apparent molecular mass of 86 kDa. However, upon addition throughout purification of a cocktail of protease inhibitors including leupeptin and antipain (bloodstream-form trypanosomes contain mainly cysteine proteases), the molecular mass of the receptor is increased to 145 kDa (compared to 130-160 kDa for the mammalian receptors: Schneider et al 1982; Wade et al 1986). Recent observations indicate that the 86 kDa fragment corresponds to the major part of the ectodomain of the LDL receptor, cleaved during isolation by an as yet undentified endoprotease. Monospecific rabbit antibodies directed against the purified LDL receptor of trypanosomes exclusively decorate the flagellar pocket and the flagellum membrane. They strongly inhibit LDL binding to trypanosomes (Fig. 4) and to rat hepatocytes, and

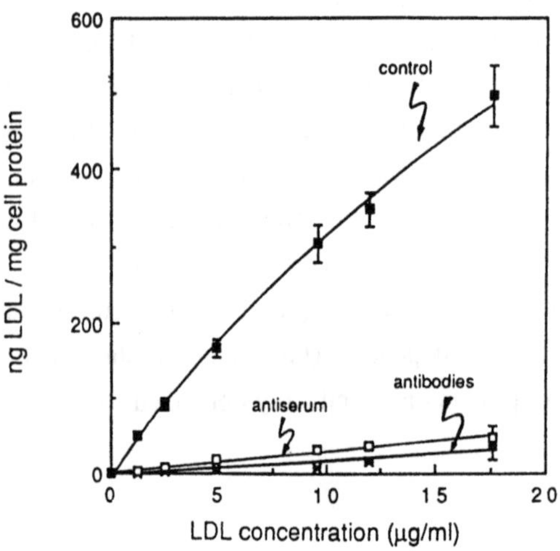

Fig. 4. Immunological inhibition of LDL binding. Trypanosomes were incubated with low concentrations of radioactive LDL together with either 10 % preimmune serum (■), 10 % total antiserum (□), or antibodies affinity-purified against the 86-kDa protein (∗). mean ± S.D., n = 3. Reproduced from Coppens et al (1988).

significantly increase the doubling time of the parasites. In the presence of complement, the antibodies induce trypanolysis in vitro.

The LDL receptor of trypanosomes bears major surface epitopes, which appears to escape antigenic variation. Indeed, the antiserum against the LDL receptor of one variant recognizes the LDL receptor of several other variants selected for the expression of different surface glycoproteins (Fig. 5). This antiserum also recognizes the host LDL receptor but evidence for epitope(s) specific of the parasite receptor has recently been obtained with the production of monoclonal antibodies.

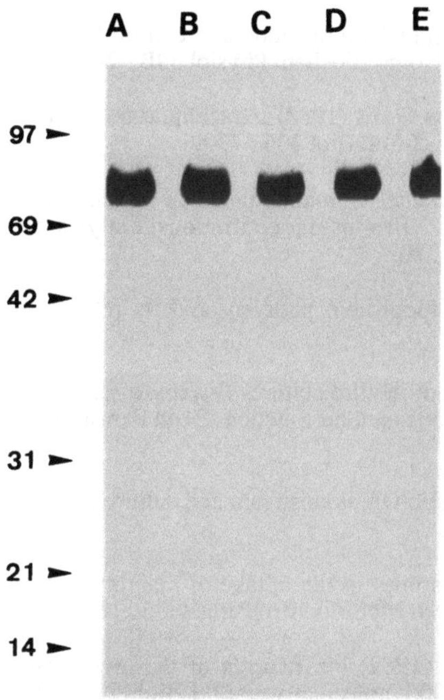

Fig. 5. Western blots of LDL receptors from *T.b.brucei* variants. Crude membrane fractions isolated from different variants in the presence of 1 mM PMSF were resolved by SDS-PAGE and immunolabeled by antibodies against LDL receptors of the variant shown at the lane A.
A: MITat 1.1a, B: MITat 1.3a, C: MITat 1.4a, D: MIT at1.5d, E: MIT at 1.8b.

PERSPECTIVES

The presence of invariant receptors for macromolecules on the surface of trypanosomes may open new avenues for the therapy of sleeping sickness. One might use their ligands as vector proteins for trypanocides, in view of a lysosomotropic chemotherapy. Alternatively, antibodies directed against these receptors could be used to interfere with the nutritional requirement of the parasites or to promote complement-mediated lysis. Moreover, the purified receptors could be good candidates for vaccines. It has been assumed so far that the only proteins recognized by the immune system of the host are the Variant Surface Glycoproteins, which cover the entire surface of the bloodstream form of *T. brucei* (Duszenko et al 1988) and that the rapid variation of these coat proteins is sufficient to allow escape from the immune response. The stability of accessible cell surface receptors now justifies to envisage immunoprophylaxis of trypanosomiasis, after further identification of the epitope(s) specific of the trypanosome receptor.

REFERENCES

Brown MS, Goldstein JL (1974) Binding and degradation of low-density lipoproteins by cultured human fibroblasts. J Biol Chem 249 : 5153-5162

Coppens I, Opperdoes FR, Courtoy PJ, Baudhuin P (1987) Receptor-mediated endocytosis in the bloodstream form of *Trypanosoma brucei*. J Protozool 34 : 465-473

Coppens I, Baudhuin P, Opperdoes FR, Courtoy PJ (1988) Receptors for the host low density lipoproteins on the hemoflagellate *Trypanosoma brucei* : Purification and involvement in the growth of the parasite. Proc Natl Acad Sci USA 85 : 6753-6757

Dixon H, Ginger CD, Williamson J (1971) The lipid metabolism of blood and culture forms of *Trypanosoma lewisi* and *Trypanosoma rhodesiense*. Comp Biochem Physiol 39B : 247-266

Duszenko M, Ivanov IE, Ferguson MAJ, Plesken H, Cross GAM (1988) Intracellular transport of a variant surface glycoprotein in Trypanosoma brucei. J Cell Biol 106 : 77-86

Fairlamb AH, Bowman IBR (1980) *Trypanosoma brucei* : maintenance of concentrated suspensions of bloodstream trypomastigotes in vitro using continuous dialysis for measurement of endocytosis. Exp Parasitol 49 : 366-380

Goldstein JL, Brown MS (1977) The Low-Density Lipoprotein pathway and its relation to atherosclerosis. Ann Rev Biochem 46 : 897-930

Herbert IV (1965) Cytoplasmic inclusions and organelles of in vitro cultures *Trypanosoma theileri* and *Trypanosoma melophagium* and some speculations on their function. Exptl Parasitol 17 : 24-40

Langreth SG, Balber AE (1975) Protein uptake and digestion in bloodstream and culture forms of *T. brucei*. J Protozool 22 : 40-53

Opperdoes FR, Van Roy J (1982) Involvement of lysosomes in the uptake of macromolecular material by bloodstream forms of *Trypanosoma brucei*. Mol Biochem Parasitol 6 : 181-190

Schneider WJ, Beisiegel U, Goldstein JL, Brown MS (1982) Purification of the low-density lipoprotein receptor, an acidic glycoprotein of 164,000 molecular weight. J Biol Chem 257 : 2664-2673

Steiger RF, Opperdoes FR, Bontemps J (1980) Subcellular fractionation of *Trypanosoma brucei* bloodstream forms with special reference to hydrolases. Eur J Biochem 105 : 163-175

Steinert M, Novikoff AB (1960) The existence of a cytostome and the occurence of pinocytosis in the trypanosome *Trypanosoma mega*. Biophys Biochem Cytol 8 : 563-569

Wade DP, Knight BL, Soutar AK (1986) Binding of low-density lipoprotein and chylomicrons remnants to the hepatic low-density lipoprotein receptor of dogs, rats and rabbits demonstrated by ligand blotting. Eur J Biochem 159 : 333- 340

Walter RD, Opperdoes FR (1982) Subcellular distribution of adenylate cyclase, cyclic AMP phosphodiesterase, protein kinases and phosphoprotein phosphatase in *Trypanosoma brucei*. Mol Biochem Parasitol 6 : 287-295.

2.6. ENDOCYTOSIS AND DRUG TARGETING

(Chairperson : Michel MONSIGNY)

The last session of the workshop was a glance at therapeutic approaches based on endocytosis, either exploiting unique features of lysosomes in the target cells, or aiming at the specific targeting of general drugs of particular cells. The first approach was illustrated for *Leishmania amazonensis* amastigotes by **Michel Rabinovitch**. He reported that lysosomotropic dipeptide esters containing a N-terminal hydrophobic amino-acid are preferentially hydrolyzed by cysteine proteinases in the unusual lysosomes of the parasite, called "megasomes". Toxicity may be due to the conversion of esters into membranolytic polymers, catalyzed by a dipeptidyl-peptidase I. Several such compounds reduce the growth of *L. amazonensis* in mouse foot pads after intralesional administration.

Drug targeting is in modern terms the equivalent to the "magic bullet" concept of Ehrlich. **Michel Monsigny** exploited the receptor-mediated endocytosis of glycoconjugate-drug complexes to target pharmacological agents and activators into macrophages. He reported several successful examples : the activation of macrophages with targeted N-acetylmuramyldipeptide, leading to the eradication of lung metastases in amurine system, and the targeted delivery of phosphonylmethoxyethyladenine to stop multiplication of herpes virus, and the targeted delivery of allopurinol riboside 5'-phosphate to kill *Leishmania* amastigotes.

As described in the chapter of Kirsten Sandvig, the most powerful cell killer agents are toxins, like the ricin A chain. **Sjur Olsnes** reported the current results obtained so far by their conjugation to monoclonal antibodies, i.e. immunotoxins, to selectively kill cancer cells. Promising results are obtained with murine models and these compounds have now entered clinical trials. Anti-T cell immunotoxins have also a potential against auto-immune diseases. The cell biology of immunotoxins remains poorly understood. Endocytosis is clearly required but the mode of translocation is still a mystery. Studies on immunotoxins emphasize that targeting to a particular cell is not sufficient, and that ensuring adequate intracellular routing is equally important.

NATO ASI Series, Vol. H 62
Endocytosis
Edited by P. J. Courtoy

LYSOSOMES OF LEISHMANIA MEXICANA SP. AS TARGETS FOR POTENTIAL THERAPEUTIC AGENTS

Michel Rabinovitch[1], Claude Ramazeilles[1], Silvia C. Alfieri[2], Vladimir Zilberfarb[1], Elliot Shaw[3], Jair Ribeiro Chagas[4] and Luiz Juliano[4]

[1] Institut Pasteur and CNRS UA 361, 25 rue du Dr Roux, 75724 Paris, France
[2] Department of Parasitology, University of Sao Paulo, Sao Paulo, Brazil
[3] Friedrich Miescher Institut, Basel, Switzerland
[4] Department of Biophysics, Escola Paulista de Medicina, Sao Paulo, Brazil.

INTRODUCTION

Hydrophobic amino acid esters and peptides disrupt lysosomes in cell-free fractions by a mechanism involving trapping by protonation, enzymatic hydrolysis and accumulation of less permeant products within the organelles (Goldman & Naider 1974; Ransom & Reeves 1983). Damage to lysosomes possibly accounts for the selective toxicity of the compounds for monocytes, NK cells and cytotoxic T cells. This toxicity may be due to ester conversion to membranolytic polymers catalysed by a dipeptidyl-peptidase I (Thiele & Lipsky 1990 a, b).

DESTRUCTION OF L. AMAZONENSIS AMASTIGOTES BY AMINO ACID AND PEPTIDE ESTERS

Amino acid and peptide esters kill *L. amazonensis* amastigotes lodged within mouse peritoneal macrophages in culture at concentrations that allow the survival of the host cells (Rabinovitch et al 1986). Amastigotes isolated from lesions are destroyed by the same compounds (Rabinovitch et al 1987). Certain L-amino acid methyl esters, such as those of Leu, Trp, Met or Phe are leishmanicidal, whereas those of Ile, Val, Gly or Ala are inactive at higher concentrations. L-amino acid amides display a similar pattern of activity but are generally less effective than the methyl esters and more toxic to the macrophages (Rabinovitch & Zilberfarb 1988). Leishmanicidal but not inactive [^{3}H]-amino acid methyl esters are rapidly hydrolysed by isolated amastigotes and the [^{3}H]-amino acid generated is retained in the parasites (Ramazeilles & Rabinovitch 1989).

NATO ASI Series, Vol. H 62
Endocytosis
Edited by P. J. Courtoy

The hypothesis that amastigote proteases are involved in parasite destruction by the esters (Alfieri et al 1987) is supported by the finding that pretreatment of infected macrophages or isolated amastigotes with the protease inhibitors antipain or chymostatin protects the parasites from damage by Leu-OMe (Alfieri et al 1988). Z-Phe-AlaCHN$_2$ and Z-Tyr-AlaCHN$_2$, specific and irreversible inhibitors of mammalian cathepsins B and L, are also protective. The activity of 25 to 33 kDa cysteine proteinases (examined after electrophoresis in gelatin-containing acrylamide gels) is inhibited in lysates prepared from living parasites pretreated with the diazomethanes (Alfieri et al 1989 and unpublished). Amastigotes of *L. mexicana* exhibit stage specific, lysosome-like organelles termed megasomes which may contain a 31 kDa cysteine proteinase (Pupkis et al 1986). The roles, if any, of megasomes and of their enzymes in the physiology of amastigotes remain to be determined. The hypothesis that megasomes are the targets of the esters (Hunter et al 1989) gained support from the findings that the organelles are acidified (Antoine et al 1988) and display clear morphological changes in parasites exposed to sublethal concentrations of Leu-OMe; these changes are prevented in amastigotes preincubated with antipain (Antoine et al 1989). Furthermore, the development of megasomes and of cysteine proteinase activity parallels the acquisition of parasite susceptibility to Leu-OMe during the intracellular differentiation of *L. amazonensis* promastigotes to amastigotes (Quintao-Galvao et al 1990). This result suggests that parasites become susceptible to the compounds only when the cysteine proteinases are packaged in megasomes. Thus, amino acid and peptide esters can be considered prodrugs activated by cysteine proteinases associated with megasomes. Although damage to the organelles may indeed cause parasite destruction, the more 'distal'mechanisms of the leishmanicidal activity of the amino acid derivatives remain to be investigated (Rabinovitch & Alfieri 1987; Rabinovitch 1989).

STRUCTURE-ACTIVITY CORRELATIONS

The activity of a series of esters on isolated amastigotes was determined by a semi-automated tetrazolium (MTT) reduction assay (Rabinovitch et al 1987). A small number of molecules was also assayed on *Leishmania*-infected mouse macrophages. Activity of amino acid esters depends on the nature of the amino acid (alluded to above) and of the ester moiety. The latter is exemplified by the series Leu-OMe (0.5 mM) < Leu-OPropyl (0.36 mM) < Leu-OPhenyl (0.33 mM) < Leu-OButyl (0.22 mM) < Leu-OIsoamyl (0.11 mM) < Leu-OBzl (0.07 mM) (the numbers are ED_{50}s for isolated amastigotes). Since the structure of Leu-OIsoamyl is related to that of Leu-Leu-OMe, it may be of interest that the ED_{50}s of the two compounds are nearly the same. Dipeptide esters are more active than amino acid esters. The activity of homodipeptide esters depends on the nature of the amino acid, and cannot be predicted from that of the esters of the constitutive amino acids; thus, activities decrease 5 fold over the series Phe-Phe-OMe (34 μM) > Val-Val-OMe > Leu-Leu-OMe > Trp-Trp-OMe > Ile-Ile-OMe (170 μM). The activity of heterodipeptide esters depends on the nature and position of the amino acids. Active dipeptide esters contain at least one hydrophobic amino acid (Leu, Trp, Phe, Ile, Val or Tyr), preferably placed in the amino-terminal position; thus, the ED_{50}s of Phe-Gly-OMe and of Gly-Phe-OMe are, respectively, 58 and 600μM (Ramazeilles et al 1990). Interestingly, when additional Leu residues are added to Leu-Leu-OMe, activity decreases. Thus, the ED_{50}s of $(Leu)_2$OMe, $(Leu)_3$OMe and $(Leu)_4$OMe are, respectively, 0.12, 0.24 and 0.20 mM. Activity increases markedly by blocking the terminal amino group by the carbobenzoxy (Z) or t-butyloxycarbonyl (BOC) moieties, showing that the weak base character is not required for destruction of the parasites. However, N-formylation or N-acetylation of the compounds reduces their leishmanicidal activity. These results are illustrated by the following series: Trp-OMe (ED_{50}, 0.88 mM); Ac-Trp-OMe (ED_{50}, 2.12 mM); Trp-Trp-OMe (ED_{50}, 0.13 mM); Boc-Trp-Trp-OMe (ED_{50}, 14 μM). Study of recto-inverso analogues of Bz-Phe-Gly-OEt indicated that increased resistance of the peptide product to enzymatic degradation enhances leishmanicidal activity (Ramazeilles et al 1990).

This is also borne out by the following series: Phe-Phe-Gly-OMe > D-Phe-Phe-Gly-OMe > Phe(R)Phe-Gly-OMe > Phe-2(-Cl-Z)Phe-Gly-OMe.

IN VIVO ACTIVITY IN AN INTRALESIONAL INJECTION MOUSE MODEL

In this assay, mice were infected in both foot pads with 2 x 10^5 *L. amazonensis* amastigotes and on the 18th day one of the lesions was injected with 50μl of a solution of the compounds to be tested. The other lesion received the diluent alone. Foot pad thickness was measured biweekly with a caliper. Ile-Leu-OMe (ED_{50}, 25 mM), Phe-Val-OMe (ED_{50}, 28 mM) or Leu-OMe (ED_{50}, 50 mM), reduced or stabilized for 1-2 weeks the thickness of the injected footpads. Thereafter the treated lesion resumed growth but the differences between injected and control pads persisted for another 4-6 weeks. The effect of the esters was equivalent or better than that of meglumine antimoniate ("Glucantime") at 1:10 dilution, which yielded an ED_{50} of 0.46 mg Sb. The number of amastigotes recovered from treated lesions was smaller by one or more orders of magnitude than that recovered from lesions injected with the diluent alone. We cannot exclude the possibility that toxicity of the peptide esters for macrophages or other host cells may have contributed to the reduction of lesion growth by the compounds. Several compounds active in vitro were either inactive in vivo (e.g. Trp-OMe) or were too inflammatory (e.g. Leu-Leu-OMe). In preliminary experiments, Z- or BOC-blocked esters at 2-4 mM concentrations were inactive. This suggests that, in the absence of trapping by protonation, the blocked compounds are not sufficiently retained within the lesions and diffuse out prior to or after hydrolysis by host and/or parasite enzymes.

FINAL COMMENTS

The idea that parasite proteases can serve as targets for chemotherapy is now pursued in many laboratories. However, very few parasite cysteine proteinases have been purified to homogeneity and sequenced (North et al 1990). In the studies discussed here leishmanicidal esters do not act as enzyme inhibitors but rather as substrates that generate molecules toxic to the parasites. A rational design of more specific leishmanicidal esters (substrates) may have to await the complete purification and/or cloning, expression and crystallographic studies of the parasite cysteine proteinases and their comparison with the homologous host enzymes. However, factors other than the specificity of parasite proteinases are involved in the toxicity of the compounds. Among them are molecular features relevant to membrane permeation, transport mechanisms, membranolytic activity, and metabolic effects of the esters and/or their hydrolysis products. The sensitivity of *L. amazonensis* to the amino acid derivatives appears to depend on the activity of cysteine proteinase(s) concentrated in megasomes. There is however, no information on lysosomes and cysteine proteinases in amastigotes of other *Leishmania* species apparently devoid of megasomes (Glew et al 1988). It is hoped that other lysosomotropic peptide derivatives may be active against more commonly found Leishmania such as *L. infantum* or *L. braziliensis*. A detectable, albeit modest, activity of L-tryptophanamide on *L. donovani* amastigotes has already been reported (Hunter et al 1989).

ACKNOWLEDGEMENTS

Research supported by CNRS, Institut Pasteur, UNDP/World Bank/WHO Special Programme for Research and Training in Tropical Diseases, the Commission of the European Communities and an INSERM/CNPq (Brazil) cooperation agreement. S.C. Alfieri is also supported by TDR, FAPESP, CNPq and BID-USP.

REFERENCES

Alfieri SC, Zilberfarb V, Rabinovitch M (1987) Destruction of *Leishmania mexicana amazonensis* amastigotes by leucine methyl ester: protection by other amino acid esters. Parasitology 95:31-41

Alfieri SC, Ramazeilles C, Zilberfarb V, Galpin IJ, Norman SE, Rabinovitch M (1988). Proteinase inhibitors protect *Leishmania amazonensis* amastigotes from destruction by amino acid esters. Mol Biochem Parasitol 29:191-201

Alfieri SC, Shaw E, Zilberfarb V, Rabinovitch M (1989) *Leishmania amazonensis*: involvement of cysteine proteinases in the killing of isolated amastigotes by L-Leucine methyl ester. Exp Parasitol 68:423-431

Antoine JC, Jouanne C, Ryter A, Benichou JC (1988) *Leishmania amazonensis*: acidic organelles in amastigotes. Exp Parasitol 67, 287-300

Antoine JC, Jouanne C, Ryter A (1989) Megasomes as targets of leucine methyl ester in *Leishmania amazonensis* amastigotes. Parasitology 99:1-9
Galvao-Quintao L, Alfieri SC, Ryter A, Rabinovitch M (1990) Intracellular differentiation of *Leishmania amazonensis* promastigotes to amastigotes: presence of megasomes, cysteine proteinase activity and susceptibility to leucine-methyl ester. Parasitology 101:7-1
Glew RH, Saha AK, Das S & Remaley AT (1988) Biochemistry of the *Leishmania* species. Microbiol Reviews 52:412-432
Goldman R, Naider F (1974) Permeation and stereospecificity of hydrolysis of peptide esters within intact lysosomes in vitro. Biochim Biophys Acta 338:224-233
Hunter CA, Macpherson LM, Coombs GH (1989) Antileishmanial activity of L-leucine ester and L-tryptophanamide, in Leishmaniasis, DT Hart, ed, Plenum Press, NY pp 741-747
North MJ, Mottram JC, Coombs GH (1990) Cysteine proteinases of parasitic protozoa. Parasitol Today 6:270-275
Pupkis MF, Tetley L, Coombs GH (1986) *Leishmania mexicana*: amastigote hydrolases in unusual lysosomes. Exp Parasitol 62:29-39
Rabinovitch M, Zilberfarb V, Ramazeilles C (1986) Destruction of *Leishmania mexicana amazonensis* amastigotes within macrophages by lysosomotropic amino acid esters. J Exp Med 163:520-535
Rabinovitch M, Alfieri SC (1987) From lysosomes to cells, from cells to *Leishmania*: amino acid esters as potential chemotherapeutic agents. Brazilian J Med Biol Res 20:665-674
Rabinovitch M, Zilberfarb V, Pouchelet M (1987) Leishmania mexicana: destruction of isolated amastigotes by amino acid esters. Am J Trop Med Hyg 36:288-293
Rabinovitch M, Zilberfarb, V (1988) Destruction of intracellular and isolated *Leishmania mexicana amazonensis* amastigotes by amino acid amides. Parasitology 96:289-296
Rabinovitch M (1989) Leishmanicidal activity of amino acid and peptide esters. Parasitol Today 5:299-301
Ramazeilles C, Rabinovitch M (1989) *Leishmania amazonensis*: uptake and hydrolysis of ^{3}H amino acid methyl esters by isolated amastigotes. Exp Parasitol 68:135-143
Ramazeilles C, Juliano L, Chagas JR, Rabinovitch M (1990) The antileishmanial activity of dipeptide esters on Leishmania amazonensis amastigotes. Parasitology 100:201-207
Ransom JT, Reeves JP (1983) Accumulation of amino acids within intracellular lysosomes of rat polymorphonuclear leukocytes incubated with amino acid methyl esters. J Biol Chem 258:9270-9275
Thiele DL, Lipsky PE (1990 a) Mechanism of L-leucyl-L-leucine methyl ester-mediated killing of cytotoxic lymphocytes: Dependence on a lysosomal thiol protease, dipeptidyl peptidase I, that is enriched in these cells. Proc Nat Acad Sci USA 87:83-87
Thiele DL, Lipsky PE (1990 b) The action on leucyl-leucine methyl ester on cytotoxic lymphocytes requires uptake by a novel dipeptide-specific facilitated transport system and dipeptidyl peptidase I-mediated conversion to membranolytic products. J Exp Med 172:183-194

GLYCOSYLATED POLYMERS AS CARRIERS FOR TARGETING ANTIVIRAL AND ANTIPARASITE DRUGS

Michel Monsigny, Annie-Claude Roche, Roger Mayer, Patrick Midoux, Eric Nègre and Edwige Bonfils
Department of Endogenous Lectin and Glycoconjugate Biochemistry, Molecular Biophysics Institute, CNRS and University, 1, Rue Haute, 45071 Orléans Cedex 02, France

INTRODUCTION

A therapeutic agent to be active must reach its target cell and more precisely the right compartment of the target cell. For a great majority of drugs, the compartment is either the cytosol or the nucleoplasm. When hydrophobic, a drug may cross the plasma membrane to enter the cytosol. However when hydrophilic, a drug is not membrane permeant and usually will enter the cytosol from an intracellular organelle. On these bases, drug targeting may enhance the efficacy of a hydrophilic drug. The rationale of drug targeting using macromolecular glycoconjugates includes the following considerations :

- a small molecular weight hydrophilic drug is usually lost *in vivo* within a few minutes, while bound to a macromolecular carrier it will stay longer in the blood,
- a macromolecular carrier containing carbohydrate moieties as a recognition signal will be recognized by cell surface receptors (lectins) with a specificity depending on the nature of the carbohydrate moiety used and on the type of lectin present on the target cell. Therefore the amount of drug at the level of the target cell will be much higher than when the free drug is used,
- a carbohydrate moiety used as recognition signal allows the binding and the internalization of the drug-carrier conjugate because membrane lectins usually mediate endocytosis,
- the drug must be attached to the carrier in such a way that it will be linked until its release in a selected organelle inside the target cell.

MEMBRANE LECTINS

Membrane lectins of animal cells were discovered by Ashwell and colleagues about 20 years ago, (see Ashwell and Harford, 1982). These authors showed that asialo-ceruleoplasmin was taken up by liver parenchymal cells within a few minutes upon injection and that many other glycoconjugates having galactose residues in a non-

NATO ASI Series, Vol. H 62
Endocytosis
Edited by P. J. Courtoy

reducing terminal position were actively endocytosed and degraded in lysosomes. Since, membrane lectins were identified on many other animal cells, including tumor cells, endothelial cells, macrophages, monocytes and lymphocytes (see for reviews Monsigny *et al.*, 1988a, Sharon and Lis, 1989). The capacity of membrane lectins to mediate receptor-driven endocytosis has been thoroughly analyzed (see for a review Wileman *et al.*, 1985).

BINDING AND UPTAKE OF GLYCOCONJUGATES

In order to determine the efficiency of glycoconjugates as carrier for targeting to a given cell, we developed a fluorescein based assay (Midoux *et al*, 1986, 1987). The fluorescence of fluorescein bound to a neoglycoprotein is partially quenched , but upon proteolytic cleavage the quenching is abolished. Furthermore as the fluorescence intensity is pH dependant, the cells are incubated in the presence of monensin at 4°C after any biological active step at 37°C in order to neutralize the acidic compartments and so to recover a maximum fluorescence intensity. Using this method, it is possible to quantitate the binding of fluoresceinylated glycoconjugates onto the cell surface , to follow the endocytotic process in acidic compartments and to determine the extent of the hydrolytic degradation. Using fluorescein-substituted neoglycoproteins, membrane lectins of various cells have been identified and their capacity to internalize their ligand was determined: the mannose and galactose receptors on macrophages (Tenu *et al.*, 1982), the glucose receptor on Lewis lung carcinoma cells (Roche *et al.*, 1983), the galactose and fucose receptors on L1210 lymphoid cells (Monsigny *et al.*, 1984b), the mannose-6-phosphate receptor on monocytes (Roche *et al*, 1985b) and an intracellular membrane mannose receptor in promonocytes (Pimpaneau *et al*, 1990). This approach is also valuable to identify membrane lectins using fluoresceinylated glycoproteins, the receptor of acid α_1-glycoprotein for instance (Pimpaneau *et al.*, 1989).

Glycoconjugates containing galactose residues in a terminal non-reducing position are efficient carriers to target drugs to the liver because they are recognized by the membrane lectins of parenchymal cells. Such glycoconjugates carrying antiviral agents (FUdR, AraA or AraC) are efficient to stop the development of virus in parenchymal cells infected by viruses (Fiume *et al*, 1986), or infected by *Plasmodium* sporozoites using primaquine(Trouet *et al*, 1982a). Using mannosylated serum albumin as carriers, we showed that, in *in vitro* experiments, linked muramyl dipeptide (MDP) was at least 100 times more active than free MDP to activate macrophages (Monsigny *et al*, 1984a) and that in *in vivo* experiments the MDP conjugates were quite active (Roche *et al*, 1985a, 1986, 1988), while free MDP was not at all.

DRUG-CARRIER LINKAGE

The main types of drug-carrier linkages which have been developed and the various drugs successfully carried with glycoconjugates are presented in table 1. The drug-carrier linkage may be :

i) cleavable in acidic medium (in the lumen of endosomes or lysosomes) as in the case of acido-labile heterobifunctional links : *maleic acid derivatives* such as *cis*-aconitic acid (Shen and Ryser, 1981; Shen, 1990) , C-ribofuranomaleic acid (Monsigny *et al*, 1988b), *phenylthiocarbamyl acidic* α*-amino acid derivatives* leading to **DALAC** (Drug-Acid-Labile-Arm-Conjugates) : *p*-benzylthiocarbamoyl-aspartic acid (Daussin *et al.*, 1988), or

ii) cleavable under reducing conditions as in the case of disulfide bridges ; gelonin-glucosylated serum albumin conjugate is 100 times more active than free gelonin killing Lewis lung carcinoma cells (Roche *et al.*, 1983), or

iii) cleavable through the action of an endosomal or lysosomal enzyme as in the case of peptidyl linkers as already described (**DAC** : Drug-Arm-Conjugate, Monsigny *et al*, 1980; Trouet *et al*, 1982a, 1982b; Delmotte, 1985).

GLYCOSYLATED POLYMERS

To improve the chemistry of the drug-glycosylated carrier conjugate and to extend the use of such carriers, we developed a new glycosylated carrier: namely poly-**L**-lysine, the ϵ-amino groups of which beeing partially acylated by δ-gluconolactone, while the other amino groups were substituted by recognition signals (carbohydrate residues for instance) **and** by drugs (Derrien *et al*, 1989). The poly-**L**-lysine based carrier is easily prepared with a high yield because the various chemical reactions are conducted in organic solvant and is highly hydrosoluble because half the lysine residues are substituted with a polyhydroxylalcanoyl moiety, such as a gluconoyl group. Surprisingly, we found that not only the activity of the drug carrier conjugate depends on the nature of the recognition signal (mannose *versus* mannose 6 phosphate, for instance) but also on the nature of the carrier itself, neutralized poly-**L**-lysine *versus* serum albumin (Petit *et al.*, 1990). Both conjugates are able to rend macrophages tumoricidal but the later induces secretion of cytotoxic factor while the former does not. The main properties of the two types of glycosylated carriers are summarized in table 2.

Table 1. **Glycoconjugates as drug carriers**

Carriers :

- Glycoproteins (orosomucoids, asialoglycoproteins, β-glucoronidase ...)
- Neoglycoproteins
- Glycosylated hydrosoluble polymers.

Rationale :

- Endogeneous lectins (membrane sugar binding proteins) are expressed in various animal cells:
    - their sugar specificity depends on the nature and on the differentiation state of the cells
    - they actively mediate endocytosis of their ligand.

Drug-carrier linkages :

- direct : - amide, ester... bonds
- indirect : - drug-arm-carrier (DAC) : peptide
    - drug-acid labile arm carrier (DALAC) :
        - *cis*-aconitic acid derivatives
        - C-ribofuranomaleic acid derivatives
        - phenylthiocarbamyl aspartic acid derivatives
        - hydrazone
    - through disulfide bridge.

Carried drugs :

- Anticancer agents: daunomycine, adriamycine, methotrexate, gelonin ...
- Biological response modifiers: N-acetylmuramyl dipeptide (MDP) and analogues
- Antiviral agents:
    - nucleoside and nucleotide analogues: fluoronucleotides, AZT, d4T, ddI, PMEA, ribavirin ...
    - Antisense oligonucleotides
- Antiparasitic agents:
    - anti-malarial : primaquine
    - anti-leishmania : allopurinol riboside, methotrexate.

Table 2. **Glycosylated macromolecular drug carriers**

Starting material	Serum albumin	Poly-**L**-lysine
Available NH_2 groups	≃ 57	≃ 190
Substituted material	Neoglycoprotein	Glycosylated polymer
Structure	globular	α-helix
Form	spherical	cylindrical
Size (nm)	d = 6	d = 3; l = 30
Mr (Da)	≃ 80,000	≃ 80,000
Yields (%)	≃ 50	≃ 95
Solubility (mg/mL)	≃ 100	> 200
Synthesis solvant	aqueous	organic
Sugar derivative	osidyl PTC	osidylphenylacetate
Sugars per molecule	≃ 25	≃ 50
Drugs per molecule	5 - 20	10 - 80
pI	≃ 4	neutral*
Biological efficacy		
Carried / free drug	up to 100	up to 100
Biodegradability	partial	total
Unspecific cell binding	±	-
Immunogenicity	+ +	-
Antigen presentation	+ +	-

* All remaining NH_2 groups are acylated
d: diameter, l: length, PTC: phenyl-isothiocyanate

GLYCOSYLATED POLYMERS AS CARRIERS OF ANTI-INFECTIOUS AGENTS

Glycosylated polymer conjugates have been substituted by various anti-infection agents. Allopurinol-ribosyl-5'-phosphate which is not active against *Leishmania* parasites grown in macrophages become quite efficient when bound through a phenyl ester linkage to a mannosylated polymer (Nègre *et al.*, unpublished data).

Antiretroviral nucleoside analogs (for a review, see Mitsuya *et al.*, 1990) have been coupled to glycosylated polymers (Roche *et al.*, 1990). Succinyl-5'-azidothymidine (Succinyl-AZT) linked to the amino groups of glycylglycyl spacers bound to the mannosylated polymer is more efficient than free AZT or free succinyl-AZT against HIV multiplication in human macrophages (unpublished results). PMEA (phosphono-methoxy-ethyladenine) which inhibits herpes virus multiplication in macrophages is at least 50 times more active than free PMEA when it is linked to the same mannosylated polymer (Midoux *et al.,* 1990). Conversely, when the drugs are linked to the recognition signal-free carrier, their activity is much lower than free drugs. These results show that low molecular weight antiviral drugs are more active than free drugs when they are actively internalized by cell surface receptors that bind the recognition signal borne by the carrier.

Antisense oligonucleotides are potential antitumor and antiviral agents (for review, see Hélène and Toulmé, 1990). They are very sensitive to exonucleases, but they can be derivatized at both 3' and 5' ends in order to avoid such degradations. Oligonucleotides (19mer) specific for target sequences of the LTR of HIV have been substituted at both ends: with a thiol group at one end and a fluorophore at the other.They are both charged and highly hydrophilic and are taken up by cells *in vitro* when they are used at relatively high concentration (usually above 1 micromolar). These oligonucleotide derivatives were linked to glycosylated polymers through a disulfide bridge; such conjugates are more efficiently taken up by cells than free oligonucleotides, but the optimal conditions to achieve a high antiviral efficiency have yet to be determined. Because oligonucleotides are quite sensitive to endonucleases, they must be released from the carrier and to cross the vesicular membrane before they reach lysosomes. Depending on the type of cells concerned, this requirement may be fulfilled; for instance the material endocytosed through the mannose-6-phosphate receptor on monocytes is not readily degraded (Roche *et al.*, 1985b).

CONCLUDING REMARKS

The poly-**L**-lysine based glycosylated polymers appear to be versatile carriers that can be substituted with various drugs leading to conjugates which are usually more active than the free drugs in *in vitro* experiments. *In vivo*, such glycosylated polymers are much less antigenic than neoglycoproteins and do not induce adverse effects even when large quantities are injected to mice. Therefore, these polymers seem to be suitable for drug targeting. However, several problems are still limiting the development of such therapeutic approach, specially at the level of the release of the drug as an active agent in the right compartment of the expected target cell.

To improve the efficiency of drug targeting, further knowledge on the intracellular trafic of endocytosed material and on the properties of intracellular vesicles are required, namely :

i) for any cell surface receptor, the precise endocytic pathway;

ii) the nature and the specificity of the hydrolases present in early endosomes, multivesicular bodies, and late endosomes;

iii) the permeability (active and/or passive) of the various endosomes, the trans-Golgi-network and of the Golgi apparatus.

Acknowledgements : This work was partly supported by grants from ANRS (Agence Nationale de la Recherche sue le Sida). A.C. Roche & P. Midoux are Directeur & Chargé de Rechèrche INSERM, respectively.

REFERENCES

ASHWELL G. & HARFORD J.(1982) Carbohydrate-specific receptors of the liver. *Ann. Rev. Biochem.* **51**, 531-554

DAUSSIN F., BOSCHETTI E., DELMOTTE F. & MONSIGNY M. (1988) *p*-Benzylthiocarbamoyl-aspartyl-daunorubicin substituted polytrisacryl. A new drug acid-labile arm carrier conjugate. *Eur. J. Biochem.* **176**, 625-628

DELMOTTE F., LESCANNE P.J., DAUSSIN F., ROCHE A.C., MIDOUX P. & MONSIGNY M. (1985) Préparation et propriétés de systèmes covalents drogue-bras-vecteur. *Actualités Chimie Thérapeutique* **12**, 121-140

DERRIEN D., MIDOUX P., PETIT C., NEGRE E., MAYER R., MONSIGNY M.& ROCHE A.C. (1989) Muramyl dipeptide bound to poly-**L**-lysine substitued with mannose and gluconoyl residues as macrophage activators. *Glycoconjugate J.* **6**, 241-255

HELENE C. & TOULME J.J. (1990) Specific regulation of gene expression by antisense, sense and anti-gene nucleic acids. Biochim. Biophys. Acta **1049**, 99-125

FIUME L., BASSI B., BUSI C., MATTIOLI A.& SPINOSA G.(1986) Drug targeting in antiviral chemotherapy. A chemical stable conjugate of 9-β-**D**-arabinofuranosyl-adenine 5'-monophosphate with lactosaminated albumin accomplishes a selective delivery of the drug to liver cells. *Biochem. Pharmacol.* **35**, 967-972

MIDOUX P., NEGRE E., ROCHE A.C., MAYER R., MONSIGNY M., BALZARINI J., DE CLERQ E., MAYER E., GHAFFAR A.& GANGEMI J.D.(1990) Drug targeting: anti-HSV1 activity of mannosylated polymer-bound 9-(2-phosphonyl-methoxy-ethyl)-adenine. *Biochem. Biophys. Res. Commun.* **167**, 1044-1049

MIDOUX P., ROCHE A.C.& MONSIGNY M.(1986) Estimation of the degradation of endocytosed material by flow cytofluorometry using two neoglycoproteins containing different numbers of fluorescein molecules. *Biol. Cell* **58**, 221-226

MIDOUX P., ROCHE A.C.& MONSIGNY M.(1987) Quantitation of the binding, uptake, and degradation of fluoresceinylated neoglycoproteins by flow cytometry. *Cytometry* **8**, 327-334

MITSUYA H., YARCHOAN R. & BRODER S. (1990) Molecular targets for AIDS therapy. *Science* **249**, 1533-1544

MONSIGNY M., KIEDA C., ROCHE A.C.& DELMOTTE F.(1980) Preparation and biological properties of a covalent antitumor drug arm-carrier (D.A.C conjugate). *FEBS Letters* **119**, 181-186

MONSIGNY M., ROCHE A.C.& BAILLY P. (1984a) Tumoricidal activation of murine alveolar macrophages by muramyl dipeptide substituted mannosylated serum albumin. *Biochem. Biophys. Res. Commun.* **121**, 579-584

MONSIGNY M., ROCHE A.C., KIEDA C., MIDOUX P. & OBRENOVITCH A. (1988a) Characterization and biological implications of membrane lectins in tumor, lymphoid and myeloid cells. *Biochimie* **70**, 1633-1649

MONSIGNY M., ROCHE A.C.& MIDOUX P.(1984b) Uptake of neoglycoproteins via membranes lectin(s) of L 1210 cells. Evidence by quantitative flow cytometry

and drug targeting. *Biol. Cell* **51**, 187-196
MONSIGNY M., ROCHE A.C.& MIDOUX M.(1988b) Endogenous lectins and drug targeting. *Ann. N.Y. Acad. Sci.* **414**, 399-414
PETIT C., MONSIGNY M.& ROCHE A.C.(1990) Macrophage activation by muramyldipeptide bound to neoglycoproteins and glycosylated polymers: cytotoxic factor production. *J.Biol.Response Modifiers* **9**, 33-43
PIMPANEAU V., MIDOUX P., DURAND G., DE BAETSELIER P., MONSIGNY M. & ROCHE A.C. (1989) Endocytosis of α_1-acid glycoprotein variants and of neoglycoproteins containing mannose derivatives by a mouse hybridoma cell line ($2C_{11-12}$). Comparison with mouse peritoneal macrophages. *Glycoconjugate J.* **6**, 561-574
PIMPANEAU V., MIDOUX P., MONSIGNY M.& ROCHE A.C.(1990) Characterization and isolation of an intracellular mannose specific receptor from the human promyelocytic HL60 cells. *Carbohyd. Res.*, in the press
ROCHE A.C., BAILLY P.& MONSIGNY M.(1985a) Macrophage activation by MDP bound to neoglycoproteins : Metastasis eradication in mice. *Invas. Metastasis* **5**, 218-232
ROCHE A.C., BARZILAY M., MIDOUX P., JUNQUA S., SHARON N. & MONSIGNY M. (1983) Sugar-specific endocytosis of glycoproteins by Lewis lung carcinoma cells *J. Cell Biochem.* **22**, 131-140
ROCHE A.C., MIDOUX P., BOUCHARD P.& MONSIGNY M. (1985b) Uptake of neoglycoproteins containing 6-phosphomannose residues by human monocytes. *FEBS Lett.* **193**, 63-68
ROCHE A.C., MIDOUX P., PETIT C., DERRIEN D., MAYER R. & MONSIGNY M. (1988) Macrophage activation by targeted biological response modifiers *in* Immunomodulators and non-specific host defense mechanisms against microbial infections, (K.N. MASIHI & W. LANGE, eds) *Adv. Bioscience* pp. 217-235. Pergamon Press, Oxford.
ROCHE A.C., MIDOUX P., PIMPANEAU V., NEGRE E., MAYER R. & MONSIGNY M. (1990) Endocytosis mediated by monocyte and macrophage membrane lectins. Application to antiviral drug targeting. *Res. Virol.* **141**, 243-249
ROCHE A.C. & MONSIGNY M. (1986) Macrophage activation by targeted biological response modifiers *Ann. Inst. Pasteur, Immunol.* 137C, 223-226
SHARON N.& LIS H.(1989) Lectins as cell recognition molecules. *Science* **246**, 227-234
SHEN W.C. (1990) Acid-sensitive dissociation between poly-lysine and histamine-modified poly-glutamate as a model for drug releasing from carriers in endosomes. *Biochim. Biophys. Acta* **1034**, 122-124.
SHEN W.C.& RYSER H.J.P.(1981) *Cis*-aconityl spacer between daunomycin and macromolecular carrier : a model of pH sensitive linkage releasing drug from a lysosomotropic conjugate. *Biochem. Biophys. Res. Commun.* **102**, 1048-1054
TENU J.P., ROCHE A.C., YAPO A., KIEDA C., MONSIGNY M. & PETIT J.F. (1982) Absence of cell surface receptors for muramylpeptides in mouse peritoneal macrophages. *Biol. Cell* **44**, 157-164
TROUET A., BAURAIN R., DEPREZ-DE-CAMPENEERE D., MASQUELIER M. & PIRSON P. (1982a) Targeting of antitumour and antiprotozoal drugs by covalent linkage to protein carriers *in* Targeting of drugs (G. Gregoriadis, Senior J. & Trouet, eds) pp 19-30, Plenum, New-York
TROUET A., MASQUELIER M., BAURAIN R.& DEPREZ-DE-CAMPENEERE D.(1982b) A covalent linkage between daunorubicin and proteins that is stable in serum and reversible by lysosomal hydrolases, as required for a lysosomotropic drug-carrier conjugate: in vitro and in vivo studies. *Proc. Natl. Acad. Sci. USA* **79**, 626-629
WILEMAN T., HARDING C. & STAHL P. (1985) Receptor-mediated endocytosis. *Biochem. J.* **232**, 1-14

IMMUNOTOXINS

Sjur Olsnes
Institute for Cancer Research,
The Norwegian Radium Hospital,
Montebello, Oslo 3,
Norway

Current cytostatic drugs have comparatively low specificity. This is a persistent problem in cancer chemotherapy and in immunosuppression, and great effort is being made to develop more selective compounds. An important line of research towards this goal is to use antibodies against defined cell surface markers as vehicles to transport poisons to malignant cells. Certain protein toxins from bacteria and plants are often used for this purpose (Olsnes and Pihl, 1986). These toxins are enzymes that enter the cytosol and inactivate catalytically vital components of the protein synthesis machinery. Although true tumor-specific antigens are rare, the expression of developmental antigens on many tumor cells may provide the required selectivity (Olsnes *et al*, 1989). This is particularly the case in lymphomas and leukemias where the cells often express antigens not present in the stem cells.

Structure of immunotoxins

The bacterial diphtheria toxin, Shigella toxin and *Pseudomonas aeruginosa* exotoxin A, and the plant toxins ricin, abrin, modeccin and viscumin are used in construction of immunotoxins. Most of these toxins consist of two functionally different polypeptides linked by a disulfide bond (Fig. 1). The one polypeptide (A) carries the enzymatic activity, whereas the other one (B) binds to receptors at the cell surface (Olsnes and Sandvig, 1988). In most cases the toxins exhibit low selectivity between cells. The concept of immunotoxins is to replace the intrinsic binding specificity of the toxin with that of an antibody against a cell surface antigen.

NATO ASI Series, Vol. H 62
Endocytosis
Edited by P. J. Courtoy

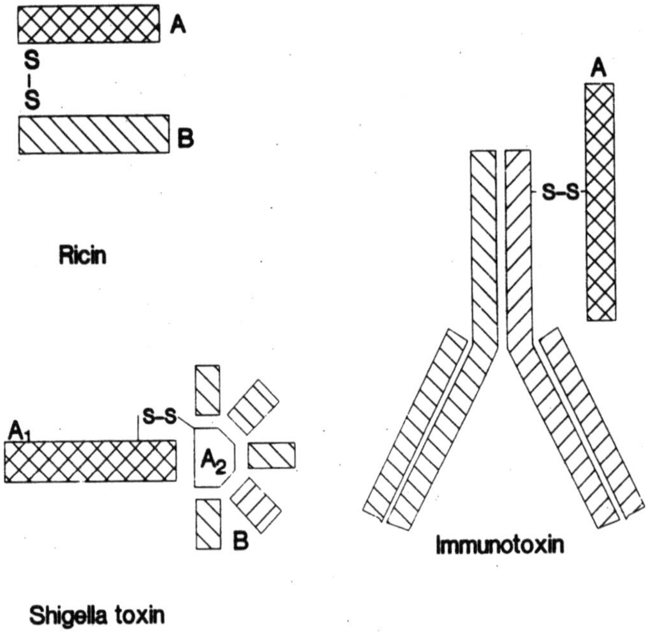

Figure 1. Schematical structure of toxins and immunotoxins.

A common strategy is to isolate the A-moiety of the toxin and link it chemically to the antibody by introduction of a disulfide bridge, but conjugates with the whole toxin have also been used. In the latter case, it is necessary to modify the B-chain to eliminate its intrinsic binding properties. Many plants contain A-chain like proteins that are not toxic as they lack the B-chain (Stirpe and Barbieri, 1986). Such proteins have also been used in immunotoxins. Several toxins are cloned, and recently a number of constructs have been made by the use of recombinant DNA technology (Bacha et al, 1988; Chaudhary et al., 1988).

Binding, uptake and intracellular routing

Immunotoxins containing monoclonal antibodies that bind with high affinity are in general more efficient than those containing low-affinity antibodies. When the binding is prevented, up to 10^6 times higher toxin concentrations are required to intoxicate cells than when binding is allowed to occur.

It appears that both proteins and lipids at the cell surface can act as productive binding sites for immunotoxins. The available evidence indicates that for a binding site to be effective in facilitating intoxication it must promote endocytosis of the toxin.

Only in the case of diphtheria toxin does translocation to the cytosol occur from the endosome. Some of the endocytosed toxin is routed back to the cell surface, whereas other molecules end up in lysosomes. A small part of the endocytosed ricin (~5%) is transported to the trans-Golgi network where it enters the biosynthetic/exocytotic route (Gonatas et al., 1977; van Deurs et al., 1986). This transport appears to be required for toxic effect.

Translocation to the cytosol

The mechanism of the final translocation across the vesicle or cisternal membrane is still not understood. The translocation process is best understood in the case of diphtheria toxin (Olsnes et al., 1988). This toxin is translocated from the endosomes, triggered by the low pH and, in contrast to ricin, there is no indication that transport to the trans-Golgi network is required for diphtheria toxin. Much less is known about the translocation of the other natural toxins and about that of immunotoxins. An acidification step appears to be involved in some, but not in all cases. NH_4Cl and monensin protect against modeccin and Pseudomonas aeruginosa exotoxin A, while the toxic effect of ricin and that of many immunotoxins is increased under the same conditions (Olsnes et al., 1989). It should be noted that even in those cases where NH_4Cl and monensin protect against a toxin or an immunotoxin, low pH may not necessarily be required for the translocation process as such. For example, the routing between intracellular organelles may be altered when the interior of acidic vesicles is neutralized.

Enhancement of the toxic effect

The sensitivity of cells to many immunotoxins can be strongly increased by NH_4Cl and by cationic ionophores such as monensin. In the presence of these compounds up to 30,000 fold increase in activity is obtained with certain immunotoxins (Casellas et al., 1984; Raso and Lawrence, 1984). More importantly, the enhancers often reduce the lag-time from the addition of toxin to cells until protein synthesis is blocked. This is particularly important when rapidly dividing cells are the targets. The reason for the potentiation is not understood.

Problems and potentials in immunotoxin therapy

The construction of immunotoxins is simple in concept, but a number of requirements must be fulfilled for a conjugate to be of therapeutic value in vivo. The conjugate must remain intact in the circulation for a considerable period of time, it must be able to penetrate through the capillary wall and into tumors and it must not bind to other cells than those intended.

After the conjugate has reached the target cell a further set of requirements must be fulfilled. To ensure high toxicity it is important that the antigen is internalized efficiently and delivered to the intracellular compartment(s) where translocation takes place.

In the construction of the immunotoxins so far made the overall structure of the conjugate has usually not been taken into consideration. By linking together either chemically or genetically two different proteins, it is likely that molecules are made that have lower stability and that are recognized as unusual by the protective mechanisms in the organism. Conceivably, the natural toxins have been polished by time to avoid these mechanisms. Therefore, the natural toxins are more active in the in vivo situation than immunotoxins.

Diphtheria toxin is toxic to most human cells, whereas murine cells are highly resistant. Therefore, diphtheria toxin can be used for targeted therapy against human tumors growing in nude mice. Raso et al. (1989) treated mice with transplanted human mesothelioma cells with injections of diphtheria toxin and obtained complete regression of the tumor even in cases of large solid tumors. In spite of the fact that the tumor cells were not highly sensitive to the toxin, a single injection of diphtheria toxin was in many cases sufficient to induce complete cure of the animal. This shows the potential of targeted toxin therapy once the problems of making stable, natural-looking immunotoxins has been solved.

Although immunotoxins were first conceived of as anticancer drugs, their potential applications reach much further. The use of anti T-cell immunotoxins in immunosuppression has already been mentioned. Conjugates with anti-idiotype antibodies to the acetylcholine receptor may eradicate B-cells producing antibodies to the acetylcholine receptor in myasthenia gravis (Bown and Krolick, 1988). Allergic conditions might be treated with IgE-toxin conjugates that bind to and destroy mast cells and basophile leukocytes. It should also be noted that the protein synthesis machinery of eukaryotic parasites is sensitive to the toxins, and that attempts have been made to form immunotoxins against such organisms (Willemez and Carlo, 1984).

It is likely that drug targeting will play an important role in future pharmacology. Studies on immunotoxins have focused on a number of problems that are inherent in all targeted therapy. In particular it draws attention to the fact that it may not be sufficient to direct a drug to the appropriate cell, but that it may be equally important to ensure correct routing inside the target cell to achieve the intended effect.

This work was supported by the Norwegian Cancer Society.

References

Bacha P, Williams DP, Waters, C, Williams, JM, Murphy, JR and

Strom, TB (1988) Interleukin 2 receptor-targeted cytotoxicity. Interleukin 2 receptor-mediated action of a diphtheria toxin-related interleukin 2 fusion protein. J Exp Med 167: 612-622

Brown, RM and Krolick, KA (1988) Selective idiotype suppression of an adoptive secondary anti-acethylcholine receptor antibody response by immunotoxin treatment before transfer. J Immunol 140: 893-898

Casellas, P, Bourrie, BJP, Gros, P and Jansen, FK (1984) Kinetics of cytotoxicity induced by immunotoxins: enhancement by lysosomotropic amines and carboxylic ionophores. J Biol Chem 259: 9359-9364

Chaudhary, VK, Mizukami, T, Fuerst, TR, FitzGerald, DJ, Moss, B, Pastan, I and Berger, EA (1988) Selective killing of HIV-infected cells by recombinant human CD4-Pseudomonas exotoxin hybrid protein: Nature, 335: 369-372

Gonatas, NK, Kim, SU, Stieber, A and Avrameas, S (1977) Internalization of lectins in neuronal GERL. J Cell Biol 73: 1-13

Olsnes, S, and Pihl, A (1986) Construction and properties of chimeric toxins target-specific cytotoxic agents. Internatl Encycl Pharm Ther Section 119: Pharmacology of bacterial toxins. (F Dorner and J Drews, eds) Pergamon Press, Oxford, pp 709-739

Olsnes, S and Sandvig, K How protein toxins enter and kill cells. In: **"Immunotoxins"** (AE Frankel, ed) Kluwer Academic Publishers, Boston 1988, pp 39-73.

Olsnes S, Sandvig, K, Petersen, OW and van Deurs, B (1989) Immunotoxins - entry into cells and mechanism of action. Imunology Today, 10: 291-295

Preijers, FWMB, Tax, WJM, De Witte, T, Janssen, A, vd Heijden, H, Vidal, H, Wessels, JMC and Capel, PJA (1988) Relationship between internalization and cytotoxicity of ricin A-chain immunotoxins. Br J Haematol 70: 289-294

Raso, V and Lawrence, J (1984) Carboxylic ionophores enhance the cytotoxic potency of ligand- and antibody-delivered ricin A chain. J Exp Med 160: 1234-1240

Raso, V and McGrath, J (1989) Cure of experimental human malignant mesothelioma in athymic mice by diphtheria toxin. J NatlCancer Inst 81: 622-627

Stirpe, F and Barbieri, L (1986) Ribosome-inactivating proteins up to date. FEBS Lett 195: 1-8

van Deurs, B, Tønnessen, TI, Petersen, OW, Sandvig, K and Olsnes, S Routing of internalized ricin and ricin-conjugates to the Golgi complex. (1986) J Cell Biol 102: 37-47

Willemez, CJ and Carlo, PL (1984) Preparation of immunotoxin for Acanthamoeba castellanii. Biochem Biophys Res Commun 125: 25-29

SUBJECT INDEX

DIRECTORY OF REGISTERED PARTICIPANTS, OBSERVERS AND CONTRIBUTORS

(The first page number of each contribution is given in parentheses)

AMAR-COSTESEC Alain, ICP & Louvain University, Medical School, Lab. Chimie Physiologique, UCL 7539, 75 av Hippocrate, B-1200-Brussels, Belgium,
T. 32/2/764-75-52,
F. 32/2/764-75-73

ANTOINE Jean-Claude (381), Institut Pasteur, Unité de Physiopathologie, 28 Rue du Dr Roux, F-75724 Paris Cx 15, France,
T. 33/1/45-68-80-00,
F. 33/1/43-06-98-35

BARTHOLEYNS Jacques, CNTS, 3 av des Tropiques - BP 100, 91943 Les Ulis Cx, France,
T. 32/41/69-07-78-71

BASTIN Philippe (475), ICP & Louvain University Medical School, Cell Biology Unit, UCL 7541, 75, av Hippocrate, B-1200-Brussels, Belgium,
T. 32/2/764-75-41,
F. 32/2/764-75-43

BEAUMELLE Bruno, URA CNRS, 530, Case 107 USTL 34095 Montpellier, Cx 2, France,
T. 33/67/14-37-41,
F. 33/67/54-30-39

BERG Trond (239, 247), University of Oslo, Institute Nutrition Research, P.O. Box 1046, Blindern, N-0316 Oslo 3, Norway,
T. 47/2/45-61-62,
F. 47/2/45-41-94

BERGERON John J.M. (151), Mc Gill University, Dept of Anatomy, 3640 University street, H3A 2B2 Montreal Québec, Canada,
T. 514/398-63-35 or 63-51,
F. 514/398-50-47

BERNHARD Wolfgang B., Weizmann Institute, Dept. Biophysics, Rehovot, Israel,
T. 972/8/34-21-11 (or 36-11),
F. 972/8/46-82-56

BIRN Henrik, University of Aarhus, Institute of Anatomy, Dept. Cell Biology, DK-8000 Aarhus C, Denmark,
T. 45/86/12-83-33,
F. 45/86/19-86-64

BOMSEL Morgane (113), University of California, Dept. Anatomy, School of Medicine, Box 0452, San Francisco CA 94143-0452, U.S.A.,
T. 1/415/476-60-47,
F. 1/415/476-48-45

BOQUET Patrice (413), Institut Pasteur, Unité des Antigènes Bactériens, 28 Rue du Dr Roux, F-75724 Paris Cx 15, France,
T. 33/1/45-68-80-00,
F. 33/1/43-06-98-35

BRODSKY Frances (343), University of California, Dept. Pharmacy, Rm S-926, Box 0446, 513 Parnassus Av, San Francisco CA 94143, U.S.A.,
T. 1/415/476-64-06,
F. 1/415/476-06-88

BUCKLAND Robin, Faculté de Médecine Alexis Carrel, CNRS, UMR 30, Immunovirologie cellulaire et moléculaire, Lyon Cx 8, France,
T. 33/78/77-86-18

BUSSON Sylvie, CNRS, Centre de Biologie Cellulaire, 67 rue Maurice Günsbourg, F-94200 Ivry-s-Seine, France
T. 31/1/46-72-18-00
F. 31/1/46-70-88-46

CARPENTIER Jean-Louis (17), Université de Genève, Institut d'histologie et d'embryologie, 1 Rue Michel Servet, CH-1211 Genève 4, Suisse,
T. 41/22/22-90-65,
F. 41/22/47-33-34

CHRISTENSEN Erik (161, 325), University of Aarhus, Institute of Anatomy, Dept. Cell Biology, 8000 Aarhus C, Denmark,
T. 45/86/12-83-33 (ext 7874),
F. 45/86/19-86-64

COPPENS Isabelle (475), ICP & Louvain University Medical School, Cell Biology Unit, UCL 7541, 75 Av Hippocrate, B-1200 Bruxelles, Belgium
T. 32/2/764-75-57,
F. 32/2/764-75-43

COURTOY Pierre J. (231, 291, 475), ICP & Louvain University Medical School, Cell Biology Unit, UCL 7541, 75, av Hippocrate, B-1200 Bruxelles, Belgium,
T. 32/2/764-75-69 (or 41),
F. 32/2/764-75-43

CUPERS Philippe, ICP & Louvain University Medical School, Cell Biology Unit, UCL 7541, 75 av. Hippocrate, B-1200-Bruxelles, Belgium,
T. 32/2/764-75-41,
F. 32/2/764-75-43

DAUTRY-VARSAT Alice (175), Institut Pasteur, Dépt. Immunologie, 25 Rue du Dr Roux, F-75724 Paris CX 15, France,
T. 33/1/45-68-85-74,
F. 33/1/45-68-86-39

DAVOUST Jean (27, 337), Centre d'Immunologie de Marseille-Luminy, Case 906, F-13288 Marseille Cx 9, France,
T. 33/91/26-94-36,
F. 33/91/26-94-30

DE BRABANDER Mark (3), Clinical Research Unit, Jan Palfijn Hospital, Lange Bremstraat 70, B-2060 Merksem, Belgium,
T. 32/3/640-28-06,
F. 32/3/646-44-79

de CHASTELLIER Chantal (375), Necker-Enfants Malades, Lab. Bactériologie, 156 rue de Vaugirard, F-75730 Paris Cx 15, France,
T. 33/1/47-83-33-03 (ext 436 or 437)
F. 33/1/45-67-53-33

de DUVE Christian (449), ICP & Louvain University Medical School, 75, av. Hippocrate, B 1200-Bruxelles, Belgium,
T. 32/2/674-75-50,
F. 32/2/674-75-73

DESBUQUOIS Bernard (141), Necker Enfants Malades, Unité 30 INSERM, 156 rue de Vaugirard, F-75015 Paris Cx 15, France,
T. 33/1/42-73-89-38

DOMURADO Dominique, Université de Compiègne, Lab. Technologie Enzymatique, BP 649, F-60206 Compiègne, France,
T. 33/44/20-99-60 (ext 2278)
F. 33/44/20-39-10

DUPREZ Véronique (175), Institut Pasteur, Dept. Immunologie, 25 rue du Dr Roux, F-75724 Paris Cx 15, France,
T. 33/41/45-68-85-71,
F. 33/41/45-68-86-39

ENRICH Carlos, Universitad Barcelona, Facultad de Medicina, Dept Biologia Cellular i Anatomia Patologica, Av. Diagonal s/n, 08028 Barcelona, Spain,
T. 34/3/411-10-72,
F. 34/3/330-29-03

EVANS William Howard, National Institute for Medical Research, The Ridgeway, Mill Hill, London NW7 1AA, U.K.,
T. 44/71/959-36-66, (ext 2157)
F. 44/71/906-44-77

FREHEL Claude (375), Necker Enfants Malades, Lab. Bactériologie, 156 rue de Vaugirard, F-75730 Paris Cx 15, France,
T. 33/41/47-83-33-03 (ext 436 or 437),
F. 33/41/45-67-53-33

FUCHS Renate (135, 301), Universität Wien, Dept. General and Experimental Pathology, Wahringerstrasse, 13, A-1090 Vienna, Austria,
T. 43/222/43-15-26 (ext 372)

GABRION Jacqueline, Université de Montpellier II, CNRS UA 1197, Case 090, Lab. Neurobiologie Endocrinologique, Place E. Bataillon, F-34060 Montpellier Cx, France,
T. 33/67/14-37-49

GEFFEN Iris, Biozentrum der Universität Basel, Dept. Biochemistry, Klingelberg-strasse 70, CH-4056 Basel, Switzerland,
T. 41/61/25-38-80 (ext 335),
F. 41/61/25-67-60

GERARD Anne, Université de Nancy I, Lab. Histologie-Embryologie II, 9 av. de la Forêt de Haye, B.P. 184, F-54505 Vandoeuvre Cx, France,
T. 33/83/56-56-56 (ext. 270 or 232)

GERLIER Denis, Ecole Normale Supérieure de Lyon, Lab. Biologie Moléculaire & Cellulaire, 46 allée d'Italie, F-69364 Lyon Cx 07, France,
T. 33/72/72-80-13,
F. 33/72/72-80-80

GEUZE Hans (169, 221), State University Utrecht Medical School, Dept. Cell Biology, AZU HO2 314 Heidelberglaan 100, NL-3584 Cx Utrecht, Netherlands,
T. 31/30/50-76-52,
F. 31/30/54-17-97

GINSEL Leo (273), University Nijmegen, Dept Cell Biology & Histology, TRIGON, Adelbertusplein 1, PO Box 9101, 6500 HB Nijmegen, Netherlands,
T. 31/80/61-43-29,
F. 31/80/54-05-25

GODELAINE Danièle, ICP & Louvain University Medical School, Lab. Chimie Physiologique, UCL 7539, 75 av Hippocrate, B-1200-Bruxelles, Belgium,
T. 32/2/764-75-60
F. 32/2/764-75-73

GONATAS Nicholas, University of Pennsylvania, School of Medicine, Dept. Neuropathology, 418 Johnson Pavillion, Philadelphia, PA 19104, U.S.A.,
T. 1/215/662-6695,
F. 1/215/898-4227

GRIFFITHS Gareth (73), EMBL, Dept. Cell Biology, Postfach 10.2209, 1 Meyerhofstrasse, D-6900 Heidelberg, Germany,
T. 49/6221/38-72-67,
F. 49/6221/38-73-16

GRUENBERG Jean (113), EMBL, Dept. Cell Biology, Postfach 10 2209, 1 Meyerhofstrasse, D-6900 Heidelberg, Germany,
T. 49/6221/38-72-88 (or 92),
F. 49/6221/38-73-06

HANSEN Steen (57), University of Copenhagen, Panum Institute, Structural Cell Biology Unit, Blegdamsveg 3, DK-2200 Copenhagen N, Denmark,
T. 45/31/35-79-00,
F. 45/31/35-61-33

HOFLACK Bernard (197), EMBL, Dept. Cell Biology, Postfach 10.2209, 1 Meyerhofstrasse, D-6900 Heidelberg, Germany,
T. 49/6221/38-72-88,
F. 49/6221/38-73-06

HOPKINS Colin (65), Imperial College, Dept. Biochemistry, Imperial College Rd, SW7 2 AZ London, U.K.,
T. 44/71/225-82-65
F. 44/71/584-75-96

HOWELL Kathryn (131), University of Colorado, School of Medicine, Dept. Cellular & Structural Biology, Campus box B-111 - 4200 East 9th Ave., Denver CO 80 262, U.S.A,
T. 1/303/270-51-53
F. 1/303/270-47-29

HUBBARD Ann (257) Johns Hopkins University, Dept. Cell Biology & Anatomy, 725 North Wolfe St, Baltimore, MD 21205, U.S.A.,
T. 1/301/955-23-33 (or 37),
F. 1/301/955-41-29 (or 433-35-89)

KAS H. Süheyla, University of Hacettepe, Faculty of Pharmacy, Dept. Pharmaceutical Technology, 06100 Ankara, Turkey,
T. 4/310-15-24,
F. 4/311-47-77

KIRCHHAUSEN Tom, Harvard Medical School, Dept. Anatomy & Cell Biology, 25 Shattuck st, Boston, MA 02115-6092, U.S.A.,
T. 1/617/432-16-84,
F. 1/617/277-27-32

KISS Anna, Semmelweis University of Medicine, Lab. I Electron Microscopy, 58 Tűzolto u, 1450 Budapest, Hungary,
T. 36/11/33-69-20
F. 36/11/14-30-64

KOSTROUCH Zdenek, Charles University, Dept. Pathology, Studnickovaz, CS-12800 Prague-2, Czechoslovakia
F. 42/2/29-17-51

KOSTROUCH Zdenek, c/o INSERM U.197, Faculté de Médecine Alexis Carrel, rue G. Paradin, F-69372 Lyon Cx 08, France

KRAEHENBUHL Jean Pierre (363) Institut Suisse de Recherche sur le Cancer, Université de Lausanne, Chemin des Boveresses, CH-1066 Epalinges s/Lausanne, Suisse,
T. 41/21/33-30-61,
F. 41/21/32-69-33

LARDEUX Bernard, Faculté de Médecine Xavier Bichat, Laboratoire de Biologie Cellulaire, 16 rue Henri Huchard, F-75018 Paris, France,
T. 33/1/42-63-84-20 (ext 434)

LEVINE Tim, University College London, Dept. Biology, Medawar building, Gower street, London WC1E 6BT, U.K.,
T. 44/71/387-70-50 (ext 3576),
F. 44/71/380-70-26

LUZIO Jean-Paul (123), University Cambridge, Addenbrooke's Hospital, Dept. Clinical Biochemistry, Hills Rd, Cambridge, CB 2 2QR, U.K.,
T. 44/223/33-67-80,
F. 44/223/21-68-62

MAGNUSSON Sigurdur (239), University of Oslo, Dept. Biology, Division Molecular Cell Biology, PO Box 1050 Blindern, N-0316 Oslo 3, Norway,
T. 47/2/45-45-97,
F. 47/2/45-46-05

MARSH Mark (355, 399), Institute of Cancer Research, Chester Beatty Laboratories, Fullam Rd, London SW3 6JB, U.K.,
T. 44/71/352-81-33,
F. 44/71/352-32-99

MATA Lucinda (161), Gulbenkian Institute of Science, Dept. Cell Biology, Apart 14, P-2781 Oeiras, Portugal,
T. 351/1/443-14-54,
F. 351/1/443-16-31

MAURICE Michèle (267), Faculté de Médecine X. Bichat, Lab. Biologie Cellulaire, 16 rue Henri Huchard, F-75018 Paris, France,
T. 33/1/42-63-84-20 (ext 431)

MELLMAN Ira (135, 309), Yale University School of Medicine, Dept. Cell Biology, 333 Cedar Street, PO Box 3333, New Haven, CT 06510, U.S.A.,
T. 1/203/785-50-58 (or 43-02),
F. 1/203/785-72-26 (or 74-46)

MILLER Karen (65), Imperial College of Science & Technology, Dept. Biochemistry, Imperial College Rd, London SW7 2 AZ, U.K.,
T. 44/71/589-51-11 (ext 4117),
F. 44/71/584-75-96,

MONSIGNY Michel (489), Université d'Orléans & CNRS, Dept. Biochimie des Glycoconjugués, 1 rue Haute, F-45071 Orléans Cx 2, France,
T. 33/38/51-55-57,
F. 33/38/69-00-94

MULLOCK Barbara (123), University Cambridge, Addenbrooke's Hospital, Dept. Clinical Biochemistry, Hills Rd, Cambridge, CB2 2QR, U.K.,
T. 44/223/33-67-82,
F. 44/223/21-68-62

MURPHY Robert (91), Carnegie-Mellon University, Dept. Biol. Sci., Fifth Ave, 4400, Pittsburgh, PA 15213, U.S.A.,
T. 1/412/268-34-80 (or 34-61),
F. 1/412/268-65-71

NATHKE Inke (343), University of California, Dept. Pharmacy, S-926, Box 0446, 513 Parnassus av, San Francisco, CA 94143, U.S.A,
T. 1/415/476-64-06,
F. 1/415/476-06-88

NIELSEN Soren (325), University of Aarhus, Institute of Anatomy, Dept. Cell Biology, DK-8000 Aarhus C, Denmark,
T. 45/86/12-83-33,
F. 45/86/19-86-64

OLSNES Sjur (497), The Norwegian Radium Hospital, Institute for Cancer Research, Ullernchausséen 70, Montebello, N-0310 Oslo 3, Norway,
T. 47/2/50-60-50,
F. 47/2/52-55-29

OPARKA Karl J., Scottish Crop Research Institute Invergowrie, Dundee, DD 2 5DA, U.K.,
T. 44/382/56-27-31,
F. 44/382/56-24-26

PACCAUD Jean-Pierre (17), Université de Genève, Département de Morphologie, CMU, 1 rue Michel Servet, CH-1211 Genève 4, Suisse,
T. 41/22/22-90-83 (or 90-85),
F. 41/22/47-33-34

PARTON Robert (317), EMBL, Dept. Cell Biology, Postfach 10.2209, 1 Meyerhofstrasse, D-6900 Heidelberg, Germany
T. 49/6221/38-72-88
F. 49/6221/38-73-06

PAULOIN Alain, CRJ-INRA, Biologie Cellulaire et Mol., Bâtiment des Biotechnologies, 78352 JOUY-EN-JOSAS Cx, France,
T. 33/1/34-65-21-21
F. 33/1/34-65-22-73

PELCHEN-MATTHEWS Anne-Gret (355, 399), Institute of Cancer Research, Chester Beatty Laboratories, Fullam Rd, London SW3 6JB , U.K.,
T. 44/71/352-8133,
F. 44/71/352-3299

PETERS Christophe B.W. (205), Universität Göttingen, Institut für Biochemie II, Gosslerstrasse 12d, D-3400 Göttingen, Germany,
T. 49/551/39-59-49,
F. 49/551/39-96-12

POUVELLE Bruno, Centre de Biophysique Moleculaire CNRS, 1A av. de la Recherche Scientifique, F-45071 Orléans Cx 2, France,
T. 33/38/63-10-04
F. 33/38/63-15-17

PRYDZ Kristian (405), Institute for Cancer Research, Dept. Biochemistry, Ullernchausséen 70, Montebello, N-0310 Oslo 3, Norway,
T. 47/2/50-60-50,
F. 47/2/50-86-92

PYPAERT Marc (99), Imperial Cancer Research Fund Labs., Cell Biology Laboratory, PO Box 123, Lincoln's Inn Fields, London WC2A 3PX, U.K.,
T. 44/71/242-02-00 (ext 3346),
F. 44/71/405-15-56

RABINOVITCH Michel (483), Institut Pasteur, Unité d'Immuno-parasitologie, 28 rue Dr Roux, F-75724 Paris Cx 15, France,
T. 33/1/45-68-80-00
F. 33/1/45-68-86-39

REGGIO Hubert, Faculté Sciences de Marseille-Luminy, Lab. Biologie de la Différentiation Cellulaire, 70 Route Léon Lachamp, Bx 901, F-13288 Marseille Cx 2, France,
T. 33/91/26-92-42
F. 33/91/26-93-86

REID Pamela (335), University of Dundee, Medical Sciences Institute, Dept Biochemistry, DD1 4HN Dundee, U.K.,
T. 44/382/231-81 (ext 4746),
F. 44/382/20-10-63

RIEZMAN Howard (467), Biozentrum Basel, 70 Klingelbergstrasse, CH-4056 Basel, Suisse,
T. 41/61/25-38-80,
F. 41/61/25-67-60

ROBENEK Horst (9), University of Münster, Institute for Arteriosklerosis Research, 3 Domagkstrasse, D-4400 Münster, Germany,
T. 49/251/83-62-36,
F. 49/251/83-62-05

ROBINSON David G. (459), Universität Göttingen, Pflanzenphysiol. Institut, Untere Karspüle 2, D-3400 Göttingen, Germany,
T. 49/551/39-78-33,
F. 49/551/39-96-12

ROBINSON Margaret S. (51), University of Cambridge, Addenbrooke's hospital, Dept. Clinical Biochemistry, Hills Rd, Cambridge CB2 2QR, U.K.,
T. 44/223/33-67-82,
F. 44/223/21-68-62

ROCHE Annie C. (489), Université d'Orléans, Dépt. Biochimie des Glycoconjugués, 1 rue Haute, F-45071 Orléans Cx 02, France,
T. 33/38/51-55-37,
F. 33/38/69-00-94

RYTER Antoinette (369), Institut Pasteur, Dept. Biologie Moleculaire, 28 rue du Dr Roux, F-75724 Paris Cx 15, France,
T. 33/1/45-68-84-95,
F. 33/1/43-06-98-35

SALAMERO Jean (337), Centre d'Immunologie Marseille Luminy Case 906 Luminy, F-13288 Marseille Cx 9, France,
T. 33/91/26-94-36,
F. 33/91/26-94-30

SANDHOFF Konrad (427), Friedrich Willem Universität, Institut für Organische Chemie & Biochemie, 1 Domagkstrasse, D-5300 Bonn 1, Germany,
T. 49/228/73-53-46 (or 77-78),
F. 49/228/73-56-83

SANDOVAL Ignacio (213), Universidad Autonoma de Madrid, Centro de Biologia Molecular, Campus de Cantoblanco , SP-28049 Madrid, Spain,
T. 34/91/397-41-57,
F. 34/91/397-47-99

SANDVIG Kirsten (57, 405), The Norwegian Radium Hospital, Institute for Cancer Research, Ullernchausséen 70, Montebello, N-0310 Oslo 3, Norway,
T. 47/2/50-60-50 (ext 9762 or 9824),
F. 47/2/52-55-59

SANSONETTI Philippe (389), Institut Pasteur, Unité Pathogénie Microbienne Moléculaire, 28 Rue du Dr Roux, F-75724 Paris Cx 15, France,
T. 33/1/45-68-83-42,
F. 33/1/43-06-98-35

SATIAT-JEUNEMAITRE Béatrice, c/o Oxford Polytechnic, School Biol. & Mol. Sciences, Gipsy lane, Headington, Oxford OX3 OBP, U.K.,
T. 44/865/81-92-40 or 41,
F. 44/865/81-90-73

SATRE Michel, Université de Grenoble, Lab. Biologie Cellulaire, CEN-G 85X, F-38041 Grenoble Cx, France,
T. 33/76/88-49-07 or 30-65,
F. 33/76/88-51-85

SCHERMAN Daniel, Institut de Biologie Physico-Chimique, CNRS, 13 rue P et M Curie, F-75005 Paris, France,
T. 33/41/43-25-26-09,
F. 33/41/40-46-83-31

SCHMID Sandra (105, 135), Scripps Research Institute, Dept. Molecular Biology-IMM-II, 10666 North Torrey Pine's Road, La Jolla, CA 92037-1093, U.S.A.,
T. 1/619/554-23-11,
F. 1/619/554-67-05

SEGLEN Per (247), The Norwegian Radium Hospital, Institute for Cancer Research, Ullernchausséen 70, N-0310 Montebello, N-0310 Oslo 3, Norway,
T. 47/2/50-60-50 (ext 9205),
F. 47/2/52-55-59

SMYTHE Elizabeth (105), Scripp's Research Institute, Dept. Molecular Biology-IMM-II, 10666 North Torrey Pines Road, La Jolla, CA 92037-1093, U.S.A.,
T. 1/619/554-82-57,
F. 1/619/554-67-05

SORKIN Alexander (181), Acad. Sciences of the USSR, Institute of Cytology, 4 Tichoretsky pr, 194064 Leningrad, USSR

SORKIN Alexander (181), c/o Vanderbilt University, School of Medicine, Dept. Biochemistry, Nashville, TN 37232-0146, U.S.A.,
T. 1/615/322-63-36,
F. 1/615/322-43-49

STOORVOGEL Willem (169), State University Utrecht Medical School, Dept. Cell Biology, Heidelberglaan 100, 3584 Cx Utrecht, Netherlands,
T. 31/30/50-65-51,
F. 31/30/54-17-97

STORRIE Brian (85), Virginia Polytechnic Inst. State University, Dept. Biochemistry & Nutrition, Blacksburg, VA 24061-0308, U.S.A.,
T. 1/703/231-64-34,
F. 1/703/231-90-70

STROUS Ger (169, 221), State University Utrecht Medical School, Dept. Cell Biology, Heidelberglaan 100, 3584 Cx Utrecht, Netherlands,
T. 31/30/50-64-76 (or 65-51),
F. 31/30/54-17-97

TAUBER Rudolf (187), Universitätsklinikums Rudolf Virchow, Institut für Klinische Chemie und Biochemie, Standort Charlottenburg, 130 Spandauer Damm, 1000 Berlin 19, Germany,
T. 49/30/30-35-35-64,
F. 49/30/30-35-20-66

TOUGARD Claude, Collège de France & CNRS, Neuroendocrinologie Cell. & Mol., URA 1115, 11 Place Marcelin Berthelot, F-75231 Paris Cx 05, France,
T. 33/41/44-27-14-29,
F. 33/41/44-27-11-09

TULKENS Paul (439), ICP & Louvain University Medical School, Lab. Chimie Physiologique, UCL 7539, 75 av Hippocrate, B-1200 Bruxelles, Belgique,
T. 32/2/762-21-36
(or 764-74-74),
F. 32/2/764-52-94 (or 75-73)

UNGEWICKELL Ernst (43), Max Planck Institute für Biochemie, Am Klopferspitz 18A, D-8033 Martinsried, Germany,
T. 44/89/85-78-22-15,
F. 44/89/85-78-37-77

VAN DEN HOVE-VAN DEN BROUCKE Marie-France, ICP and University of Louvain, Medical School, Hormone and Metabolic Research Unit, UCL 7529, 75 av Hippocrate, B-1200 Bruxelles, Belgique,
T. 32/2/764-75-23 or 75-29,
F. 32/2/762-74-55

VAN DER SMISSEN Patrick (231), ICP & Louvain University Medical School, Cell Biology Unit, UCL 7541, 75 av Hippocrate, B-1200 Bruxelles, Belgique,
T. 32/2/764-75-46,
F. 32/2/764-75-43

VAN DEURS Bo (57, 405), University of Copenhagen, Panum Institute, Structural Cell Biology Unit, Blegdamsveg 3, DK -2200 Copenhagen N, Denmark,
T. 45/31/35-79-00,
F. 45/31/35-64-52

VAN MEER Gerrit (281), State University Utrecht Medical School, Dept. Cell Biology, AZU HO2 314 Heidelberglaan 100, NL-3584 Cx Utrecht, Netherlands,
T. 31/30/50-64-80,
F. 31/30/54-17-97

VEGA-PALACIOS Miguel A. (213), Universidad Autonoma de Madrid, Centro de Biologia Molecular (Modulo CXV), Cantoblanco, Madrid 28049, Spain,
T. 34/91/397-41-57
F. 34/91/397-47-99

VIDAL Michel J., CNRS 530 URA, Dept. Biologie-Santé, USTL Case 107, Bât 24, Place E. Bataillon, F-34095 Montpellier CX 5, France,
T. 33/67-14-37-43

VON FIGURA Kurt (205), Georg-August Universität Göttingen, Institut für Biochemie II, Gosslerstrasse, 12d, D-3400 Göttingen, Germany,
T. 49/551/39-59-48,
F. 49/551/39-96-12

WATTIAUX Robert (231, 433), FUNDP, Lab. Chimie Physiologique, 61 rue de Bruxelles, B-5000 Namur, Belgique,
T. 32/81/22-90-61,
F. 32/81/23-03-91

WATTIAUX-DE CONNINCK Simone (231, 433), FUNDP, Lab. Chimie Physiologique, 61 rue de Bruxelles, B-5000 Namur, Belgique,
T. 32/81/22-90-61,
F. 32/81/23-03-91

WATTS Colin (335), University of Dundee, Medical Sciences Institute, Dept. Biochemistry, Dundee DD1 4HN, Scotland - U.K.,
T. 44/382/23-181 (ext 4233),
F. 44/382/20-10-63

ZAMOYSKA Rose, University College London, Dept. biology, Medawar building, Gower street, London WC1E 6BT, U.K.,
T. 44/71/387-70-50 (ext 3529),
F. 44/71/380-70-26

NATO ASI Series H

Vol. 1: **Biology and Molecular Biology of Plant-Pathogen Interactions.** Edited by J. A. Bailey. 415 pages. 1986.

Vol. 2: **Glial-Neuronal Communication in Development and Regeneration.** Edited by H. H. Althaus and W. Seifert. 865 pages. 1987.

Vol. 3: **Nicotinic Acetylcholine Receptor: Structure and Function.** Edited by A. Maelicke. 489 pages. 1986.

Vol. 4: **Recognition in Microbe-Plant Symbiotic and Pathogenic Interactions.** Edited by B. Lugtenberg. 449 pages. 1986.

Vol. 5: **Mesenchymal-Epithelial Interactions in Neural Development.** Edited by J. R. Wolff, J. Sievers, and M. Berry. 428 pages. 1987.

Vol. 6: **Molecular Mechanisms of Desensitization to Signal Molecules.** Edited by T. M. Konijn, P. J. M. Van Haastert, H. Van der Starre, H. Van der Wel, and M. D. Houslay. 336 pages. 1987.

Vol. 7: **Gangliosides and Modulation of Neuronal Functions.** Edited by H. Rahmann. 647 pages. 1987.

Vol. 8: **Molecular and Cellular Aspects of Erythropoietin and Erythropoiesis.** Edited by I. N. Rich. 460 pages. 1987.

Vol. 9: **Modification of Cell to Cell Signals During Normal and Pathological Aging.** Edited by S. Govoni and F. Battaini. 297 pages. 1987.

Vol. 10: **Plant Hormone Receptors.** Edited by D. Klämbt. 319 pages. 1987.

Vol. 11: **Host-Parasite Cellular and Molecular Interactions in Protozoal Infections.** Edited by K.-P. Chang and D. Snary. 425 pages. 1987.

Vol. 12: **The Cell Surface in Signal Transduction.** Edited by E. Wagner, H. Greppin, and B. Millet. 243 pages. 1987.

Vol. 13: **Toxicology of Pesticides: Experimental, Clinical and Regulatory Perspectives.** Edited by L. G. Costa, C. L. Galli, and S. D. Murphy. 320 pages. 1987.

Vol. 14: **Genetics of Translation. New Approaches.** Edited by M. F. Tuite, M. Picard, and M. Bolotin-Fukuhara. 524 pages. 1988.

Vol. 15: **Photosensitisation. Molecular, Cellular and Medical Aspects.** Edited by G. Moreno, R. H. Pottier, and T. G. Truscott. 521 pages. 1988.

Vol. 16: **Membrane Biogenesis.** Edited by J. A. F. Op den Kamp. 477 pages. 1988.

Vol. 17: **Cell to Cell Signals in Plant, Animal and Microbial Symbiosis.** Edited by S. Scannerini, D. Smith, P. Bonfante-Fasolo, and V. Gianinazzi-Pearson. 414 pages. 1988.

Vol. 18: **Plant Cell Biotechnology.** Edited by M. S. S. Pais, F. Mavituna, and J. M. Novais. 500 pages. 1988.

Vol. 19: **Modulation of Synaptic Transmission and Plasticity in Nervous Systems.** Edited by G. Hertting and H.-C. Spatz. 457 pages. 1988.

Vol. 20: **Amino Acid Availability and Brain Function in Health and Disease.** Edited by G. Huether. 487 pages. 1988.

NATO ASI Series H

Vol. 21: **Cellular and Molecular Basis of Synaptic Transmission.**
Edited by H. Zimmermann. 547 pages. 1988.

Vol. 22: **Neural Development and Regeneration. Cellular and Molecular Aspects.**
Edited by A. Gorio, J.R. Perez-Polo, J. de Vellis, and B. Haber. 711 pages. 1988.

Vol. 23: **The Semiotics of Cellular Communication in the Immune System.**
Edited by E.E. Sercarz, F. Celada, N.A. Mitchison, and T. Tada. 326 pages. 1988.

Vol. 24: **Bacteria, Complement and the Phagocytic Cell.**
Edited by F.C. Cabello und C. Pruzzo. 372 pages. 1988.

Vol. 25: **Nicotinic Acetylcholine Receptors in the Nervous System.**
Edited by F. Clementi, C. Gotti, and E. Sher. 424 pages. 1988.

Vol. 26: **Cell to Cell Signals in Mammalian Development.**
Edited by S.W. de Laat, J.G. Bluemink, and C.L. Mummery. 322 pages. 1989.

Vol. 27: **Phytotoxins and Plant Pathogenesis.**
Edited by A. Graniti, R.D. Durbin, and A. Ballio. 508 pages. 1989.

Vol. 28: **Vascular Wilt Diseases of Plants. Basic Studies and Control.**
Edited by E.C. Tjamos and C.H. Beckman. 590 pages. 1989.

Vol. 29: **Receptors, Membrane Transport and Signal Transduction.**
Edited by A.E. Evangelopoulos, J.P. Changeux, L. Packer, T.G. Sotiroudis, and K.W.A. Wirtz. 387 pages. 1989.

Vol. 30: **Effects of Mineral Dusts on Cells.**
Edited by B.T. Mossman and R.O. Bégin. 470 pages. 1989.

Vol. 31: **Neurobiology of the Inner Retina.**
Edited by R. Weiler and N.N. Osborne. 529 pages. 1989.

Vol. 32: **Molecular Biology of Neuroreceptors and Ion Channels.**
Edited by A. Maelicke. 675 pages. 1989.

Vol. 33: **Regulatory Mechanisms of Neuron to Vessel Communication in Brain.**
Edited by F. Battaini, S. Govoni, M.S. Magnoni, and M. Trabucchi. 416 pages. 1989.

Vol. 34: **Vectors as Tools for the Study of Normal and Abnormal Growth and Differentiation.**
Edited by H. Lother, R. Dernick, and W. Ostertag. 477 pages. 1989.

Vol. 35: **Cell Separation in Plants: Physiology, Biochemistry and Molecular Biology.**
Edited by D.J. Osborne and M.B. Jackson. 449 pages. 1989.

Vol. 36: **Signal Molecules in Plants and Plant-Microbe Interactions.**
Edited by B.J.J. Lugtenberg. 425 pages. 1989.

Vol. 37: **Tin-Based Antitumour Drugs.** Edited by M. Gielen. 226 pages. 1990.

Vol. 38: **The Molecular Biology of Autoimmune Disease.**
Edited by A.G. Demaine, J-P. Banga, and A.M. McGregor. 404 pages. 1990.

Vol. 39: **Chemosensory Information Processing.** Edited by D. Schild. 403 pages. 1990.

Vol. 40: **Dynamics and Biogenesis of Membranes.**
Edited by J.A.F. Op den Kamp. 367 pages. 1990.

Vol. 41: **Recognition and Response in Plant-Virus Interactions.**
Edited by R.S.S. Fraser. 467 pages. 1990.

NATO ASI Series H

Vol. 42: **Biomechanics of Active Movement and Deformation of Cells.**
Edited by N. Akkaş. 524 pages. 1990.

Vol. 43: **Cellular and Molecular Biology of Myelination.**
Edited by G. Jeserich, H. H. Althaus, and T. V. Waehneldt. 565 pages. 1990.

Vol. 44: **Activation and Desensitization of Transducing Pathways.**
Edited by T. M. Konijn, M. D. Houslay, and P. J. M. Van Haastert. 336 pages. 1990.

Vol. 45: **Mechanism of Fertilization: Plants to Humans.**
Edited by B. Dale. 710 pages. 1990.

Vol. 46: **Parallels in Cell to Cell Junctions in Plants and Animals.**
Edited by A. W. Robards, W. J. Lucas, J. D. Pitts, H. J. Jongsma, and D. C. Spray. 296 pages. 1990.

Vol. 47: **Signal Perception and Transduction in Higher Plants.**
Edited by R. Ranjeva and A. M. Boudet. 357 pages. 1990.

Vol. 48: **Calcium Transport and Intracellular Calcium Homeostasis.**
Edited by D. Pansu and F. Bronner. 456 pages. 1990.

Vol. 49: **Post-Transcriptional Control of Gene Expression.**
Edited by J. E. G. McCarthy and M. F. Tuite. 671 pages. 1990.

Vol. 50: **Phytochrome Properties and Biological Action.**
Edited by B. Thomas and C. B. Johnson. 337 pages. 1991.

Vol. 51: **Cell to Cell Signals in Plants and Animals.**
Edited by V. Neuhoff and J. Friend. 404 pages. 1991.

Vol. 52: **Biological Signal Transduction.**
Edited by E. M. Ross and K. W. A. Wirtz. 560 pages. 1991.

Vol. 53: **Fungal Cell Wall and Immune Response.**
Edited by J. P. Latgé and D. Boucias. 472 pages. 1991.

Vol. 54: **The Early Effects of Radiation on DNA.**
Edited by E. M. Fielden and P. O'Neill. 448 pages. 1991.

Vol. 55: **The Translational Apparatus of Photosynthetic Organelles.**
Edited by R. Mache, E. Stutz, and A. R. Subramanian. 260 pages. 1991.

Vol. 56: **Cellular Regulation by Protein Phosphorylation.**
Edited by L. M. G. Heilmeyer, Jr. 520 pages. 1991.

Vol. 57: **Molecular Techniques in Taxonomy.**
Edited by G. M. Hewitt, A. W. B. Johnston, and J. P. W. Young. 420 pages. 1991.

Vol. 58: **Neurocytochemical Methods.**
Edited by A. Calas and D. Eugène. 352 pages. 1991.

Vol. 59: **Molecular Evolution of the Major Histocompatibility Complex.**
Edited by J. Klein and D. Klein. 522 pages. 1991

Vol. 60: **Intracellular Regulation of Ion Channels.**
Edited by M. Morad and Z. Agus. 261 pages. 1992.

Vol. 61: **Prader-Willi Syndrome and Other Chromosome 15q Deletion Disorders.**
Edited by S. B. Cassidy. 277 pages. 1992.

NATO ASI Series H

Vol. 62: **Endocytosis.**
Edited by P. J. Courtoy. 547 pages. 1992.

MIX
Papier aus verantwortungsvollen Quellen
Paper from responsible sources
FSC® C105338

FSC
www.fsc.org

If you have any concerns about our products,
you can contact us on
ProductSafety@springernature.com

In case Publisher is established outside the EU,
the EU authorized representative is:
Springer Nature Customer Service Center GmbH
Europaplatz 3, 69115 Heidelberg, Germany

Printed by Libri Plureos GmbH
in Hamburg, Germany